Studies in the History of

Volume 1

The Engineering
of Medieval Cathedrals

*the assistance of M. M. Chrimes, A.W. Skempton,
N.A.F. Smith and R.J.M. Sutherland*

Twelve volumes are to appear from the Autumn of 1997:

1. **The Engineering of Medieval Cathedrals**
    Edited by LYNN T. COURTENAY, University of Wisconsin

2. **Masonry Bridges and Viaducts**
    Edited by TED RUDDOCK, University of Edinburgh

3. **Land Drainage and Irrigation**
    Edited by SALVATORE CIRIACONO, University of Padova

4. **Dams**
    Edited by DONALD C. JACKSON, Lafayette College

5. **Water Supply and Public Health Engineering**
    Edited by DENIS SMITH, formerly University of East London

6. **Port and Harbour Engineering**
    Edited by ADRIAN JARVIS, Merseyside Maritime Museum and
    Liverpool University

7. **The Civil Engineering of Canals and Railways before 1850**
    Edited by MIKE CHRIMES, Institution of Civil Engineers

8. **The Development of Timber as a Structural Material**
    Edited by DAVID T. YEOMANS, Liverpool University

9. **Structural Iron, 1750–1850**
    Edited by R.J.M. SUTHERLAND, Consultant, Harris and Sutherland
    London

10. **Structural Iron and Steel, 1850–1900**
    Edited by ROBERT THORNE, Editor, *Construction History Journal*

11. **Early Reinforced Concrete**
    Edited by FRANK NEWBY, formerly F.J. Samuely and Partners, London

12. **Structural and Civil Engineering Design**
    Edited by WILLIAM ADDIS, University of Reading

Studies in the History of Civil Engineering
General Editor: Joyce Brown

Volume 1

The Engineering
of Medieval Cathedrals

edited by
Lynn T. Courtenay

Routledge
Taylor & Francis Group

LONDON AND NEW YORK

First published 1997 by Ashgate Publishing

2 Park Square, Milton Park, Abingdon, Oxon OX14 4RN
711 Third Avenue, New York, NY 10017, USA

Routledge is an imprint of the Taylor & Francis Group, an informa business

First issued in paperback 2016

British Library CIP Data
The Engineering of Medieval Cathedrals. – (Studies in the History of Civil
    Engineering; 1).
    1. Cathedrals – History.   2. Cathedrals – Design and construction – History.
    I. Courtenay, Lynn T.
    690.6'6'0902

US Library of Congress CIP Data
The Engineering of Medieval Cathedrals/ edited by Lynn T Courtenay.
    p. cm. – (Studies in the History of Civil Engineering ; v. 1).
    Includes bibliographical references and index.
    ISBN 0–86078–750–8 (hardbound)
    1. Cathedrals – Europe – Design and construction – History.
    2. Middle Ages.   I. Courtenay, Lynn T. (Lynn Towery), 1943–
    II. Series.
    TH4221.E54   1997
    690'.664'0940902—dc21   97–29054
                                                          CIP

ISBN 978-0-86078-750-1 (hbk)
ISBN 978-1-138-26806-7 (pbk)

Transfered to Digital Printing in 2012

STUDIES IN THE HISTORY OF CIVIL ENGINEERING – VOL. 1

Contents

Acknowledgements

The chapters in this volume are taken from the sources listed below, for which the editors and publishers wish to thank their authors, original publishers or other copyright holders for permission to use their material as follows:

Chapter 1: Robert Mark, 'Robert Willis, Viollet-le-Duc and the structural approach to Gothic architecture', *Architectura*, 7, no. 2 (Munich, 1977), 52–64. Copyright © 1977 by Deutscher Kunstverlag GmbH, Munich.

Chapter 2: J. Heyman, 'On the rubber vaults of the Middle Ages and other matters', *Gazette des Beaux-Arts*, 5th ser. 51 (Paris, 1968), 177–188. Copyright © 1968 by the *Gazette des Beaux-Arts*.

Chapter 3: Lon R. Shelby, 'The geometrical knowledge of medieval master masons', *Speculum*, 47 (Cambridge, MA, 1972), 395–421. Copyright © 1972 by the Medieval Academy of America.

Chapter 4: Robert Branner, 'Villard de Honnecourt, Reims, and the origin of Gothic architectural drawing', *Gazette des Beaux-Arts*, 6th ser. 61 (Paris, 1963), 129–146. Copyright © 1963 by the *Gazette des Beaux-Arts*.

Chapter 5: J.H. Harvey, 'The tracing floor of York Minster', *Friends of York Minster*, 40th annual rep. (York, 1968), 1–8. Copyright © 1968 J.H. Harvey and The Friends of York Minster.

Chapter 6: Lon R. Shelby and Robert Mark, 'Late Gothic structural design in the 'instructions' of Lorenz Lechler', *Architectura*, 9, no. 2 (1979), 113–131. Copyright © 1972 by Deutscher Kunstverlag GmbH, Munich.

Chapter 7: Eric Fernie, 'The ground plan of Norwich Cathedral and the square root of two', *Journal of the British Archaeological Association*, 129 (Oxford, 1976), 77–86. Copyright © 1976 by Eric Fernie.

Chapter 8: Jean Bony, 'The stonework planning of the first Durham master', *Medieval Architecture and its Intellectual Context: Studies in Honour of Peter Kidson*, ed. E. Fernie and P. Crossley (London, 1990), 19–34. Copyright © 1990 by Hambledon Press.

General Editor's Preface

Joyce Brown

Civil engineering has a long history. Although practitioners did not begin to describe themselves as 'civil engineers' until the eighteenth century, the origins of their work lie in the construction skills of the ancient world and in the works of military engineers. Shortly after British civil engineers had formed themselves into a professional body, the Institution of Civil Engineers in London, their role was described in 1828 by Thomas Tredgold in these words:

> Civil engineering is the art of directing the great sources of power in Nature for the use and convenience of man.[1]

He thus neatly encapsulated what the civil engineer's job is about – the provision of the means to give society what it wants for its survival.

As time has gone by, the civil engineer's skills have had to diffuse over an ever-widening field and at the same time become more specialised. In all periods he has had to be the ingenious 'fixer', but more than a mere artisan – a point observed in the Institution's motto, *scientia et ingenio*.

The history of civil engineering is a fascinating one. Certainly 'the great sources of power in Nature' are not always directed without a certain amount of opposition by them, and the provision of the basic requirements – habitation, water supply, main drainage, harbours, bridges, places of worship, transport systems by land and water – has called for inventiveness and sometimes for daring and courage.

Yet interest in the history of civil engineering is not of long standing, and it seems fair to state that most of the important work has been done in the last forty years. With a few notable exceptions, most of the major books on the subject have appeared in that era; many specialist societies with their own transactions have been founded; and many of the journals in which articles might be published have come into existence. While archaeologists and architectural historians have long made their contribution to this subject, their ranks have been swelled by a large number of serious scholars – including engineers themselves – now interested in recording and evaluating the engineering achievements of earlier times. In particular, there is a lively interest in the study of our industrial past, at least in part because some of the evidences of it are already disappearing from the landscape.

Our intention in creating the series of which this volume forms part is to provide, through the reproduction of important contributions to this subject, an invaluable reference collection for its study. The series encompasses many different branches of engineering from early times to the beginning of this century; its

---

[1] *Minutes of Council, Institution of Civil Engineers*, 4 January 1828.

perspective is global; and the chosen articles have an international authorship, to the extent that this can be achieved for an essentially English-language series. The introduction to each volume has allowed the volume editor to set the papers selected in the context of the whole history of the subject and its historiography, while the provision of references and a bibliography will enable the reader to go further into the study of the topic.

This series will have succeeded if it gives civil engineers some unexpected insights into their own craft and other readers a new way of looking at engineering structures.

* * *

The volume editors were chosen for their particular knowledge and expertise in the field. They have worked with enthusiasm and good will to put together collections of articles intended to be both informative and stimulating, and I am grateful to them for their efforts. Where possible, they have incorporated the work of the best scholars, while at the same time giving preference to articles often obscurely published and not readily available. The learned journals are represented, as also are conference proceedings and essay chapters contributed to books.

In all of this, I have been much assisted by my Advisory Panel. Emeritus Professor A.W. Skempton (Imperial College) has made himself freely available to me for consultation on many aspects of this venture. Mike Chrimes (Librarian of the Institution of Civil Engineers), Dr. Norman Smith (Emeritus Reader, Imperial College) and James Sutherland (formerly a Senior Partner in Harris and Sutherland, London, consulting engineers) have all contributed valuable advice based on their extensive knowledge of particular fields. The idea for the series was developed in discussion with Dr John Smedley of Ashgate, and Publisher of the Variorum Collected Studies Series, who has given me much personal support and the benefit of his experience.

I am grateful also for the help given by the staff of several libraries, notably Kay Crooks and Susannah Parry in the library of the Department of Civil Engineering, Imperial College, and Mike Chrimes and staff of the library of the Institution of Civil Engineers, London. Many friends have given me encouragement and useful advice; in particular, I want to thank Paula Kahn and Mark Baldwin.

Formerly Department of Civil Engineering
Imperial College, London

Introduction

Lynn T. Courtenay

Part One – Historiography and Scope

As early as 1842, engineer, historian, and Professor of Natural and Experimental Philosophy at Cambridge, Robert Willis remarked:

> It becomes . . . a curious and interesting subject of inquiry to trace from examination of the structures themselves, what geometrical methods were really employed in setting out the work, and how the necessity of these methods gradually arose...For the forms and proportions of every [medieval] structure are so entirely dependent upon its construction and derived from it, that unless we thoroughly understand these constructions, and the methods and resources which governed and limited them, we shall never succeed in obtaining the master key to their principles . . . [1]

Willis is speaking generally to an audience (the Royal Institute of British Architects) interested in designing architecture in the Gothic idiom, yet his fascination with the means by which medieval buildings were constructed, or 'engineered', has generated a quest for the methods of early builders that has continued to the present. Given this keen curiosity in a man of considerable erudition, it is not surprising that Willis should turn his attention not only to the buildings themselves, but also to available documentary sources, one of the most important being a newly published Parisian manuscript of the thirteenth century known as the *Sketch Book* of Villard de Honnecourt; Willis translated the French edition of Jean-Baptiste Lassus in 1859.[2]

In his seminal paper 'On the construction of vaults in the Middle Ages', Willis cites two important sources: Philibert de l'Orme's *Premier Livre de l 'Architecture* (1567), a treatise of eleven books on technical subjects such as vaults, geometry and stereometry, and a second treatise, *Nouvelles Inventions pour bien bastir* (1578).[3]

---

[1]   Willis, 'On the construction of vaults . . . ', 2.

[2]   This manuscript (Bibliothèque National, Paris, MS fr. 19093), which has addenda by two later hands (Magister 2 and 3) is an autographed compilation of sketches and commentary by an alleged architect, Villard de Honnecourt. Lassus' edition appeared in French in 1858 as: *Album de Villard de Honnecourt: Architecte du XIIIe siècle, manuscrit publié en facsimile* (Paris, 1858) and was reprinted photographically in 1976 (Paris: Léonce Laget). Willis' edition (the only English version until 1959) is: *Facsimile of the Sketch Book of Wilars de Honecort with commentaries and Descriptions by M.J.B.A. Lassus and M.J. Quicherat: Translated and Edited with Many Additional Articles and Notes by the Rev. R. Willis* (London, 1859). An observation Willis made, accepted by Villard scholars, is that the text was added after the drawings. The standard edition is that of Hahnloser (1972, see Bibliography); on the various editions, see Barnes, *Villard de Honnecourt . . .* xli–lv.

[3]   Philibert de l'Orme, *Traités d'architecture . . .* , ed. Jean-Marie Pérouse de Montclos (Paris: Léonce Laget, 1988); H. Mitchell, 'An unrecorded issue of Philibert Delorme's *Le premier tome de*

Willis regarded Philibert as an educated 'architect' but more importantly, as a master mason trained in the medieval craft tradition. Combining Philibert's descriptions and instructions with his own keen observation of buildings, Willis used Philibert's terminology to describe vaulting forms and commented particularly on technical aspects of construction such as the *tas de charge*, that is, the horizontal stone coursing (often a single block) from which the vault ribs spring.[4] Willis' keen desire to examine buildings from above the vaults or during repairs provided him with opportunities to discover archaeological evidence of the construction process. Willis' documentary research combined with his meticulous observations of buildings represent an enduring contribution to the understanding of medieval architecture.

As Robert Mark's paper (chapter 1) points out, though Willis was trained in engineering mechanics, his writings are not concerned with the structural issues that fascinated Viollet-le-Duc. Willis' object of inquiry remained historically rooted in the design and construction processes of medieval builders. Nonetheless, as engineers Robert Mark and Jacques Heyman maintain, there are important, indeed essential, questions about how structures behave and respond to their loads, especially over time. How and why do medieval cathedrals and great churches stand or fall? What did the master masons of the cathedrals understand and intend?

In the 1960s engineers began using modern tools of analysis to answer these questions, or as J.E. Gordon's delightful book title puts it *Structures – or Why Things Don't Fall Down*. Whether undertaken in relation to restoration or purely from curiosity, engineering investigations are now a recognised component in the interpretation of early buildings. Midway in this generation of scholarship, Robert Branner in a review essay of 1973, remarked, 'At the present moment the most fundamental and "hottest" question in the literature concerns the structural analysis of the High Gothic French (or German) cathedral'.[5] Clearly by the 1990s what was a 'hot' topic in 1973 has been absorbed into the literature. One has only to look at the number of conference sessions (e.g. in *Avista Forum*[6]) and the growth of the periodical literature to realise that a technological awareness of architecture, history, and material culture is a significant component in current historical thinking and research.

From the 1960s to the 1980s a vigorous group of scholars, encompassing engineers, art historians, and historians of technology and science inaugurated a

l'architecture, annotated by Sir Henry Wotton', *Journal of the Society of Architectural Historians*, 53 (1994), 20–9; and S. Sanabria, 'From Gothic to Renaissance stereotomy: the design methods of Philibert de l'Orme and Alfonso de Vandelina', *Technology and Culture*, 30 (1989), 266–99.

4   Willis, 'On the construction of vaults . . . '. 6–8.

5   R. Branner, 'Gothic architecture', *Journal of the Society of Architectural Historians*, 32 (1973), 331–2.

6   AVISTA, the Association Villard de Honnecourt for the Interdisciplinary Study of Medieval Technology, Science, and Art (Haverford College, Haverford, PA, USA) was founded in 1985.

seminal era marked by Paul Frankl's monumental historiographic essay, *The Gothic* (1960), followed in 1961 by John Fitchen's *The Construction of Gothic Cathedrals, a Study of Medieval Vault Erection*. These pioneering studies (including Conant's work at Cluny and von Simson's contextual approach to Gothic architecture[7]) have profoundly shaped medieval scholarship. These were followed by important contributions by, *inter alios*, Jacques Heyman, Robert Mark and Lon Shelby (see chapters 1, 2, 3 and 12).

The date, 1960, thus emerges as a logical point of departure for this collection of papers which present different approaches to engineering aspects of cathedrals and great churches, mainly in England and France. A major portion of the volume has been selected to encompass primary sources that reveal building methods, the constructional process, and archaeological observations pertinent to engineering issues. Several papers provide historical background about medieval builders and individual churches in order to complement the more analytical studies of engineers. The contributions cover medieval cathedrals across a considerable chronological period and in different regions, thus dating 'medieval' in the manner of L.F. Salzman, namely, from the Norman cathedral of Durham 'down to 1540'. In a period of rapid economic, intellectual, liturgical and technical expansion, as in northern Europe between *ca.* 1050 and 1280, medieval builders mastered the techniques of stone vaulting and pushed building technology to unprecedented heights. They built at an astonishing pace and volume, leaving to the European architectural legacy the cathedrals and great abbey churches of the twelfth and thirteenth centuries, such as Durham, Norwich, St Denis, Notre-Dame-de-Paris, Bourges, Amiens, Salisbury, Lincoln and Beauvais. The period between 1130 and 1270 frequently defines the essential 'Gothic achievement'.[8] Nonetheless, 'Gothic' building continued, vaulting techniques were refined, and the social and educational position of craftsmen improved as they achieved guild status and privileges in an increasingly urban society. Moreover, the late medieval flowering of architecture in Germany and especially in the Bohemian capital of Prague is undeniable, just as Brunelleschi's vast ribbed dome of Florence Cathedral (completed 1436) is also part of the medieval engineering legacy. Hence papers representing 'The Late Gothic' and alternatives to French High Gothic have been included (see chapters 6, 13, 14 and 15).

Part Two – Medieval Builders and Their Methods

Part Two concentrates on primary sources of the period and their interpretation, that

---

[7] Conant, *Cluny: les églises et la maison . . .* ; von Simson, *The Gothic Cathedral*.

[8] See, for example, Heyman (chapter 11 in this volume); Kimpel et al., *Die gotische Architektur in Frankreich, 1130–1270*.

is, medieval builders, their design methods, and the mechanical arts. These papers confirm the understanding that medieval 'architects' were designers, builders, geometricians and *working craftsmen* whose skills embraced those of civil engineers, architectural surveyors, purveyors of materials, and inventors of machines and mechanical devices for construction.

The master mason was usually the primary figure responsible for the plan and construction, although there are documented instances of clerical or royal input.[9] A master mason was a hired salaried professional who headed a hierarchical work force consisting of his foreman, fellow masons, journeymen, and apprentices as well as his chief collaborators, the master carpenter and master smith, in addition to common labourers, diggers, and carters. The 'team' thus included skilled professionals often imported from elsewhere in addition to local labourers who might be hired on the spot.[10] The Work (*Opus*) was generally supervised by a clerical administrator, who provided the liaison between the cathedral chapter or patron and the *masons' lodge*, that is, the workmen's temporary headquarters. As various building accounts reveal, master builders received high wages and belonged socially to a superior artisan class. Within this organizational framework, it is critical to bear in mind that cathedrals and great abbeys were designed and built for the performance of the liturgy whose ritual ideology and music influenced design and structural forms.[11]

The complementary papers by Shelby, Branner, and Shelby and Mark (chapters 3, 4, and 6) present a representative survey of many of the important documentary sources relating to the building process. Harvey's paper (chapter 5) on the plaster tracing floor of York Minster, examines an actual production site, *viz.* the upper room of the Chapter House vestibule, and its incised drawings. The markings on the tracing floor, discussed by Harvey and referred to in Branner's article, illustrate the large and important category of incised 'project drawings' or full-scale working drawings that provided the basis for the templates used to cut stones to a particular size and shape. These survivals of the actual design process have received considerable attention in the scholarly literature and most are conveniently catalogued by Wolfgang Schöller.[12]

Distinction between full-scale working drawings found at numerous sites and the unique small parchment drawings contained in the famous manuscript of Villard de Honnecourt is vital to the interpretation of medieval design process. Because

---

[9] For examples, see Fergusson, 'The builders of Cistercian monasteries . . . ; and R. Sundt, '*Mediocres domos et humiles habeant fratres nostri*: Dominican legislation on architecture and architectural decoration in the thirteenth century', *Journal of the Society of Architectural Historians*, 46 (1987), 394–407.

[10] See Shelby's review article of John James, 'The contractors of Chartres', *Gesta*, 20, 1 (1981), 173–8.

[11] See von Simson, *The Gothic Cathedral*, on music and proportions, *passim*.

[12] Schöller, 'Ritzzeichnungen . . .', see Bibliography.

Villard's sketches represent the only surviving example in the thirteenth century of such annotated drawings, Shelby and others doubt whether this manuscript represents a genre of mason's manuals or 'lodge books' that were actually used. Thus many scholars question whether Villard was actually a practising mason, or rather, a well-educated amateur, or perhaps a goldsmith.[13]

Shelby's paper on the geometrical knowledge of medieval masons (chapter 3) raises fundamental issues and provides a broad view of the concepts and historical background to the teaching of geometry in the Middle Ages. He surveys the major scholastic literature, as well as the non-scholastic sources, in search of the geometry actually learned and practised by medieval builders. He argues convincingly that the *modus operandi* of medieval builders was based on 'constructive geometry', i.e. the use of instruments (compass, square, plumb bob, etc.) and the physical manipulation of simple geometric forms as demonstrated in fifteenth century treatises by Lechler and others. None of the geometric constructions depended upon mathematical proofs in Euclidean terms but were rather 'rules of thumb'.

In a second paper co-authored by Shelby and Mark on the German master mason, Lorenz Lechler (chapter 6), we are able to see the 'constructive' method applied to the projection of vault ribs from a plan or '*reiung*'.[14] Significantly, what has long been regarded as the mysterious and carefully guarded 'secret' of the medieval mason can be understood as nothing more than the prescriptive stages needed to project an elevation from a plan.[15] Lechler employed very few actual measurements or mathematical calculations; rather, he used experiential knowledge of materials and simple ratios based on the building's span, i.e. rule-of-thumb proportioning that in his view would ensure safety.

One of the most provocative sections in Shelby's paper concerns Magister 2's addenda to Villard's *Sketchbook* (also mentioned by Branner, chapter 4, 68). This is the question of the precise meaning of Magister 2's assertion that: *Totes ces figures sunt estraites de geometrie* [all these figures are drawn from geometry], an ambiguous statement that is open to a number of interpretations.[16] Does he imply that his 'figures', or exercises, came from treatises on geometry? Or, did they derive from project drawings or manuals in a masons' lodge? Did he intend the sketches as an *aide memoire* based on geometrical knowledge learned orally? Shelby concludes from the evidence available that Magister 2's '*geometrie*' is best interpreted as the

---

[13] The literature on Villard de Honnecourt and his 'career' is extensive and best accessed via Barnes, *Critical Bibliography* and thereafter in *Avista Forum* (see note 6). See also Bechmann, *Villard de Honnecourt* . . .

[14] This paper incorporates contributions of Müller; cf. also the other papers by Bucher and Müller in the Bibliography.

[15] Shelby, 'The "secret" of the medieval masons', 201–19.

[16] On Magister 2's comments, see Shelby, chapter 3 in this volume, 30; cf. Branner, chapter 4 in this volume, 68.

geometry learned first hand and not from manuals. But, is there a middle ground between texts on geometry and hands-on experience?

Shelby's interpretations of Magister 2's enigmatic comment can now be placed in a wider context provided by scholarship in the history of science, in particular the work of Stephen Victor and of Elspeth Whitney.[17] For example, Magister 2's illustrated 'problems' – 'measuring the height of a tower'; 'finding the width of a river without crossing it'; and, 'finding the width of a distant window' – are all included in treatises on practical geometry, but Victor points out that 'The drawings, even if only a shorthand, show an understanding of practical application which surpasses that of the practical geometries'.[18] Thus we can suggest that first-hand knowledge was the most critical part of a master mason's education above what he may have gleaned from treatises.

In the context of medieval scientific attitudes towards technology, Elspeth Whitney's study is pertinent to the status of the geometrical and mechanical arts and also to the issue of theory and practice raised in the context of Magister 2. Specifically, Whitney examines a large sample of medieval texts to see how schoolmen described the distinctions between theory and practice in the mechanical arts.[19] Whitney argues for a general rise in the status of mechanical arts and their practitioners in the twelfth and thirteenth centuries, in contrast to the traditional Graeco-Roman ambivalence towards manual labour. Her evidence suggests that expertise in *ars mechanica*, now classified as distinct from speculative science, achieved a moral purpose in the thirteenth century. By using his hands and intellect for technical invention, the craftsman contributes towards improving mankind's lapsed and sinful condition as he progresses towards salvation. Whitney claims that this view (which one might argue actually places boundary conditions on technology as a hand maiden to soteriology) created a 'modern' attitude towards technical progress, fostered by influential thinkers like Robert Kilwardby, Hugh of St Victor, and Roger Bacon. Thus, she maintains, by 1300 medieval scholars, having assimilated both Aristotelian and Arabic philosophy, had integrated the classical heritage into a soteriological view of the mechanical arts and craftsmanship.[20] The attitudes of the writing intelligentsia, whose works were probably known to clerical patrons, thus suggest the essential appeal of a moralised technology, i.e. similar to Abbot Suger's view of the importance of creating inspirational buildings and liturgical spaces that would enhance mankind's progress towards the Divine, as

[17] Victor, *Practical Geometry* . . . ; Whitney, 'Paradise restored . . . ', 1–169. See also Ovitt, *The Restoration of Perfection* . . .

[18] Victor. *Practical Geometry*, 71.

[19] The influential texts concerning the status of the mechanical arts by Gundissalinus of Segovia is quoted at some length by Shelby (chapter 3, 34–5) and by Victor, *Practical Geometry*, 8–10, who explains the 'mutually beneficial' relationship between theory and practice in an epistemological context. Cf. Whitney, 'Paradise restored . . . ', 132–4.

[20] Whitney, 'Paradise restored . . . ', 60, 73, 125–7.

exemplified in the splendid churches lavished with colour, stained glass, and rich furnishings, supported by ecclesiastical, urban pride and liturgical pageantry.

Turning to what the *building fabric* reveals in respect to foundations, plan, measurement, geometry and proportions, the papers of Fernie (chapter 7) and Bony (chapter 8) represent recent scholarship dedicated to understanding the rationale of medieval design by analysing precisely-dimensioned plans, elevations, and details (especially piers). Similarly, the careful recording of moulding profiles (i.e. what was produced by the mason's template) and the database assembled by Richard Morris[21] requires a revision of the assumption that profiles of architectural members were used, or re-used *en masse*. The study of metrology and recent computerization of medieval dimensions has, however, inevitably led to debate as to whether the original designer actually used the particular system extracted from the measurements, i.e. did he design according to Pythagorean proportions of harmony, with a square (*ad quadratum*), triangle, pentagon, the golden section, radius of a circle, or a combination of these?[22]

Using Norwich Cathedral as a primary example, Fernie postulates that the ratio of one to root two is one of the most common proportions used in medieval buildings,[23] but, to what extent does it apply beyond the particular buildings Fernie has studied? And should not one always examine alternative hypotheses in the arduous task of translating dimensions into geometric schemes of medieval masons? Ultimately one may find whatever one is looking for, though scholars like Fernie are careful to point out deviations.

The paper by Jean Bony (chapter 8) combines a careful assessment of the original stonework and plan of the east end of Durham Cathedral (AD 1093–1104) with detailed mensuration and attention to both quantitative and qualitative aspects of the masonry (also discussed in Fernie). This is an important contribution to the understanding of design procedures in Anglo-Norman construction, and like the work of Fernie, Bony bases his assessment on a meticulous survey of the fabric to deduce work procedures and modes of thinking – an approach in recent scholarship also represented in the provocative study of John James, 'The contractors of Chartres'.[24] Bony concludes that the advanced planning and elegant proportioning evident in the first Durham campaign constitute 'an essential new document on the intellectual history of the late eleventh century' – not to mention

[21] Morris, 'Mouldings . . .', 239–47.
[22] For examples see Kidson; Sanabria; Neagley; von Simson (1962), 21–2 in the Bibliography; and M.T. Davis, 'Plan design in Gothic architecture: Beauvais Cathedral and Erbach Abbey', *Avista Forum*, 8, 1 (1994), 5–7.
[23] Fernie shows how the dimensions of the cathedral complex at Norwich bear out the 1:1.4142 proportions and traces this system from the overall plan down to individual members, e.g. the major and minor piers. See Fernie, *An Architectural History of Norwich Cathedral*, 94–100 and Appendix 3, pp. 205–6.
[24] J. James, 'The contractors of Chartres', *Architectural Association Quarterly*, 4, 2 (1972), 42–53.

the importance of Durham as the first entirely rib-vaulted cathedral in northern Europe.[25]

The short article by Bonde et al. (chapter 9) draws attention to geophysical conditions and foundation systems, necessary considerations in the design of large-scale buildings. In this field, the soils engineer Kerisel has made a lengthy study of medieval foundations.[26] The importance of superimposed loads on foundations is discussed in the context of towers and spires, specifically at Salisbury Cathedral (Part Four).

Part Three – Vaults and Their Support: Structural Archaeology and Engineering Analysis

Part Three presents major engineering studies of medieval cathedrals using a variety of analytical approaches. It has been the engineer's quest to determine how and why certain structural forms work, and beyond that, to make speculative interpretations about what the original builders intended. This section intentionally incorporates controversy among engineers, but it is also designed to appeal to the general reader. To this end, some engineering definitions are provided to facilitate an appreciation of the more technical papers. For example, the distinction between *statically determinate* and *statically indeterminate* structures is particularly useful in understanding why engineers can disagree radically and is pertinent to the considerable difficulties that arise in interpreting historic structures.

A *statically determinate* structure has *only* the members required for stability. All of its member internal forces can be accounted for from equations of static equilibrium, i.e. the sum of all forces acting on any structural element must equal zero. *Statically indeterminate* structures contain extra (*redundant*) members; in solving problems of indeterminacy, engineers must have (along with the basic equations for equilibrium) information relating to support fixity in the structure (e.g. the displacement, if any, at the ends of the supports of a statically indeterminate two-hinged masonry arch) as well as any individual member's resistance to deformation under loading. Thus statically indeterminate structures demand far more complex analysis than determinate ones, particularly when displacements are time-dependant, such as the relative motion (displacement) in the end supports of such a masonry arch that occur over a long period of time; or, in the much-debated

---

[25] Bony, chapter 8 in this volume, 131. On Durham's quasi-industrial production and its proportional differences from Norwich, see E. Fernie, 'Design principles of early medieval architecture as exemplified at Durham Cathedral', in *Engineering a Cathedral*, ed. M. Jackson (London, 1993), 150–2.

[26] Kerisel, 'Old structures . . . ', Appendix 2: Structures and foundations of Cluny III – equilibrium conditions, 480–1. See also S. Bonde, C. Maines and R. Richards, 'Soils and foundations', in Mark, *Architectural Technology . . .* , 36–46 especially.

failure of the choir of Beauvais Cathedral whose high vaults stood for twelve years before their collapse (see below). Statically determinate structures are not sensitive to small changes in support conditions; however, statically indeterminate structures (e.g. a cracked Gothic vault) have an advantage in that they may continue to carry load and remain stable even after the complete failure of an individual element.

Theory, even for the simpler statically determinate structures, was employed only rarely until the later nineteenth century. Yet complex structural behaviour was probably understood empirically by early builders, who observed their structures carefully during construction and made adjustments, such as increasing abutment or thickening members. Gothic builders, however, could only resolve static issues by tested 'rules of thumb' as described in Lechler's advice on wall thickness and abutment in relation to vault spans (chapter 6, 101–4). Today statically indeterminate structures are generally analysed using numerical (computer) modelling. This powerful tool of analysis has largely superseded physical modelling including *photoelastic* modelling.[27] Another approach, taken, by Heyman, for example, is *limit analysis* in which a pattern of failure (that eliminates the structural redundancies) is presumed for an indeterminate structure so that it can be simply treated as a determinate one.[28]

The Case of Beauvais (1272–1284)

The choir of Beauvais, discussed in three papers in this volume with additional comments appended to Heyman's paper (chapter 11), provides a vivid if controversial example of the application of engineering principles to structural archaeology and analysis. Moreover, these papers, taken with Mainstone's classic static analysis of the dome of Florence Cathedral (chapter 15), represent the principal types of investigation by twentieth-century engineers.

The range of analytical methods already mentioned accounts for the fascinating

---

[27] In photoelastic modelling, small-scale models are tested under scaled loadings that simulate the forces acting on a full-scale building; the stress measurements are derived from optical interference patterns in the model; see S.P. Timoshenko, *History of Strength of Materials* (New York, 1983), 249, 271 and 351; and S.P. Timoshenko, *Theory of Elasticity* (3rd edn., New York, 1970), 150–67. Mark explains the rationale for its application to historic structures: 'Research at Princeton during the 1960s demonstrated that tests on small-scale plastic models could be used to predict internal forces within complex reinforced-concrete structures subjected to normal in-service loadings, even though concrete is notoriously inelastic, compositionally non-homogeneous, and subject to tensile micro-cracking. It was this experience with concrete structures that led to the idea that masonry buildings might be amenable to investigation using small-scale models, but only if the masonry acted . . . as a monolith [in compression]. Mark, *Light, Wind, and Structure*, 24.
[28] Heyman, 'The stone skeleton', 252–60. It is also of interest in this context to note Heyman's remark with reference to hinging in the voussoir arch: 'The actual elasticity of stone will, in practice (though not admitted in theory) close up some cracks, and introduce a degree of indeterminacy into the construction', 259.

variety of interpretations of the collapse of the choir of Beauvais, whose 48-metre high vaults stood proudly for twelve years. In November of 1284, however, they failed sufficiently to require major repairs and prevent use of the choir until 1339.

Murray's paper has been chosen to open the Beauvais debate because it contains detailed archaeological information and cites the relevant historical sources. Thus it provides not only a structural assessment but also a context into which the discussion of Heyman's paper by Skempton and Mainstone can be placed (chapter 11). Using a retrospective approach, with the repaired cathedral as a 'model' for what went wrong, Murray looks at the repairs and points out the essential fact that it was *only* in the central straight bay of the choir (bay 7, Murray's fig. 1) that the vertical elements of the aisles and intermediate buttresses were entirely rebuilt; significantly he notes (chapter 10, 157) that Benouville's influential drawing (1891) is of a bay that did *not* fail (cf. Viollet's similar section, chapter 11, Heyman's fig. 12). The fourteenth-century builders, however, took no chances and interposed extra piers throughout the choir, replaced the original quadripartite system with sexpartite vaults, and substantially reinforced the intermediate buttresses of the straight bays of the choir.

An early explanation of the failure was that of Viollet-le-Duc, who argued that the slenderness of the twin colonnettes below the abutment of the lower flying buttress at triforium level was the critical design flaw.[29] Both Heyman and Kerisel accept his sequential collapse mechanism; as Kerisel states, the problem leading to the collapse was 'parallel flexible members of different moduli of deformation' (one member being the colonnette and the second the main pier supporting the vault).[30] We do not know whether the colonnettes pin-pointed by Viollet were actually *en délit,* members added after the main piers were in place. However, according to Murray's archaeological investigation, the main piers were not entirely rebuilt, as such a mode of failure might suggest. Nonetheless, those who repaired the choir must have been concerned about the load taken by the main piers, since intermediate piers were added. In the discussion following Heyman's paper, Skempton and Mainstone stressed the exceptional slenderness of the main piers and the wide original bays.

Murray proposes that design problems existed in the configuration of the aisle supports of the superstructure, *viz.* the eccentric positioning of the intermediate buttresses carried only partly above the main aisle piers (*porte-à-faux* construction).

---

[29] E.E. Viollet-le-Duc, *Dictionnaire*, IV, trans. G.M. Huss, *Rational Building* (New York, 1895), 238. Cf. the comments by Mark, *Experiments . . .* , 58–72.

[30] Kerisel, see note 26, 449–50. Kerisel's reference to flexible members of different 'moduli of deformation' is confusing in that both colonnette and main supporting wall are of stone; their relative deformability is a function of the depth of the section and not of their modulus of elasticity. Kerisel notes the excellent foundations provided for Beauvais, which leads him to reject any suggestion of displacement in that part of the structure.

He argues that this asymmetry was a key point of weakness in the central bay of the choir, i.e., the only bay that did not receive additional support either from the transepts or from the adjacent bays of the hemicycle. This explanation has been rejected by Heyman, and Kerisel, but Murray's observations, supplemented by Mark's photoelastic model, demonstrate the strength of this interpretation, particularly since Mark's analysis takes into account the cumulative effect of wind-loading and cracking in the area of the *porte-à-faux* construction.[31]

Ironically, the later repairs can support several interpretations, since the intermediate buttresses in the straight bays were rebuilt and reinforced, but also the load on each primary support was reduced by the auxiliary piers effectively taking a portion of the load of the high vaults. It is probable that the builders could not pinpoint what triggered the collapse and thus took no chances with either the buttressing system or the piers.

The failure of the Beauvais choir was, however, only partial. Despite the history of this ill-fated building, whose crossing tower also collapsed in 1573, much of the original thirteenth century fabric still stands, and current investigation of the roof carpentry demonstrates that a considerable portion of the early carpentry above the high vaults remained intact.[32] Failures inevitably excite curiosity and a keen desire to find causes that remain elusive. Yet, the rebuilt choir of Beauvais Cathedral still stands, having survived both wars and tempest.

The Rib and Medieval Vaults

The gradual shift in the eleventh century from main-span timber trusses to stone vaults in the primary areas of the choirs and naves of medieval cathedrals initiated far more complex systems of construction and abutment. In Romanesque cathedrals and great churches, such as Cluny III, barrel vaults, which behave statically under deadload like a series of planar arches along a longitudinal axis, were initially employed for nave high vaults and were probably introduced for acoustical reasons.[33] This vault type, however, had several major disadvantages in that it required full timber centering, continuous abutment at the haunches, and massive

---

[31]   Cf. Mark (chapter 1), 4–5 and Mark, *Experiments* . . . , 70–1; cf. Wolfe and Mark, 'The collapse . . . ', 462–76; and R. Mark and R.S. Jonash, 'Wind loading on Gothic structure', *Journal of the Society of Architectural Historians*, 29, 3 (1970), 222–30.

[32]   Personal communication, June 1995. Dr Patrick Hoffsummer, University of Liège, Centre de Recherches Archéologiques, Laboratoire de Chrono-écologie.

[33]   Taking into account the major qualitative contribution of resonance and reverberation to medieval liturgical music, the difference between the smooth, curved surfaces of stone vaults versus planked timber ceilings or open roofs is of profound acoustical significance. Andrew V.V. Tallon, *Medieval Architectural Acoustics: An Inquiry into Some Aspects of the Relationship between Ecclesiastical Music and Architecture in Medieval France* (MA Thesis, University of Paris, IV, La Sorbonne, Faculté d'Art et d'Archéologie, 1992).

piers and walls for support. Hence, the clerestory could not be easily opened to fenestration without the use of thick-wall construction and additional abutments, as in the large nave of the eleventh-century priory of St Etienne at Nevers.[34]

Unlike the simpler barrel vault, the groin vault, widely employed in early medieval crypts, galleries, and aisles had the advantage of concentrating the forces (i.e. vertical, horizontal, longitudinal and lateral reactions) to specific points of concentrated abutment (e.g. at the corners of a square bay). Hence, the areas between supports and buttressing could be opened and used to provide large clerestory windows to light the interior space. Formed by the intersection of orthogonal semi-circular or pointed barrel vaults, groin vaults are more complex structures whose precise behaviour is difficult to determine, since the distribution of forces can vary according to how the vault was built and how it was supported.

In the late eleventh century ribbed vaults appeared in Lombardy and at Durham, and these were to gain widespread popularity from the twelfth century onwards.[35] The rib vault has all of the structural advantages of groin vaults with the additional bonuses: (1) they are easier to construct; (2) involve less centering; and (3) produce the enormous aesthetic appeal of linear articulation and visual cohesiveness associated with the Gothic style. On the other hand, as will become apparent, the behaviour of these complex, 'statically indeterminate' structures has engaged engineers in considerable debate.

A primary issue that has occupied engineers and architectural historians is the function of the ribs in Gothic and proto-Gothic vaulting systems. The question of whether the rib, which articulates and imposes a pattern on the vault surface, is structural or decorative, or both, has a considerable history of opinion, much of which is discussed by Frankl and, later, by Fitchen.[36]

The debate about early vaults like those at Durham Cathedral continues to be raised in the literature. Both Thurlby and James, for example, argue that the Romanesque ribs at Durham are primarily decorative rather than structural, though their reasons are different.[37] Thurlby, citing the research of Alexander et al. (chapter 12), as well as the archaeological evidence of timber centering, notes the importance of the ribs during construction, as they were used to support centering planks for the

---

[34] Nevers, constructed between 1068 and 1098, is a unique example of an eleventh-century barrel-vaulted, wide nave with a clerestory and internal quadrant arches in the gallery. See Marie-Thérèse Zenner, *Saint-Étienne de Nevers: un ancien prieuré de Cluny dans le Nivernais* (Varennes, 1995); idem., *Methods and Meaning of Physical Analysis in Romanesque Architecture, a Case Study: Saint-Étienne in Nevers* (Doctoral dissertation, Bryn Mawr College, 1994; Ann Arbor, University Microfilms International, 1994, No. 9425215).

[35] The dominance of the four-part ribbed vault is explained in Taylor and Mark, 'The technology of transition . . . ', 579–87.

[36] Frankl, *The Gothic* . . . ; and Fitchen, *The Construction* . . .

[37] Thurlby, 'The purpose of the rib . . . '; and John James, 'Rib vaults in Italy', *Avista Forum*, 6, 2 (1992/3), 5–6.

webbing. He concludes, however, that the ribs are decorative and function mainly to articulate the groins, a theme that has been expanded by Hoey who concentrates on aesthetic rather than structural issues and rejects Bilson's 'evolutionary' approach to vault articulation.[38] Structural discussions, which tend to distinguish structure from ornamentation, invariably refer to earlier seminal papers presented in this collection, e.g., Heyman (chapter 2) and Alexander et al. (chapter 12), the latter devoted to the distribution of forces in the French Gothic vault, as exemplified by the choir of Cologne Cathedral.

In the late 1960s the analysis of the ribbed vault as a shell structure was taken up by engineers Heyman in England and Curcio in Buenos Aires.[39] In this context, Heyman applied 'plastic design' concepts, originally developed to evaluate the safety of steel and reinforced concrete framed structures, to the examination of the structural mechanics of historic masonry vaults and domes.[40] Engineers contemplating the problem of forces in a thin shell structure have arrived at quite different ideas about the necessity of reinforcing or stiffening ribs within the vault shell, especially the diagonal groins of the quadripartite vault. Briefly, Heyman continues to argue that the rib is important structurally to reinforce the groins which constitute planes of weakness in the shell of the vault. Abraham and later Curcio, suggest that the forces of the vault shell (naturally thicker near the groins) remained the same with or without the ribs and that the structure was too complex to be analysed as a planar arch.[41]

In an effort to resolve this controversy, the forces in a typical large-scale, statically indeterminate, high Gothic vault were examined by *combined* photoelastic and finite element modelling (Alexander et al., chapter 12). According to this analysis, using the vaults of Cologne Cathedral choir as a model, the force trajectories follow the shortest path to the supports, i.e. to the piers and the *tas-de-charge*, and not to the groins reinforced by ribs. In other words, the entire web-and-rib shell is self-supporting, and the ribs *per se* have no structural function in the completed vault. The compressive forces in the masonry shell radiate upward from the *tas-de-charge* and are fairly evenly distributed throughout the structure (see chapter 12 in this volume, fig. 8).

Though not engaged in similar model analysis, Mainstone tends to concur with

---

[38] L. Hoey, 'A problem in Romanesque aesthetics: the articulation of groin and early rib vaults in the larger churches of England and Normandy', *Gesta*, 35, 2 (1996), 156–76; Bilson, 'Durham Cathedral . . . ', 101–60.

[39] Heyman, 'The stone skeleton'; L.C. Curcio, *Estudio y reflexiones sobre estructuras medievales y equilibrio de la Catedral de Reims* (Buenos Aires, 1967).

[40] J. Heyman, *Beams and Framed Structures* (Oxford and New York, 2nd edn. 1974), especially Chapter III, 'Plastic Beams and Frames', 88–136. Cf. Heyman, 'The stone skeleton', and chapter 2 in the volume.

[41] P. Abraham, *Viollet-le-Duc et le rationalisme médiéval* (Paris, 1935); Curcio, *Estudio e reflexiones* . . . (see note 39).

Mark's view. Both Mainstone and Mark agree that the Gothic ribbed vault has the essential properties and advantages of a groined vault, but is easier to construct. Thus rib and webbing together act as a single shell structure, each portion carrying its own weight. Given the way in which vaults were constructed, however – ribs before web – Mainstone suggests the diagonal ribs may actually take a good deal of the load,[42] but agrees that the rib's function is primarily *constructional*, acting as permanent centering – clearly of critical importance as buildings became larger, higher and thinner.

The question remains, however, whether or not the Cologne model (analysed as an indeterminate structure) can be taken as representative of the force distributions in rib vaults in general, since engineers dispute whether simplified models can accurately reflect the structural behaviour of real masonry vaults, despite the fact that the same type of analysis is used for the design of contemporary large-scale structures. The objection, arguably applicable to the study of any actual building, is that the real structure must be abstracted to a degree that some find unrealistic.[43] Yet all engineering models (and applications of theorems) must make certain assumptions to facilitate calculations.

Regardless of their structural role and of the unknown views of early Gothic builders, moulded ribs projecting from vault surfaces produce the decorative features that have so encouraged scholars to interpret Gothic architecture as an expressive art form. In the later Middle Ages designers appear to have delighted in the textural effects of complex, multi-ribbed canopies of wood and stone. Late Gothic vault forms composed of complex jointed masonry, including rib-and-panel construction with *liernes* (short connecting ribs) and *tiercerons* (additional ribs rising from the springing of the vault), were conceived much like decorative tracery with linear patterns pervading wall surfaces, vaults, and windows, as for example, at King's College, Cambridge, described in Walter Leedy's paper 'Fan vaulting' (chapter 13).

Italian Gothic

The last two papers in this part (chapters 14 and 15) draw attention to the long-standing interest in Brunelleschi's dome of Florence Cathedral (Santa Maria del Fiore) and the *Expertise of Milan*, a report of vehement debates among French, German, and Italian architects which took place between about 1391 and 1399.[44]

---

[42] Mainstone, *Developments in Structural Form*, 129.

[43] For the simplification rationale used to permit 'elastic modelling' at Exeter Cathedral, see E.W. Mander, 'Some structural studies of Exeter Cathedral', *The Structural Engineer*, 73, 7 (1995), 105–10.

[44] The *Expertise of Milan* is published in *Annales della fabbrica del Duomo di Milano dall'origine al presente*, I (Milan, 1877). See Frankl, *The Gothic . . .* , 62–86; and the classic discussion in J. Ackerman, '*Ars sine scientia . . .* ', 84–111; reproduced in W. Addis, ed. *Structural and Civil Engineering Design*, vol. 12 in this series (forthcoming).

They also reflect recent scholarly interest in 'Italian Gothic' *per se*, particularly buildings of the Cistercians and Mendicants.[45]

Citing the large Dominican church of Santa Maria Novella in Florence (1246–ca. 1470) as an archetypal example of the Italian-gothic structural system, Smith's paper stresses the statical 'rationale' of the domical ribbed vault and its abutment. She challenges the traditional view of Ackerman and others who used the *Expertise of Milan* to build a case for the inability of Italian architects to understand 'gothic' structure and statics.[46] Smith's paper (concentrating on central Italy) thus brings us chronologically and structurally to Santa Maria del Fiore (begun in 1294 and completed apart from the dome in *ca.* 1417).

Using the dome of the Roman Pantheon as a point of departure, Mainstone's paper (chapter 15) on the equally vast dome of Florence Cathedral (180 ft/55 m high and 138 ft/42 m in span) examines Brunelleschi's specifications, the construction process, and in particular, the design strategies of reinforcement to control tensile forces during construction without the use of conventional centering (chapter 15, 241–243). For example, the construction in horizontal courses was critical and allowed Brunelleschi to incorporate circular 'rings' within the two-metre thickness of the inner dome of masonry, each course of which was to be self-supporting once the mortar had dried. To enhance vertical integration as the dome inclined inward towards the apex, Brunelleschi devised his famous, brick *spinapesce* (herring-bone) bond that spirals upwards within the masonry to keep the bricks upright as the mortar cured. Considering the geometry of the octagonal dome and its materials, Mainstone's analytical account brings us closer to an understanding of this highly complex structure, through his critical discussion of the evolution and experimental nature of Brunelleschi's design in the light of earlier precedents; these have been highlighted in a subsequent paper.[47]

By successfully combining an octagonal ribbed cloister vault and a double-shell dome, Brunelleschi's great cupola can be regarded as the *summa* and synthesis of medieval and ancient building technology in central Italy. The great dome, built between 1418 and 1436, continues to interest engineers and historians of technology, especially because of its ingenious construction. Moreover, the dome's hybrid nature has led Mainstone and others to inquire as to exactly what kind of

---

[45] The articles by M. Trachtenberg, 'Gothic/Italian "Gothic": towards a redefinition', *Journal of the Society of Architectural Historians*, 50 (1991), 22–37; and David Gillerman, 'The evolution of the design of Orvieto Cathedral, *ca.* 1290–1310', *Journal of the Society of Architectural Historians*, 53, 3 (1994), 300–21, are indicative of this interest, as are the reprints of L.H. Heydenreich's *Architecture in Italy, 1400–1500* (Yale, 1996); and W. Lotz's, *Architecture in Italy, 1500–1600* (Yale, 1995). On mendicant architecture see P. Heliot, 'Sur les églises gothiques des ordres mendiants en Italie centrale', *Bulletin Monumental*, 130 (1972), 231–5; and Sundt, 'Mediocres domos . . . ', see note 9.

[46] Smith, chapter 14 in this volume, 230–233; see also Heyman (chapter 2, 23; chapter 11, 169–70; and Ackerman, '*Ars sine scientia . . .* ', see note 44.

[47] Mainstone, 'Brunelleschi's Dome'.

structure it is: an octagonal vault (like that of the nearby Florentine Baptistry), or a true dome?

In his subsequent paper, Mainstone affirms that Brunelleschi constructed the cupola as if it were a circular dome, despite its octagonal form. Historians of science, such as Gustina Scaglia, have investigated the mechanics and machinery of building, which Brunelleschi personally designed and supervised.[48] While the specific sources of Brunelleschi's ideas continue to elude scholars, it is now clear that the cupola of Santa Maria del Fiore, though resembling an octagonal vault, behaves like a true dome with the rings functioning to eliminate bending so that horizontal forces are mainly circumferential. Nonetheless, as Mainstone notes, the dome has cracked substantially, perhaps in response to temperature changes or localised tensile stresses in the masonry. When this cracking began is still unclear (see chapter 15, 254), but as the fissures propagate, their patterns of behaviour will no doubt provide engineers with further insights.[49]

Part Four – Roofs, Towers and Spires

The papers in Part Four examine the most and the least visible structural components of medieval great churches in northern Europe. On the one hand are the imposing towers, spires, flying buttresses, and high-pitched roof gables, and on the other, the invisible roof carpentry, internal iron and timber reinforcement, and the foundation systems that bear these loads.

The compelling desire in the later thirteenth century to build enormous towers and spires extending hundreds of feet beyond the roof apex required innovative engineering solutions to cope with their demands upon the original fabric. Challenges of height and wind load often called for strategic reinforcement in timber, iron and masonry, such as the dramatic strainer arches in the crossings of Wells and Salisbury cathedrals, and the ingenious integration of the high roofs and flèche at Notre-Dame de Paris (chapter 18). Since the addition of large towers inevitably exceeded the loads on walls, piers and foundations anticipated by the original builders, the stability of the entire fabric was often put at risk. The papers in this part address these structural dynamics and the developments these super-structures fostered in structural carpentry, masonry and ironwork.

Fitchen's perceptive article (chapter 16), though now in part out-dated by

---

[48]   G. Scaglia, 'Building the cathedral in Florence', *Scientific American*, 264 (1991), 66–72; she stresses Brunelleschi's debt to Roman technology as well as his considerable mechanical and mathematical skills.

[49]   See also F. Toker, 'Excavations below the Cathedral of Florence, 1965–74', *Gesta*, 14 (1975), 17–36; and more recently *Rapporto sulla situazione del complesso strutturale cupolabasamento dell cattedrale di Santa Maria del Fiore in Firenze* (Pistoia, 1985), cited in Mark, *Architectural Technology . . .* , 170.

subsequent research, makes the essential observation that there is a critical structural relationship between cathedral roofs, gutteral walls, and masonry abutments. As he illustrates, the careful placement and design of these interactive structures is important both for conditions of static equilibrium and, especially, for the live loads of wind forces acting unpredictably on tall structures – an area of inquiry that has since been explored by Robert Mark and others.[50] When writing in 1955, Fitchen commented on a seeming lack of interest by architects and engineers in the consequences of wind and thermal variations for timber and masonry (chapter 16, 267-69 and 271, fn. 21). Hence this paper signals an area of investigation that has only recently come to fruition.

Roofs

Fitchen embraces the rationalist view of Viollet-le-Duc, that all major architectural elements are functional. He argues convincingly for a structural rationale for the addition and placement of the upper tier of flyers in French High Gothic cathedrals as necessary abutment for the roof. But, typical of this era of scholarship, the actual carpentry is not a primary interest, and it is clear that Fitchen has limited knowledge of medieval roofs. This leads him, for example, to the incorrect belief that tiebeams function 'in compression' to counteract the inward pressure of the 'crown thrust' of the upper flyers against the thin clerestory walls (chapter 16, 271). But, had he observed the carefully-braced, slender tiebeams of the nave of Notre-Dame de Paris (width to length ratio of 1:60), he would probably have concluded that these members are in tension and that the triangulated form of the roof and its seating counteracts overturning at the top of the parapet wall. Nonetheless, Fitchen's observations about the general increase in roof pitch and height (and thus weight), wind pressure, leeward suction, and vortex action are an early contribution to understanding this realm of cathedral engineering. The detailed study (chapter 17) of the high roofs of Lincoln complements Fitchen's simplified description of a typical Gothic roof. Moreover, the Lincoln carpentry, whose construction sequence has been established by dendrochronology,[51] exhibits many of the characteristic features of northern European medieval roofs.

The earliest surviving cathedral roofs date to the late twelfth century,[52] though

---

[50] For example, Mark and Jonash, 'Wind loading . . . ', see note 31; and R. Mark and W.W. Clark, 'Gothic structural experimentation', *Scientific American*, 251, 5 (1984), 176–85.

[51] Dendrochronology (tree-ring dating) is the science of dating timber by the matching of ring widths of individual timbers with the widths of a previously dated master ring sequence; M.G.L. Baillie, *Tree-ring Dating and Archaeology* (London, 1982).

[52] Essential literature on roofs includes C.A. Hewett, *English Cathedral and Monastic Carpentry* (Chichester, 1985); and for France, Henri Deneux, 'L'évolution des charpentes du XIe au XVIIIe siècle', *L'Architecte*, 4 (1927), 49–68; a recent summary and bibliography is in L.T. Courtenay, 'Timber roofs and spires', in Mark, *Architectural Technology . . .* , 182–231.

evidence for more archaic carpentry survives in masonry scars, tie-beam sockets, and re-used timbers incorporated into later construction, as in the western bays of the Lincoln Angel Choir (chapter 17, 287). The wealth of datable early Gothic carpentry surviving at Lincoln from the period of 1192 to *ca.* 1250 thus makes these roof structures of exceptional interest for this vital period of cathedral building.

The eastern roofs, above St Hugh's Choir, the Angel Choir, and the eastern transept, all have a notably steep pitch (60 to 70 degrees) and thus reveal the medieval carpenters' desire to emphasise the verticality of the structure, in keeping with aesthetic tendencies that became apparent in England and northern France during the thirteenth century. Such a roof inclination, which nearly always requires a lead covering, is characteristic of High Gothic roofs in north-western Europe, in contrast to the typically lower pitches of 35 to 50 degrees which preceded them.[53] As Fitchen noted, the steeper pitch has the structural consequence of reducing horizontal forces at the wall head, but this is countered by the considerable increase in materials, weight, and the 'sail' area of the roof subject to wind loading.

As well as pitch, the Lincoln roofs illustrate some of the major developments in medieval carpentry. For example, the use of a tiebeam at the base of every rafter couple has been abandoned in favour of intermittent base ties. This practice substantially reduced the need for long, expensive timbers spanning the width of the structure and was thus a major step in economy, as well as decreasing the overall weight of the framing.

A second important structural development that relates roof and wall and pertains to overall support conditions (often designed by close collaboration of mason and carpenter) is the development of the rafter foot in conjunction with the anchoring of the tie-beams firmly to the wall head by trenching them over two or even three longitudinal wall plates often braced from below (as they are in the nave of Notre-Dame de Paris).[54]

Third, the Lincoln roofs (chapter 17, especially Figs. 5b and c, and Fig. 8) exhibit some of the means available to carpenters to stiffen the roof frame in the transverse plane, principally based on diagonal bracing, triangulation and the development of joints specialised to take either compression or tension. What is lacking at Lincoln, typical of roofs prior to *ca.* 1250, is effective longitudinal bracing to prevent axial deformation (*racking*). An early form of this type of bracing, as seen at Notre-Dame de Paris, involved the insertion of longitudinal members (*roof plates*) at various levels; this was eventually followed by the addition

---

[53] For typical early examples see P. Hoffsummer, *Les Charpentes de toitures en Wallonie. Typologie et dendrochronologie* (Namur, 1995), 121–7.

[54] For rafter-foot development and techniques for seating a high roof on a thin parapet wall see L.T. Courtenay, 'Where roof meets wall', *Science and Technology in Medieval Society*, ed. Pamela Long (New York, 1985), 109–17.

of substantial axial connecting members (*purlins*) in contact with the primary rafter couples, or principal trusses.

Spires

Aesthetically and symbolically, the tall, slender spires of Gothic cathedrals express the spiritual aspirations of an age of piety as well as its technological mastery. The construction of these remarkable stone and timber structures called for unique skills and techniques. Furthermore, their vulnerability to lightning and tempest, coupled with their invariably light-weight construction, has required equally ingenious and elegant engineering solutions to their reinforcement, repair, and support. Surprisingly, these remarkable symbols of medieval piety and urban pride have received little attention from an engineering perspective.

Two exceptions are included here (chapters 18 and 19). The first examines the slender spire (or 'flèche') erected over the crossing of Notre-Dame de Paris. This structure of timber and lead survived from its construction in the early thirteenth century until the ninetenth century, when it was removed by a cathedral architect and subsequently rebuilt by Viollet-le-Duc. Considering the usual fate of medieval timber spires surmounted by iron crosses, for example those of Reims, Arras, and Amiens, which had long since perished from violent storms or fire ignited by bolts of lightning, it is extraordinary that such a perilous structure rising to an overall height of 250 ft/77 m could survive for 600 years. It is also fortunate that the early Gothic timberwork was recorded by several architects well *before* Viollet-le-Duc, thus facilitating the interpretation of Viollet's reconstruction as incorporating the same design principles used in seating the original flèche and integrating it with the adjoining roofs.[55]

The second paper (chapter 19) concerns the masonry crossing tower and spire of Salisbury Cathedral, the tallest medieval structure in Britain, rising to *ca.* 404 ft/123 m at the apex. The main concern of this paper is the construction sequence. Using a combination of historical, archaeological, and stylistic evidence, Tatton-Brown concludes that both tower and spire were built in rapid campaigns beginning *ca.* 1300–1315. This revises the conventional construction date of 1338 and is consistent with the stylistic dating of the extensive ballflower ornamentation.[56] Tatton-Brown also argues that the initiation of the project probably belongs

---

[55] Additional drawings of the carpentry of Notre-Dame prior to Viollet-le-Duc are included in Jean Charles Krafft, *Traité sur l'art de la charpente théorique et pratique*, 5 vols. (Paris, 1819–21); these were only discovered after the publication of chapter 18.

[56] Richard Morris' detailed study of mouldings and ornament on the tower, buttresses, internal arches, and spire yield the most convincing dates and reinforce the interpretations of Tatton-Brown. See R.K. Morris, 'The style and buttressing of Salisbury Cathedral tower', in *Medieval Art and Architecture at Salisbury Cathedral*, ed. L. Keen and T. Cocke (Leeds, 1996), 46–58; T. Tatton-Brown, 'The archaeology of the spire of Salisbury Cathedral', ibid., 59–67.

to the late thirteenth century, i.e. prior to 1297, after which date 'all six deans of the cathedral were non-resident foreigners'.[57]

Structurally, the most remarkable features of the Salisbury tower cum spire are the original timber scaffold within the spire, and the medieval ironwork, particularly the extraordinary iron 'machine' by which much of the weight of the scaffold is suspended from the capstone of the spire. The scaffold functions as an immense, centralised dead-weight, stabilising the whole spire against wind loading.

Strengthening the Structure   The ever taller towers and spires added ubiquitously to great churches in the later Middle Ages imposed extra stresses both on the visible structure and on its foundations. The collapse of towers and spires was by no means uncommon; Beauvais, Chichester, Ely and Norwich are but some among the many recorded examples. Although the immediate cause of such failures was in general violent storms, contributory factors were probably movement and cracking caused by insufficient foundations, differential settlement over time, or changes in water-table and soil conditions. This is well illustrated by the recent evidence from York Minster, where extensive cracking due to inadequate foundations and large differential settlement threatened imminent collapse of the central tower.[58]

In this context, A.W. Skempton's 'Report' on the foundations of the crossing piers at Salisbury is instructive. Significantly, the archaeological evidence reveals the exceptional care the medieval builders took to minimise the weight of the tower and octagonal spire.[59] The original pier foundations, which provided for a low lantern tower rising only just above roof level, comprised a 30-inch/75 cm thick course of bedded stone and a layer of compacted flints, resting on natural gravel whose bearing capacity is estimated at 16.5 ton/ft^2/1800 kPa.[60] As constructed in *ca.* 1240, the dead load on each pier is calculated at 550 tons, transmitting a force per unit area of 3.1–3.9 ton/ft^2/330–420 kPa. However, the successive addition of the two-stage tower, additional flying buttresses, octagonal masonry spire, and crossing vault increased the total load to 1,750 tons on each pier, thus augmenting the pressure more than threefold to 9.8–12.5 ton/ft^2/1050–1340 kPa. As Skempton comments, 'Today no builder or engineer would be permitted to build the Salisbury spire on its original foundations, yet the foundations of the main crossing piers

---

[57]   Cf. T. Tatton–Brown, 'The tombs of the two bishops who built the tower and spire of Salisbury Cathedral', *Wiltshire Archaeological and Natural History Magazine*, 88 (1995), 134–7.

[58]   Dowrick and Beckmann, 'York Minster . . . '; Phillips, *Excavations at York Minster . . .*

[59]   A.W. Skempton, unpublished report to the Dean and Chapter, 1983.

[60]   For unknown reasons, the bedded stone course beneath the south-west pier was only half this thickness, leading to its bearing capacity being considerably reduced. Unlike the others, this pier has settled significantly since precise levels were first taken in 1743.

remain stable after many centuries, albeit with what would now be considered unacceptable margins of safety (1.3–1.7).'[61]

Medieval builders had no means of carrying out such precise calculations; nor could they undertake the massive reinforcements used by modern engineers to correct foundation problems – unless, indeed, they were rebuilding after a catastrophic failure. Thus, in adding towers and spires, Gothic builders could rely only on faith, prayer and invocation of the holy relics of the church's patron saints to secure the adequacy of their predecessors' foundations. However, for observable danger signs such as cracks, buckling, or fractures, in crossing piers or adjacent masonry walls, they could use traditional methods of repair to provide strengthening vaults, struts, wall buttresses, and internal masonry arches, such as the inverted 'strainer arches' at Wells Cathedral, or those added as primary reinforcement to the main crossing at Salisbury.[62]

At Wells Cathedral dramatic 'strainer arches' were erected in the 1330s by William Joy in response to major observable distress to the crossing piers, occasioned by the load of the new tower of 1315. Recent calculations of the additional load on these piers, relative to the sinking of the foundations at the west end and the resistance to the shear stresses involved, suggest that the actual effectiveness of these arches is minimal, in contrast to the steeply-pitched arched buttressing provided within the wall of the nave.[63]

In contrast to the situation at Wells, careful measures were taken at Salisbury to strengthen the base of the tower at the onset of construction by means of additional flying buttresses, internal masonry arches, and iron ties with carpentry-inspired joints. In addition to the ingenious linking of the capstone and timber scaffold (discussed above), the top of the new tower was also girded with an interlocking system of lead-sheathed iron ties buried within the masonry, while the thin masonry of the spire was later ringed with ten iron bands.[64]

Though it would have been impossible for the medieval masons to pre-determine mathematically the effects of the superimposed loads of the Salisbury tower and spire, recent scholarship and archaeological evidence indicate that these contingencies were seriously taken into account from the beginning. Thus the intuitive measures of reinforcement seen at Salisbury bear witness to the

[61] Personal communication, January 1996. Professor A.W. Skempton, Department of Civil Engineering, Imperial College, London.

[62] There are four pairs of internal arches at Salisbury; those spanning the nave, i.e. the west face of the crossing and the east face beneath the tower, are primary and date to the early fourteenth century as do those spanning the choir and retrochoir of the eastern transept. Both the main crossing and eastern transept received additional arches at a later date; Morris, 'The style and buttressing . . . ', see note 56, 51–4 and especially Fig.1, E.

[63] Mark *Experiments in Gothic Structure*, 82–6.

[64] The dating and nature of all the iron reinforcements, both original and later, are examined in detail in Reeves et al., 'Iron reinforcement . . . '.

engineering skills of medieval masons, smiths and carpenters. Moreover, the analysis of this aspect of their work may be taken as representative of much of the contents of the present volume in that it draws attention to the creativity of cathedral engineers, the named or unknown builders, who continued to respond energetically to the successes, failures, and architectural vision of their age. These papers also underscore the complexity of the buildings involved and the almost miraculous ability of these 'indeterminate' structures to endure.

Select Bibliography

Ackerman, J., 'Ars sine scientia nihil est, Gothic theory of architecture at the Cathedral of Milan, Art Bulletin, 31 (1949), 84–111.

Addis, W., Structural Engineering: The Nature of Theory and Design (Chichester, 1990).

Barnes, Carl F., Jr., Villard de Honnecourt, the Artist and his Drawings: a Critical Bibliography (Boston, 1982).

Bechmann, R., Villard de Honnecourt: la Pensée technique au XIIIe siècle et sa communication (Paris, 1991).

Bessac, Jean-Claude, 'Outils et techniques spécifiques du travail de la pierre dans l'iconographie médiévale', in Pierre et métal dans le bâtiment au Moyen Age, ed. O. Chapelot and P. Benoit (Paris 1985), 169–84.

Bilson, J., 'Durham Cathedral: the chronology of its vaults', Archaeological Journal, 79 (1922), 101–60.

Bony, Jean, French Architecture of the Twelfth and Thirteenth Centuries (Berkeley, Los Angeles and London, 1983).

Branner, Robert, Gothic Architecture, 10th edn. (New York, 1988).

Bruzelius, C. A., The Thirteenth Century Church at St Denis (New Haven, CT and London, 1985).

—, 'The construction of Notre-Dame in Paris', Art Bulletin, 69 (1987), 540–69.

Bucher, F., 'Medieval architectural design methods', Gesta, 11 (1972), 37–57.

Clark, W., and R. Mark, 'The first flying buttresses: a new reconstruction of the nave of Notre-Dame de Paris', Art Bulletin, 66 (1984), 47–65.

Clasen, K. H., Deutsche Gewölbe der Spätgotik (Berlin, 1958).

Conant, J. K., Cluny: les églises et la maison du chef d'ordre (Cambridge and Macon, 1968).

Crosby, Sumner McK., 'Some uses of photogrammetry by the historian of art', in Etudes d'art médiéval offerts á Louis Grodecki, ed. P. Crossley et al. (Paris, 1981), 119–28.

Dabas, M., C. Stegeman and A. Hesse, 'Cases, prospection géophysique dans la cathédrale de Chartres', Bulletin de la Société Archéologique d'Eure-et Loir, 36, (1993), 5–25.

Dowrick, D. J., and P. Beckmann, 'York Minster structural restoration', *Proceedings of the Institution of Civil Engineers*, 49, supplement (1971), 93–156.

du Colombier, P., *Les chantiers des cathédrales* (Paris, 1973).

Fergusson, Peter, 'The builders of Cistercian monasteries in twelfth century England', *Studies in Cistercian Art and Architecture*, II, ed. M.P. Lillich (Kalamazoo, 1984), 14–29.

—, *Architecture of Solitude: Cistercian Abbeys in Twelfth-Century England* (Princeton, NJ, 1984).

Fernie, Eric, *An Architectural History of Norwich Cathedral* (Oxford, 1993).

Fitchen, J., *The Construction of Gothic Cathedrals: a Study of Medieval Vault Erection* (Chicago, 1961).

Frankl, Paul, *The Gothic: Literary Sources and Interpretations through Eight Centuries* (Princeton, 1960).

Goodman, D., 'Ground-penetrating radar simulation in engineering and archaeology', *Geophysics*, 59, 2 (1994), 224–32.

Gordon, J. E., *Structures or Why Things Don't Fall Down* (Harmondsworth, 1978).

Graefe, Rainer, ed., *Zur Geschichte des Konstruierens* (Stuttgart, 1989).

Hahnloser, H. R., *Villard de Honnecourt: Kritisch Gesamtaugabe des Bauhuttenbuches ms fr 19093 der Pariser National Bibliotek* (Vienna, 1935; Graz, 1972).

Harvey, J. H., *The Medieval Architect* (New York, 1972).

—, *English Medieval Architects: a Biographical Dictionary Down to 1550*, 2nd edn. (Gloucester, 1984).

Hewett, C. A., *English Cathedral Carpentry* (London, 1974).

—, *English Cathedral and Monastic Carpentry* (Chichester, 1985).

Heyman, J., 'The stone skeleton', *International Journal of Solids and Structures*, 2 (1966), 249–79.

—, *The Stone Skeleton: Structural Engineering of Masonry Architecture* (Cambridge, 1995).

—, *Arches, Vaults and Buttresses: Masonry Structures and their Engineering* (Aldershot, 1996).

Jackson, M., ed., *Engineering a Cathedral: Proceedings of the Conference 'Engineering a Cathedral', Durham, 1993* (London, 1993).

James, J., 'The rib vaults of Durham Cathedral', *Gesta*, 22, 2 (1981), 135–46.

—, *Chartres: The Masons Who Built a Legend* (London, 1982).

Kerisel, J., 'Old structures in relation to soil conditions', *Géotechnique*, 25, 3 (1975), 433–83.

Kidson, Peter, 'A metrological investigation', *Journal of the Warburg and Courtauld Institutes*, 53 (1990), 71–97.

Kimpel, D., and R. Suckale, 'Le dévelopement de la taille en série', *Bulletin Monumental*, 135 (1977), 195–222.

Kimpel D., R. Suckale and A. Hirmer, *Die gotische Architektur in Frankreich, 1130–1270* (Munich, 1985). French edn., *L'architecture gothique en France, 1130–1270*, trans. Françoise Neu (Paris, 1990).

Knoop, D., and G. P. Jones, *The Medieval Mason* (Manchester, 1967).

Kubler, G., 'A late Gothic computation of rib-vault thrusts', *Gazette des Beaux-Arts*, 26 (1944), 135–48.

Kusaba, Yoshio, 'Some observations on the early flying buttresses and the choir triforium of Canterbury Cathedral' *Gesta*, 28, 2 (1989), 75–89.

Lalbat, C., M. Gilbert and J. Martin, 'De la stéréometrie médiévale: la coup des pierres chez Villard de Honnecourt (II)', *Bulletin Monumental*, 145 (1987), 387–406.

Leedy, Walter, Jr., *Fan Vaulting: a Study of Form, Technology and Meaning* (London, 1980).

Mainstone, R. J., 'Brunelleschi's Dome', *Architectural Review*, 162 (1977), 155–66.

—, *Developments in Structural Form* (Cambridge, MA, 1983; first edn. 1975).

Mark, R., *Experiments in Gothic Structure* (Cambridge, MA, 1982).

—, *Light, Wind, and Structure* (Cambridge, MA, 1990).

—, ed., *Architectural Technology up to the Scientific Revolution* (Cambridge, MA, 1993).

Mark, R., John F. Abel, and K. O'Neill, 'Photoelastic and finite-element analysis of a quadripartite vault', *Experimental Mechanics*, 13 (1973), 322–9.

Morris, R. K., 'Mouldings and the analysis of medieval style' in *Medieval Architecture and its Intellectual Content: Studies in Honour of Peter Kidson*, ed. E. Fernie and P. Crossley (London and Ronceverte, 1990), 239–47.

Müller, Werner, *Grundlagen gotischer Bautechnik: Ars sine scientia nihil est* (Munich, 1990).

Murray, S., *Beauvais Cathedral: Architecture of Transcendence* (Princeton, 1989).

—, *Notre-Dame, Cathedral of Amiens: The Power of Change in Gothic* (Cambridge and New York, 1996).

Neagley, Linda E., 'Elegant simplicity; the Late Gothic plan design of St Maclou in Rouen', *Journal of the Society of Architectural Historians*, 74, 3 (1992), 395–422.

Ovitt, George, *The Restoration of Perfection: Labor and Technology in Medieval Culture* (New Brunswick, NJ, 1987).

Pevsner, Sir Nikolaus, *An Outline of European Architecture* (first pub. London, 1943).

—, *Some Architectural Writers of the Nineteenth Century* (Oxford, 1972).

Phillips, Derek, *Excavations at York Minster: the Cathedral of Archbishop Thomas of Bayeux*, II (London, 1984).

Praeger, F. D. and G. Scaglia, *Brunelleschi: Studies in his Technology and Inventions* (Cambridge, MA, 1970).

Raguin, V., K. Brush and P. Draper, ed., *Artistic Integration in Gothic Buildings* (Toronto, 1995).

Recht, R., 'Sur le dessin d'architecture gothique', *Etudes d'art médiéval offerts à Louis Grodecki*, ed. P. Crossley et al. (Paris, 1981), 233–50.

Reeves, John, W. G. Simpson and P. Spencer, 'Iron reinforcement of the tower and spire of Salisbury Cathedral', *Archaeological Journal*, 149 (1992), 380–406.

Saalman, H., *Filippo Brunelleschi: The Cupola of Santa Maria del Fiore* (London, 1980).

Salzman, L. F., *Building in England Down to 1540* (Oxford, 1952).

Sanabria, S., 'Metrics and geometry of Romanesque and Gothic St Bénigne, Dijon', *Art Bulletin*, 62 (1980), 518–32.

Sauerlander, Willibald, *Das Jahrhundert der grossen Kathedralen: 1140–1260* (Munich,1990).

Scheller, R. W., *A Survey of Medieval Model Books* (Haarlem, 1963).

Schöller, Wolfgang, 'Ritzzeichnungen: Ein Beitrag zur Geschichte der Architekturzeichnung im Mittelalter', *Architectura*, 19 (1989), 31–61 (includes catalogue, 47–61).

Shelby, L. R., 'The education of medieval English master masons', *Medieval Studies*, 32 (1970), 1–26.

—, 'The "secret" of the medieval masons', in *On Pre-Modern Technology and Science: Studies in Honor of Lynn White, Jr.,* ed. Bert Hall and D. C. West (Malibu, 1976), 201–19.

Sundt, R., 'The Jacobin Church of Toulouse and the origins of its double-nave plan', *Art Bulletin*, 71 (1985), 185–207.

—, 'From half to full palmier: factors contributing to the final chevet design of Toulouse's Jacobin Church', *Avista Forum* 9, 2 (1996), 7–15.

Taylor, W., and R. Mark, 'The technology of transition: sexpartite to quadripartite vaulting in High Gothic architecture', *Art Bulletin*, 64, 4 (1982), 579–87.

Thurlby, M., 'The purpose of the rib in the Romanesque vaults of Durham Cathedral' in *Engineering a Cathedral,* ed. M. Jackson (London, 1993), 64–76.

Trachtenberg, M., *The Campanile of Florence Cathedral* (New York, 1971).

Ungewitter, George G., *Lehrbuch der gotischen Konstructionen*, 2 vols. (Leipzig, 1901/1903).

Van der Meulen, Jan, 'Die Kathedrale im Verfall: Chartres und die Expertise of 1316' in *Akten des XXV Internationalen Kongresses für Kunstgeschichte* (Vienna, 1985), 53–64.

Victor, Stephen K., ed., *Practical Geometry in the High Middle Ages: 'artis cuius libet' and the 'Pratike de geometrie'* (Philadelphia, PA., 1979).

Viollet-le-Duc, E. E., *Dictionnaire raisonné de l'architecture française du XIe au XVIe siècle*, 10 vols. (Paris, 1854–1868).

von Simson, Otto, *The Gothic Cathedral* (2nd edn. revised, New York, 1962).

—, 'The Cistercian contribution to architecture' in *Monasticism and the Arts*, ed. T. G. Verdun (Syracuse, 1984), 127–37.

Whitney, Elspeth, 'Paradise restored, the mechanical arts from antiquity through the thirteenth century', *Transactions of the American Philosophical Society*, 80, 1 (1990), 1–169.

Wiemer, W., and G. Wetzel, 'A report on data analysis of building geometry by computer', *Journal of the Society of Architectural Historians*, 53, 4 (1994), 448–601.

Wilcox, R. P., *Timber and Iron Reinforcement in Early Buildings* (London, 1981).

Willis, Robert, 'On the construction of vaults in the Middle Ages', *Transactions of the Royal Institute of British Architects*, 1, 2 (1842), 1–69; reprinted R. Willis, *Architectural History of Some English Cathedrals*, II (Chichley, 1972).

Wolfe, M., and R. Mark, 'The collapse of the Beauvais vaults in 1284', *Speculum*, 51 (1976), 462–76.

1

Robert Willis, Viollet-le-Duc and the structural approach to Gothic architecture

Robert Mark

Following soon after its genesis as a romantic-literary movement, the Gothic revival kindled a number of serious investigations of the organization and construction of medieval building. The bulk of these early studies have been so far surpassed by modern scholarship that they remain merely as interesting curiosities, but a few mid-nineteenth-century writers have avoided this fate and are still being consulted and frequently quoted in our time. By far the most important of these for English medieval architecture is Robert Willis (1800–1875), and for French architecture, Eugène Viollet-le-Duc (1814–1879). Willis's most valuable contributions comprise a paper published in 1842 on the construction of medieval vaults and the series of papers derived from lectures on the archi-

tectural history of individual English cathedrals that he delivered during the years 1842–1863[1]. Viollet's major works are his monumental, ten-volume encyclopedia, *Dictionnaire Raisonnée de l'Architecture Française du XIᵉ au XVI Siècle*, ori-

* The research for this paper was largely carried out while the writer was a recipient of a National Endowment for the Humanities Fellowship in 1973–1974 when he was extended the courtesy of the use of the Library of the Society of Antiquaries, London.

[1] Robert Willis, "On the Construction of the Vaults of the Middle Ages", *Trans. Royal Institute of British Architects*, Vol. 1, Part ii, 1842, pp. 1–69. This article and all of Willis's known studies of individual cathedrals have been reproduced in a two-volume set: Robert Willis, *Architectural History of Some English Cathedrals*, Chicheley 1972–1973.

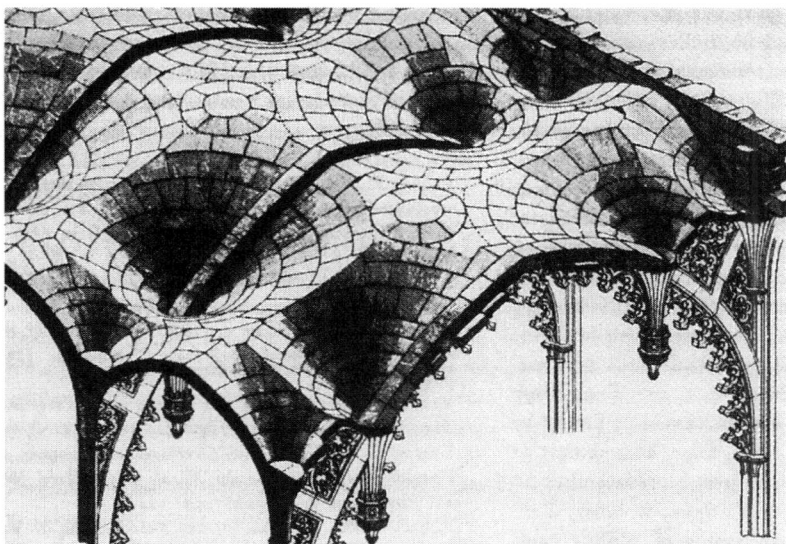

1. Sketch of the vaults over the Henry VII Chapel at Westminster by Willis

ginally published between 1854 and 1861, and his two-volume manifesto, *Entretiens sur l'Architecture*, published in 1863 and 1872, which first appeared in English translation in 1875 [2].

Both Willis and Viollet are usually depicted as *functionalists* or interpreters of historic architecture with emphasis on the construction and performance of the building fabric. Willis was one of the first to view Gothic architecture in this way. Indeed, while the opening of Pugin's *True Principles* has often been taken as a seminal statement of functionalism in medieval architecture and its application to contemporary architecture ("There should be no features about a building which are not necessary for convenience, construction, or propriety" [3]), seven years previously Willis had stated about Gothic construction: "Every member, nay, almost every moulding is a sustainer of weight, and it is by this ... that the eye becomes satisfied of the stability of the building" [4].

Figure 1, an often-reproduced Willis drawing of the vaults over the Henry VII chapel at Westminster Abbey, is an example of his singular ability to penetrate the complexities of Gothic construction. Similar, but usually less elaborate, drawings of construction details abound in Viollet's *Dictionnaire*. Hence, it may not be surprising that Reyner Banham could infer that Willis was Viollet's mentor. Banham writes in his *Theory and Design in the First Machine Age*: "... there were a number of particular predisposing causes that helped to guide the mainstream of development [of the theory of Modern architecture] into the channels through which it flowed in the [nineteen] Twenties. These predisposing causes were all of nineteenth-century origin, and may be grouped under three heads ... [the second of these is] the Rationalist or structural approach to architecture ... of English extraction, from Willis, but elaborated in France by Viollet-le-Duc ..." [5]. That some communication between Willis and Viollet took place is well documented – but not enough, I think, to justify Banham's chauvinism. A translation of Willis's vault construction article appeared in the *Revue Générale de l'Architecture* in 1842, the same year as the

publication of the original English version, and Viollet, in Vol. IV of the *Dictionnaire*, used sketches of vault surfaces that are reminiscent of Willis's drawings and he made reference to the translation [6]. Willis, in his annotated *Villard de Honnecourt Sketchbook*, published in 1859, reproduced a Viollet illustration of a Reims Cathedral chapel [7]. However, more important than these few cross-references are the basic differences of both writers in their interpretation of the Gothic as a structural example. Willis, a keen observer, attempted to sort out characteristic geometric patterns: "In the oldtime, one style alone was practiced in each period, and a few simple [geometric] rules were sufficient for the purpose. The change or improvements of one or more of these rules introduces new features and new characters, but which still are alone employed as long as they last, and until they are in turn superseded. But we, imitators of all styles, must have more comprehensive and flexible rules, capable of imparting to our works the characters of every age in turn" [8]. Viollet, convinced that many of the elements of the mature Gothic were derived from the unique demands of construction or structural stability, took a further step which distinguished him from all his contemporaries, including Willis, and which has led Sir John Summerson to characterize him as "the last great theo-

[2] Eugène E. Viollet-le-Duc, *Discourses on Architecture*, translated into English by Henry Van Brunt, Boston 1875, 1881 (originally published in Paris in 1863 – Vol. I. 1872 – Vol. II).

[3] A. N. W. Pugin, *The True Principles of Pointed or Christian Architecture*, London 1841, p. 1.

[4] Robert Willis, *Remarks on the Architecture of the Middle Ages, especially of Italy*, Cambridge 1835, p. 20.

[5] Reyner Banham, *Theory and Design in the First Machine Age*, New York 1967, p. 14.

[6] Eugène E. Viollet-le-Duc, *Dictionnaire Raisonée de l'Architecture Française du XI au XVI Siècle*, Vol. IV (construction), Paris 1875, p. 121 ff.

[7] Robert Willis, *Facsimile of the Sketch Book of Villard de Honnecourt*, from G. B. Lassus, translated and annotated, London 1859.

[8] Willis, "Vaults", pp. 23–24.

rist in the world of architecture" [9]. Instead of geometric rules, he sought for universal principles in historic construction which might then be applied to current architectural design (it was no accident on Viollet's part that his writings became one of Banham's "predisposing causes" behind the Modern movement). He realized that the laws of structural mechanics must apply to all architecture at all times: "We have various opinions respecting the method of expressing our ideas in architecture . . . but we are all agreed as to the rules dictated by good sense and experience and by the inexorable laws of statics . . . it is not so much the forms of art that we must teach our youth as these invariable principles" [10].

Viollet's crisp, modern view of a fundamental tenet of modern architectural design contrasts sharply with that of Willis. The great irony in this comparsion is that the structural approach as an architectural doctrine might have been expected from Willis, who was a prominent engineer and a fellow of the Royal Society, instead of from a man who came primarily from a literary background, was self-trained as an architect, and was not particularly knowledgeable about the science of structural mechanics. Some further information on the experience of both Willis and Viollet, and comparsions of their comments on specific buildings with the results of recent structural investigations are offered here to clarify further their different viewpoints and to help explain this irony.

Eugène Emmanuel Viollet-le-Duc, Restorer and Architectural Theorist

According to Summerson, Viollet's ability to "think his way through the romantic attraction of style to a philosophic point of view applicable to all buildings at all times" derived from a conscious "independence of mind" [11]. His informal training in architecture did not stem from any lack of interest in the subject, but rather from a decision to remain free of the formalism of a proffered École des Beaux-Arts education. The major influence on Viollet's formative years seems to have come from his uncle, Eugène Délécluze, a political radical, artist (he had been a student of Jacques-Louis David), and art critic. Délécluze was in the habit of holding weekly salons which gathered such literary lights as Prosper Mérimée, the author of Carmen. Since the movement to restore the ancient French monuments was also fueled by romantic literature (Victor Hugo's Hunchback of Nôtre-Dame was published in 1831), it was not entirely a trick of fate that led two habitués of Délécluze's literary salons, one of whom was Mérimée, to be appointed senior officers of the Commission des Monuments Historiques at its formation in 1837. In the following year, Viollet, then 24, was nominated to a Commission post, and in 1840 he was sent to repair the abbey church of Vézelay. Fame came in 1844 when he and the architect Lassus won a competition to restore Nôtre-Dame in Paris. This was followed over the next two decades by numerous commissions for restoration, including the cathedrals of Amiens, Chartres, Reims and Clermont-Ferrand. With this experience, Viollet became the foremost authority on medieval construction and, as an admirer of what he perceived as the products of a secular, intellectual movement, he persuasively argued that many of the High Gothic architectural elements, such as the vault ribs, flying buttresses and pinnacles, were originally derived from structural exigency. What set Viollet apart from his contemporaries, awash in the Gothic revival movement, was his proposition that modern architectural elements might be derived in the same way from the newly available materials of the industrial age. He had in mind iron, not the reinforced concrete, steel and glass of the Modern Movement, but his ideas were not to be dependent upon the use of any particular material. Because this hypothesis took on such importance, by the turn of the century (well after

9 John Summerson, *Heavenly Mansions*, New York 1963, p. 135.

10 Viollet-le-Duc, *Discourses*, p. 144 ff.

11 Summerson, *op. cit.*, p. 193 ff. Further biographical details may also be found here as well as in Nikolaus Pevsner, *Some Architectural Writers of the Nineteenth Century*, Oxford 1972.

2. *Beauvais Cathedral. Mechanical
Diagram of the Buttressing
by E. E. Viollet-le-Duc*

3. *Photoelastic-interference Patterns
in Model of Reconstructed Beauvais
Section under Simulated Wind Loading*

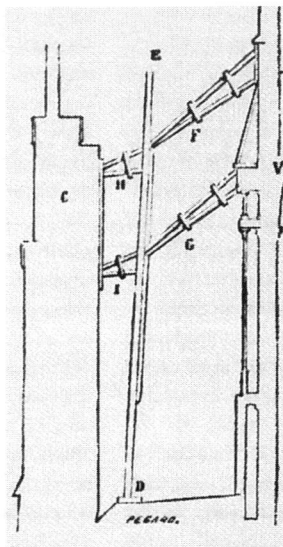

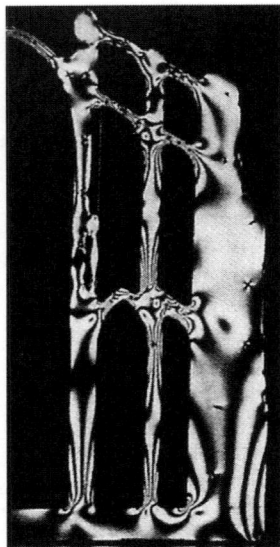

Viollet's death) his work became the target of considerable controversy among architectural historians[12]. His restoration was criticized as being insensitive to particular periods of medieval construction and it was pointed out that his few contemporary projects were far behind the expectations of his theory. But the strongest arguments came from antagonists such as Pol Abraham (1883–1966) who implied that if Viollet's technical analysis of the Gothic could be shown to be wrong, then the whole concept of structural rationalism might fall. Abraham accused Viollet of employing "romanticized mechanics" to prop up a "subjective" thesis[13], and in the light of his more modern analysis, inconsistencies in Viollet's reasoning did show up. But Abraham, trained as an architect, also got into difficulties of analysis; one result has been a lively debate on the technical issues which has been carried on up to our own time.

A sample of Viollet's analysis of structural form and its pitfalls is provided by part of a recent study of possible causes of the collapse of the vaults of Beauvais Cathedral[14]. Although he was not directly associated with it as restorer, the unsurpassed vaulting height (48 meters) of Beauvais and the fact that the high vaults collapsed in 1284, twelve years after completion of the choir, could not have escaped Viollet's attention. From his observation of its existing fabric, he reconstructed a portion of the pre-collapse choir section, and then cited its buttressing system as an example of rational design (for which he was criticized by Abraham "as reducing the most lyric architecture ... to the level of the narrowest utilitarianism"[15]). Viollet's analysis is based on his observation that the upper portion of the intermediate flying buttress support is just balanced on the lower section so that it will

[12] Paul Frankl, *The Gothic, Literary Sources and Interpretations Through Eight Centuries*, Princeton 1960, p. 799 ff.
[13] Pol Abraham, *Viollet-le-Duc et le Rationalism Médiéval*, Paris 1934, p. 102.
[14] Maury I. Wolfe and Robert Mark, "The Collapse of the Vaults of Beauvais Cathedral in 1284", *Speculum*, Vol. LI, No. 3, July 1976, pp. 462–476.
[15] Abraham, *op. cit.*

tend to incline towards the clerestory and hence help to resist the thrust of the high vault. To explain this notion, he employs a mechanical model with a system of inclined props [16]. But for the support to function as he suggests (fig. 2), hinges must be assumed just above the side aisle (point D on his sketch), as well as at the points of attachment to the flying buttresses. And since hinging in masonry construction is tantamount to its cracking, this explanation of the mechanical action of the overhanging supports could just as easily be taken as a demonstration of the structure's weakness rather than its rationality. This weakness was underscored by our investigation which was based on a structural analysis of the reconstructed Beauvais choir section using a small-scale photoelastic model (fig. 3) [17]. Under combined, scaled, live (wind) and dead (self-weight) loadings, the model revealed the presence of bending and consequently high tensile stresses at the base of the intermediate flying-buttress support – just at the point where Viollet should have placed his hinge at the base of the inclined prop of figure 2. A combination of wind and dead weight loadings would have produced cracks at this location on both sides of the intermediate support, and this, we believe, may have led to the collapse of the vaults. Viollet ought not to be faulted severely for his inference, since the structural forces within the Beauvais buttress system are not intuitively obvious, and he has correctly noted that the pier buttresses do not receive as much horizontal thrust from the vaults (and the effect of wind on the high roof) as might be expected. His mechanical model was not badly chosen, and if he had interpreted the model in a different way, it could have helped him to sense the cause of the vault failure. Furthermore, there is no evidence that this type of mechanical model was employed by Willis in any of his considerations of historic structures.

Robert Willis, Engineer and Architectural Historian

Willis appears to have been a kind of beneficent Jekyll and Hyde. His penetrating historical research was somehow accomplished while he pursued

a brilliant technical career. Educated at Cambridge, elected a fellow of the Royal Society in 1830, and appointed to the chair of Jacksonian Professor of Natural and Experimental Philosophy at Cambridge in 1837, his principal scientific work was in kinematics, the study of the motion of mechanisms. Through this he became "the first Cambridge professor to win an international reputation as a mechanical engineer" [18]. Although André-Marie Ampère is credited with originally defining statics and kinematics as separate branches of elementary mechanics in 1829, it was not until the publication of Willis's *Principles of Mechanisms* in 1841 that the field of kinematics was given a systematic, analytical basis (the Anglicized, standard term for this branch of mechanics is from Willis) [19]. Statics of structures was not Willis's speciality, but he must have been familiar with the subject as he was in charge of the first documented, dynamic structural testing of beams (for railway bridges, published in 1848 in the Iron Commissioner's Report) and the subsequent development of mathematical theory for deflection under dynamic loading [20].

The qualities which distinguished Willis's scientific methodology are echoed in Sir Nikolaus Pevsner's appraisal of his historical approach: "Willis

---

16 Viollet-le-Duc, *Dictionnaire* IV, p. 177 ff.
17 Details of the photoelastic modeling approach may be found in Robert Mark, "The Structural Analysis of Gothic Cathedrals: A Comparison of Chartres and Bourges", *Scientific American*, Vol. 227, Nov. 1972, pp. 90–99.
18 T. J. N. Hilken, *Engineering at Cambridge University 1783–1965*, Cambridge University Press 1967.
19 Eugene S. Ferguson, "Kinematics of Mechanisms from the Time of Watt", *Bulletin of the U.S. Museum*, 228: Paper 27, Museum of History and Technology, Washington, Smithsonian Institution, 1962, pp. 209–213.
20 Published as an Appendix to the account of the tests of the Conway and Britannia bridge sections by Fairbairn and Hodgkinson in the *Report of the Commissioners Appointed to Inquire into the Application of Iron to Railway Structures*, London 1849. See also Stephen P. Timoshenko, *History of Strength of Materials*, New York 1953, pp. 173–178.

was much ahead of others in precision of description, clarity of thought and expansion of general theory ... standard of insight and meticulous accuracy which has never been surpassed ... the greatest observer and indefatigable collector of data, both visual and literary"[21]. In fact, Pevsner credits Willis and his engineering colleague at Cambridge, William Whewell (1794–1866), with pioneering the systematic study of medieval buildings. (Whewell published one important architectural work on German medieval churches in 1830, but did not continue a lifelong preoccupation with architectural history as did Willis.)

Another side of Willis also has some bearing on our inquiry. He was ordained as deacon and priest of the Church of England in 1827, and at the formation in 1839 of the Cambridge Camden Society, he and Whewell were made vice-presidents. The Society, constituted at the height of the Revival movement in England, professed to promote the study of medieval architecture. However, it soon became an instrument for criticism of new church styles with the ultimate intent of reintroducing Catholic ritual to the Anglican church. Willis resigned from the Society in 1841 in protest that it had been improperly converted "into an engine of polemic theology"[22], which suggests what we would expect: that his interests were more architectural than religious. Possibly, too, he was opposed to the direction of the Society's dogma.

Willis's interpretation of the master builder's relation to the structural mechanics of medieval buildings is probably best summed up by the closing statement from his essay on vaults: "The necessary limits of a paper of this kind have prevented me from introducing several topics which may appear to belong to the question in hand. Thus, I have said nothing respecting mechanical principles, and have confined myself to form and management [i.e., geometry and construction]. But it appears to me from examination of the works of the Middle Age architects, that the latter considerations had an infinitely greater influence upon their structures than the relations of pressure, then very little understood, and about which they made

manifest and sometimes fatal errors"[23]. This description of the medieval architect's understanding of internal stress distributions is no doubt correct. However, this should not conceal the fact that certain design decisions were taken in response to the action of forces within the large buildings. Willis's decision not to grapple with structural questions might be put down to prudence, since the structural action of vaulting is so complex[24]; but he did not seem to take any great interest in the structural rationale of simpler elements either. For example, in his comprehensive history of Canterbury Cathedral, he compared sections of the choir as it was before and after the fire of 1174 (fig. 4)[25]. The post-fire section, largely the work of William of Sens until his near-fatal fall from construction scaffolding in 1179, embodies unusual, light flying buttresses slung low over the side aisle roof. If Willis is correct in showing their existence before this campaign of building was finished in 1185, then these flyers must be among the oldest now extant. The modern view of the development of flying buttresses is that they were employed for the first time in the nave of Nôtre-Dame at Paris shortly before 1180[26]; however, these were dismantled with the thirteenth-century campaign of rebuilding so that the original disposition of the Paris flying buttresses remains something of a mystery. The work on the Canterbury choir begun under the direction of a French master familiar with the work at Paris could, therefore, be a link to the earliest flying-buttress system. It would be unfair to have expected Willis, well over a century ago, to have known this; but we should expect some comment

[21] Pevsner, op. cit., pp. 52, 54, 59.
[22] Pevsner, op. cit., p. 124.
[23] Willis, "Vaults", pp. 68–69.
[24] See Robert Mark, John F. Abel and Kevin O'Neill, "Photoelastic and Finite-element Analysis of a Quadripartite Vault", Experimental Mechanics, Vol. 13, No. 9, Aug. 1973, pp. 322–329.
[25] R. Willis, The Architectural History of Canterbury Cathedral, Oxford 1845, p. 72 ff.
[26] Robert Branner, Gothic Architecture, New York 1961, p. 27.

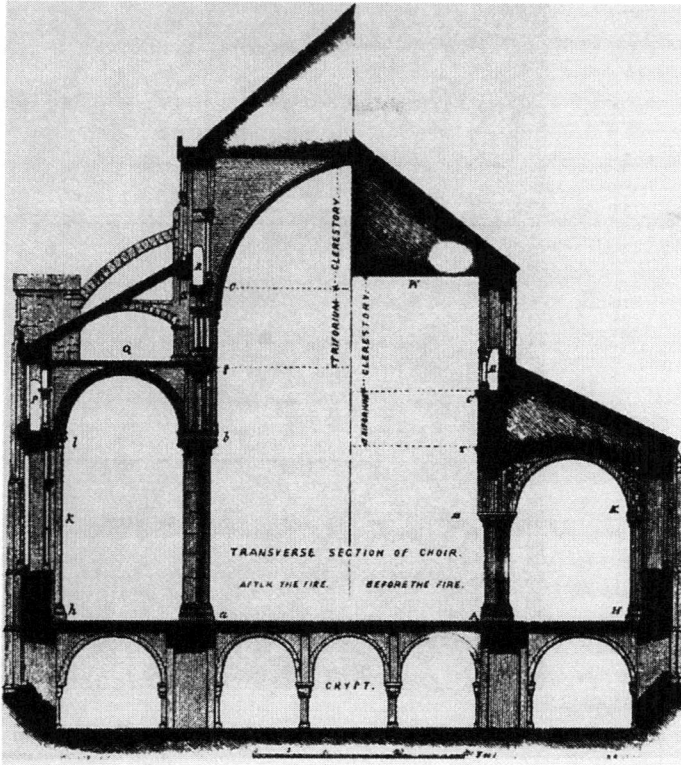

4. *Canterbury Cathedral. Pre and Post 1174 – fire Cross-sections of the Choir by Willis*

from him about the use, or at least the appearance, of this important technical device at Canterbury – yet there is none. The absence is felt particularly when one considers that for Viollet the flying buttress was the culminating example of structural need redefining the whole aesthetic of medieval building.

It is not intended here to demean Willis's imposing historical work, but rather to illustrate that, although he applied his keen sense of geometry for describing the construction and form of medieval buildings, he paid rather less attention to the work of their structure. A final example to illustrate this point is his remarks on the tower reinforcement at Wells.

The Inverted Arches of Wells Cathedral

In the beginning of the third decade of the fourteenth century, following what had become the style of major English churches, a great tower and spire were raised at Wells over the central crossing piers [27]. The additional weight of the tower caused distress to the earlier construction which required that the crossing be reinforced; part of that reinforcement consists of three sets of inverted arches across the nave and both transept arms (fig. 5). The

[27] E. A. Freeman, *The History of the Cathedral Church of Wells*, London 1870, p. 118.

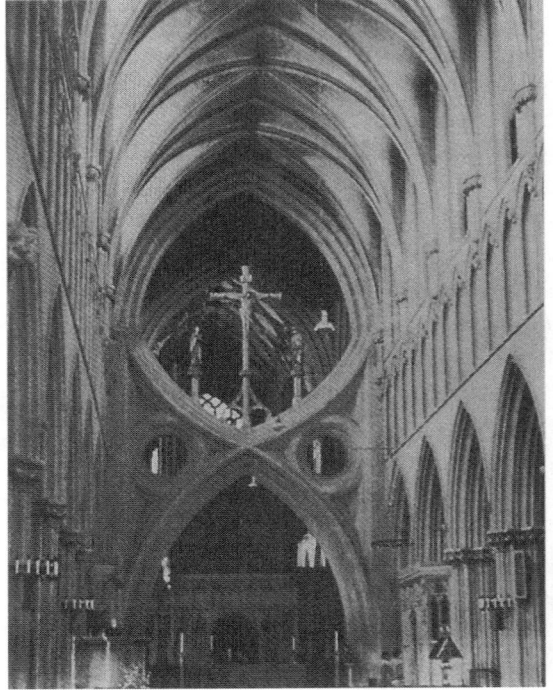

5. Wells Cathedral Interior. Inverted Arches Across the Nave

unique visual impact of these structures has made them a much-discussed element of the building. What structural role, if any, do these inverted arches play?

To begin, modern histories of the cathedral (after 1870) were consulted and it was found that all incorporated the substance of two lectures delivered by Willis at Wells in 1851 and in 1863. Unfortunately they were never transcribed; there remains only a summary of the second lecture [28]. From Wells records, Willis traced convocations of the diocese that authorized various campaigns of construction and repair to the building. According to the summary (f.n. 28) he indicated that the "convocation (of 1338) was summoned because the church of Wells is so enormously fractured and deformed ('enormiter confracte ... et enormiter defor-

mate') that its structure can only be repaired, and with sufficient promptitude by the common counsel and assistance of its members." The result of this meeting was that "the great piers of the tower are cased and connected by a stone framework which is placed under the north, south and west tower-arches, but not under the east." He further notes that "the original, high, narrow windows had been fortified with later insertations, by way of bonding and stiffening the structure endangered by the sinking of its piers below, and producing on the outside a singular mosaic of styles in which late canopy and panel work is inserted in the earlier openings.

---

[28] *Proceedings of the Somersetshire Archaeological and Natural History Society*, Vol. XII, 1863–64, pp. 14 to 22.

6. *Wells Cathedral Interior. Wall Buttress in the Nave*

These works probably occupied many years, and were added from time to time as fresh symptoms of failure exhibited themselves, although the first alarm is indicated by the convocation of 1338." There is no indication here that Willis applied any structural theory to verify his premise about the role of the inverted arch "connectors" between the tower piers. In fact, the premise can be refuted by an analysis based on some observations of pier settlements and alterations of the building fabric, estimating the weight for the early tower and its supporting structure, and by simply evaluating the structural capacity of the tower supports.

The most important observation in the first instance is that the settlement is largely a problem of the western crossing piers only and of these, the settlement is worst of all in the north-west pier. String courses above the arcading in the nave and western walls of the transepts deform downward

near the crossing piers, and the tower itself as well as the nave piers leans westward. Further, buttressing which partially blocks off clerestory window and triforium openings in bays adjacent to the tower has been added to the nave and western transept walls. If the inverted arches had indeed been intended to tie together offending piers, the existence of three arches (across the nave and each transept arm) rather than four points also to the problem of the western piers. We many then consider the structural performance of the principal supports added after 1338, the wall buttresses and inverted arches, as they might have acted to prevent this western settlement. Unlike the inverted arches, which must be the most striking feature of the church interior, the wall buttresses are unobtrusive (figs. 6, 7). Their construction, particularly their coursing, indicates that they were conceived to function as very steeply sloped flying buttresses carrying mainly vertical loading from the tower to the adjacent piers. As with any flying buttress, they also contain horizontal components of force and these must be adequately resisted at the buttress ends, at the top of the buttresses by the crossing arches, and at their base by the gallery floors and walls. Although the buttresses are relatively slight, their form, which places constituent stones primarily in compression, is most efficient for masonry. On the other hand, the form of the inverted arches, trussed by stone rings within the spandrels, does not permit similar buttressing. The appearance of the inverted arches has drawn comment about their being "strange flying buttresses", but in a structural sense, this is not truly the case. Instead of the efficient compression action of flying buttresses, these rigid structures would transmit the pier loading across their spans by shearing action which tends to slide rather than compress constituent stones and thus produces tensile cracking (the supporting action of the structural elements is depicted in fig. 8).

The relative effect of both supporting structures may be derived from an estimate of their strength. The minimum cross-sectional area of the (two) wall buttresses supporting each western pier is about four

7. Wells Cathedral. Wall Buttresses Below Central Tower

square feet, and the cross sectional area at the center of the inverted arches (this section is shown in
fig. 9) is eighteen square feet. The masonry may be
assumed to accept 1000 pounds per square inch in
compression without distress and an average shear
stress across a section of 50 pounds per square inch
(or equivalently 50 psi in tension, for which this is
a generous estimate). Assuming stable supports, the
maximum load that can be carried by each structural element is found by multiplying these stresses

by the corresponding cross-sectional area. The resulting maximum shear through the inverted arches
is then found to be 65 tons and the maximum vertical component of the wall buttress compression is
270 tons, giving the ratio of their relative ability to
support the pier as being somewhat less than one to
four.

These resisting forces may be compared with an
estimate of 2500 tons for the total dead-weight
loading of the tower, fourteenth-century spire and

the crossing piers (exclusive of the footings) made from observations of the existing structure and from drawings given by one of the building's surveyors[29]. Considering a fourth of this total acting on each crossing pier, the wall buttresses could, if called upon, carry almost half of the pier loading, while the arches across the transepts could relieve the western piers by only about ten percent. The inverted arches do not appear to have been called upon to transmit major loads between the crossing piers, as no significant distress is evident at their centers. Their other functional role could be to brace the piers laterally below the crossing arches in a manner similar to that achieved at Salisbury Cathedral with similar but lighter structure. However, the relatively low height of Wells (the springing of the crossing arches begins at a level of only about 15 meters above the floor) should preclude the need for this additional lateral bracing.

In any discussion of the utility of the pier reinforcement, a typical characteristic of soils should also be taken into account. With stable subsoil conditions (e.g., with the water table level remaining constant), the rate of foundation settlement is usually far greater immediately after new construction and then decreases, often becoming imperceptible after about 10 to 20 years. In consequence it may well be that almost all the possible damage to the building from the uneven settlements was already realized by 1338, and that the reinforcement had little effect at that time. However, it could have helped to prevent further settlement during later alterations which added still more weight to the central tower. Hence, the buttressing placed within the western walls, though much less in evidence than the inverted arches and implied by the Willis summary to play a secondary role, was in fact a prudent response to the "fracture and deformation" of the building; it is most unlikely that the inverted arches have similar effect.

The analysis employed here was very simple, within the realm of an engineer of Willis's time. Admitting that the estimates of masonry strength would not have been readily available to Willis, nevertheless it must be concluded that he did not seriously consider the building in terms of the mechanics of its structure, in spite of the fact that the question posed was so overwhelmingly structural. Some explanation for this may be found by reviewing Willis's cultural milieu and comparing it with Viollet's.

The Structural Paradox

In the first place, we note that Willis's work at Cambridge was not primarily concerned with statical analysis. Both Willis and Whewell were important figures in the establishment of the English university tradition of applied mechanics and the introduction of analytic approaches to problem solving. In effect, they subdivided the field into the specialties of statics (as a forerunner of civil engineering studies) and kinematics (a forerunner of mechanical engineering) with Whewell contributing to the former and Willis mainly to the latter. However, even the most intense specialization in another technical area would not seem a convincing reason for a man of Willis's talent to shun questions of statics of historic structures.

A better explanation is suggested by taking into account Willis's architectural environment, particularly the important structural differences between the major English and French High Gothic churches. Excepting Westminster Abbey, which is more French than any of the others in structure (and was not dealt with by Willis), all of the English cathedrals had internal vessel heights well below thirty meters. The extremely high-vesseled church, a thirteenth-century French phenomenon first seen at Chartres and Bourges and then at Reims, the choir of Le Mans, Amiens, Metz and culminating with the choir of Beauvais, crossed French borders to Germany (Cologne), Italy (Milan) and Spain (Palma, Majorca), but not to England. (The cathedrals of Cologne, Milan and Palma rank two, three and four following Beauvais in height of Gothic

[29] Charles Nicholson, "Construction and Design", *RIBA Journal*, Vol. XIX, 1912, pp. 627–628. See also John Britton, *The History and Antiquities of the Cathedral Church of Wells*, London 1824, p. 98.

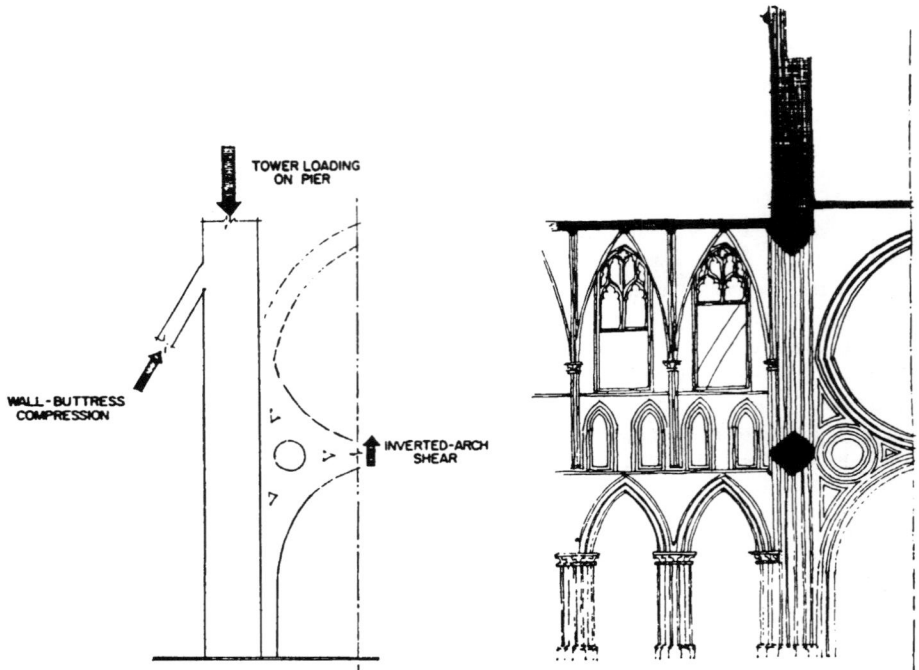

8. *Wells Cathedral. Simplified Mechanical Diagram of Inverted Arch and Wall Buttress Forces acting on a Western Pier*

9. *Wells Cathedral. Section Drawing showing Disposition of Wall Buttress and Inverted Arch by W. Shellman*

building, while even the high vaults of Westminster Abbey are lower than those of any of the other buildings cited here). It was all very well for Francis Bond to make the observation that "the practical, English [Gothic] builder avoided . . . mischief of rain, frost and storm" by eschewing the exposed flying buttress [30], but it must be realized that unlike their French counterparts, the classical English buildings were sufficiently heavy and low in height so that they could maintain stability without flying buttresses. Indeed, flyers were only later added on many English churches as required for repairs or, more often in the nineteenth century, for stylistic reasons. In this environment, it is easier to accept that Willis's attention might have been diverted from structural questions. On the other hand, the technical options for the French Gothic builders

were fewer. New, critical problems of stability of the high vaults and the effect of wind loads on lofty roofs had to be met. There were also economic pressures to lighten the fabric of these huge buildings in order to reduce the expense of transporting stone from the quarries and of shaping and setting it into place. There should be no doubt that technical exigencies provided strong impetus in the development of the French High Gothic form, and these were the buildings that Viollet chose as his examples of structural rationalism.

Finally, and perhaps most important were the different intellectual climates of nineteenth-century Britain and France. In Britain the Gothic revival

30 Francis Bond, *Gothic Architecture in England*, London 1905, p. 371.

63

was *backward looking*. The rampant industriali- zation of that country brought forth a campaign of reform: an attempt to revive a romanticized version of the medieval structure of society, as well as of architecture. Willis, at Cambridge, was at its epicenter. Although he resigned from the vice-pres- idency of the Camden Society, the influence of the revival movement must have carried over, for he too was caught up in the notion that an important product of his research was the accurate *reproduc- tion* of medieval construction. In the year following the resignation, Willis introduced his study of vaults by observing that "independent of the value of such investigations to the history of the science of construction, the [technical] knowledge ... would greatly assist us in the imitation of the works of each period" [31].

As in England, the Revival movement in France had its beginnings in literature, but unlike England, the major issues by the mid-nineteenth century were architectural: what was to be the course of contem- porary French design? As might be expected, Viollet took up the cause of defending the ideas of the Gothic, presumably with the faculty of the Ecole des Beaux-Arts pressing for eclecticism on the other side [32]. The lines were drawn; conditions were ripe for the presentation of a new, overwhelming thesis and Viollet was the man of the hour with the in- tellect and the strong conviction from his construc- tion experience, if not the technical ability of a Ro- bert Willis, to transcend the controversy. Perhaps he succeeded all too well. His theory is so perva- sive that the meaning of the structural approach is usually taken as being synonymous with functional- ism (à la Pugin) in modern architectural literature, and this has tended to mask the uniqueness of Viollet's contribution.

[31] Willis, "Vaults", p. 2.
[32] Georg Germann, *Gothic Revival in Europe and Bri- tain: Sources, Influences and Ideas*, Cambridge 1972, p. 135 ff.

2

On the rubber vaults of the
Middle Ages and other matters

J. Heyman

FRANKL's massive study [1] of Gothic is a complete survey "of what has been thought and written about the phenomenon of Gothic as a whole since Suger" [2]. As such, the study tends towards the historian's, or perhaps historiographer's, approach, in that the prime question posed is "What did contemporary architects think they were doing?", coupled with discussion of what art historians thought that contemporary architects thought they were doing.

An equally valid prime question is "What did contemporary architects actually do?". This is discussed thoroughly, but as a secondary matter, by Frankl; he gives a large bibliography, but did not have the benefit of the basic work of Fitchen [3], published in 1961. As an architect, Fitchen makes an exhaustive study of the techniques that must have been used by mediaeval builders, and throws much light on what might be called purely historical matters. Equally, however, Fitchen illuminates what might be called purely engineering topics. In his account of the constructional methods actually used by mediaeval architects, Fitchen is forced to consider also the way in which masonry behaves when assembled into the Gothic structure.

Thus a third prime question, to be answered by an engineer, as the first two have been answered by a historian and an architect, is "How does a stone structure actually behave?". A recent attempt [4] to answer some of the problems posed by this question is phrased very largely in mathematical terms. For example, the mechanics of the rib vault is studied, and the main forces are determined; no reference is made, however, to a particular problem (to be discussed below) which has exercised art historians in the past, of whether or not the ribs "carry".

The purpose of the present paper is to present some findings on mediaeval architecture derived from engineering considerations, that is, from structural analysis, and to give some specific indications of how these findings can be applied to some of the problems of the art historian. The three questions of how an actual structure (a) was designed, (b) was built, and (c) actually behaves cannot, in the final analysis, be answered independently. But concentration on the third question, of structural analysis, is only marginally relevant to the study of how a particular cathedral was designed and built. It must be borne in mind, therefore, that the engineering approach, while leading to some valid and interesting conclusions, will throw very little light on the ratio-

cinations of a mediaeval architect. In particular, there will be a temptation to conclude
that the mediaeval architect *must* have thought
in the same way as a modern engineer ; to this
temptation, Frankl provides a complete and
effective answer.

*
* *

Frankl also is forced on occasion to consider
the structural analysis of Gothic cathedrals,
and, insofar as he reports on published work,
his comments seem fair and pertinent. On the
structural topics on which information is lacking, however, and on which Frankl hazards
an opinion, he is very often wrong. It comes
as something of a shock to find, embedded in
a text of utmost clarity and considerable poly-
mathy, one or two simple mistakes which
appear almost deliberately wrong-headed [5].

An example may be found in Frankl's discussion of the *rationalisme* of Viollet-le-Duc [6].
"Every form has its practical purpose. The
pinnacles are an important example of this,
'which by ther weight give the pier buttresses
all the stability necessary to support the thrust
of the flying buttresses'. From this thesis,
traceable to Sir Christopher Wren, one would
have to conclude that the flying buttress would
give way and consequently the nave vaults collapse if the pinnacles were to be removed.
Paradoxically, according to this thesis, the
more the piers are weighted down vertically by
a superstructure, the thinner they can be
made". [7]

That these conclusions appear paradoxical is
due to a confusion between stability and
strength. Frankl seems to be thinking that
the weight of a pinnacle will cause distress to
a pier, perhaps causing crushing of the masonry ; but Viollet-le-Duc's word "fixeté" is
clear, and has been adequately translated by
Frankl as "stability". No mention is made of
what would be called in modern times the
stresses in the masonry.

Some simple calculations reveal the astonish-
ingly small values of stress that can exist, generally, in masonry structures. A significant
parameter expressing the strength of a particular stone was often used by nineteenth century
engineers : The height to which a prismatic
column could be built before crushing due to

FIG. 1.—Nave of Lichfield. The large pinnacles help to reduce
the size of the main buttresses.

its own weight occurred at the base [8]. Taking
a medium sandstone of weight $2\,000$ kg/m³
and crushing strength say 400 kg/cm², the
column can be erected to a height of $2\,000$ m
before crushing at the base [9]. Put another way,
the addition of an extra meter of superstructure
of the same diameter as the pier supporting
that superstructure will cause an increase of
stress in the pier of $0,2$ kg/cm² ; this figure is
to be compared with a crushing strength of
400 kg/cm².

The cross-sectional dimensions of the pier
did not enter the calculations. If a pier of a
certain height has its cross-sectional area doubled, then the weight of the pier will also be
exactly doubled, and the stress due to self-
weight remains unaffected.

A pinnacle will usually be tapered, and, in
any case, be of smaller diameter than the supporting pier, so that a pinnacle of total height
5 m may be expected to increase the stress in
the pier by considerably less than 1 kg/cm².
In practical terms, the presence of an additional pinnacle will have negligible effect on the
overall *strength* of the supporting masonry, but
it will have some slight beneficial effect on the
stability of that masonry.

ON THE RUBBER VAULTS OF THE MIDDLE AGES, AND OTHER MATTERS   179

As a numerical example, suppose that a main buttress is of uniform cross-section and 10 m high, so that the stress level at the base is, in the absence of a pinnacle, 2 kg/cm². If this stress is increased by the pinnacle to 3 kg/cm², then, roughly speaking, the horizontal component of the thrust delivered by the flying buttress may be increased in the ratio 3/2 with the same safety factor against overturning of the main buttress. The precise increase depends on the placing of the pinnacle; the pinnacle will be more effective if it is placed as near to the flying buttress as possible, as at Lichfield, Fig. 1.

Now the horizontal thrust from the Lichfield flying buttress [10] is of the order of 3 000 kg, so that the pinnacle might be thought of as counteracting a horizontal force of about 1 000 kg. Alternatively, the presence of the pinnacles has enabled the main buttresses at Lichfield to have the relatively slender dimensions of Fig. 1.

It is certain that the addition of a pinnacle must improve the stability of the buttress as a whole, although the effect, even for the relatively massive pinnacles of Lichfield, may be quite small. The action can be seen from the diagram of Fig. 2. The inclined thrust P acting from the flying buttress tends to rotate the whole main buttress about the outer edge A; this tendency to instability is counteracted by the opposing couples due to the weight W of the main buttress and the additional weight w of the pinnacle. Such "hinging" is a key to the structural analysis of masonry, and will be elaborated below. It was the main form of failure envisaged by Coulomb [11] in 1773, but the idea was, curiously, neglected until about 25 years ago, when it served as the basis for the development of a completely new form of structural analysis, the so-called plastic theory [12].

The general stabilizing effect of a pinnacle was well understood as early as 1843; Moseley's book [13] of that date discusses lines of thrust in masonry, "the stability of a pier or buttress surmounted by a pinnacle", and, specifically, "The Gothic buttress" [14]. Now Moseley makes continual reference to French work and records his indebtedness to Dupin, Morin, Navier, and others, and above all, to Poncelet, whose (pirated) *Mécanique Industrielle* had been published in 1839 in Belgium [15]. Ponce-

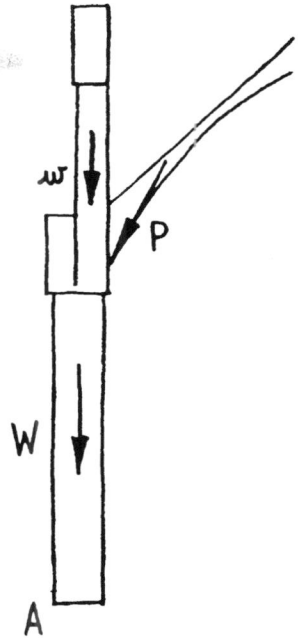

FIG. 2.—Forces acting on a main buttress (Lichfield). The buttress will be *stable* if it does not overturn about the point A.

let does not deal specifically with buttresses in this book, but had discussed similar problems elsewhere [16], and he was, at this time, regarded as one of the leading engineers in France [17].

It is inconceivable that Viollet-le-Duc was ignorant of the work of these French engineers, who were, according to Moseley, well ahead of their English colleagues; yet Moseley was including in 1840 in his lectures to the undergraduates of King's College, London, "... the theory of piers, walls supported by counterforts and shores, buttresses, walls supporting the thrust of roofs and the weights of the floors of dwellings, and Gothic structures..." [18]. It is difficult to accept Pol Abraham's sneer, reproduced by Frankl [19], "...that neither flying buttresses nor pinnacles were necessary. Many a French cathedral had none, and acquired them only when restored by Viollet-le-Duc".

Yet there is some excuse for the doubts about the efficacy of pinnacles. It seems to the eye, and this is confirmed by calculation, that

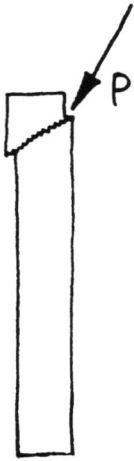

FIG. 3.—Sliding failure
at the top of a buttress
in the absence
of a pinnacle
(Lichfield, schematic).

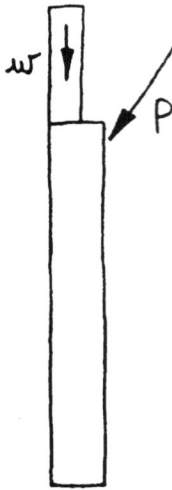

FIG. 4.—Sliding failure
prevented by the weight
of a pinnacle.

shows Viollet-le-Duc's cross-section of Reims [23]; Fitchen [24] has given reasons for the use of two rather than a single flying buttress.

* *
* *

Coulomb considered both sliding and hinging failure in his original paper. The study of *frictional* materials leads on directly to the science of soil mechanics, of which Coulomb was the father. For masonry, however, he remarks that "friction is often so high for the materials used in arch construction, that the voussoirs can never slide one on another" [25]. For the remainder of this article the danger of sliding failure will be disregarded, so that mechanisms of collapse are limited to those involving "hinging".

a pinnacle can have little effect on the *overall* stability of a pier. Abraham is right when he says that it is a fine pinnacle that weighs a hundredth of the total weight on a pier [20].

The main effect of a pinnacle is, indeed, not one of overall stability, but is localized at the head of the pier. Fig. 3 shows a main buttress without a pinnacle, and it will be seen that, if the force P delivered by the flying buttress departs too much from the vertical, there will be a tendency to sliding failure. To avoid such failure, the simplest expedient is to add a relatively small weight w, Fig. 4, so that the frictional force is increased along a potential line of slip. Moseley, in his article "The conditions necessary that the stones of the pier may not slip on one another" [21], shows that the calculations are extremely simple.

The weight w in Fig. 4 necessary to prevent slip of the main buttress at the point where it receives the flying buttress can thus be determined easily; it turns out that the line of action of w is immaterial. Thus there would be no objection, from this point of view, to placing the pinnacle towards the *outside* of the main buttress, where its small effect on *overall* stability would be even further diminished [22]. Fig. 5

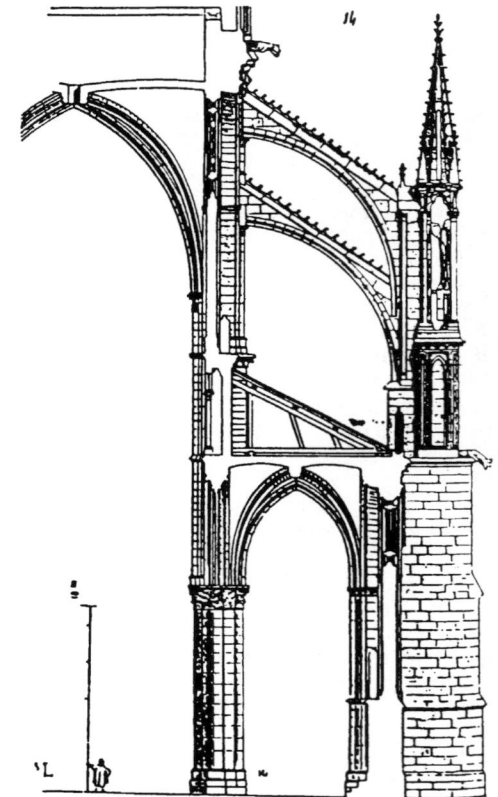

FIG. 5.—Massive pinnacle at Reims (Viollet-le-Duc).

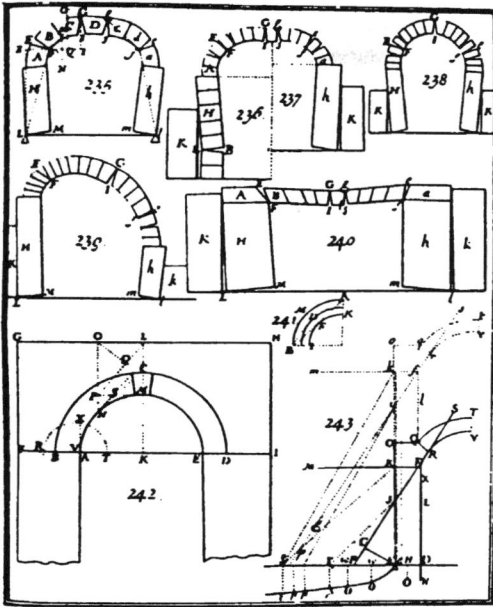

FIG. 6.—Danisy's tests on model plaster arches; collapse is by "hinging" of the voussoirs.

has cracked in *three* places. Such cracking is completely harmless. Since, in fact, no masonry can be designed to fit *exactly* between given abutments, such cracking is inevitable; indeed, it might be said that the cracked state is the natural state of masonry, although the cracks may be so small as to be invisible, or be closed by the elasticity of the stone. In Fig. 8 (*b*),

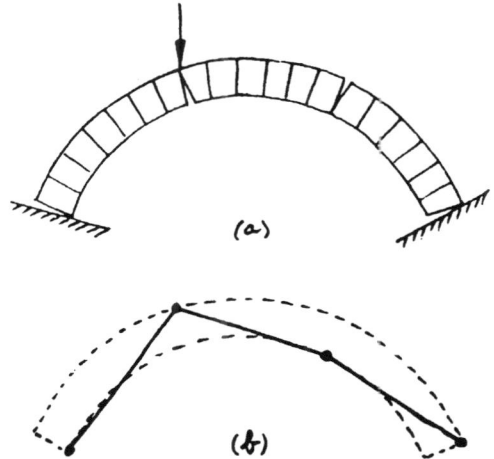

FIG. 7.—Collapse mechanism for a voussoir arch; four hinges are necessary.

The idea of a mechanism of collapse goes back at least 200 years. Frézier [26] quotes tests made by Danisy in 1732, and his Figs. 235-240, reproduced here in Fig. 6, record the results of model experiments on voussoir arches. Fig. 7 (*a*) shows such an idealized arch subjected, in addition to its own weight, to a single point load sufficient in magnitude just to cause collapse; the masonry has cracked in four places, permitting the development of a "four bar chain" mechanism, Fig. 7 (*b*).

The arch of Fig. 7 *must* develop four hinges in order to collapse; three or fewer hinges will not permit a mechanism motion. The locations of the hinges are not arbitrary, but serve to define the line of thrust within the masonry. The development of *four* hinges represents the *limit state* at which it is only just possible to find a position for the line of thrust lying *wholly within* the arch.

Thus, in Fig. 8 (*a*), an arch spans between two abutments which have spread slightly under the thrust of the arch. In order to accommodate itself to the increased span, the arch

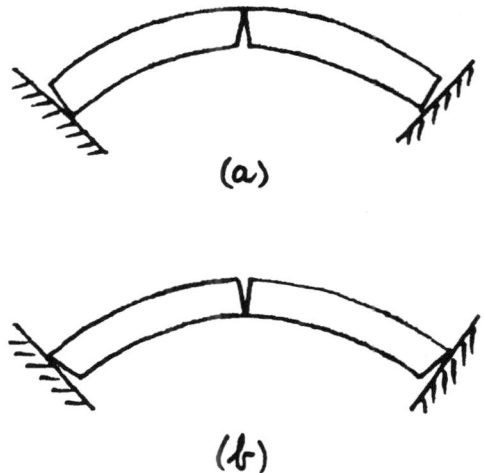

FIG. 8.—Voussoir arch fitted between abutments of slightly the wrong span; (*a*) too large, and (*b*) too small. In either case the arch cracks in three places.

the abutments are too close, and a different configuration of cracks is necessary. Again, however, the arch is as completely stable as it was in the uncracked state.

These statements can be proved rigorously [27] using the techniques of modern structural analysis developed for the plastic theory of steel frames [28]. The masonry considered is not limited to voussoir construction; it is necessary only to assume (conservatively) that an assemblage of stones is incapable of resisting tension. This assumption is equivalent to ignoring the relatively weak tensile strength of mortar, or to considering dry construction. In particular, it is the fact that masonry can develop completely harmless cracks that has ensured the survival of stone structures, and it is simple to demonstrate, for example, that small accidental settlements of the foundations of a stone structure will never, of themselves, promote collapse of that structure.

All this was understood with complete insight by Choisy: "The ribbed vault is, as it were, flexible and deformable; the points of support can settle, the piers lean, it will follow their movements" [29]. Frankl offers as comment: "When an expert like Choisy said that, it is no wonder that whole generations of architects and art scholars calmly continued to live and theorize under Viollet-le-Duc's influence, although they must all have known that Gothic transverse arches, ribs and vaults were not made of rubber" [30].

*
**

A structural engineer, looking at a Gothic cathedral, will see, not a massive array of nave piers, but the skeletal structure formed by the centre-lines of those piers; not a thick vault, but a thin doubly-curved sheet spanning between the mathematical centre-lines of the ribs. He will determine primary structural forces for this conceptual model, in which the real structural elements are replaced by a vanishingly thin skeleton, and for which the lines of action of the forces must, of necessity, coincide with the skeleton. It may be necessary, in the final analysis, to take account of the fact that the actual structure has a finite thickness; in most cases, however, such an adjustment to the calculations is trivial, or even meaningless [31].

Piers may be reduced to the skeletal form of lines, straight for vertical thrusts, curved if there is also a horizontal thrust; the web of a vault, however, must be idealized by a two-dimensional doubly-curved shell [32]. Now thin shell theory, or membrane theory, has been developed extensively in recent years [33], and can be used to determine primary vault forces. For example, the stresses in a smoothly curving thin shell subjected to its own weight are of the order $R\rho$, where R is the local radius of curvature of the shell and ρ is the density of the material [34]. Before putting numbers into this formula, it may be noted that the value of stress is independent of the thickness of the shell; exactly as for the self-crushing pier, doubling the shell thickness will double the gravity forces, but the resisting area is also doubled.

A doubly curved shell has two radii of curvature, and both values are, in general, needed to determine stresses. For the order-of-magnitude calculations, however, a single value will suffice; a pointed vault over a nave of 15 m span might have a radius R of about 10 m. Taking the density of stone again as $\rho = 2\,000$ kg/m³, the product $R\rho$ becomes 2 kg/cm², to be compared with a crushing strength (for medium sandstone) of 400 kg/cm².

The ambient stress in the shell is so low that the strength of the masonry is again irrelevant. Thus the vault webs were constructed "of a light stone, if it could be had" [35], and "rubble vaults with thick mortar joints were... the rule" [36]. The weakness of tufa, as at Canterbury, or of the mortar in the joints, is of no consequence.

These remarks apply only to *smoothly* curving shells. At shell intersections (groins) there exist very large stress concentrations, and the intersections must be reinforced. A crease in a shell is, in fact, a line of weakness; a *reinforced* crease, however, confers rigidity to the whole shell structure. There is, of course, conflicting evidence as to the role of the ribs, some of which seem very small (e.g. Reims); Porter cites the evidence of Ourscamp [37], where the ribs stand intact although the webs have fallen, and of Longpont, Aisne [38], where the webs stand although the ribs have fallen. Porter uses both pieces of evidence to reinforce his thesis that ribs were constructional devices, useful to avoid formwork during erection, and he is right to point to the fact that the rib vault can "...be applied not only to quadripartite rectangular spaces but equally well to irregular plans, to the trapezoidal compartments of an ambulatory or annular gallery, to the semi-circular vault of an apse, or to the polygonal vault of a chevet" [39]. However, Fitchen [40] objects to Porter's preoccupation solely with economic considerations, and he produces convincing arguments [41] that "...the rib's function was not only the erectional one of determining the contour of the vault panels but also the erectionally simplifiying aesthetic one of covering the joint along the groin, which was made as simple to build as possible".

Thus the rib, besides having a structural purpose as reinforcement for potentially weak "creases" at the groins, also enables the vaulting compartments to be laid out more easily, enables a good deal of formwork to be dispensed with, and covers ill-matching joints at the groins, As a bonus, the rib has been thought to be aesthetically satisfying, and all of these functions might be thought of as the "function" of the rib.

From the point of view of structural analysis, however, the functional test is simple; a sharp crease in a shell will lead to a large stress concentration and hence should be reinforced with a rib. Thus in a quadripartite nave vault formed by intersecting barrels (pointed or semi-circular) with a level soffit, there will be no creases in the shell at the nave walls or at the position of the transverse arches. Under these circumstances, then, neither the wall arches (formerets) nor the main transverse arches carry anything but their own weights.

By contrast, the deep diagonal creases, increasing in sharpness towards the haunches of the vault, require reinforcement, and the diagonal ribs emerge as the effective structural members carrying the whole vault. These conclusions seem identical to those arrived at by Marcel Aubert [42], as quoted by Kubler [43]. In particular, if the soffit of the vault is not level, as when each vaulting bay is strongly domed, for example, then a crease will appear at the transverse arches, and Aubert "...also arrived at the conclusion that vaults of domical form require heavy transverse arches..." [44].

The evidence of Longpont, Aisne, where the vaults survived the fall of the diagonal ribs, requires explanation, which can best be done

numerically. A 20 cm thick quadripartite vault, slightly pointed, may be used to span a rectangular bay 15 m by 7,5 m [45]; such a vault, together with its ribs, will weigh about 80 000 kg and the main shell will be stressed to about 1,5 kg/cm². The maximum force in a diagonal rib is about one quarter of the weight of a vaulting bay, i.e. about 20 000 kg. Accepting Yvon Villarceau's factor of safety of 10 on a crushing strength of 400 kg/cm², a rib working at 40 kg/cm² must have a cross-sectional area of 500 cm² if it is to carry a total force of 20 000 kg. Thus a rib of 25 cm width and 20 cm depth would seem adequate.

These calculations have been made for the *skeletal* structure, but it will be appreciated that a 25 cm by 20 cm rib can be thought of as contained *within* the 20 cm thickness of the main shell severies. If then, a vault of these dimensions is built without ribs, or is built with ribs that subsequently fall, the above analysis indicates that the primary shell stress of about 1,5 kg/cm² will increase very sharply in the neighbourhood of the groins to about 50 kg/cm². There will therefore be a high stress concentration, but it may be that the crushing strength of the stone is so high (400 kg/cm²), the diagonal intersection of the severies sufficiently regular, and the mortar sufficiently thin and strong, that collapse of the vault does not occur. In this case the vault will have succeeded, so to speak, in the attempt to construct its own ribs, and the structural engineer, viewing a ribless vault, will always see a diagonal skeleton defining the primary lines of force [46].

Conversely, the structural engineer's eye will be irritated by the main transverse arches across the nave. A supreme example of this irritation is to be found in the fan vaulting at King's College Chapel, Fig. 9. The nave was originally designed for a quadripartite rib vault, but was fan-vaulted by John Wastell in 1512-1515. The clumsy intersections of the fans, and the strong transverse arches, structurally meaningless, combine to arrest the eye bay by bay, instead of permitting a smooth progress down the nave. The small ribs on the fans, being applied to an otherwise smooth shell, do not of course carry any force, but serve to define visually the shape of the vault.

As a further comment on the mechanics of the rib vault, reference may perhaps be made

184

FIG. 9.—King's College Chapel (fan vault by John Wastell). The heavy transverse arches are a survival from the original intention to provide a rib vault.

ming the horizontal components from the diagonal ribs, a simple calculation shows that the horizontal thrust at the position of a main transverse arch is about three times as great as that at an intermediate arch [49]. If flying buttresses were used at every bay, main and intermediate, the former would transmit about three times the thrust of the latter. Thus it would be perfectly possible to dispense with the intermediate flying buttresses, providing it could be ensured that "the thrust of the intermediate part of the vault would be taken up by the strip buttresses" [50].

This brief discussion hardly answers the question of whether or not Notre-Dame was originally built with the intermediate flying buttresses omitted. However, it has shown that it would be *possible*, under certain circumstances, for them to be omitted.

***

The emphasis given above to the low stress levels encountered in large portions of the masonry of Gothic structures does not mean that stresses are *everywhere* low. Indeed, the example of the rib vault indicated much higher stresses in the ribs. Similarly, nave piers carry much more than their own weight; the whole "intention" of Gothic is the collection of the vertical forces from the vault and great roof into slender columns, the walls becoming non-structural sheets of glass.

Nevertheless, even the piers have working stresses which are probably an order of magnitude less than the crushing strength of the stone. The piers and vault ribs are probably the most highly stressed members at about one-tenth of the crushing strength; then might come the flying buttresses, and finally the main buttresses, self-supporting walls, and vault webs at about one-hundredth of the crushing strength. The gross overloading of a portion of the structure, for example by the attempted construction of a tower or high spire, or the bad design of a detail, involving bending (and hence tension) rather than pure compression of the masonry, might lead to a failure directly related to the strength of the stonework. In normal circumstances, however, the conclusion

to Conant's discussion [47] of the sexpartite vaults of Notre-Dame, Paris. This article is concerned, *inter alia,* with a discussion of whether or not the original flying buttress system was applied to alternate bays only. Fig. 10 shows, schematically, two complete bays (i.e. four nave bays) of a sexpartite vault (after Fitchen). The creases in the shell indicate, as before, those ribs which carry; as before, the diagonal ribs have high forces, and the *main* transverse arches carry merely their own weight [48]. The *intermediate* transverse arches, however, *do* carry.

A rib force has, of course, horizontal and vertical components; the vertical components load the nave piers, and the horizontal components are conducted through the flying buttresses to the main external buttresses. Sum-is that *stability* rather than strength is the over-riding design criterion.

Now *stability* of a masonry structure may be ensured by the proper *proportioning* of the various parts of the structure; the thickness of a vault web should be a certain fraction of the span (and a function of the curvature), the

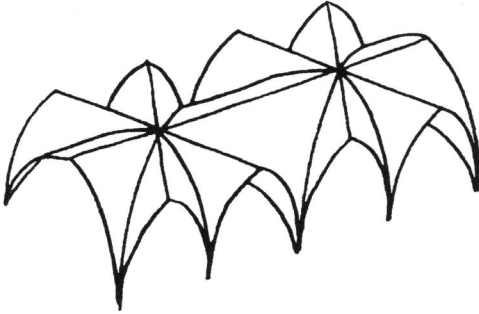

FIG. 10.—Schematic sexpartite vault (after Fitchen).

diameter of a nave pier should bear a certain relation to the pier height, the breadth of a main buttress should be a certain fraction of its depth, both being related to its height, and so on. Rules like these are essentially *nume-rical*, and Frankl's brilliant exposure of the sources from the Gothic period makes it very clear that it was precisely numerical rules of proportion that concerned mediaeval architects. The fragmentary rules given in Villard's ma-nuscript can be reduced to proportions; the great discussion of 1392 at Milan concerned the notions of *ad quadratum* and *ad triangulum*, and was resolved on advice from a mathema-tician (Stornaloco); the equally famous set of criticisms of Milan by Mignot in 1399 were phrased in terms of proportion.

However such numerical rules were cons-tructed, mediaeval architects were right to insist on them. A knowledge of statics might have enabled them to construct *better* rules, but they should still have been rules of the same sort, since the safety of their structures depends on stability and not on strength, and stability depends on proportion.

Even the simplest (and most basic) problem

of statics was not solved until the end of the sixteenth century; the parallelogram of forces is usually attributed to Simon Stevin, 1586 [51]. Without any clear notion of force, or of its line of action [52], mediaeval architects had no option but to construct their rules in terms of proportions. For any problem of design for which they had no rule they had either to guess or to make a model.

The use of models is well attested [33], not only to solve geometrical problems (i.e. pro-blems of stereotomy) but also to simulate the actual structure. For example, the brick and plaster model of S. Petronio, Bologna was over 18 m long, which is of enormous scale. Such a model can be used with complete confi-dence to check the stability of the whole or of any part of the real structure, since questions of stability, depending as they do on relative proportions, *can* be scaled. By contrast ques-tions of strength cannot be scaled in this way; the square-cube law ensures that a scale model will always be misleadingly strong.

So long as architects were prepared to expe-riment, both on models and on paper, progress was possible. Mignot's great cry "Ars sine scientia nihil est" sounds to our modern ears like the initiation of a new age of architecture, when practical experience would be firmly reinforced by theory. That it meant nothing of the sort is shown by Mignot himself, who, at the end of the fourteenth century, was work-ing without that seemingly effortless and in-tuitive mastery exemplified by the twelfth and thirteenth century architects of the Ile-de-France. His *credo* implies in reality that, in future, any quarrel between practice and theory would be resolved in favour of theory; and by theory he means that *scientia*, that pedantry of the written word, contained in the mysteries of the Lodges, whose significance, however empirical, eluded more and more Mignot and his successors. Villard's manu-script, added to with reverence by Magister 2 and Magister 3, and guarded with secrecy in the Lodge, would come in time to have an almost Biblical authority (or the authority, equally stultifying, of a modern Code of Prac-tice), and is, perhaps, a symbol of that internal decay which ensured the death of Gothic at the Renaissance.

186

NOTES

1. P. FRANKL, *The Gothic, Literary Sources and Interpretations through Eight Centuries*, Princeton, 1960.

2. FRANKL, *op. cit.*, p. v.

3. J. FITCHEN, *The Construction of Gothic Cathedrals, A Study of Medieval Vault Erection*, Oxford, 1961.

4. J. HEYMAN, *The Stone Skeleton, International Journal of Solids and Structures*, vol. 2, N° 2, April 1966.

5. FRANKL, despite his apparent objectiveness, was clearly influenced by the work of Pol ABRAHAM : *Viollet-le-Duc et le rationalisme médiéval*, Paris, 1934. Abraham demonstrates some correct (and some incorrect) structural analysis, but the whole of his long essay is distorted by his desire to prove his paradoxes that, of the two *essences* of Gothic, the rib vault and the flying buttress, the rib does not carry the vault and the flying buttress does not thrust. Discussing the problem of the mechanics of the vault, FRANKL states (*op. cit.*, p. 812) : "Certainly it is interesting, but we should do better to wait until the physicists have agreed among themselves." Here the use of the word "physicists" (? Einstein) is revealing; Frankl should have written (and should have been thinking of) the word "engineers". Again, there is a certain arrogance in the statement (FRANKL, *op. cit.*, p. 818) : "It remains an inescapable duty of the builder, the technician, the architect, as *ingeniere*, to know structural mechanics so that his work will not collapse; but the art historian, who wants to understand the architect as artist, comes back to Fiedler's formula: Art exists for the eye."

6. FRANKL, *op. cit.*, p. 569.

7. The inner quotation is from VIOLLET-LE-DUC, *De la construction des édifices religieux en France, Annales archéologiques*, 1845, p. 339: "...qui par leur poids, donnent aux contre-forts toute la fixeté nécessaire pour maintenir la poussée des arcs-boutants."

8. See, for example, Yvon VILLARCEAU, *L'établissement des arches de pont*, Institut de France, Académie des Sciences, Mémoires présentés par divers savants, 1854, vol. 12, p. 503. Yvon Villarceau advocated a factor of safety of 10 on the height of the self-crushing column; that is, nominal stresses in a masonry structure should be limited to one-tenth of the crushing strength of the stone.

9. Crushing strengths may be found in standard handbooks; for sandstone, a range of values between 150 and 900 kg/cm² is given by G. G. UNGEWITTER, *Lehrbuch der Gotischen Konstruktionen*, 2 vol., Leipzig 1901 (Tauchnitz).

10. HEYMAN, *op. cit.*

11. C. A. de COULOMB, *Essai sur une application des règles de maximis et minimis à quelques problèmes de statique, relatifs à l'architecture*, Mémoires de Mathématique et de Physique présentés à l'Académie Royale des Sciences, 1773, p. 343. The rotation of a pier about its base was envisaged by Viollet-le-Duc, who used such words as *"articulation"* and *"base-rotule"*.

12. An account of the development of the plastic theory for steel framed structures is given by Sir John BAKER, M. R. HORNE, and J. HEYMAN, *The Steel Skeleton*, vol. 2, *Plastic Behaviour and Design*, Cambridge 1956. See also J. HEYMAN, *Beams and Framed Structures*,

Oxford (Pergamon Press) 1964. The use of the word « plastic » here implies the opposite of elastic, taking elastic in the usual sense of the capability of resuming an original shape after deformation. Thus a plastic material has the capability of permanent deformation. A discussion of the two words elastic/plastic is given by Pol Abraham, *op. cit.*, p. 17 f.

13. H. Moseley, *The mechanical principles of engineering and architecture*, London, 1843.

14. *Ibid.*, articles 302, 303, p. 422.

15. J. V. Poncelet, *Mécanique Industrielle*, Liège, 1839.

16. Moseley, *op. cit.*, p. 458 n.

17. Biographical information and discussion of the history of structural analysis can be found in S. P. Timoshenko, *History of Strength of Materials*, McGraw-Hill 1953, and in H. Straub, *A History of Civil Engineering*, London 1952 (being a translation of *Die Geschichte der Bauingenieurkunst*, Basle 1949).

18. Moseley, *op. cit.*, p. xiv.

19. Frankl, *op. cit.*, p. 807. See Abraham, *op cit.*, p. 88 n.

20. Abraham, *op. cit.*, p. 90 : "Quand son poids atteint le centième de la charge totale, c'est qu'il est déjà un beau pinacle."

21. Moseley, *op. cit.*, article 292, p. 411.

22. In concluding that the main *raison d'être* of a pinnacle may be found in the consideration of local rather than overall conditions, it is perhaps significant that Viollet-le-Duc *did* use the word *fixeté* rather than the word *stabilité* which he uses freely elsewhere. (See note 7 above.)

23. E. E. Viollet-le-Duc, *Dictionnaire raisonné de l'architecture française du XI° au XVI° siècle*, Paris 1858-68, 10 vols : Article "Cathédrale" vol. 4, p. 318, Fig. 14.

24. J. Fitchen, *A comment on the function of the upper flying buttress in French Gothic architecture*, *Gazette des Beaux-Arts*, ser. 6, vol. 45, 1955, p. 69.

25. Coulomb, *op. cit.*, p. 380. "Le frottement est souvent assez considérable dans les matériaux que l'on emploie à la construction des voutes, pour que les différens voussoirs ne puissent point glisser l'un contre l'autre."

26. A. F. Frézier, *La théorie et la pratique de la coupe des pierres et des bois pour la construction des voûtes et autres parties des batimens civils et militaires, ou traité de stéréotomie à l'usage de l'architecture*, 3 vols, Strasbourg and Paris, 1737-9.

27. Heyman, *The Stone Skeleton*. The proof of the stability of an arch in its cracked state is completely at variance with the intuitive belief of Pol Abraham that such cracks were "signes avant-coureurs de la ruine de l'arc." (*op. cit.*, p. 8). Indeed, Abraham uses the fact that cracks are visible in flying buttresses to "prove"

that flying buttresses cannot transmit thrusts (*op. cit.*, p. 84), yet he was well aware (p. 100 n) that such cracks had often been developed on the original decentering, had been stuffed with mortar, covered with plaster, and had false joints painted.

28. See, for example, Baker, *et. al.*, *op. cit.*

29. A. Choisy, *Histoire de l'architecture*, Paris 1899, vol. 2, p. 270 : "...la voûte nervée est pour ainsi dire flexible et déformable : les points d'appui peuvent tasser, les piles se déverser, elle en suivra les mouvements."

30. Frankl, *op. cit.*, p. 577.

31. Much time and effort can be spent by engineers in refining what might be called "crude" calculations, when it can be shown by the theorems of plastic design that certain classes of simple calculation give safe estimates of the actual behaviour of the structure.

32. The two dimensional shell may, in turn, sometimes be "sliced" with advantage into a series of independent elements. This technique seems first to have been used, with brilliant success, by Poleni, in his analysis of the dome of St. Peter's, Rome. See G. Poleni, *Memorie istoriche della Gran Cupola del Tempio Vaticano*, Padua 1748. Pol Abraham's rolling ball following a line of steepest descent (*op. cit.*, p. 18 ff) is an equally brilliant anthropomorphic way of defining independent slices of a vault.

33. The most useful book on shell theory for the purpose of the analysis of the Gothic structure (indeed, perhaps the most useful book on shell theory *tout court*) is W. Flugge : *Stresses in shells*, Berlin 1961 (Springer). Flügge does not deal with Gothic vaults, but his equations for the analysis of ribbed segmental domes can be adapted readily.

34. By "order" is meant that the calculation is subject to error within a factor of about 2, but is correct to within a factor of 10. Thus if a stress is determined as of the order $2 \, kg/cm^2$, the actual stress, which may vary from point to point, will be much greater than $0,2 \, kg/cm^2$ and much less than $20 \, kg/cm^2$.

35. Francis Bond, *Gothic Architecture in England*, London 1906, p. 303.

36. Fitchen, *The Construction of Gothic Cathedrals...*, p. 91.

37. Arthur Kingsley Porter, *The Construction of Lombard and Gothic Vaults*, Yale University Press, 1911, p. 8.

38. *Ibid.*, p. 16.

39. *Ibid.*, p. 25.

40. Fitchen, *op. cit.*, p. 105.

41. *Ibid.*, p. 111. Pol Abraham finally concludes (*op. cit.*, p. 59 f) that the rib's function was geometrical, enabling masons who knew nothing of stereotomy to first define the web intersections and then warp the vault webs to suit, rather than *vice versa*: "...la nervure a été le moyen commode à la fois constructif et esthétique de la voûte à pénétrations généralisées."

188

42. Marcel AUBERT, *Les Plus Anciennes Croisées d'Ogives, Bulletin Monumental*, XCIII, 1934, pp. 6-67 and 137-237.

43. George KUBLER, *A Late Gothic Computation of Rib Vault Thrusts, Gazette des Beaux-Arts*, series 6, vol. 26, July-Dec. 1944, pp. 135-148. The first three pages of this article summarize the issues joined in the 1930's over Viollet-le-Duc's "rational" interpretation of the structure of the rib vault. The title of Kubler's article is slightly misleading. The sixteenth century text (by Rodrigo Gil) under discussion does not in fact compute rib vault thrusts, but gives a series of rules for determining the dimensions of wall buttress to support such a vault; it is a modern deduction that the buttress is required to absorb the vault thrust. The rules, incidentally, while perhaps working empirically within a certain limited range, seem fairly nonsensical, since the required answers, a set of lengths, are computed by taking the square root of another set of lengths, which is a dimensional absurdity.

44. KUBLER, *op. cit.*, p. 136.

45. HEYMAN, *op. cit.*

46. These conclusions about the mechanics of the quadripartite rib vault differ only in one important particular from those of Pol Abraham. He also concludes that the formerets and transverse ribs do not carry, and that the diagonal arches support the vault. ABRAHAM insists, however (*op. cit.*, p. 29), that the diagonal ribs are non load-carrying appendages, and that the main force is transmitted within the shell itself: "Il faut bien insister sur ce fait que, dans la réalité, tout se passe nécessairement dans l'épaisseur des maçonneries et que, par suite, il se constitue spontanément de culée à culée un arc diagonal incorporé."

47. Kenneth John CONANT, *Observations on the Vaulting Problems of the Period 1088-1211, Gazette des Beaux-Arts*, series 6, vol. 26, July-Dec. 1944, pp. 127-134.

48. At Notre-Dame, there is some warping down of the vault severies at the main transverse arches, so that a large proportion of the total vault weight will in fact be carried by these.

49. Rosenberg also quotes this factor of 3. See G. ROSENBERG, *The Functional Aspect of the Gothic Style, Journal of the RIBA*, vol. 43, 1936, pp. 273-290 and 364-371.

50. CONANT, *op. cit.*, p. 132.

51. STRAUB, *op. cit.*, p. 60 f.

52. These notions were correctly identified by Stevin, but twentieth century architects still have difficulty with forces. BOND, for example, in his discussion of the action of pinnacles (*op. cit.*, p. 363) confuses horizontal and vertical forces, and Pol ABRAHAM (*op. cit.*, p. 10) feels compelled to make a small digression to explain the difference between vertical, horizontal, and inclined forces.

53. FRANKL (*op. cit.*) has several references to the construction of models. See e.g. p. 299 for details of the model of S. Petronio, Bologna, 1390.

3

The geometrical knowledge
of mediaeval master masons

Lon R. Shelby

DURING the past one hundred and fifty years numerous scholars have searched for the geometrical canons which supposedly were used by master masons in the design and construction of mediaeval churches. But in this search for one of the keys to an understanding of mediaeval architecture, these scholars have seldom asked themselves what was the actual character and content of the geometrical knowledge which a mediaeval master mason might have been expected to possess. This paper attempts to answer that question.

The idea that geometry played a fundamental role in the mediaeval masons' craft is by no means the invention of modern scholars; it was a notion commonly held by mediaeval masons themselves. The thirteenth-century French master mason, Villard de Honnecourt, touched on the matter in the "Preface" to his *Sketchbook*: "Villard de Honnecourt greets you and bids all those who work with the devices found in this book to pray for his soul and to remember him. For in this book one will find good advice concerning the proper technique of masonry and the devices of carpentry. You will also find the technique of drawing — the forms — just as the art of geometry requires and teaches it."[1] In another place Villard commented, "Here begins the technique of the forms of drawing, just as the art of geometry teaches them for working more easily. And on other sheets are those of masonry."[2] Although he does not quite say it, the implication seems to be that "on other sheets will be found the technique of the forms (*li force des trais*) of masonry, likewise as taught by the art of geometry."

An even stronger assertion of the essential role of geometry in the masons' craft was made around 1400 by the unknown author — probably a cleric — who compiled a historical introduction to the "Articles and Points of Masonry" which set forth the customs and regulations pertaining to the masons' craft in England at this time. Although this introduction contains overlays of "clerical" learning,

I am grateful to the Graduate School Office of Research and Projects of Southern Illinois University, Carbondale, for a grant which made possible the research for this paper. I should also like to thank Mr Alan Cohn and Miss Kathleen Eads of Morris Library at S.I.U. for their constant help in procuring books for my research through purchase or interlibrary loans. Finally, a special word of thanks is due Dr Stephen Victor of Simon Fraser University for several bibliographical leads and to Prof. John Hoag of the University of Colorado and Prof. Harold McFarlin of S.I.U. for their critiques of the paper while it was in draft form.

1 H. R. Hahnloser, ed., *Villard de Honnecourt* (Vienna, 1935), p. 11: "Wilars de Honecort v(os) salue (et) si proie a tos ceus qui de ces engiens ouverront, c'on trovera en cest livre q(u)'il proient por s'arme (et) qu'il lor soviengne de lui. Car en cest livre puet o(n) trover grant consel de le grant force de maconerie (et) des engiens de carpenterie, (et) si troveres le force de le portraiture, les trais, ensi come li ars de iometrie le (com)ma(n)d(e) (et) ensaigne."

2 *Ibid.*, p. 91: "Ci comence li force des trais de portraiture si con li ars de iometrie les ensaigne. por legierem(en)t ouvrer. (et) en l'autre fuel s(un)t cil d(e) le maconerie." What Villard could have meant by the "technique of the forms of masonry as taught by the art of geometry" will become clear, it is hoped, in the course of this paper.

it is patently founded on the traditions of the craft itself, and thus it provides a valuable insight into the masons' perceptions of the history and character of their craft.[3] The author begins with a review of the seven liberal arts, but he quickly singles out for special consideration geometry, which he defines in the old-fashioned way — following Isidore of Seville — as the measure of the earth. He then propounds the significance of geometry for the handicrafts:

Marvel you not that I said that all science lives all only by the science of geometry. For there is no artifice nor handicraft that is wrought by man's hand but it is wrought by geometry. . . . For if a man works with his hands he works with some manner [of] tool, and there is no instrument of material things in this world but it comes of the kind of earth and to earth it will return again. And there is no instrument, that is to say, a tool to work with, but it has some proportion more or less. And proportion is measure, [and] the tool or the instrument is earth. And geometry is said [to be] the measure of earth, wherefore I may say that men live all by geometry.[4]

Having established the connection between geometry and work with tools, the author proceeds to say "that among all the crafts of the world of man's craft, masonry has the most notability and most part of this science [of] geometry, as it is noted and said in historial, as in the Bible and in the Master of Stories."[5] He next turns to the origins of geometry and masonry and recounts several versions drawn from these various sources. The one of greatest interest here is the story of Euclid, who, according to the author, had been a clerk of Abraham during the latter's sojourn in Egypt. Indeed, it was Abraham who had taught the science of geometry to Euclid, who in turn taught it to the Egyptians.

Then this worthy clerk Euclid taught them to make great walls and ditches to hold out the water [of the Nile]. And he by geometry measured the land and departed it in divers parts, and made every man to close his own part with walls and ditches, and then it became a plenteous country. . . . And they took their sons to Euclid to govern them at his own will, and he taught to them the craft [of] masonry and gave it the name of geometry because of the parting of the ground that he had taught to the people.[6]

The chief value, for our present purposes, of this quaint and garbled account of the historical person Euclid, and of the origin and meaning of Euclidean geometry, is that it reveals the connotative significance which the word geometry had acquired for masons by 1400; in their view, masonry was a craft which historically went back to Abraham through Euclid, and which had originally been founded on that preeminent science of the handicrafts — geometry. For mediaeval masons Euclid had virtually become an eponymous hero of the craft, and the word geometry had become synonymous with masonry. When a word has acquired such rich and special meanings, we must beware of misreading it when

---

[3] See Douglas Knoop, G. P. Jones, and Douglas Hamer, eds., *The Two Earliest Masonic MSS. The Regius MS. (B.M. Bibl. Reg. 17 A1); The Cooke MS. (B.M. Add MS. 23198)*, "Publications of the University of Manchester," No. CCLIX (Manchester, 1938), pp. 7–10, for discussion of the authorship and literary sources for the "Cooke MS.", to which I refer.

[4] *Ibid.*, pp. 73, 75. I have rendered this and later quotations from this text in modern spelling and punctuation.

[5] *Ibid.*, p. 75.

[6] *Ibid.*, pp. 95, 97.

Geometrical Knowledge of Mediaeval Master Masons   397

used by those who have given it these meanings. That is to say, neither "Euclid" nor "geometry" may have meant to mediaeval masons what today we mean by Euclidean geometry.

To reconstruct the geometrical knowledge of mediaeval master masons, we must first consider the kind of education which these men would normally have obtained. Since in a previous study I have done this in detail for English master masons, only a summary review will be provided here.[7] It does not appear that literacy was a necessary accomplishment for a mason to become a master of his craft, for clerks were readily available to provide whatever reading and writing skills might be needed in the transaction of business and the keeping of records in building construction. On the other hand, there is evidence that from the thirteenth century onward at least some master masons learned to read and write; this seems to be simply part of the larger story of the increasing literacy of the laity in the later Middle Ages. Literacy in the vernacular languages could be acquired in a variety of ways — formal and informal — but literacy in Latin normally meant that one had attended a grammar school, where indeed the major thrust of the studies was to teach the young scholars to read, write, and speak Latin. Therein lies an important key to the problem of the geometrical knowledge of mediaeval master masons. While grammar school teachers normally paid lip service to the *quadrivium,* more often than not arithmetic, geometry, music, and astronomy received little attention in the actual curriculum. Thus even the lad who completed several years of study in a grammar school would have had little or no contact with Euclidean geometry, even in the diminished form in which it had come down through the early mediaeval encyclopedias, anthologies, and textbooks.

Only in the higher levels of formal schooling, that is, at some of the more renowned monastic and cathedral schools, the *studia generalia,* and later, the universities, could a student find geometry as a regular part of the curriculum.[8] But given the social, economic, and professional circumstances of mediaeval master masons, it may be inferred that a young man who wished to become a master in this craft would not have pursued such higher studies — and I have seen no evidence which contradicts this inference. Conversely, a young man who had studied at a university would expect to find career opportunities outside the building crafts themselves; he might become a clerk of the works, but not a master mason.

[7] Lon R. Shelby, "The Education of Medieval English Master Masons," *Mediaeval Studies,* xxxii (1970), 1–26.

[8] See Pearl Kibre, "The Quadrivium in the Thirteenth Century Universities (With Special Reference to Paris)," *Arts libéraux et philosophie au moyen âge,* "Actes du quatrième congrès international de philosophie médiévale (1967)," (Montréal, 1969), pp. 175–191; and James Weisheipl, "The Place of the Liberal Arts in the University Curriculum during the XIVth and XVth Centuries," *ibid.,* pp. 209–213. Unfortunately, in this Congrès' monumental program on the mediaeval *artes liberales,* virtually no attention was given to the place of the *trivium* and *quadrivium* in mediaeval grammar schools. The same is true for the compendium of studies in Josef Koch, ed., *Artes Liberales von der Antiken Bildung zur Wissenschaft des Mittelalters,* "Studien und Texte zur Geistesgeschichte des Mittelalters," Bd. v (Leiden, 1959).

398   *Geometrical Knowledge of Mediaeval Master Masons*

It thus appears that whatever geometrical knowledge a mediaeval master mason might have possessed, he had not gotten it from formal schooling.[9] On the other hand, there were informal means for masons to acquire this knowledge — such as conversations with clerical building patrons that could amount to a kind of tutoring. And the really determined mason who was literate could teach himself geometry by studying the mediaeval treatises on the subject. But the deep conviction of the masons that "geometry" was the basis of their craft suggests that these informal — and probably quite exceptional — avenues do not satisfactorily explain how the rank and file of the masons, and even the master masons, acquired their knowledge of geometry. A far more probable answer is to be found in the masons' education into the traditions of their craft, whereby the technical knowledge required in design and construction was transmitted from father to son, from master to apprentice, from learned journeyman to those who were less learned in the craft traditions.[10] Since the geometry of the masons was an essential part of that technical knowledge, mediaeval master masons would normally have acquired their geometrical knowledge in the same way that they acquired the rest of their knowledge and skill in building — by mastering the traditions of the craft.

Those traditions were by and large transmitted orally from one generation of masons to the next; consequently, the vast bulk of the technical knowledge upon which mediaeval building and architecture were based disappeared with the dying of those oral traditions at the close of Gothic building in Europe. In view of this, the task of reconstructing the geometrical knowledge of mediaeval master masons would appear to be hopeless. Fortunately, however, near the end of the Middle Ages a few German master masons wrote little books on some of the technical aspects of their craft, and from these we can get a fairly substantial picture of the geometry of the masons. But before we turn to these late fifteenth-century documents we must give attention to that solitary and crucially important thirteenth-century *Sketchbook* of Villard de Honnecourt.

As noted above, Villard several times referred to the *ars de iometrie* as the basis for his technique of *portraiture*, and he implied that it was also the basis for the craft of *maconerie*. But since we have recognized the loaded character of the word geometry for mediaeval masons, we must ask, what precisely did Villard mean by *iometrie*? And what did his follower, Magister 2, mean by *geometrie* when he inscribed the sentence, "Totes ces figures sunt estraites de geometrie," on one of the pages devoted to *maconerie*? It has generally been assumed that Villard and Magister 2 meant "practical geometry" in the traditional mediaeval sense;

[9] This conclusion, based on my detailed study of the sources for education and the building crafts in mediaeval England, still corresponds to the views of Sigmund Günther on the mathematical education of laymen in the Middle Ages, set forth long ago in his *Geschichte des mathematischen Unterrichts im deutschen Mittelalter* (Berlin 1887), pp. 286–335; for geometry in particular see pp. 326–328. I am not aware of any detailed study of the mathematical education of laymen in mediaeval France or Italy.

[10] Shelby *op. cit.* pp. 18–23.

Geometrical Knowledge of Mediaeval Master Masons   399

Magister 2's assertion has therefore been taken to mean, "All of these figures are drawn from [a treatise on practical] geometry."[11]

Certainly mediaeval scholars showed a considerable interest in practical geometry, as evidenced by the numerous extant treatises on the subject. If the geometry of the masons was even approximately equivalent to this practical geometry — if Magister 2 did copy his examples from a mediaeval *practica geometriae* — then we clearly have a wealth of material from which to reconstruct the geometrical knowledge of mediaeval masons, in addition to the *Sketchbook* of Villard and the booklets by late mediaeval German masters.

This is a critical problem for the subject of this paper. In order to deal with the specific meaning of Magister 2's assertion, and to provide a general framework for a discussion of the content of the geometry of mediaeval masons, it will be necessary at this point to make an excursus into the mediaeval traditions in geometry through the thirteenth century, with particular attention on the development of mediaeval practical geometry.[12] This does not require us to enter the vexed and much-discussed questions regarding the translations of Euclid's *Elements* and their dissemination in mediaeval Europe. Suffice it to say that during the early Middle Ages there were in circulation Latin translations (by

[11] Hahnloser *op. cit.*, pp. 196–197, 240–242, 254–259; and before him, Victor Mortet, "La mesure de la figure humaine et le canon des proportions d'après les dessins de Villard de Honnecourt, d'Albert Durer et de Léonard de Vinci," *Mélanges offerts à M. Émile Chatelain* (Paris, 1910), pp. 367–371.

[12] For the subject of mediaeval practical geometry the general histories of mathematics (F. Cajori, 1894; S. Günther, 1908; H. G. Zeuthen, 1912; D. E. Smith, 1923; J. E. Hofmann, 1953; H. W. Eves, 1953) and even the histories of geometry (M. Chasles, 1837; J. L. Coolidge, 1940; H. W. Eves, 1963) are of little value, with the exceptions of the following: Moritz Cantor, *Vorlesungen über Geschichte der Mathematik*, 3rd. ed. (Leipzig, 1907; rpt. Stuttgart, 1965), I, 821–878 and II, 35–53; Gina Loria, *Storia della matematiche*, Vol. I, *Antichità-Medio Evo-Rinascimento* (Turin, 1929), pp. 233–260; 393–396; and A. P. Juschkewitsch, *Geschichte der Mathematik im Mittelalter* (Basel, 1964), pp. 338–346 and 384–387. However, the best surveys of mediaeval practical geometry are two articles by Paul Tannery, "La géométrie au xiᵉ siècle," *Mémoires scientifiques*, ed. J. L. Heiberg *et. al.* (Paris and Toulouse, 1922), V, 79–102 [rpt. from *Revue générale internationale, scientifique, litteraire et artistique*, No. 15 (1897), 343–357]; and "Histoire des sciences: Géométrie," *Mémoires scientifiques* (Paris and Toulouse, 1930), X, 37–59 [rpt. from *Revue de synthèse historique*, II (1901), 283–299]. Also helpful is Victor Mortet, "Note historique sur l'emploi de procédés matériels et d'instruments usités dan la géométrie pratique au moyen âge (xᵉ–xiiiᵉ siècles)," *Congrès international de philosophie*, 2nd. session (Geneva, 1904), pp. 925–942. But the history of mediaeval geometry is yet to be written. The last two decades of the nineteenth century saw a spate of activity by German scholars — H. Weissenborn, S. Günther, J. L. Heiberg, and especially M. Curtze — and French scholars — C. Henry, V. Mortet, and P. Tannery — in publishing mediaeval texts and monographs on detailed problems. Curtze had undertaken to write a history of mediaeval geometry that would synthesize this work, but the project remained unfinished at his death in 1903. The death of Tannery in 1904, of Mortet in 1914, and the coming of W. W. I (which saw the demise of several German journals devoted to the history of mathematics) virtually brought to a close this concerted interest in mediaeval geometry. Fortunately, since W. W. II interest in the subject has been revived by Marshall Clagett, John Murdoch, Guy Beaujouan, Roger Baron, and H. L. L. Busard, as will become evident from the numerous citations to their work in this present study. We may also welcome into the field Prof. Murdoch's student, Stephen Victor, whose Harvard dissertation (1971) contains an edition and translation of a late twelfth-century *practica geometriae* which begins "Artis cuiuslibet consummatio. . . ."

Boethius, "Pseudo-Boethius," and others) of portions of the *Elements* which transmitted to western students some of the definitions, postulates, axioms, and propositions of Euclid, but without the Euclidean proofs.[13] Then in the twelfth century the entire *Elements* was translated into Latin from Arabic versions by at least three Latin scholars.[14]

Geometry came to mediaeval Europe not only by way of Euclidean fragments. Since late antiquity it had been recognized as one of the seven liberal arts, so that it received proper obeisance — but short shrift — in the handbooks of Martianus Capella, Cassiodorus Senator, and Isidore of Seville.[15] Furthermore, early mediaeval scholars who wanted a more detailed knowledge of practical geometry than was provided in the handbooks and "Boethian" excerpts from Euclid could turn to the treatises on surveying written by the Roman *agrimensores*. Fragments from these works of Frontinus, Hyginus, Balbus, Nipsus, Epaphroditus, Vitruvius Rufus, and others were preserved in the famous "Codex Arcerianus," a very early (sixth or seventh century) manuscript known to have been in the monastery at Bobbio in the tenth century.[16] But it is doubtful that early mediaeval scholars were interested in or capable of doing much more than copying these passages on geometry from the handbooks, the Euclidean excerpts, and the agrimensorial treatises. Not until the time of Gerbert of Reims (c. 940–1003) was a western Latin scholar able to understand these sources sufficiently to attempt a geometrical treatise on his own. To be sure, the compilation which came to be known as the *Geometria Gerberti* was not all written by Gerbert. Again, it is not necessary here to enter the complex questions regarding the authorship of this work.[17] Our purpose will be served by noting the achievement represented

---

[13] The best guide to the mediaeval MSS and to the modern editions and literature on the texts is Menso Folkerts' introductory chapters to his edition of one of the "Pseudo-Boethius" geometries: *"Boethius" Geometrie II: Ein mathematisches Lehrbuch des Mittelalters,* "Boethius: Texte und Abhandlungen zur Geschichte der exakten Wissenschaften," Bd. IX (Wiesbaden, 1970), pp. 3–107.

[14] Marshall Clagett, "The Medieval Latin Translations from the Arabic of the *Elements* of Euclid, with Special Emphasis on the Versions of Adelard of Bath," *Isis,* XLIV (1953), 16–30. Cf. John E. Murdoch, "The Medieval Euclid: Salient Aspects of the Translations of the *Elements* by Adelard of Bath and Campanus of Novara," *Revue de synthèse,* LXXXIX (1968), 67–74. On a twelfth-century translation of the *Elements* from Greek into Latin see *idem,* "Euclides Graeco-Latinus: A Hitherto Unknown Medieval Latin Translation of the *Elements* Made Directly from the Greek," *Harvard Studies in Classical Philology,* LXXI (1966), 249–270.

[15] *Martiani Capellae de Nuptiis Philologiae et Mercurii,* Lib. VI, ed. F. Eyssenhardt (Leipzig, 1866), pp. 194–254; *Cassiodori Senatoris Institutiones,* Lib. II, cap. vi, ed. R. A. B. Mynors (Oxford, 1937), pp. 150–153; *Isidori Hispalensis Episcopi Etymologiarum sive Originum Libri XX,* Lib. III, cap. viii–xiv, ed. W. M. Lindsay (Oxford, 1911). Cf. William H. Stahl, *Roman Science: Origins, Development, and Influence to the Later Middle Ages* (Madison, 1962), pp. 173, 177, 206–209, 216–218.

[16] For over a century modern scholars have wrestled with the complex problems of establishing and interpreting the texts of the *agrimensores.* For a guide to the literature on the subject and an interesting argument that the monastery of Corbie "was the gromatic and geometric capital of the mediaeval world," see B. L. Ullman, "Geometry in the Mediaeval Quadrivium," *Studi di bibliografia e di storia in onore di Tammaro de Marinis* (Verona, 1964), IV, 263–285.

[17] Nicolaus Bubnov, ed., *Gerberti postea Silvestris II papae opera mathematica* (Berlin, 1899; rpt. Hildesheim, 1963), pp. 310–313, summarized the scholarship on the problem and explained his own separation of the authentic *Geometria Gerberti* (edited on pp. 48–97) from the *Geometria incerti auctoris*

Geometrical Knowledge of Mediaeval Master Masons 401

in this compilation by Gerbert and his eleventh-century successors, namely, the mastery of the Euclidean excerpts and the agrimensorial treatises at least to the extent that the authors could restate and even attempt to move beyond these sources in the formulation of geometrical problems. In reworking these materials the *Geometria Gerberti* provided the prototypes for the two main approaches to geometry in the High Middle Ages. The more strictly mathematical approach eventually found its way into a rather small corner of the university curriculum as the study of Euclid's *Elements*; practical geometry, on the other hand, became a subject of common interest to both Schoolmen and craftsmen, and treatises on it in Latin and in the vernacular languages continued to be written throughout the remainder of the Middle Ages.

But the formal distinction between "theoretical" and "practical" geometry did not appear in the Latin West until the twelfth century, when it was introduced by Hugh of St. Victor in a short treatise on *Practica geometriae*.[18] "The entire discipline of geometry is either theoretical, that is, speculative, or practical, that is, active. The theoretical is that which investigates spaces and distances of rational dimensions only by speculative reasoning; the practical is that which is done by means of certain instruments, and which makes judgments by proportionally joining together one thing with another."[19] Having separated theoretical from practical geometry, Hugh then developed a tripartite division of the latter into *altimetria, planimetria,* and *cosmimetria.* "Altimetry is that which investigates heights and depths. . . . It is called planimetry when one seeks to find the extent of a plane. Cosmimetry however takes its meaning from the word cosmos. Cosmos in Greek means the world; hence cosmimetry is the measurement of the world, that is to say, it concerns the measurement of circumference, as in the motion of a heavenly sphere and of other heavenly circles, or in the globe of the earth and many other things which nature has placed in the round."[20]

(edited on pp. 317–364). Tannery remained unconvinced of Bubnov's solution, however, and insisted that there were at least three authors involved in the compilation of this work. See his "Géométrie au xiᵉ siècle," pp. 99–101; and "Histoire des sciences: Géométrie," pp. 49–51.

[18] Roger Baron, "Sur l'introduction en Occident des termes 'geometria theorica et practica'," *Revue d'histoire des sciences et de leurs applications,* VIII (1955), 298.

[19] *Hvgonis de Sancto Victore opera propaedevtica: Practica geometriae, De grammatica, Epitome Dindimi in philosophiam,* ed. Roger Baron. "Publications in Mediaeval Studies," XX (Notre Dame, 1966), p. 16: ". . . omnis geometrica disciplina aut theorica est, id est speculatiua, aut practica, id est actiua. Theorica siquidem est que spacia et interualla dimensionum rationabilium sola rationis speculatione uestigat, practica uero est que quibusdam instrumentis agitur et ex aliis alia proportionaliter coniciendo diiudicat."

[20] *Ibid.,* p. 17: "Hinc namque altimetria dicta est quod sublime siue profundum uestigat. . . . Planimetria appellata uidetur quando porrectionem secundum planum persequitur. Cosmimetria autem ab eo quod et cosmus nomen accepit. Cosmus enim grece mundus dicitur, et inde cosmimetria dicta est quasi mensura mundi, ea uidelicet que circumferentiam metitur, quam in ambitu celestis spere et reliquorum circulorum celestium nec non in globo terre, multorum etiam aliorum que natura in orbem disposuit, consideramus." Cf. Hugh's definitions of altimetry, planimetry, and cosmimetry in his *Didascalicon,* Lib. II, cap. 14 [Migne, *Patrologia Latina,* CLXXVI, col. 757; and *The* Didascalicon *of Hugh of St. Victor,* trans. Jerome Taylor (New York, 1961), p. 70]. Damien van den Eynde, *Essai*

402   *Geometrical Knowledge of Mediaeval Master Masons*

In short, Hugh's *Practica geometriae* is a Schoolman's textbook on surveying, both terrestrial and celestial. Hugh mentions in his "Prologue" that he is not attempting something new, but is simply pulling together material scattered through older works, in order to smooth the way for students interested in such matters. But what he did in effect was to establish a genre of scholastic treatises and to give to the genre its basic framework in the distinctions between altimetry, planimetry, and cosmimetry. Henceforth the mediaeval treatises on *practica geometriae* would normally follow the path marked out by Hugh, except that cosmimetry in time was transformed into stereometry. Thus mediaeval practical geometry as reflected in the treatises on the subject was confined to surveying and metrology, which meant that other applications of geometry to the world of practice remained outside the ken of the authors and readers of the *practicae geometriae*.

It might have been otherwise if these authors had taken up the suggestions on practical geometry which Dominicus Gundissalinus, the twelfth-century Spanish philosopher and translator, introduced into his schematization of knowledge, *De divisione philosophiae*. In this work Gundissalinus leaned heavily on the classification of the sciences developed by the tenth-century Arabic scholar, al-Farabi.[21] But he also appears to have been influenced by Hugh of St. Victor, for his discussion of geometry reflects a blending of terms and ideas from both the *Practica geometriae* and al-Farabi's *De scientiis*.[22] Following the latter's distinction between the theoretical and the practical sciences, Gundissalinus placed mathematics within the theoretical branch. He then divided mathematics into seven arts, one of which was geometry.[23] He further divided each of these arts into its

---

sur la succession et la date des écrits de Hugues de Saint-Victor, "Spicilegium Pontificii Athenaei Antoniani," xii (Rome, 1960), pp. 47–48, suggests that Hugh wrote the passage on geometry in the *Didascalicon* before he wrote the *Practica geometriae;* Roger Baron, "Note sur la succession et la date des écrits de Hugues de Saint-Victor," *Revue d'histoire ecclésiastique*, lvii (1962), 110, argues to the contrary. In either case, Hugh remains the source for this fundamental structuring of mediaeval *practicae geometriae*.

[21] Gundissalinus himself translated al-Farabi's *De scientiis* and *De ortu scientiarum*. The former has been published in al-Farabi, *Catálogo de las ciencias*, ed. and trans. Ángel Gonzalez Palencia, 2nd. ed. (Madrid, 1953), pp. 87–115 (on geometry, see pp. 98–99); the latter in al-Farabi, *Über den Ursprung der Wissenschaften (De ortu scientiarum)*, ed. Clemens Baeumker. "Beiträge zur Geschichte der Philosophie des Mittelalters," Bd. xix, Heft 3 (Münster, 1916), pp. 17–24 (on geometry, p. 18).

[22] Since my purpose is simply to outline the development of mediaeval *practicae geometriae* I shall not attempt here to sort out precisely what Gundissalinus owed to al-Farabi, to Hugh of St. Victor, and to his own thinking. A beginning in this task has been made by Roger Baron, "Note sur les variations au xii° siècle de la triade géométrique: Altimetria, Planimetria, Cosmimetria," *Isis*, xlviii (1957), 31–32; but see R. W. Hunt's earlier attempt to find some of the intellectual roots of the *De divisione philosophiae* in the works of Thierry of Chartres: R. W. Hunt, "The Introductions to the 'Artes' in the Twelfth Century," *Studia Mediaevalia in honorem admodum Reverendi Patris Raymundi Josephi Martin* (Bruges, n.d. [1948]), pp. 86–93.

[23] "Mathematica quoque uniuersalis est, quia sub ea continentur septem artes, que sunt arismetica, geometria, musica et astrologia, sciencia de aspectibus, sciencia de ponderibus, sciencia de ingeniis." Dominicus Gundissalinus, *De divisione philosophiae*, ed. Ludwig Baur. "Beiträge zur Geschichte der Philosophie des Mittelalters," Bd. iv, Heft 2–3 (Münster, 1903), pp. 31–32.

theoretical and practical aspects, and in good scholastic fashion he developed distinctions between the theoretical and the practical in terms of genus, species, parts, artificer, instruments, office, purpose, etc. Some of the distinctions which he made concerning geometry are particularly pertinent to our present study and worth quoting at length.

There are three species of practical geometry: altimetry, planimetry, cosmimetry. That science by which one considers lines, surfaces, and bodies in height is called altimetry, that is to say, the science of measuring altitudes; in planes it is called planimetry, that is, the science of measuring any plane surface; in depth, it is called cosmimetry or the science of measuring solids. . . . The purpose of theory is to teach something. The purpose of practice is to do something. . . . The artificer of theory is the geometer, who clearly knows all parts of geometry and can teach it. His instrument is the demonstration. . . . The artificer of practice is he who uses [geometry] in working. There are two kinds of these, namely, surveyors and craftsmen. Surveyors are those who measure the height and depth and plane surface of the earth. Craftsmen are those who exert themselves by working in the constructive or mechanical arts — such as the carpenter in wood, the smith in iron, the mason in clay and stones, and likewise every artificer of the mechanical arts — according to practical geometry. Each indeed forms lines, surfaces, squares, circles, etc., in material bodies in the manner appropriate to his art. These many kinds of craftsmen are distinguished according to the different materials in which and out of which they work. Any one of these thus has his proper materials and instruments. The instruments of the surveyors are the foot, palm, cubit, stadium, perch, and many others. Those of the carpenters are the axe, adze, broadaxe, string, and many others. Those of the smith are the anvil, shears, hammer, and many others. Those of the masons are the string, trowel, plumb bob, and many others. . . . The office of practical geometry is, in the matter of surveying, to determine the particular dimensions by height, depth, and breadth; in the matter of fabricating, it is to set the prescribed lines, surfaces, figures, and magnitudes according to which that type of work is determined. The goal is either the certification of dimensions, or money and praise for the completion of the work.[24]

It is clear from this passage that Gundissalinus followed Hugh of St. Victor's distinction between theoretical and practical geometry, but he significantly

[24] *Ibid.*, pp. 107–110: "Species quoque practice sunt tres: altimetria, planimetria, cosmimetria. Sciencia enim, qua considerat lineas superficies et corpora in altum, altimetria dicitur, scilicet sciencia de mensura altitudinis. qua uero in planum, dicitur planimetria i.e. scientia de mensura alicuius planiciei. qua uero in profundum, dicitur cosmimetria quasi sciencia de mensura solidi. . . . Finis enim theorice est aliquid docere. finis uero practice est aliquid agere. . . . Artifex uero theorice est geometer, qui plane nouit omnes partes geometrie et eam docet. Instrumentum eius est demonstracio. . . . Artifex uero practice est, qui eam operando exercet. Duo autem sunt, qui eam operando exercent, scilicet mensores et fabri. mensores sunt, qui terre altitudinem uel profunditatem uel planiciem mensurant. fabri uero sunt, qui in fabricando siue in mechanicis artibus operando desudant, ut carpentarius in ligno, ferrarius in ferro, cementarius in luto et lapidibus et similiter omnis artifex mechanicarum arcium secundum geometriam practicam. Ipse enim per semetipsum format lineas, superficies, quadraturas, rotunditates et cetera in corpore materie, que subiecta est arti sue. horum autem fabrorum multe species esse dicuntur secundum diuersitatem materiarum in quibus et ex quibus operantur. quorum unusquisque sicut habet materiam propriam sic et instrumenta propria. mensorum enim instrumenta sunt pes, palmus, cubitus, stadium, pertica et multa alia. carpentariorum uero securis et ascia, dolabrum et linea et multa alia. ferrariorum uero incus, forficus et malleus et multa alia. cementariorum uero linea, trulla, perpendiculum et multa alia. . . . Officium uero practice est uel in mensurande de alto, profundo et lato certam dimensionem reddere, uel in fabricando statutas lineas, superficies, figuras et magnitudines secundum quod species operis exigit non excedere. Finis eius est uel certitudo dimensionis, uel merces et laus de consummacione operis."

404   *Geometrical Knowledge of Mediaeval Master Masons*

broadened Hugh's definition of the latter. Whereas Hugh had limited his analysis of practical geometry to terrestrial and celestial surveying, Gundissalinus recognized that this "art" was used by *fabri* as well as *mensores*. Indeed, his perception of the essence of the geometry of the craftsmen was incisive; as we can see from the quotation above, that geometry consisted of the manipulation of "lines, surfaces, figures, and magnitudes" in the materials, with the instruments, and according to the rules appropriate for each craft. It is unfortunate that Gundissalinus' recognition of the *geometria fabrorum* was not followed up by mediaeval authors of *practicae geometriae*, for even a Schoolman's rendering of the practice of geometry by the building crafts would be a welcome source of information on a subject for which there is so little direct evidence.

There is virtually a conscious exclusion of the geometry of the craftsmen from the *Practica geometriae* completed by the now-famous Leonardo Pisano in 1220. Perhaps because this was an advanced treatise on mathematics, Leonardo did not feel obliged to begin with the usual distinctions between theoretical and practical geometry. Instead, he set down definitions for many of the terms and concepts with which he would be dealing, and then he launched forth on an eight-part treatise concerned with technical problems in geometry, arithmetic, trigonometry — and surveying. He devoted many pages to this latter subject in his large book, but they differed radically from the agrimensorial treatises and previous mediaeval works on surveying. Unlike the authors of those treatises, Leonardo did not merely provide rule-of-thumb procedures in surveying; he offered demonstrations of the mathematical correctness of the procedures which he described.[25] Thus Leonardo turned the mediaeval distinction between theoretical and practical geometry on its head, for in his *Practica geometriae* he demonstrated with "theoretical" geometry the proofs of his propositions. But he paid a price for his mathematical sophistication, since his book does not appear to have been widely read or used by others interested in practical geometry.

It was not only the sophistication of Leonardo's mathematics that put his book beyond the reach of craftsmen; unless they had at least a pretty thorough grammar school education, they would have been unable to understand the Latin in which it was written. The same of course would have been true of Hugh of St. Victor's *Practica geometriae*. While Latin was a great boon to scholarship because of its universality, it did constitute a stumbling block for those laymen who did not know the language, but who wanted direct access to at least some of the learning of the Schoolmen. The increasing literacy of the laity in the

---

[25] Leonardo made his intentions clear in his preface: "Rogasti amice dominice et reverende magister, ut tibi librum in pratica geometrię conscriberem; igitur amicitia tua coactus, tuis precibus condescens, opus iam dudum inceptum taliter tui gratia edidi, ut hi qui secundum demonstrationes geometricas: et hi qui secundum uulgarem consuetudinem, quasi laicalj more, in dimensionibus uoluerint operari super .viij. huius artis distinctiones, que inferius explicantur, perfectum inueniant documentum." *Scritti di Leonardo Pisano*, Vol. II, *Practica geometriae ed opuscoli*, ed. B. Boncompagni (Rome, 1862), p. 1. At another point he wisecracked: "Sed hec talis inuestigatio non est operando ab agrj mensoribus, qui secundum uulgarem modum procedure uolunt. . . . Sed ut ipsi, qui secundum geometricam scientam operarj desiderant leuius quam dictum sit. . . ." (p. 95)

Geometrical Knowledge of Mediaeval Master Masons 405

vernacular languages in the later Middle Ages produced a certain amount of popular pressure for learned treatises to be written or translated into the vernacular. This desire extended to mathematical works as early as the thirteenth century, as may be noted from the introductory comment in a versified algorism dating from that period. "The two clerks who translated the computus into French are urged by many people to undertake the task of putting algorism into French, just as they did the computus. . . ."[26] Indeed, the number of treatises on "applied mathematics" composed in Latin and in the vernacular tongues from the thirteenth century onward reveals that here was a common ground of interest for Schoolmen and laymen alike. For an example one may cite another algorism in French, dating from c. 1275, but more pertinent to the present study is the French *Pratike de geometrie* to be found in the same manuscript.[27]

The anonymous author begins his *Pratike de geometrie* with this prefatory statement: "We shall commence a work on the practice of geometry, which we shall divide in three parts. In the first part we shall teach how to find the measurement of plane surfaces; in the second, how to find the measure of heights and depths and of large measures; in the third, how to find the details of geometry and astronomy appropriate to the two preceding parts."[28] One recognizes immediately the customary tripartite division of practical geometry into planimetry, altimetry, and cosmimetry — although the author appears a bit uncertain about what he is to do with that third section. He does not in fact follow his own outline very well, as may be ascertained from a brief review of the contents of the treatise.

First there is a short description of how to use the astrolabe in calculating the length of a straight line, for examples, the distance across a woods or a river, or the height of a tree or steeple. Next comes an explanation of how to find the area of various geometrical figures — the circle, square, pentagon, hexagon, heptagon, and a number of different triangles. This is followed by exercises in finding "surpluses," that is, the difference between the areas of a circle and of a square which inscribes it, and vice versa. The author then turns to some practical surveying problems, such as how to find the number of acres in a field, the number of messuages of a given size in a given area, the number of messuages in a round city. The last of the "geometrical" problems concerns the measurement of volumes of various containers, such as the hogshead and tun. The little treatise closes with

[26] Translation by E. G. R. Waters, "A Thirteenth Century Algorism in French Verse," *Isis*, XI (1928), 55.

[27] Charles Henry, "Sur les deux plus anciens traités français d'algorisme et de géométrie," *Bulletino di bibliografia e di storia delle scienze matematiche e fisiche*, XV (1882), 53–70, edited both little treatises. Victor Mortet, "Le plus ancien traité français d'algorisme," *Bibliotheca Mathematica*, Ser. 3, IX (1908–09), 60–63, re-edited the algorism and promised a new edition of the *Pratike de geometrie*, but the latter never appeared in print. Dr. Stephen Victor informs me that he is preparing a new edition; this is certainly desirable, for Henry's edition is quite unsatisfactory, as Mortet noted long ago.

[28] Henry, *op. cit.*, p. 55: "Nous commencerons une oeure soi le pratike de geometrie la ke le nous deuiserons en .3. parties. En la premiere partie ensengerons nous a trouer le mesure des planetes. En le seconde a trouer le mesure des hautes ches et des profondeces et des crasses mesures en la tierce a trouer les minuces de gyometrie et dastronomie couignables as .IJ. parties deuant."

406   *Geometrical Knowledge of Mediaeval Master Masons*

several pages devoted to methods of calculating exchanges of money from one system to another.

This description may serve as an index of the contents of practical geometry as perceived by the anonymous author, and perhaps by those "many people" anxious to obtain works like this in the French language. In terms of applied geometry, clearly the author's main interests were directed towards surveying and metrology — the latter meaning the techniques of measuring the volume of various kinds of containers.[29] With the exception of the tunmaker, he scarcely bothers with the application of practical geometry to the crafts concerned with the mechanical and constructive arts. Only in one passage does he deal directly with a problem that might arise in building construction; there he provides a somewhat confused formula for finding the volume of a round column.[30] But even this formula probably did not derive as much from mediaeval as from classical practice, for it seems to be little more than a pale reflection of the fragment from antiquity, *De geometria columnarum et mensuriis aliis*, which had gotten included in some of the mediaeval manuscripts preserving the agrimensorial treatises and other fragments of Roman practical geometry.[31]

The *Pratike de geometrie* brings our review of mediaeval practical geometry back to Villard's *Sketchbook*, for the two books were approximately contemporary, they were both written in the Picard dialect, and they both were concerned with the application of geometry to practical problems. But generally speaking, the problems which interested Villard and Magister 2 were quite different from those which caught the attention of the author of the *Pratike de geometrie*, for the latter was concerned only with surveying and metrology, whereas the pages of the *Sketchbook* were primarily devoted to problems of *portraiture, maconerie*, and

[29] This thirteenth-century French author was thus carrying on the classical tradition of handbooks and treatises on metrology which applied practical geometry and arithmetic to the work-a-day problems of weights and measures. For the classical texts, see Friedrich Hultsch, ed., *Metrologicorum scriptorum reliquiae* (Leipzig, 1864–66), I, 179–355 and II, 48–146. Even the section on money-changing in the *Pratike de geometrie* was within the classical tradition, for this was an important branch of ancient metrology: Friedrich Hultsch, *Griechische und Römische Metrologie*, 2nd. ed. (Berlin, 1882), pp. 162ff.

[30] Henry, *op. cit.*, p. 61: "Se tu ueus trouer le coube dun piler reont tu troueras laire par le moitie de son dyametre en sa circonference. Keure laire sor le lonc la somme fera la coube du piler che pues prouer ausi ke deuant."

[31] Victor Mortet edited the fragment and made a detailed historical and philological analysis of the text in "La mesure des colonnes à la fin de l'époque romaine d'après un très ancien formulaire," *Bibliothèque de l'École des Chartes*, LVII (1896), 277–324; he followed this with a study of early mediaeval texts which showed the continuing literary influence of these classical formulae for mensuration and proportions of columns: "La mesure et les proportions des colonnes antiques d'après quelques compilations et commentaires antérieurs au xiie siècle," *ibid.*, LIX (1898), 56–72. The formula for calculating the volume of a column was included in the *Geometria incerti auctoris*: Bubnov, *op. cit.*, p. 360. The traditional character of this problem is indicated by the fact that the simplified formula for its solution — very similar to that given in the *Pratike de geometrie* — was included in the section on stereometry in the *Tractatus quadrantis* composed by Robertus Anglicus sometime before 1276: Paul Tannery, ed., "Le Traité du quadrant de Maître Robert Anglès (Montpellier, xiiie siècle). Texte latine et ancienne traduction grecque," *Mémoires scientifiques*, v, 188 [rpt. from *Notices et extraits des manuscrits de la Bibliothèque Nationale*, xxxv, pt. 2 (1897), 561–640].

Geometrical Knowledge of Mediaeval Master Masons 407

carpenterie. To be sure, Villard and Magister 2 did include a few problems in surveying — how to measure the width of a river or of a window from a distance, or the height of a tower. (Figs. 1 and 2) The first and third of these were traditional problems of *altimetria* and *planimetria* in the agrimensorial and mediaeval literature on surveying. Indeed, Villard's demonstration of how to measure a tower could have served as an illustration for the description of this technique in the *Geometria Gerberti*:

The geometer devises a right-angle triangle composed of a base and vertical of the same number; the proportion of the hypotenuse is not considered, since in determining height by means of a right-angle triangle, it is thought to be quite useless. The device is carried by the surveyor along the base plane to the point where — with the eye placed at ground-level — the summit of the height to be investigated can be seen at the top of the vertical. Then, from the place where the view has been taken, one measures the base plane to the foot [of the thing which is being measured], and however far it is, that is the height.[32]

Conceivably, Villard got his knowledge of the technique directly from the *Geometria Gerberti*, although by his time it could well have become a rule-of-thumb practice handed down through the oral traditions of the craft. The traditional character of this rather crude technique is indicated by its survival in an English version of the fifteenth century, when surveying techniques had become even more refined than in the thirteenth century.[33] I think it doubtful that Villard extracted the technique directly from the *Geometria Gerberti*, but even if he did, it is important to note what he passed over in selecting this particular technique. For in fact this is only one, and the simplest, of several techniques for measuring heights which are described in the *Geometria Gerberti*. The others measure by means of a mirror, a shadow, a staff, a string and arrow, and an astrolabe, and some of the procedures are fairly complex.[34] Since neither Villard nor Magister 2 gives evidence of using either the astrolabe or the surveyor's quadrant, one suspects that they were not abreast of the advances in surveying which had come to Europe with the introduction of Arabic learning and mathematical instruments.[35]

[32] Bubnov, *op. cit.*, p. 327–328: "Componatur a geometra orthogonium basi cathetoque ejusdem numeri compositum, hypotenusae vero proportio praetermittatur, quae ad altum investigandum in hoc orthogonio prorsus inutilis judicatur. Compositum autem tandiu per planum a mensore trahatur, donec oculo humi apposito per catheti summitatem summitas altitudinis investigandae cernatur. Qua visa, a loco, cui visus inhaeserat, planities ad radicem usque metiatur; et quanta fuerit, tanta altitudo dicatur." This passage is from the eleventh-century *Geometria incerti auctoris*. Cf. Maximilian Curtze's edition of this section of the *Geometria Gerberti* in "Die Handschrift No. 14836 der Königl. Hof- und Staatsbibliothek zu München," *Abhandlungen zur Geschichte der Mathematik*, VII (1895), 84–95.

[33] J. O. Halliwell, ed., *Rara Mathematica* (London, 1839), pp. 27–28, transcribed this English document from the British Museum MS Lansdowne 762, fol. 23b. Leonardo Pisano also included the technique in his *Practica geometriae* (pp. 202–203 of Boncompagni's edition), but in typical fashion he provided a detailed mathematical explanation of the technique, as well as variations on it with the use of similar but non-equilateral triangles.

[34] See, for example, the procedure, "Ad rem inaccessibilem nobis altioribus metiendam," Bubnov, *op. cit.*, pp. 328–330.

[35] On mediaeval surveying instruments, see Maximilian Curtze, "Über die im Mittelalter zur Feldmessung benutzten Instrumente," *Bibliotheca Mathematica*, N.F., x (1896), 65–72; and R. T. Gunther, *Early Science in Oxford* (Oxford, 1921–23; rpt. London, 1967), I, 333–340. For knowledge

The question of the source of Villard's technique for measuring the height of a tower brings us back to the problem of Magister 2's assertion that "Totes ces figures sunt estraites de geometrie." He could hardly have meant that *all* of these practical problems of the masons' craft were drawn from some mediaeval treatise on practical geometry. As we have seen, these treatises were almost entirely confined to problems of surveying and metrology, and their authors showed little concern for the application of geometry to the building and mechanical crafts. If not from some *practica geometriae*, were the examples of practical problems taken from another shop-manual of the masons' craft, as suggested by Prof. Robert Branner?[36] In spite of the ingenuity with which Branner argues this possibility, it remains an *argumentum e silentio*. Apart from the *Sketchbook* itself, there is not a whit of evidence for the existence at this time of other shop-manuals of the masons' craft, let alone a continuing tradition of such books of which Villard's is the only survivor.[37]

There are so many puzzling problems in Villard's *Sketchbook* — and they have elicited such elaborate and diverse explanations — one is tempted to posit as a fundamental exegetical principle the rule that the simplest explanation is *ipso facto* the best. Therefore, instead of assuming that Magister 2 copied the shop problems on folios 39 and 40 from some other book — of whatever kind — let us begin with the fact that he inserted additional examples of practical problems of the masons' craft into those pages which Villard had already set aside for *maconerie*. Let us then assume that the figures and texts were not necessarily meant to be self-explanatory, and that Magister 2 inserted them as reminders, as memory tags, for other masters or journeymen who would orally explain the details of the problems and their solutions to apprentices and fellow masons of the craft. In brief, let us assume that the *Sketchbook* is what it appears to be, namely, an exceptional literary record of some of the oral traditions of the masons' craft. These simple assumptions help to explain why the *Sketchbook* contains so many unresolved, and perhaps in some cases unresolvable, puzzles; they also make it unnecessary to set up further puzzles about the relationship of this document to a completely unknown — and perhaps never existent — literary genre of the thirteenth century.

What then of Magister 2's assertion about the source of his figures? Within the context of an oral rather than written tradition, the statement would mean that these figures were drawn from the "science" of geometry upon which all the handicrafts are based, and which "masonry has the most notability and most part of," to borrow a phrase from the "Articles and Points of Masonry." And what is the content of that geometry? Since it is not that of the treatises on *practica geo-*

---

and use of the astrolabe and quadrant by the authors of twelfth- and thirteenth-century *practicae qeometriae*, see *Hvgonis de Sancto Victore opera propaedevtica*, pp. 25ff.; Henry, "Anciens traités français," p. 55; Leonardo Pisano, *Practica geometriae*, pp. 204ff.; and Tannery, "Traité du quadrant," pp. 151ff.

[36] "A Note on Gothic Architects and Scholars," *Burlington Magazine*, XCIX (1957), 372–375.

[37] As suggested by Paul Frankl, *The Gothic: Literary Sources and Interpretations through Eight Centuries* (Princeton, 1960), pp. 47–48.

Geometrical Knowledge of Mediaeval Master Masons   409

metriae, and since there are extant no other previous or contemporary shop manuals of the masons' craft, Magister 2's statement does not direct us outward to other literary documents, but inward to the *Sketchbook* itself, as the primary source for reconstructing the content of the geometry used by master masons of this period.

Concentrating first on the pages devoted to *maconerie,* we may enumerate the kinds of problems which Villard and Magister 2 proposed to solve with the geometry of the masons' craft. Of the thirty-eight problems on these three pages, four figures (39 1, m, 40 1, and 41 b of Hahnloser's edition) deal with techniques of surveying, while another three (39 j, p, and q) essentially belong to other crafts. Five figures (39 e, k, n, and 41 a, c) deal with larger problems of designing and setting out masonry work at full scale. The remainder are more strictly speaking stereotomical problems of setting out and cutting various stones used in the masons' repertoire of architectural forms: columns, springers, pendants, cusps, and above all, voussoirs of different kinds. Several generations of scholars have entertained themselves in trying to sort out the puzzles which these figures and cryptic comments present. In recent years Professor Branner and I have published a series of articles on these figures, and while we have sometimes differed on points of detail, our approach has been similar in seeking the simplest possible explanation of the techniques of the masons' craft to which the drawings refer.[38] Because of these studies, it will not be necessary in this paper to go into details of the stereotomical techniques of mediaeval masons as illustrated in Villard's *Sketchbook.* But we may note the general conclusion which has emerged from these studies, namely, that stereotomical problems were solved by mediaeval masons primarily through the physical manipulation of geometrical forms by means of the instruments and tools available to the masons. These were rule-of-thumb procedures, to be followed step by step, and there were virtually no mathematical calculations involved.

We may thus characterize the practical geometry of Villard's *Sketchbook* more precisely as constructive geometry, by means of which technical problems of design and building were solved through the construction and physical manipulation of simple geometrical forms — triangles, squares, polygons, and circles.

This reconstruction of the geometry of Villard and Magister 2 is based on the stereotomical problems of the pages devoted to *maconerie;* but once having defined it, one easily recognizes its application to other problems in the *Sketchbook.* For instance, the surveying techniques noted above now reveal themselves entirely as physical procedures. To find the height of a tower, a mason sets up an

---

[38] Robert Branner, "Three Problems from the Villard de Honnecourt Manuscript," *Art Bulletin,* xxxix (1957), 61–66; *idem,* "Villard de Honnecourt, Archimedes, and Chartres," *Journal of the Society of Architectural Historians,* xix (1960), 91–96 (with additional discussion in the "Letters to the Editor," *ibid.,* xx (1961), 143–146; *idem,* "Villard de Honnecourt, Reims and the Origin of Gothic Architectural Drawing," *Gazette des Beaux-Arts,* Ser. 6, lxi (1963), 129–146; Lon R. Shelby, "Medieval Masons' Tools. II. Compass and Square," *Technology and Culture,* vi (1965), 236–248; *idem,* "Setting Out the Keystones of Pointed Arches: A Note on Medieval *Baugeometrie,*" *ibid.,* x (1969), 537–548; and *idem,* "Mediaeval Masons' Templates," *Journal of the Society of Architectural Historians,* xxx (1971), 140–154.

instrument in the shape of a right-angle isosceles triangle at a distance from the tower which allows the line of sight along the hypotenuse to strike the top of the tower. The distance from the instrument to the tower will then be the height of the tower. Very simple, very physical, and very non-mathematical. Although the procedure was based on a theorem of similar triangles, the mason did not have to know that; all he had to do was to follow the procedure correctly to obtain the height of the tower. One cannot tell from the drawing whether Villard himself understood the geometrical principle which underlay the procedure. In this respect, the difference between the "content" of the masons' geometry and even that of the *Pratike de geometrie* becomes evident, for though the latter also used simplified procedures, these did involve some mathematical calculations, as well as simple mathematical demonstrations of the correctness of the procedures described.

The very same kind of constructive geometry is to be found in those pages of the *Sketchbook* devoted to *portraiture*, where Villard taught the method of drawing according to the *ars de iometrie*. Again it is a matter of constructing simple geometrical forms that provided the framework into which, or around which, the drawing was devised. Professor Panofsky clearly perceived the essence of this constructive geometry in his classic study on "The History of the Theory of Human Proportions as a Reflection of the History of Styles":

What the French architect Villard de Honnecourt wants to transmit to his *confrères* as the "art de pourtraicture" is a "méthode expéditive du dessin" which has but little to do with the measurement of proportions, and from the outset ignores the natural structure of the organism. Here the figure is no longer "measured" at all, not even according to head- or face-lengths; the schema almost completely renounced, so to speak, the object. The system of lines—often conceived from a purely ornamental point of view and at times quite comparable to the shapes of Gothic tracery — is superimposed upon the human form like an independent wire framework. The straight lines are "guiding lines" rather than measuring lines. . . .[39]

The *ars de iometrie* of Villard and Magister 2 was thus one and the same throughout the *Sketchbook*, whether it was applied to problems of drawing the faces and bodies of men and animals (Hahnloser, pls. 35–38), or to calculating the height of a tower (40 1), or to devising the shape of the keystones for third- and fifth-point arches (40 c, d),[40] or to delineating the plan of one level of a tower of Laon Cathedral (18 a).[41] In brief, the "art of geometry," as far as these masons were concerned, was "the technique of the forms" (*li force des trais*), just as Vil-

[39] Erwin Panofsky, *Meaning in the Visual Arts* (Garden City, N. Y., 1955), p. 83.

[40] See my article, "Setting Out the Keystones," especially pp. 541–545.

[41] See Walter Ueberwasser, "Nach Rechtem Masz: Aussagen über den Begriff des Maszes in der Kunst des XIII.-XVI. Jahrhunderts," *Jahrbuch der preussischen Kunstsammlungen*, LVI (1935), 259–261 and Abb. 7; and Maria Velte, *Die Anwendung der Quadratur und Triangulatur bei der Grund- und Aufrissgestaltung der gotischen Kirchen* (Basel, 1951), pp. 53–55 and Taf. VIII; both of whom apply the technique of quadrature to Villard's plan of the tower, though Ueberwasser begins with the outermost measurements, including the buttresses, while Velte's scheme excludes the buttresses. Neither scholar may be precisely correct in reconstructing the procedures employed by Villard, but their studies strongly suggest that Villard did use some technique of manipulating squares to produce the main features of the tower and the lines of his drawing.

Geometrical Knowledge of Mediaeval Master Masons 411

lard himself stated in his "Preface." That technique — the manipulation of geometrical forms — was what I have called the constructive geometry of mediaeval masons.

Because so many of the details of the *Sketchbook* apparently were intended to be explained orally, we have had to reconstitute this constructive geometry of Villard and Magister 2 by inference and deduction from the mere hints which their drawings and comments provide, and from comparison of the contents of the *Sketchbook* with other kinds of practical geometries compiled through the thirteenth century. These conditions make our case for the geometrical knowledge of master masons somewhat circumstantial for the period of High Gothic architecture. Fortunately, we enter upon much solider ground when we come to Late Gothic, thanks to several little books written by German master masons in the late fifteenth and early sixteenth centuries.

The earliest of these booklets to be printed were the two by Matthias Roriczer, *Büchlein von der Fialen Gerechtigkeit* and *Geometria deutsch*, the first published in 1486, the second a year or two later.[42] For our present purposes the *Geometria deutsch* is particularly interesting, since here is a treatise specifically devoted to "geometry" by a known mediaeval master mason. The booklet, or better still, pamphlet — for it contains only twelve pages — consists of lettered figures and brief explanations of the solution of nine problems: how to construct a right angle, a pentagon, a heptagon, and an octagon; how to find the length of the circumference of a circle; how to find the center of a circle with only part of the circumference known; how to construct a square and a triangle which have the same areas; how to set out the moldings and the finials for a gable; how to set out the plan of a gable.

It is not clear at first reading just what was Roriczer's intention in compiling this "German Geometry," since it is a somewhat incongruous agglomeration of simple geometrical problems along with techniques of architectural design and construction. One is reminded of the heterogeneous character of Villard's *Sketchbook* and tempted to comment that, whatever may have been the relationships between Gothic architecture and scholasticism, when mediaeval master masons did write books, they certainly did not reveal that penchant for systematic literary organization characteristic of scholastic treatises. This is not really surprising when one places these little books within the context of an oral rather than a written tradition. The Schoolmen taught from books, and in turn their own scholastic treatises were shaped by the techniques which they developed for teaching from books. Mediaeval masons did not teach from books,

---

[42] Although neither of these booklets has received a critical edition, both have been reprinted several times: the *Büchlein von der Fialen Gerechtigkeit* by Karl Heideloff (Nürnberg, 1844), A. Reichensperger (Trier, 1845), Karl Schottenloher (Regensburg, 1923), and Ferdinand Geldner (Wiesbaden, 1965); the *Geometria deutsch* by Heideloff (Nürnberg, 1844), Sigmund Günther, "Zur Geschichte der deutschen Mathematik im fünfzehnten Jahrhundert," *Zeitschrift für Mathematik und Physik, Historisch-literarische Abtheilung*, xx (1875), 1–14; and Geldner (Wiesbaden, 1965). It was Geldner who established (pp. 70–71) that Roriczer wrote and published both works, contrary to the prevailing views that the author of the *Geometria deutsch* was either unknown or that he was a certain Hans Hösch.

412   *Geometrical Knowledge of Mediaeval Master Masons*

but rather from memory and from experience in the techniques of the craft. When they did decide to describe some of these techniques in written words and illustrations, they had no established literary forms to follow. Consequently, it appears that they too wrote as they taught — by piling up one description after another of the particular rules and procedures of the craft, with little of that scholastic concern for placing these particulars into some sort of systematic framework.[43]

But if most masons acquired their technical knowledge and skills through oral transmission of the craft traditions, at least some master masons must have had access to knowledge not carried by those traditions. Whether this knowledge came directly from a literary source, or from someone else who had access to that source, the knowledge itself could easily and quickly lose its literary moorings and become a part of the oral traditions of the craft, if it proved useful in design and construction.

Roriczer's two little books provide a fascinating insight into the masons' thought processes in this respect. In his *Booklet on the Correct Design of Pinnacles*, which is entirely concerned with architectural design problems, he cites as the source for his design technique the Junkers of Prague, that is, the famous fourteenth-century Parler family of master masons from Prague.[44] Since there is no reference to a previous book on the technique which Roriczer himself is describing, one presumes that he had gotten his knowledge of the technique by the usual oral transmission of craft traditions. On the other hand, for his *Geometria deutsch* Roriczer may have used a fifteenth-century treatise on geometry, *De inquisicione capacitatis figurarum*, which he does not cite, although there are some striking parallels between passages in it and some of his own geometrical exercises. It is uncertain who was the author of *De inquisicione*, but the manuscript in which the treatise survives was compiled by a certain Magister Reinhard de Vurm before the middle of the fifteenth century. In 1457 it was in the possession of Johannes Fleckel, who carried it with him to Vienna when he entered the Dominican Order. At Fleckel's death the Dominican house sold the manuscript to Burchard Keck, a citizen of Salzburg, and upon Keck's death it passed to the Bibliotheca Regiae in Munich.[45] Since Roriczer spent his career as a master mason in southern Germany, with positions at Eichstätt, Nürnberg, Munich, and Regensburg, he

[43] While I continue to be intrigued by the central thesis of Professor Panofsky's *Gothic Architecture and Scholasticism* (Cleveland, 1957), it does seem that he was carried away by his argument in his discussion (p. 87) of the "scholastic disputation" between Villard de Honnecourt and Pierre de Corbie regarding the plan of an ideal chevet. Likewise, it seems to me, he has overdrawn (pp. 23–26) his portrait of the education and learning, not to speak of social and professional status, of the mediaeval architect. On the other hand, it might be profitable to rethink his thesis in terms of the argument of this present paper.

[44] The handiest guide to the enormous literature on the Parlers is Otto Kletzl's article on the family in Thieme-Becker, *Allgemeines Lexikon der Bildenden Künstler* (Leipzig, 1932), xxvi, 242–248.

[45] These facts about the history of the manuscript are known from inscriptions within the manuscript itself. See Maximilian Curtze's introduction to his edition of the text, " 'De Inquisicione Capacitatis Figurarum.' Anonyme Abhandlung aus dem fünfzehnten Jahrhundert," *Abhandlungen zur Geschichte der Mathematik*, Heft VIII (1898), 31–32.

Geometrical Knowledge of Mediaeval Master Masons 413

might have had access to the *De inquisicione*, or at least to someone who knew the contents of the treatise. That someone may well have been the Bishop of Eich-stätt, Wilhelm von Reichenau, to whom Roriczer dedicated his booklet on pin-nacles. In the dedication Roriczer mentioned that the bishop was a lover and patron of the "free art of geometry," and that they had discussed the subject together many times.[46] Incidentally, it is just this kind of rapport between the ecclesiastical patron and his master mason that must have provided the input of a great deal of "clerical" learning into the craft traditions of the Middle Ages.

The procedures which Roriczer used to construct a heptagon and an octagon are precisely the same as those in the *De inquisicione*.[47] In both cases the proce-dure was strictly by constructive geometry, that is, the manipulation of compass and straightedge to inscribe a heptagon within a circle and an octagon within a square. In neither case is the construction mathematically demonstrated to be correct.

On the other hand, Roriczer's method of determining the length of the circum-ference of a circle differed in important ways from that given in the *De inquisi-cione*. The solution to the problem had long been available in the Archimedean treatise, *On the Measurement of the Circle*, which in the twelfth century Gerard of Cremona had translated into Latin from an Arabic version. Throughout the rest of the Middle Ages European scholars copied, commented on, expanded, con-tracted, and otherwise transformed the *De mensura circuli*, with its three theorems concerning the area and circumference of the circle.[48] It is the third theorem which concerns us here: "Every circumference of a circle exceeds three times its diame-ter by an amount less than one seventh and more than 10 parts of 71 parts of the diameter."[49] Here is Roriczer's version of Archimedes' theorem:

Whoever wishes to make a circular line straight, so that the straight line and the circular

[46] Matthäus Roriczer, *Das Büchlein von der Fialen Gerechtigkeit. Faksimile der Originalausgabe Regensburg 1486* und Matthäus Roriczer, *Die Geometria deutsch: Faksimile der Originalausgabe Regens-burg um 1487/88*, ed. Ferdinand Geldner (Wiesbaden, 1965), pp. [14], 45–46. Bishop Wilhelm was also a noted patron of architecture and was engaged in building projects in Eichstätt, Regensburg, Ulm, and Ingolstadt. See Geldner's comments on him in *ibid.*, p. 63.

[47] Roriczer, *Geometria deutsch*, pp. [34–35], 57–58; Curtze, "De Inquisicione," pp. 55–56. Roriczer's relatively complex procedure for constructing a pentagon (pp. [32–33], 56–57) is not contained in the *De inquisicione*, nor have I been able to find a literary source or parallel to it in the *practicae geometriae* already cited or in those compiled after 1300. For the latter I have used the following editions: *Geo-metria Culmensis. Ein agronomischer Tractat aus der Zeit des Hochmeister Conrad von Jungingen (1393–1407)*, ed. Hans Mendthal (Leipzig, 1886); "Die 'Practica Geometriae' des Leonardo Mainardi aus Cremona," in Part II of *Urkunden zur Geschichte der Mathematik im Mittelalter und der Renais-sance*, ed. Maximilian Curtze, *Abhandlungen zur Geschichte der mathematischen Wissenschaften*, XIII (1902), 342–434; H. L. L. Busard, ed., "The Practica Geometriae of Dominicus de Clavasio," *Archive for History of Exact Sciences*, II (1962–66), 524–575; and Maximilian Curtze, ed., "Der Tractatus Quadrantis des Robertus Anglicus in Deutscher Übersetzung aus dem Jahre 1477," *Abhandlungen zur Geschichte der Mathematik*, IX (1899), 45–63.

[48] In his monumental work, *Archimedes in the Middle Ages*, Vol. I, *The Arabo-Latin Tradition* (Madi-son, 1964), Marshall Clagett has edited, translated, and commented on the numerous manuscripts of this mediaeval tradition.

[49] Clagett's translation, *ibid.*, p. 49, of Gerard's Latin translation as edited on p. 48.

414   *Geometrical Knowledge of Mediaeval Master Masons*

are the same length: Make three circles next to one another and divide [the diameter of] the first circle in seven equal parts, with the letters as shown h : a : b : c : d : e : f : g. As far as it is from h to a, set a point behind it and there put an i. Thereby, as far as it is from i to k, equally as long is the circular line of one of the three circles which stand next to each other as shown in the attached figure.[50] (Fig. 3)

One could hardly ask for a more revealing example of the constructive geometry of mediaeval masons. Finding the circumference of a circle (for columns, pillars, round towers, etc.) must have been a relatively common requirement of mediaeval building. By following the steps prescribed by Roriczer, any mason could, with only his compass and straightedge, "construct" a solution to the problem, without knowing either the Archimedean theorem or the proofs pertaining to this theorem. The *Geometria deutsch* thus clearly reveals how mediaeval masons approached geometrical problems which would seem to require some mathematical calculations, yet they managed to avoid those calculations through step-by-step manipulations of their working tools. This point can be emphasized by comparison of Roriczer's solution with that given in the *De inquisicione:*

Given the diameter, to find the circumference of a circle: Let it be that the circle is *a b* and the diameter *a b* of the circle is given as 14. Triple the diameter and it becomes 42. If you add to the product 1/7 of the said diameter, that is to say 2, there will be produced [the number] 44, which is the circumference of the circle. This is made clear by [theorem] 7 of the geometry of the three brothers.[51]

Immediately one recognizes that the author of *De inquisicione* has rendered the Archimedean theorem in the form of an arithmetical calculation, whereas Roriczer has presented it as a geometrical construction. Indeed, the language of Roriczer's formula suggests that he was hardly thinking of this as a mathematical problem. He seems to have been visualizing certain geometrical forms — a circle and a straight line — and asking himself, "How do you make a straight line that is as long as a circle is round?" This mental "set," it seems to me, differs importantly from the geometer's question, "Given the diameter, how do you find the circumference of a circle?" Secondly, one notices the merely prescriptive character of Roriczer's formula: if you want to solve *that* problem, then do it *this* way. He feels no compulsion to demonstrate the correctness of his prescription. On the other hand, the author of *De inquisicione* refers the reader to the proof of the theorem in the *Geometria trium fratrum.*[52] While this is rudimentary mathematical

---

[50] Roriczer, *Geometria deutsch*, p. [86]: "Wer ain gerunden riss scheitgerecht machen wil das der scheitgerecht ris vnd das gerund ain leng sey So mach drew gerunde neben ain ander vnd tail das erst rund in siben gleiche tail mit den puchstaben verzaichnet .h: .a: :b .c. :d. :e· :f· .g: Darnach als weit vom .h· in das :a. ist da sez hindersich ain punckt da sez ain :i: Darnach als weit von dem :i pis zv dem :k: ist Gleich so lanck ist der runden riss ainer in seiner rundens der drey neben ain ander sten des ain figur hernach gemacht stet."

[51] Curtze, "De Inquisicione," p. 37: "*Datae dyametri circumferenciam circuli invenire.* Esto, ut sit circulus *a b*, et dyametrus eius data, verbi gracia 14, sit *a b*. Quam dyametrum tripla, et proveniunt 42. Producto si 1/7 dyametri praedictae, scilicet 2, addideris, 44, quae sunt circuli circumferencia, producuntur. Patet per 7^{am} geometriae trium fratrum."

[52] This latter work, known variously in the Middle Ages as the *Librum trium fratrum* or the *Verba filiorum Moysi filii Sekir*, was composed by three ninth-century Arabic mathematicians who were brothers. Translated by Gerard of Cremona, it provided further treatment of Archimedes' theorems

Geometrical Knowledge of Mediaeval Master Masons   415

reasoning, it at least shows the author's recognition that geometrical theorems do have to be demonstrated.[53]

Another example of Roriczer's non-mathematical approach may be seen in the following formula for what is an acceptable sort of Euclidean problem:

> Whoever wishes to make a square and a triangle so that the square and the triangle each contain as much as the other: then make an [equilateral] triangle, that is, a : b : c. Divide c b in three equal parts, that is, [set points at] d and e. Then make a square out of [the line] c e [which] will become f g. Thereby the square contains equally as much as the triangle, as the example herewith shows.[54] (Fig. 4)

But the illustration merely shows the configuration of the two geometrical forms. Roriczer has certainly not proved that the two forms contain the same areas; he has only asserted that they do. In fact, the areas are only approximately the same — the area of the square is 4.000, while that of the triangle is 3.6742.

Again it is instructive to compare Roriczer's formula with the related one in *De inquisicione*:

> Let there be triangle a b c. [Fig. 5] I make line d e equal to and equal distance from b e and I join b d and e c. According to Euclid I, 41, the parallelogram b c e d is double the triangle a b c; therefore, half of parallelogram b c g f is equal to triangle a b c. We must now seek the "quadratic side" of this parallelogram b c g f.[55] To side b c I add, in a continuous and straight line, c h equal to g e; having made the diameter b h, from its center I draw the semicircle b k h, and I produce c k. I say, therefore, that line c k is the "quadratic side" of parallelogram b c g f, and by consequence, of triangle a b c.[56]

It will be noted that the author of *De inquisicione* has also "constructed" a solution to the problem, but in so doing, he has relied on two Euclidean propositions. The first half of his solution depends on Proposition 41 of Book I of the *Elements*, which he dutifully cites. But the second half, in which he finds the square equal in area to the parallelogram, is based on Book II, Proposition 14, which for some

---

concerning the circle, sphere, and cone, as well as other geometrical and arithmetical problems. Gerard's translation has been edited and translated into English in Clagett, *Archimedes*, I, 238–355.

[53] For most theorems in the *De inquisicione* the author made a stronger effort to do so on his own. The demonstration of the Archimedean theorem for the circumference of the circle is rather sophisticated, and it is not really surprising that the author shied away from it. For the proofs of Archimedes and of the Three Brothers, see Clagett, *Archimedes*, I, 48–55, 264–279. On the other hand, a brief *Tractatus de quadratura circuli* — traditionally attributed to Campanus de Novara, but authorship and date uncertain — contains a solution of the problem by means of "constructive geometry" that is similar to the one used by Roriczer. For the text, see *ibid.*, pp. 590–93.

[54] Roriczer, *Geometria deutsch*, p. [38]: "Wer machen wil ain firung vnd ain driangel das die firung vnd der driangel yediichs als vil in im helt als das ander So mach ain driangel das ist :a: :b: :c: tail vom .c. pis zv dem :b: in drew gleiche tail das ist .d. .e. Darnach mach ain firung aus dem :c: :e: wirt .f. .g. So helt dy firung gleich als vil in als der driangel des ain exempel hernach gemacht stet."

[55] That is, the side of a square equal in area to the parallelogram b c g f.

[56] Curtze, "De Inquisicione," pp. 56–57: "Sit trigonus a b c. Ducam lineam d e aequalem et aeque distantem b e et coniungam b d et e c; ergo per quadragesimam primam EUCLIDIS parallelogrammum b c e d est duplum ad trigonum a b c, ergo medietas parallelogrammi b c g f est aequalis trigono a b c. Huius ergo parallelogrammi b c g f quaeratur latus tetragonicum sic. Lateri b c adiungam in continuum et directum lineam c h aequalem g e, et facta dyametro b h et centro in medio eius circinabo semicirculum b k h et producam c k: dico igitur, quod linea c k est latus tetragonicum parallelogrammi b c g f, et per consequens trianguli a b c."

416   *Geometrical Knowledge of Mediaeval Master Masons*

reason he does not cite; consequently, his demonstration is left hanging a bit. Nevertheless, there are important differences between his and Roriczer's constructions. In the first place, the procedure in *De inquisicione* produces a geometrically correct solution, and not just an approximate one like Roriczer's. Secondly, this procedure applies to *any* triangle, and not just the equilateral triangle of Roriczer's formula. Finally, the author of *De inquisicione* assumes that the reader will understand his solution by reference to one of the Euclidean propositions on which the solution is based. Roriczer makes no such assumption about the mathematical interests or abilities of the reader; he simply says to do it this way. Whether Roriczer himself would have understood the Euclidean propositions is a moot question; certainly he did not require that understanding of his readers.

A final point of comparison between the *Geometria deutsch* and the *De inquisicione*: if Roriczer did have access to the latter, or to someone who knew its contents, he borrowed from the treatise only those formulae which could be entirely expressed in terms of constructive geometry, and he avoided all those others which required some kind of Euclidean mathematical reasoning. This point brings out an interesting contrast between the geometry of mediaeval masons and classical Greek geometry as developed by Euclid in the *Elements*. Both began with the same presupposition, namely, that all geometrical constructions must be possible with the use of a few simple tools or instruments. But Euclid accepted that presupposition as a theoretical restriction in order to prescribe the limits within which he would rigorously develop his arguments. Mediaeval masons, on the contrary, made that presupposition as a practical necessity because they lacked the ability in mathematical reasoning to push beyond the mere manipulation of their tools into the Euclidean world of definitions, postulates, axioms, propositions, and proofs. For Euclid the construction of a geometrical figure with compass and straightedge was merely a part — and not an absolutely necessary part — of his mathematical exercise; there remained the more difficult and important task of demonstrating the mathematical correctness of the construction.[57] For Roriczer and his fellow masons, such a construction was, geometrically or mathematically speaking, the end of the exercise; the next task was not to prove its mathematical correctness, but to transform the geometrical construct into an architectural form in stone.

Some of the ways in which mediaeval masons transformed geometrical figures into architectural forms were described and illustrated in three other small treatises by late mediaeval German master masons. The best known of these is Roriczer's *Büchlein von der Fialen Gerechtigkeit*. In his dedication of the booklet to Bishop Wilhelm von Reichenau, Roriczer explained that since every art has its

---

[57] On the formal structure of Euclidean propositions and the place of geometrical constructions within these, see Thomas Heath, *A History of Greek Mathematics* (Oxford, 1921), I, 370–371. On the Euclidean techniques of construction with the straightedge and the "collapsing compass," see Julian L. Coolidge, *A History of Geometrical Methods* (Oxford, 1940), pp. 43–45; and Howard W. Eves, *A Survey of Geometry* (Boston, 1963), I, 183–184.

Geometrical Knowledge of Mediaeval Master Masons 417

own materials, forms, and measures (shades of Dominicus Gundissalinus!), he wished to set forth the basic principles of these for the art of masonry, as founded on the art of geometry. "I have [tried], with the help of God, to make clear this previously mentioned art of geometry and to explain, in the first place, the beginning of drawn-out stonework — how and in what measure it arises out of the fundamental basis of geometry through the manipulation of compasses and [how it] should be brought into the correct proportions."[58]

The booklet is in fact entirely devoted to the technique of setting out the groundplan and elevation of a pinnacle, but the procedure of moving step by step is the same as that in the *Geometria deutsch*. Roriczer begins, "If you wish to draw the groundplan for a pinnacle, according to the stonemason's art and with the correct geometry, then begin by drawing a square, as it is shown here with the letters a, b, c, d."[59] Inside this square Roriczer inscribed a second square at a forty-five degree angle to the first, and inside the second he inscribed a third square in the same manner. (Fig. 6) He then rotated the second to make the sides of all three squares parallel. Having obtained the basic outline of the groundplan of the pinnacle by this manipulation of three squares, Roriczer proceeded to determine the details of both the plan and the elevation in a step-by-step elaboration of this manipulative technique. The entire process required 234 separate steps, which he illustrated with eighteen figures, the last three showing the completed geometrical design for the groundplan and elevations. (Fig. 7) Roriczer finished his instructions with these comments: "Thereupon, place the cap of the pinnacle on the body of the pinnacle and erase all drawing lines, so that there remain only correct lines which are necessary for the pinnacle. Accordingly, the figure is called a correct pinnacle, drawn from the groundplan."[60] In short, this is the famous technique of quadrature, whereby the elevation is derived from the square, the basic geometrical figure of the groundplan. But it cannot be too strongly emphasized that the entire operation consisted simply of the manipulation of geometrical figures through a long series of carefully prescribed steps, or that it was quite devoid of mathematical formulae and calculations. What Roriczer did in his little book was to provide in writing the detailed exposition of a particular design problem and its solution.[61] The literary form of this exposition

[58] Roriczer, *Büchlein*, p. [14]: "Hab ich mit der hilff gotes ettwas berurter kunst der geometrey zuerleutern Und am ersten dasmale den anefang des aussgezogens stainwerchs wie vnd in welcher mass das auss dem grunde der geometrey mit austailung des zirckels herfurkomen vnd in die rechten masse gebracht werden solle Zuerclern. . . ." Cf. John Papworth's translation of Roriczer's dedication in Elizabeth Holt, ed., *A Documentary History of Art*, Vol. i, *The Middle Ages and the Renaissance* (Garden City, N. Y., 1957), pp. 96–97.

[59] Roriczer, *Büchlein*, p. [15]: "Wilt dv ain grvndt reyssen zw ainer vialen: nach stainmezischer art: avss der rechten geometrey So heb an vnd mach ain virvng als hernach bezaichnet ist mit den puchstaben :a:b:c:d·"

[60] *Ibid.*, p. [29]: "Darnach so sez den risen der fialen auf den leib der fialen vnd du al tail riss noher so pleibt nur dy rechten riss dy noturftig sein in der fialen Darnach so haist dj figur ain rechte fialen aus gezogen auss dem grunt."

[61] This is not the place to pursue the question of why Roriczer and other contemporary master masons felt inclined to publish these descriptions of the technique of quadrature. I hope later to provide a more intensive study of the problem of the "secret" of the mediaeval masons.

418   *Geometrical Knowledge of Mediaeval Master Masons*

suggests that it is nothing more than a written record of the kind of oral teaching that was traditional in the craft. It is perhaps not unlike the exposition which a master mason might have given his apprentice in explaining some of the sketches and curt comments of Villard and Magister 2.

The same design problem and pretty much the same technique for resolving it were set forth in another *Fialenbüchlein* by Roriczer's contemporary, Hans Schmuttermayer.[62] In a prefatory statement Schmuttermayer provides a definition of geometry and its use in building construction that is very close to the point of view expressed in the "Articles and Points of Masonry" and to the content of constructive geometry as expounded in this present paper. Schmuttermayer explains that he is writing the book

... for the instruction of our fellowmen and all masters and journeymen who use this high and free art of geometry, in order that their feelings, speculations, and imaginings can, with thought, be better subjected to the correct rules of measured stonework and take root. Fundamentally, this art is freely and truly planted and founded on the center point of the circle, together with its circumference of correctly set point and construction. [I explain these matters,] not for my own reputation, but more to praise the fame and reputation of our forerunners, the inventors of this high art of building construction, which from the beginning has had its true base in the level, set-square, triangle, compass, and straightedge, but which is now pursued with greater subtlety, higher understanding, and deeper reckoning. Thus have I, Hans Schmuttermayer of Nürnberg, correctly shown the art of such measured stonework in the square and round [parts] of the pinnacle, the gable, and the pillar, with all of the things belonging to these, according to the new as well as the old art. ... I have not discovered these things by myself, but have learned them from many other great and famous masters, such as the Junkers of Prague, and Masters Ruger and Nicholas of Strasbourg, who for the most part brought this new art to light, along with many others.[63]

Schmuttermayer's technique for designing a pinnacle was basically the same as that used by Roriczer, but he clarified the procedure somewhat by separating the inscribed squares, placing them in a row, and giving them individual letter-symbols for easy reference. (Fig. 8) Then in the customary manner of carefully prescribed procedures, he transferred the length of the sides of these various squares to establish the dimensions and outlines of the essential features of the pinnacle.

---

[62] The little pamphlet was edited by A. Essenwein, "Hans Schmuttermayer's Fialenbüchlein,' *Anzeiger für Kunde der Deutschen Vorzeit*, xxviii (1881), 73–78.

[63] *Ibid.*, p. 73: ". . . vntterweysung vnnserm nachsten vñ allē maisteren vñ gesellen die sich diser hohen vñ freyen kunst der Geometria geprauchen ir gemute speculirung vnd ymaginacion dem warē grunt des maswercks pass zuuntterwerffen nach gedencken vnd ein zu wurtzeln. Auch fundamentlicher die art so auss dem Centrum des zirckels mitsamt seines vmbschweiffs warer saczung punct vñ austeylung dest freyer vñ warhafftiger eingepflanczt vnd gegrundt werden. Vnd nit vmb meiner eygen Ere willen. Sunder mer zupreyse rum vnd lob der altten vnnser vorgeer seczer vñ vinder diser hohē kunst des pauwercks die auss der wage. winckelmoss. triangel. zirckel. vñ linial. vrsprunglichē iren warē grunt habē. vñ nu mit der scherff. subtilitet. hoher synne. vñ tieffer rechnũg. yecz ersucht ist. Hyrumb hab ich Hanns schmuttermayer von Nurmberg die art solichs maswercks. virung. rotund. der violn. winperg. vñ der pfeyler mit aller irer zugehorungē auff die new mitsamt der alttē art gerecht gemacht. ... Vnd hab solichs auss mir selber nit erfunden. sunder von vil andern grossen berumbtē maisteren. Als die Junckhern von prage. Maister ruger. Niclas von straspurgk. Der dan am mainsten die new art an das licht gepracht mitsamt vil andern genomen."

Geometrical Knowledge of Mediaeval Master Masons   419

The result was a "geometrical construct" of the pinnacle which he then trans-formed into an architectural drawing showing the details of moldings, crockets, and finial. (Fig. 9) In these same figures Schmuttermayer illustrated the applica-tion of this technique to the design of a gable or canopy. His illustrations, which are not nearly so well known as those of Roriczer, reveal more explicitly the use of "geometry" by mediaeval master masons in manipulating geometrical forms to produce the framework in which they could then develop their own individual architectural forms. Comparison of Schmuttermayer's gable with the one produced by Roriczer in the *Geometria deutsch* (Fig. 10) shows that while their technique of designing by constructive geometry was basically the same, the end results — in both overall composition and formal details — were distinctively different.

The development of variations on a theme constitutes a fundamental charac-teristic of the constructive geometry of mediaeval masons, and it is virtually stated as a matter of principle in another little book by a German master mason named Lorenz Lechler. Written in 1516 for the instruction of his son Moritz into the techniques and skills of the masons' craft, this small work bears no title and is generally referred to simply as Lechler's *Unterweisung*.[64] Because of the variety of design and construction problems with which Lechler concerned himself, his treatise is more interesting than the two booklets on pinnacles by Roriczer and Schmuttermayer. Lechler gave particular attention to the design of the templates used in cutting stones for window mullions, tracery, vault ribs, bases and capitals of pillars, and moldings for gables, pinnacles, and buttresses. Since I have pub-lished a detailed study of Lechler's prescriptions for designing templates by the technique of quadrature, it will be necessary here to review his system only very briefly.[65] It was based on a modular unit determined by the choir wall of the church: "Take the wall thickness of the choir, whether it be small or large, then draw two squares through one another; therein will you find all templates, just as you will find them drawn in this book."[66] As can be seen from Fig. 11, these two squares of the same size were inscribed over each other at a forty-five degree angle. One of these was divided into nine squares, and the centermost square then provided the frame for inscribing three small squares in the manner already made familiar by Roriczer and Schmuttermayer. This grid of inscribed squares deter-mined the length and breadth of large and small mullions. (Fig. 11 bottom) An-other combination of circles and squares within this same framework provided the dimensions of the cross-sections for large and small vault ribs. Finally, by manipulating compasses and straightedge on the corners of the original overlap-ping squares, Lechler obtained the geometrical forms of large and small vaulting shafts, along with their bases and capitals. Though by now it sounds like a refrain,

[64] The unique text of Lechler's book is contained within a late sixteenth-century MS copy (Cologne, Historisches Archiv, Handschrift Wf. 276*, fols. 41–56ᵛ); it was published in an unsatisfactory edition by August Reichensperger, *Vermischte Schriften über christliche Kunst* (Leipzig, 1856), pp. 133–155.

[65] Shelby, "Mediaeval Masons' Templates," pp. 147–153.

[66] Handschrift Wf. 276*, fol. 44 (Reichensperger's edition, p. 135): "Item so nimb die mauer dickhe, von dem khor, er sey khlein oder gros so reiss Zwo fürung durch einander, darinen fintestu alle bredter, wie du dan alhie in desem Buech gerisen findest. . . ."

420   *Geometrical Knowledge of Mediaeval Master Masons*

one must repeat that throughout Lechler's prescriptions there were no mathematical calculations involved, and the entire procedure was accomplished by manipulating geometrical figures. Even the numerical ratios of seven to five which keep cropping up in Lechler's prescriptions turn out to be inevitably determined by his geometrical technique of inscribing squares.[67]

One of the more significant points which emerges from a study of Lechler's technique for designing templates is that while it was certainly prescriptive, it was not rigidly restrictive. The present study helps to see why this was the case. The constructive geometry of mediaeval masons was prescriptive in that it consisted of carefully prescribed steps which the masons were taught to follow. But since they were scarcely concerned with mathematical preciseness or correctness, those steps could be altered at will. That is to say, there were no logical or mathematical rules which they were obligated to follow; they were restricted only by their own skill and inventiveness in manipulating geometrical forms with the tools at their disposal, and by their willingness, or unwillingness, to change the prescriptions which had been handed down to them through the craft traditions.

* * *

Mediaeval masons insisted that their whole craft was based on the "art and science of geometry." It has been the purpose of this paper to reconstruct the character and content of the geometrical knowledge of mediaeval master masons from the few literary remains of the masons themselves. As reconstructed from these writings, this geometry scarcely resembles either the classical geometry of Euclid and Archimedes, or the mediaeval treatises on *practica geometriae*. Mathematically speaking, it was simple in the extreme; once it is recognized that there was virtually no Euclidean-type reasoning involved, the way is cleared for understanding the kind of geometrical thinking which the masons did employ. This non-mathematical technique I have labeled constructive geometry, to indicate the masons' concern with the construction and manipulation of geometrical forms. It becomes evident that the "art of geometry" for mediaeval masons meant the ability to perceive design and building problems in terms of a few basic

[67] Lechler's use of geometrical constructions and simple arithmetical ratios calls to mind the mathematical calculations of the sixteenth-century Spanish master mason, Rodrigo Gil de Hontañon. See George Kubler, "A Late Gothic Computation of Rib Vault Thrusts," *Gazette des Beaux-Arts*, 6th Ser., xxvi (1944), 138–148. Rodrigo's arithmetical computations are considerably more sophisticated than anything to be found in Villard or the German master masons, but his geometrical formulae for determining the thickness of the buttresses for a barrel vault appear to be within the traditions of the constructive geometry of mediaeval masons. See his text and illustrations in the only modern, and unsatisfactory, edition by Ricardo de Mariátegui, "Compendio de Arquitectura," *Arte en España*, vii (1868), 174–176. Rodrigo was also knowledgeable in at least some aspects of Renaissance theory and practice of architecture, and this, combined with the textual problems of his writings being transmitted only through a seventeenth-century architectural treatise, raises several questions which have to be sorted out before one can make a confident statement on the extent to which his mathematical computations reflect the practice and knowledge of mediaeval masons. I have had the opportunity to study briefly Prof. John Hoag's very fine, but unpublished, dissertation, "Rodrigo Gil de Hontañon: His Work and Writings" (Yale, 1958). Since this study provides a firm basis for the resolution of some of these problems, it is to be hoped that Prof. Hoag will soon put into print the results of his research.

Geometrical Knowledge of Mediaeval Master Masons   421

geometrical figures which could be manipulated through a series of carefully pre-scribed steps to produce the points, lines, and curves needed for the solution of the problems. Since these problems ranged across the entire spectrum of the work of the masons — stereotomy, statics, proportion, architectural design and drawing — the search by modern scholars for the geometrical canons of mediaeval archi-tecture is appropriate enough, so long as we keep clearly in mind the kind of geometry that was actually used by the masons. The nature of that geometry suggests that these canons, when recovered, will not be universal laws which will at last provide *the key* to mediaeval architecture; rather, they will be particular procedures used by particular master masons at particular times and places.

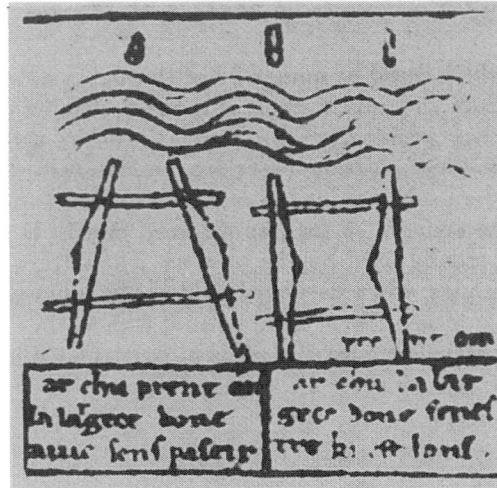

FIG. 1. Surveying technique, from Villard de Honnecourt's *Sketchbook* (ed. Hahnloser, Pl. 39 1, m).

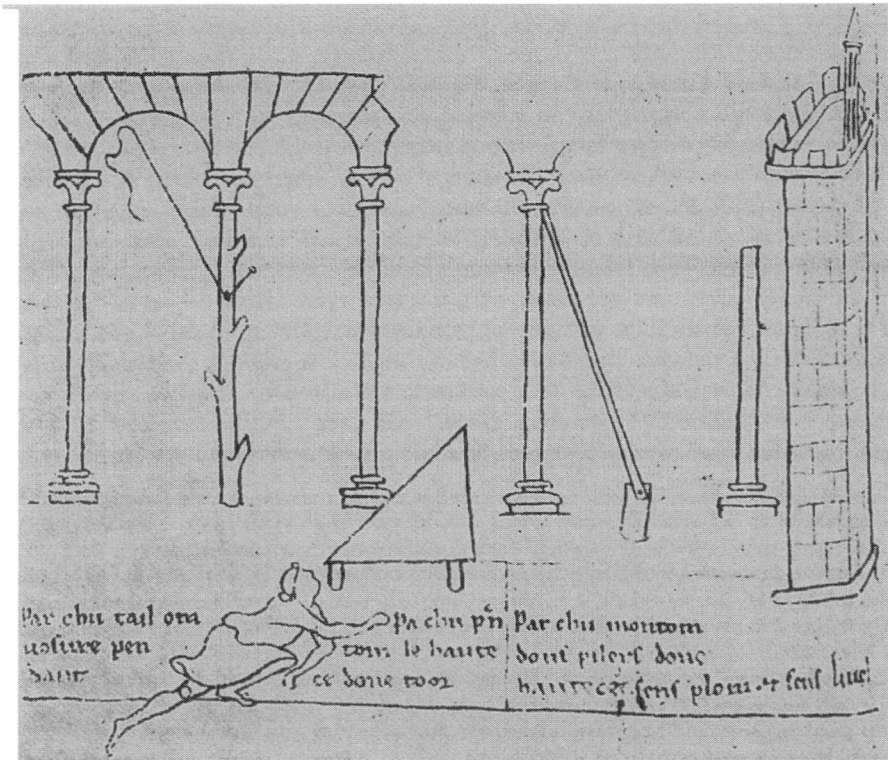

FIG. 2. Surveying technique, from Villard de Honnecourt's *Sketchbook* (ed. Hahnloser, Pl. 40 1).

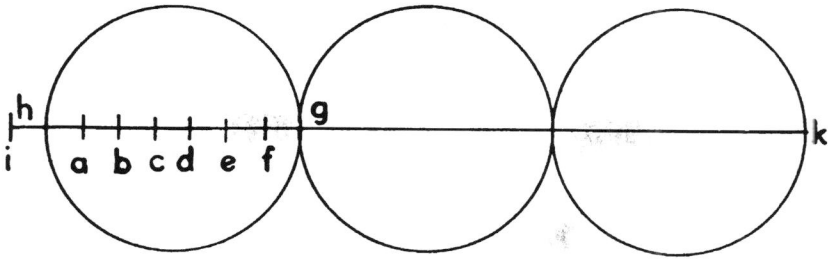

FIG. 3. Finding the circumference of a circle, redrawn from Roriczer's *Geometria deutsch*.

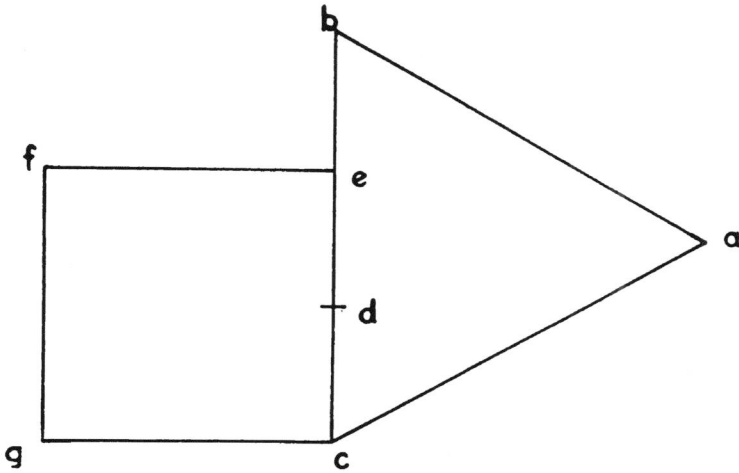

FIG. 4. Finding a square and triangle with the same area, redrawn from Roriczer's *Geometria deutsch*.

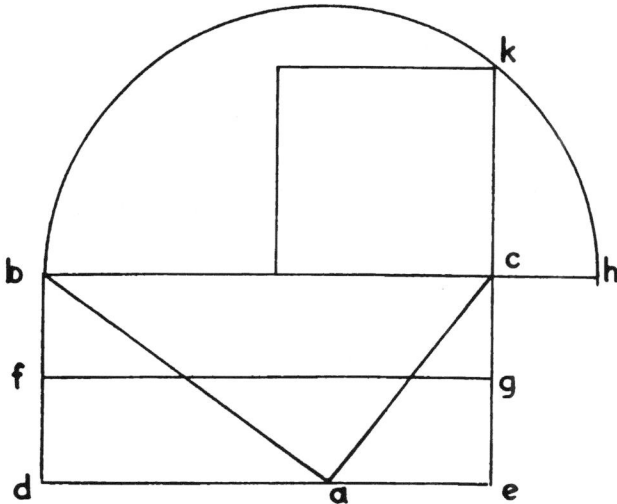

FIG. 5. Finding a square and triangle with the same area, redrawn from the anonymous *De inquisicione capacitatis figurarum*.

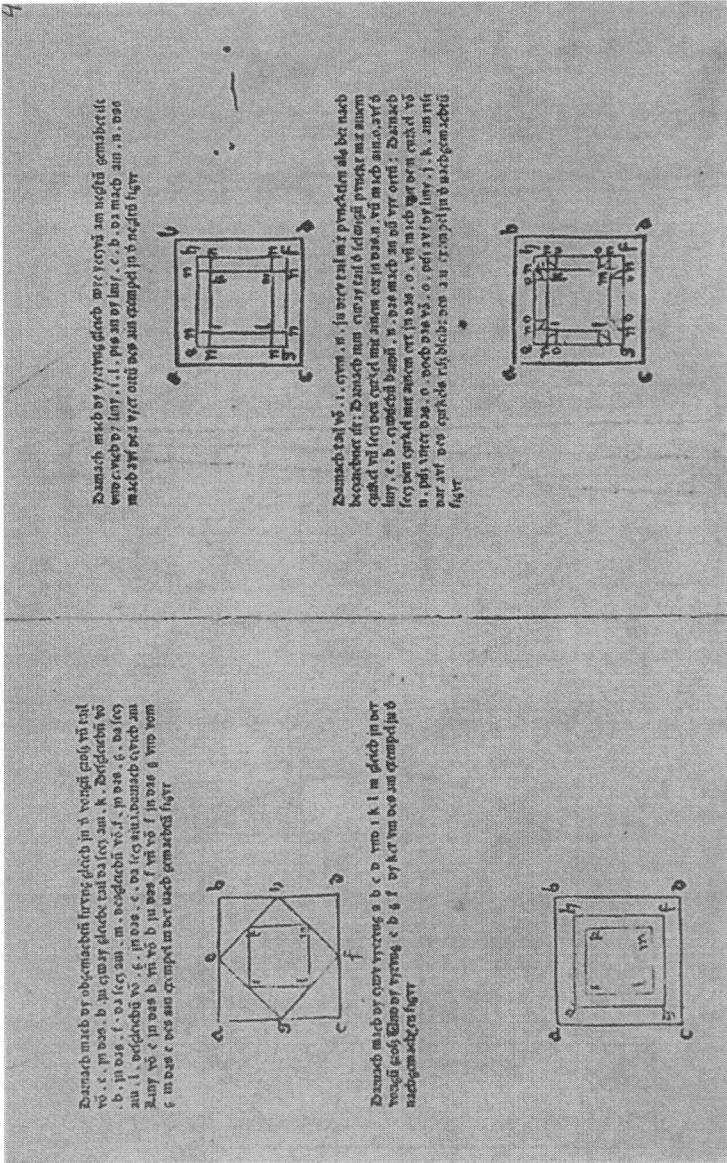

Fig. 6. Setting out the plan of a pinnacle, from Roriczer's *Büchlein von der Fialen Gerechtigkeit* (Würzburg, Universitätsbibliothek, I.t.q. XXXX, fols. 3ᵛ–4).

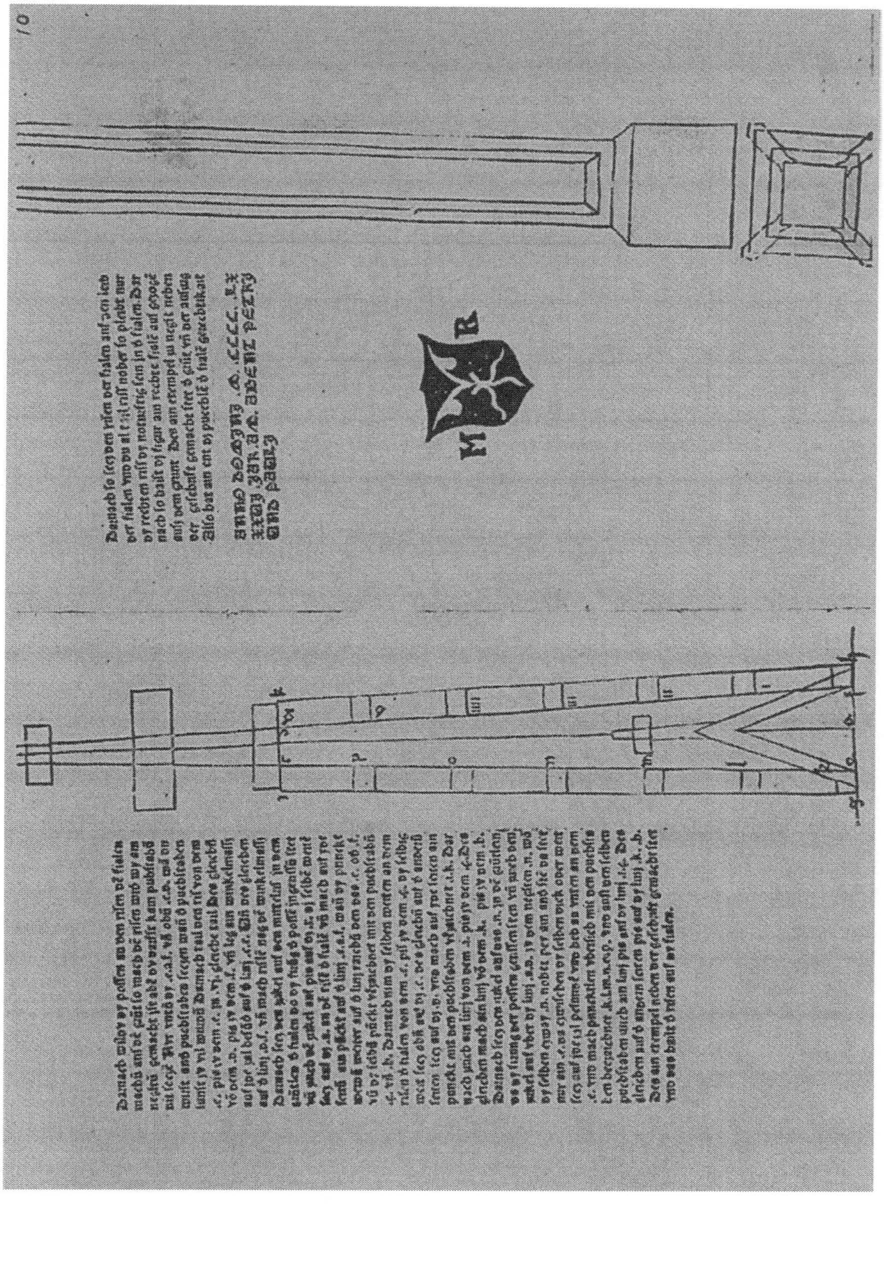

Fig. 7. Setting out the elevation of a pinnacle, from Roriczer's *Büchlein von der Fialen Gerechtigkeit* (Würzburg, Universitätsbibliothek, I.t.q. XXXX, fols. 9v–10).

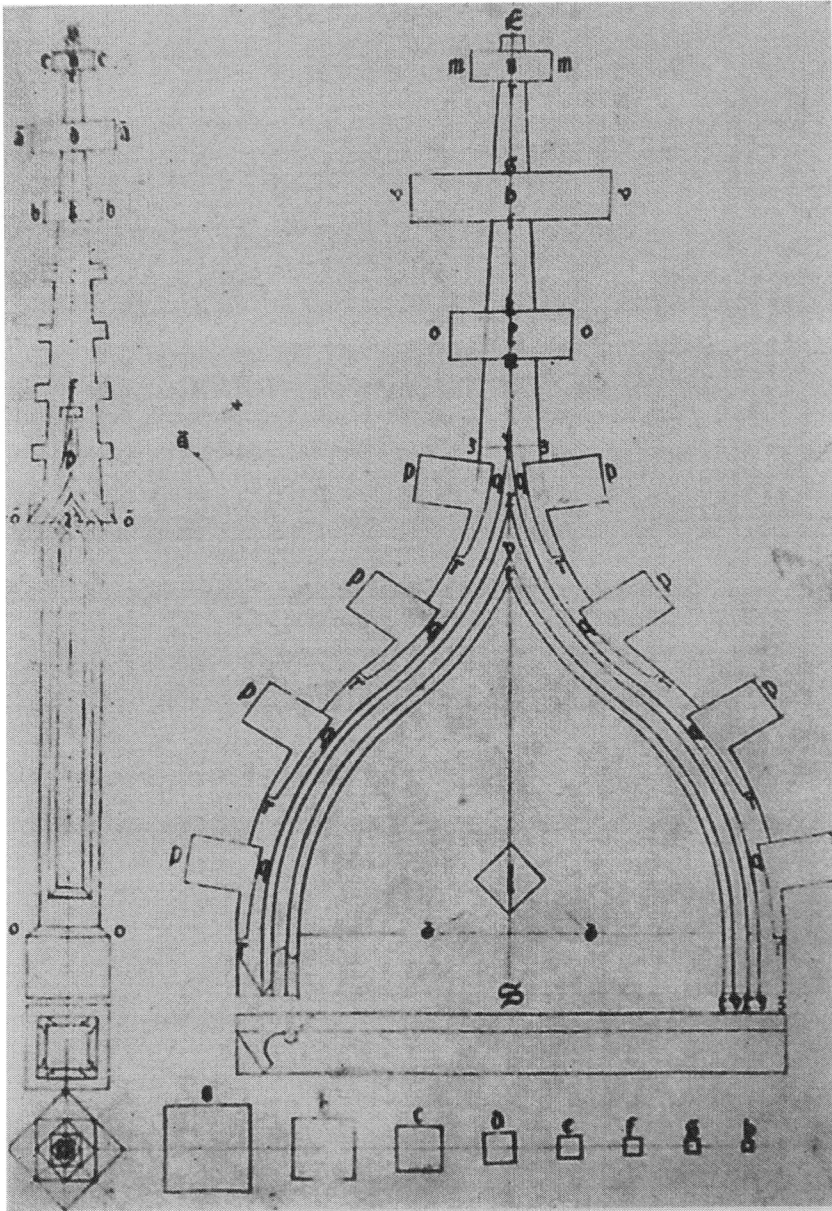

Fig. 8. Geometrical construction of a pinnacle and gable, from Schmuttermayer's *Fialenbüchlein* (Nürnberg, Germanisches Nationalmuseum, Bibliothek, No. 86,045, fol. 1ᵛ).

FIG. 9. Architectural rendering of a pinnacle and gable, from Schmuttermayer's *Fialenbüchlein* (Nürnberg, Germanisches Nationalmuseum, Bibliothek, No. 36,045, fol. 6).

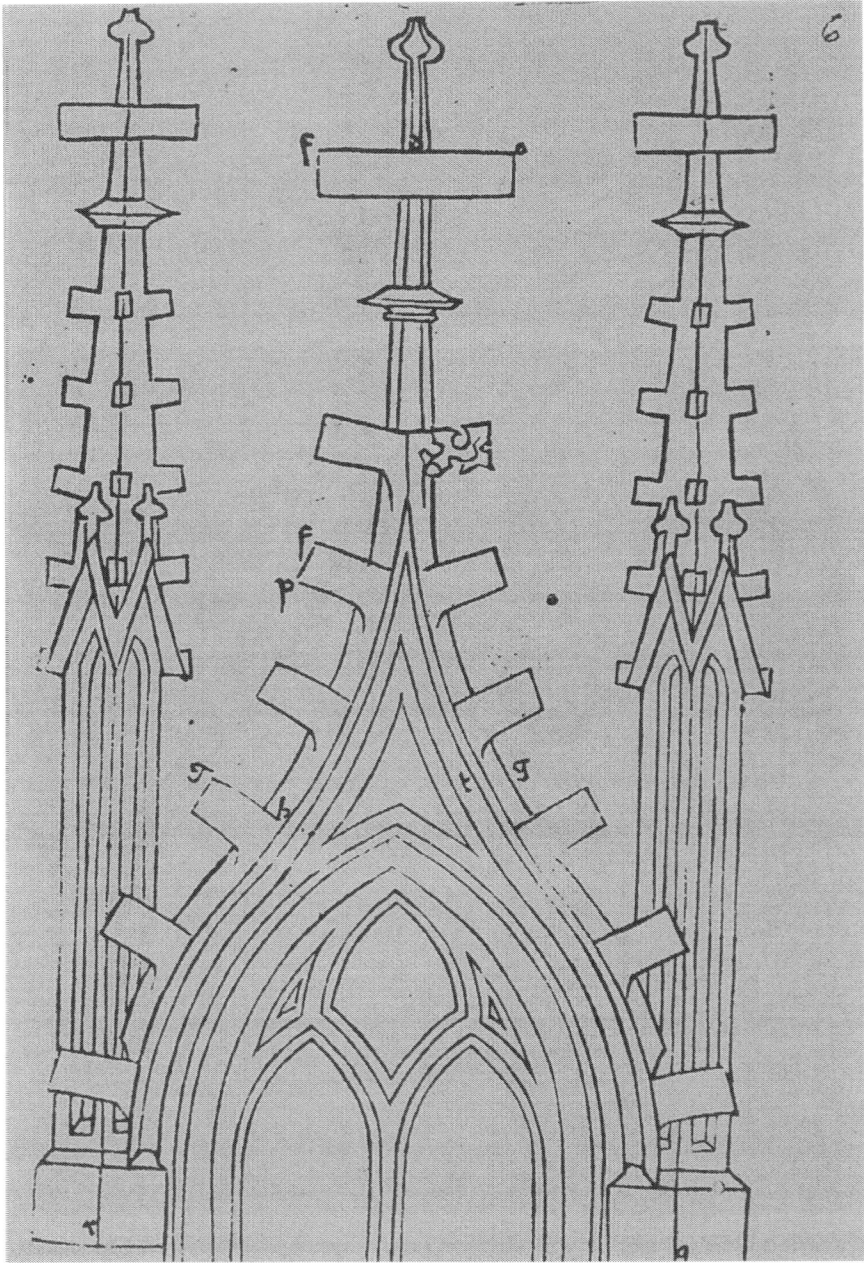

FIG. 10. Geometrical construction of a gable, from Roriczer's *Geometria deutsch* (Würzburg, Universitätsbibliothek, I.t.q. XXXX, fol. 6).

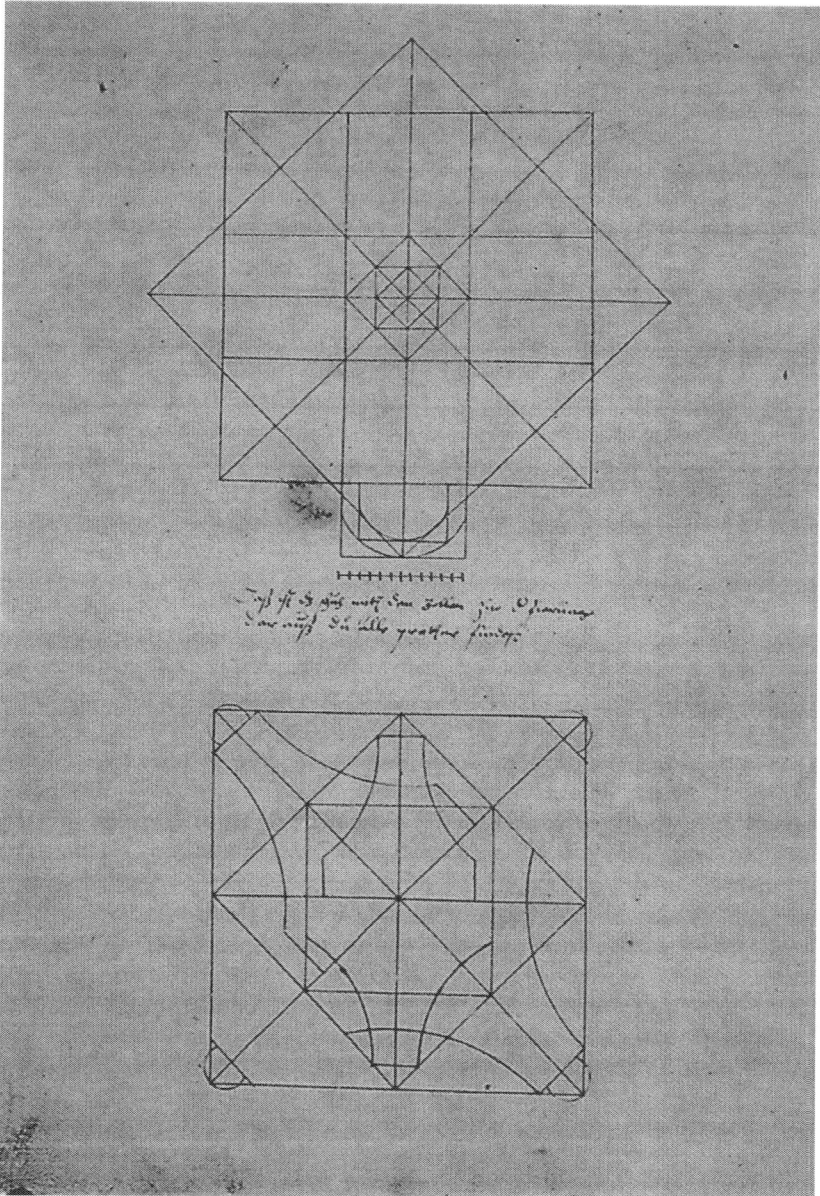

FIG. 11. Geometrical construction of a template for a window mullion, from Lechler's *Unterweisung* (Cologne, Historisches Archiv, Handschrift Wf. 276*, fol. 42ᵛ).

4

Villard de Honnecourt, Reims, and the origin of Gothic architectural drawing

Robert Branner

FROM its inception Gothic architecture revealed a strong preoccupation with linearity. This embraced not only the surface pattern of the building but also the very fabric of the structure, such as the piers, the buttresses and the ribs of the vault [1]. In comparison, Romanesque and earlier architectures appear bulky and resistant to articulation. The contrast between the styles probably reflects differences in the procedure of design as well as in conception, for it was Gothic, with its emphasis on the linear, that seems to have called modern architectural drawing into being.

Project drawings—those from which buildings could be constructed—were made by Roman and Byzantine architects, but contrary to current belief, it is highly questionable whether they were used in the early mediaeval west [2]. Architectural drawings were of course made throughout the Middle Ages, but prior to the Gothic period they always seem to have been programmatic or simply representational in nature. The plan of St. Gall, for instance, contains the layout of an entire monastery but does not inform the builder how to solve any technical problems [3], and the "perspective" renderings in manuscript illuminations were undoubtedly all made from edifices that had already been terminated [4]. . Neither of these is the kind of drawing that could be used in the actual process of building. This was probably unnecessary at the time, however, since the technique of construction was still relatively simple and in all likelihood it was the patron who informed the "architect" how he wanted the finished monument to look. Even if buildings such as Hildesheim, Cluny and Fontenay were designed on arithmetical, or, more likely, on geometrical principles, no matter how subtle or complex, the schemes could be worked out and the buildings erected from them without the intermediary of drawings [5]. Arithmetical calculations require numbers, not lines, and the great advantage of geometrical schemes lies in the fact that they can be directly developed at full scale on the basis of their proportions. Stakes, cords and simple instruments served in the layout of the plan, measuring rods were probably used for the elevation, and simple "rules of thumb" based on long masonic experience provided solutions to whatever stereotomical problems may have arisen [5a].

There are other reasons for the absence of project drawings prior to the thirteenth century. The ground-plan, for example, was probably thought out in his head by the architect and then laid out at full scale on the ground, so that the intermediate stage of a small-scale drawing was unnecessary [6]. It is likely that Jean d'Orbais, who was represented in the labyrinth of Reims Cathedral tracing a plan with a great compass, actually laid out the plan of the chevet at full scale in some such manner [7]. The elevation was undoubtedly also thought out, and many of the changes observable in the upper stories of mediaeval monuments, in addition to their archaeological value in revealing the direction and progress of the work, also seem to be indications of "improvements" that were effected after the original design had been put into work and could actually be seen. Although it may be difficult for us to imagine nowadays, when sketchbook and pencil are the architect's vade mecum, the habit of thinking out the design, even down to the details, was perfectly normal when there was no strong tradition of drawing, and several texts from the early thirteenth century indicate that the mental procedure was still well known at that time [8]. This explains why the master was indispensable to the work. When he left the shop, no one else could finish the building according to

FIG. 1.—Reims G 661, sheet E.

THE ORIGIN OF GOTHIC ARCHITECTURAL DRAWING   131

his design because he had literally taken it with him. The famous and often-quoted comment of the mid-thirteenth-century Dominican friar, Nicolas de Biard, "The master-masons...say to the others, 'Cut it for me here,'" has a meaning beyond the social and economic one usually given, for it also implies that the architect was still the only one to

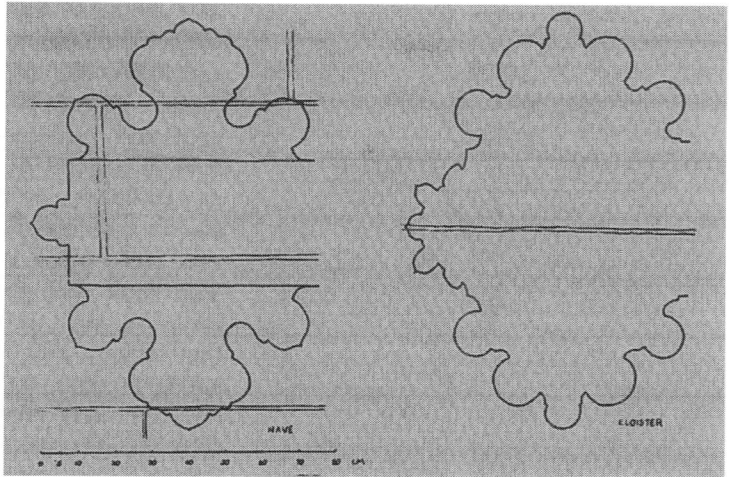

FIG. 2.—Noirlac, rejected springers for cloister vaults.

know the order and purpose of the various operations [9]; and Gervase of Canterbury tells us that when William of Sens was injured in 1179, he continued to direct the work from his bed, commanding "what must be done first and what afterwards [10]." It was probably for such a reason that some masters were bound by contract not to leave the area until their work was complete [11].

This is in sharp contrast to the procedure that we follow today. The architect now inspects the site, draws up plans for the work and gives them to a foreman or contractor to execute. Such a process became a possibility only when the design was first recorded in the form of drawings, and it does not seem to have become standard practice until the later thirteenth century. The tracing-house, for instance—the "lodge" where the projects were drawn and preserved—is first mentioned in an English text of 1274 [12]. Moreover two of the extant thirteenth-century sets of measured drawings, one palimpsested in manuscript G 661 of the archives at Reims (about 1240-1260) and the other preserved at Strasbourg (about 1275 ff.), were not devoid of problems for the builder [13]. Reinhardt and Fels have shown that the man who first undertook to execute Strasbourg B in stone made two fundamental errors in reading the drawing [14], and the row of clearstory windows on Reims E (fig. 1) could scarcely have served as a guide to construction, since the tracery varies in thickness and in some places the curves even intersect one another [15]. Fortuitous as they may seem, these problems nevertheless suggest that parchment drawings were still a relatively new idea in the process of planning and construction in the later thirteenth century.

The reasons for the appearance of project drawings are not difficult to find. The

132   GAZETTE DES BEAUX-ARTS

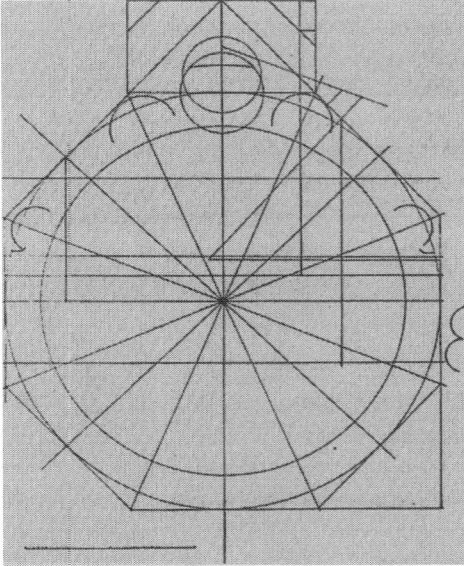

FIG. 3.—Saint-Quentin, rejected pier plan (?).

idea seems to have been rooted in certain technical developments and to have grown from the use of full-scale drawings of parts. The 1190's and early 1200's saw profound changes in Gothic style, not only in general design and structure, such as can be seen at Chartres and Reims, but also in a series of rapid technical advances that were made in these very cathedrals. The vault springer *en tas de charge*, for instance, with horizontal beds and with the various ribs cut from single stones, seems to have been invented at Chartres, and the first distinct steps toward the creation of bar tracery were taken at Reims. Both required careful stereotomy that had to be worked out in detail beforehand, and this could be done most simply by making accurate, "measured" drawings at full scale. Early examples of such drawings, engraved into stone, are the

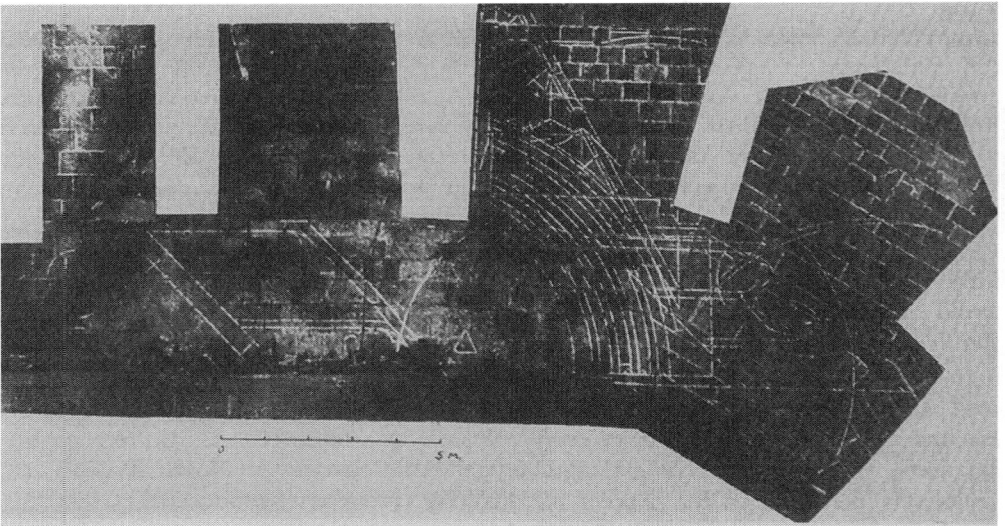

FIG. 4.—Clermont-Ferrand, Cathedral, terrace over north chevet collateral (montage).

THE ORIGIN OF GOTHIC ARCHITECTURAL DRAWING 133

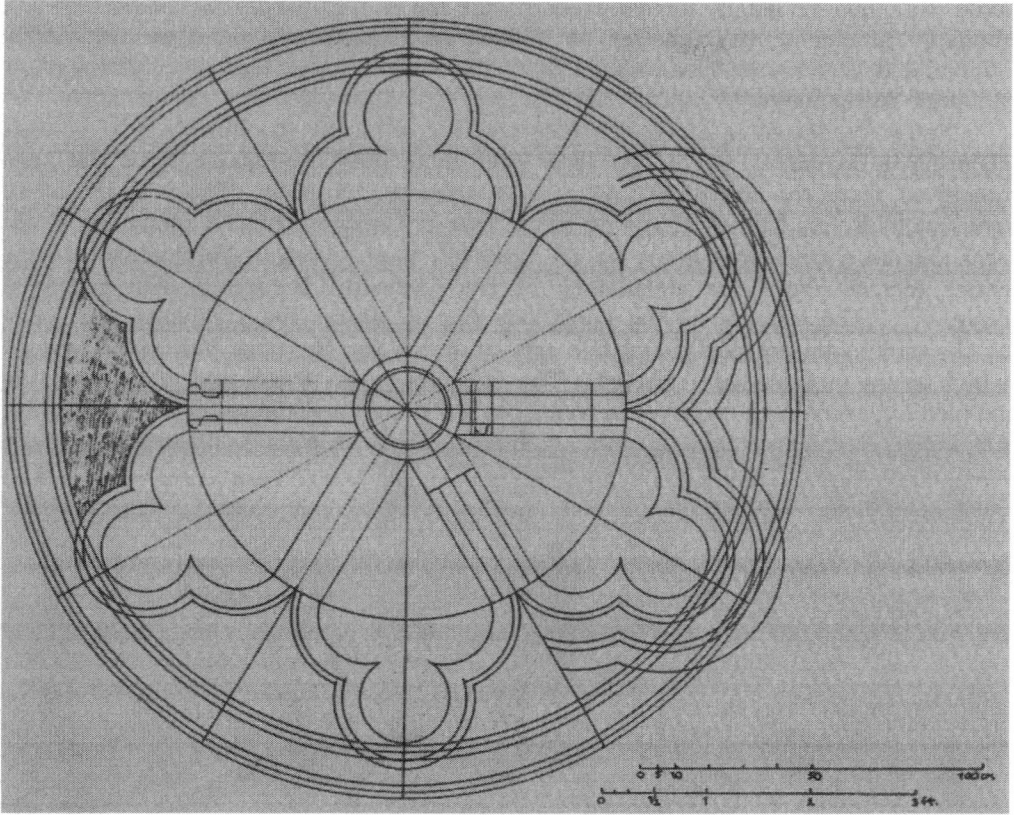

FIG. 5.—Saint-Alban's Cathedral, rose window from north aisle of nave.

stencils for vault departures found by Willis at Southwark (c. 1220-1230), the spirals from the south transept porch at Chartres which were most likely used in laying out the keystones of the arches (c. 1225), and the designs for windows in the western tribune at Soissons (c. 1235)[16]. The technique seems very quickly to have become a standard part of the Gothic building process and it remained so for several centuries[17]. Interestingly enough, some of the extant designs seem to be rejected ones, for instance the two late-thirteenth-century plans of springers for the cloister vaults at Noirlac (fig. 2) and a geometrical construction at St. Quentin that was probably intended for a pier (fig. 3)[18]. Rejected plans suggest that a design was developed by stages; and the fact that such plans were also engraved into stone at full scale, like the final design from which the template could be cut, in turn suggests that tentative, working drawings were not made at reduced size. The only small-scale drawings of details known from before the end of the thirteenth century are those of Master 2, the follower of Villard de Honnecourt, and they are merely abbreviated sketches endeavoring to show the various steps that were followed at full scale in the lodge [19].

Full-scale drawings of considerable portions of Gothic monuments, that is, ones differing in degree but not in character from those mentioned above, have also been preserved from the thirteenth and early fourteenth centuries. Those in the south transept of Reims Cathedral and on the terrace at Limoges are well known [20]. Probably the most prodigious group of all, however, is at Clermont-Ferrand, where the entire terrace above the ambulatory of the Cathedral is covered with engraved designs [21]. The montage of the north side (fig. 4) shows the archivolts and gable of the north transept portal and two sets of flyers for the nave buttresses, one of which seems to have been rejected. The calculation of thrust and abutment, the geometrical procedures employed in laying out the design and the assemblage of stones can all be determined from this set of drawings. It is equally important to note that the procedure is still alive.

There is a rose drawn on the plaster of the nave-wall of Saint Alban's Cathedral, probably by Lord Grimthorpe's foreman only a century ago

FIG. 6.—Reims, Cathedral, choir buttresses, and Villard's view (p. 64) (left).

THE ORIGIN OF GOTHIC ARCHITECTURAL DRAWING   135

(fig. 5), and there are drawings on the south wall of the nave of Lausanne, from the time of Viollet-le-Duc, as well as a set on the pavement of the crypt at Bourges, the latter made in the 1930's [22].

Thus if the full-scale, engraved drawing was invented in response to technical developments, to the need for greater precision in stereotomy and even in general layout, then it is possible that the small-scale, parchment drawing first came into being as a kind of corollary, in response to a growing habit of linear expression. Before the implications of this statement can be discussed, however, it will be necessary to take up the question of when and where project drawings may first have been used in Gothic architecture.

The oldest extant project drawings are those in the Reims palimpsest, which were made in or near Reims about 1240-1260. Since Lassus studied the Villard de Honnecourt manuscript a hundred years ago, however, it has been a commonplace

of mediaeval scholarship to assume that project drawings were also used at Reims Cathedral when that monument was begun in 1210 [23]. Villard is supposed, by most authors, to have copied them into his sketchbook and the divergences between manuscript and monument are taken to mean that the former represents early designs that were altered before or during the course of construction [24]. But a careful comparison of the two not only shows this was not the case, it also suggests that project drawings were not employed at all at Reims, at least during the early stages of work there.

Villard's drawings of Reims comprise a window from the side-aisle of the nave (p. 20), interior and exterior elevations of a

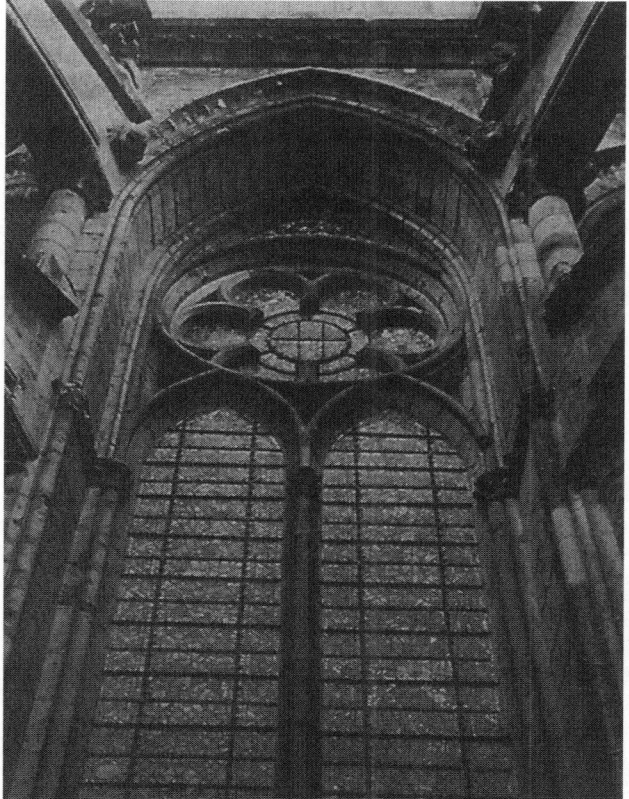

FIG. 7.—Reims, Cathedral, chevet clearstory window, détail.

radiating chapel (pp. 60-61) and of the nave (p. 62), a page of templates and pier-plans (p. 63) and one showing the flying buttresses of the choir (p. 64). The last-mentioned drawing has long been noted to differ from the real flying buttresses of the Cathedral (fig. 6): the proportions are elongated, the form of the *culées* differs, the triforium passage has been omitted, and so on. Since Henri Deneux showed that the present flying buttresses are not the ones originally planned, there might seem to be additional reason for assuming that Villard's drawing represents an older state [25]. But one factor alone is perhaps sufficient to indicate that this could not have been the case. The window capitals in the drawing are located only slightly below the "cornice," so that there is no place for the famous clearstory rosette (fig. 7), and the rosette, of course, was always a major feature of the window design at Reims from the very inception of work. If this page had been copied from a project drawing in the *chantier*, the rosette would surely have been present and in its proper place. The drawing therefore seems to be in error on this point, and a similar explanation can be offered for the many other differences from the monument, no matter how outstanding they seem to be at first glance. A comparison of the right portion of the drawing with the building suggests that Villard may have begun the

FIG. 8.—Reims, Cathedral, Villard's views (pp. 60-61).

THE ORIGIN OF GOTHIC ARCHITECTURAL DRAWING 137

image there, that is, on the interior of the monument, and that when he went outside, he realized that he had not left enough room for the upper flight of the buttresses. His solution to this problem, simply shifting the buttresses downward,
would be considered nothing short of irresponsible on the part of any master mason.

The remainder of the drawings are quite close to the monument, and the differences can definitely be ascribed to simple error. The radiating chapel (pp. 60, 61),
for example, was drawn in "perspective" with certain elements exaggerated, such
as the bases and plinth or the *crétiaus* [26], and others, such as the vaults, omitted in
order to give a more complete view of the elevation [27] (fig. 8). In this set, the
errors are limited to the design and *découpage* of the tracery and to the form of the
bases and the extreme buttresses [28]. Omissions and exaggerations are also found in
the elevations of the nave on page 62. Here, the errors comprise the additions of
a dado arcade and of *crétiaus* at the upper edge of the aisle roof, and incorrect
tracery forms (fig. 9). The drawing on page 20 is like an enlarged detail of page 62
(although it undoubtedly was made at a different time), for it shows a window of the
aisle of the nave seen from the inside and not partly hidden by the piers. But the
fuller view betrays an error that was noted long ago by Lassus: the capitals for
the tracery and for the vault departures
are placed at the same level, whereas in
the building they are consistently at different levels [29] (fig. 10).

There is only one case where Villard
represents an older form that was
rejected early during the construction of
Reims, and that is the plan of the pier
between the radiating chapels (p. 63d) [30].
But the figure is part of a group of
details including mullion plans, rib profiles and other pier plans, which he must
have copied from templates kept in the
chantier. In fact, he specifically identifies
a number of the figures as templates, or
"molles." There can therefore be little
question of the source for page 63d
being a project drawing [31].

The obvious conclusion is that Villard drew the elevations inaccurately
from the Cathedral of Reims itself. It
is strange, when one comes to think of
it, that he has ever been considered an
accurate draftsman, for his drawings

FIG. 8 a.—Reims, Cathedral, radiating chapel.

of Laon, Chartres and Lausanne all contain errors. If project drawings had been used at Reims, it seems safe to assume that Villard would have copied them or given some positive indication of their existence, a point to which we shall return shortly. As it is, however, it would appear that small-scale drawings were not used at Reims, at least between the start of construction in 1210 and the time of Villard's visit (probably about 1240-1245 [32]). Indeed, Villard's sketchbook, far from proving the existence of such drawings, is the best evidence for the reverse.

If project drawings were not used at Reims, which had one of the most modern and best organized *chantiers* of the time, then when and where were they first used? We must again turn to Villard for the answer, for he indicates that drawings were used at the Cathedral of Cambrai, undoubtedly from the beginning of the High Gothic campaign there. The text on page 28, next to the ground-plan of Cambrai, is revealing in this respect: "See here the plan of the chevet of My Lady Saint Mary of Cambrai just as it rises from the ground; earlier in this book you will find the interior and exterior elevations, and the whole manner of the chapels and also of the walls and the manner of the flying buttresses [33]," and on page 60 he again indicates that he saw the start but not the termination of the chapels: "Those of Cambrai will be the same (as those at Reims), if they are made properly [34]." It is amply evident from these phrases that Villard was at Cambrai at the very beginning of the High Gothic work, probably about 1220 [35], and equally clear that he could copy down both the main elevation and that of the flying buttresses, even though the radiating chapels had only just been begun. The implication that he must have seen project drawings seems inescapable.

Since the Cathedral of Cambrai has disappeared, together with all such hypothetical drawings, it is difficult to evaluate the evidence accurately. But certain inferences can be made. In his drawings of Reims, with the single exception of page 61, Villard consistently portrayed a more advanced form of window tracery than was actually employed in the metropolitan Cathedral [36]. This suggests that he went from Cambrai to Reims, from the younger shop to the older one. Villard also seems to have modelled his drawings of Reims on those of Cambrai. This is suggested by the fact that the list of Cambrai drawings on page 28 corresponds precisely with the preserved set of Reims on pages 60-64. The Reims drawings may therefore give us a rather precise idea of the lost ones from Cambrai. At the time, a group of three or four drawings was perhaps all that was needed: inner and outer main elevations, flying buttresses of the choir and a view or two of a chapel [37]. These must of course be understood as in the nature of a supplement, rather than a replacement, to the normal procedures of construction, which comprised such standard items as templates and pegs and string.

If drawings were employed at Cambrai, one would expect to find some corroboration of the procedure in shops that were affiliated with it, such as the one at the Liebfrauenkirche at Trier. There is in fact a small-scale plan engraved into the wall

FIG. 9.—Reims, nave elevations, and Villard's views (p. 62).

of a spiral staircase at Trier (fig. 11), but Herr Eichler has recently demonstrated that it is not a copy of an early parchment project for the church [38]. The nave and transept are considerably wider in the plan than in the monument, and the vaulting of the chapels is completely novel. For precisely the same reasons. it is difficult to see in it, with Herr Eichler, "the reproduction of an older idea for the general form of the edifice." The Trier plan is most likely a graffito, inscribed by an apprentice before the monument was completed. Graffiti of this sort seem constantly to have been made since the thirteenth century, for instance at Dommartin [39] or at Auxerre (fig. 12), the latter dating from the nineteenth century. If the Trier plan does not confirm our hypothesis regarding Cambrai, however, at least it suggests that the habit of drawing was becoming more widespread in the second quarter of the thirteenth century [40].

The oldest extant project drawings, those in the Reims palimpsest (figs. 1, 13, 14), strengthen the suggestion that the technique was becoming increasingly popular in the mid-thirteenth century, without regard to one particular shop tradition. The Reims drawings are stylistically unrelated to Cambrai or its affiliates, but they show close connections with monuments deriving from Amiens, particularly St. Nicaise at Reims. While it would be incorrect to define mediaeval shop traditions too sharply, these tend to stand out more distinctly in the first half of the thirteenth century than before or after. Villard himself, for instance, recorded data almost exclusively from High Gothic shops: from Cambrai and Reims, and from Chartres, the fountain-head of the movement; and when he went to Laon, the only part of the Cathedral he chose to record was one of the towers with incontestably Chartrain piers, which must have been designed by a master coming directly from the Beauceron Cathedral. In fact, the only non-High Gothic sites he seems to have examined were Vaucelles, Meaux and Lausanne. But it is equally significant that Villard visited only one branch of the Chartrain family, to the exclusion of the others. He gives no indication that he ever saw Amiens, for instance, which was begun in 1220, or any of the Parisian shops active in the 1230's, such as Notre-Dame, Royaumont or St. Denis [41], and he seems never to have crossed the city of Reims to Hugh Libergier's shop at St. Nicaise (1231 ff.). Such family ties may give greater significance to the appearance of project drawings in a *chantier* closely related to St. Nicaise, for they suggest that Cambrai should be considered, not the "creator" of architectural drawings in the fullest sense of the word, but merely the first of a number of north-French shops that came to use them within a relatively short span of time.

With the exception of Villard's views of the chapel at Reims and consequently perhaps also the original working drawings of the chapels at Cambrai, all the project drawings preserved from the thirteenth century are flat, orthogonal projections, and this flatness, while it is inherent in the technique of drawing. also corresponded to a particular development that can be discerned in Gothic design about 1230. After the nave of Amiens and the chevet of Beauvais had been begun, speculation on the configurations of space and mass waned and interest began to focus on surface patterns. The elevation was articulated into a complex design of mullions, arches and oculi; stories formerly distinct from one another were merged and the effects of screens and of subtle planar recessions were explored. Amiens played an important rôle in this early Rayonnant development, which can be charted in such buildings as St. Nicaise and St. Denis. The architecture of these monuments, both inside and out, was one that for the first time *could* be composed on parchment and then translated into stone. Its complexities of form also suggest that the design was seldom

arrived at with the first try. The master who made a project drawing first, however, found it easier to appreciate and hence to correct the design before construction
began, and in this respect it is worth noting that the Reims palimpsest façades are
both incomplete and seem, like Strasbourg A, to have been in the nature of trials [42].
Finally, the technique of parchment drawing may even have exerted an influence on
the avant-garde movement of the 1230's and 1240's, for which it had a much more
profound meaning than it did for the Cambrai of 1220. Be this as it may, from
that time on project drawings were more and more frequent in Gothic shops, until
eventually they became a standard step in the process of building, and they have
remained so ever since.

FIG. 10.—Reims, Cathedral, window from nave aisle (right), and Villard's view (p. 20) (left).

RÉSUMÉ : *Villard de Honnecourt, Reims et l'origine du dessin d'architecture gothique.*

Bien qu'employé par les Romains et les Byzantins, le dessin d'architecture servant à la construction semble avoir disparu pendant le haut moyen âge, pour être réinventé au XIIIᵉ siècle sous l'impulsion des procédés gothiques. Ceux-ci ont d'abord produit l'épure, dessin de détail à grandeur réelle, et ensuite le dessin à petite échelle sur parchemin. Villard de Honnecourt, dans ses vues de la cathédrale de Reims, nous conserve un témoin des plus anciens dessins gothiques qui furent utilisés non pas à Reims, semble-t-il, mais à Cambrai. La confrontation de l'album de Villard avec Notre-Dame de Reims montre clairement qu'il a dessiné d'après nature, d'après le bâtiment déjà construit ; mais son choix de

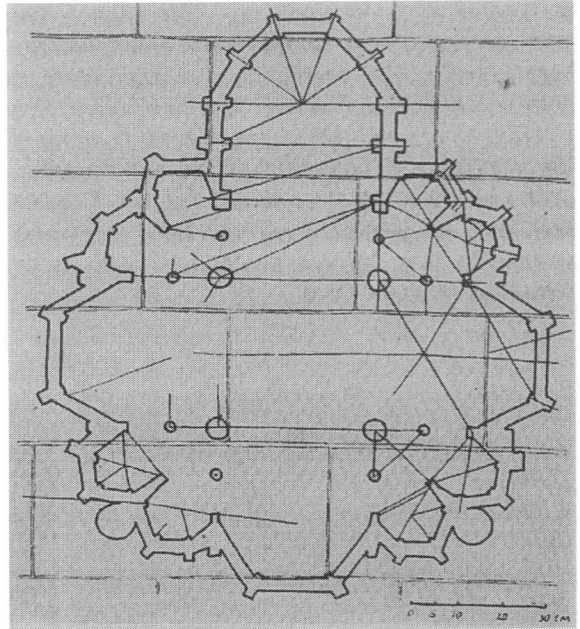

FIG. 11.—Trier, Liebfrauenkirche, engraved plan, after Eichler (augmented).

coupes et de vues, par contre, semble être inspiré de ce qu'il a pu voir sur le chantier de Cambrai. A partir de 1230 environ, le dessin devint de plus en plus fréquent, accompagnant le développement d'une architecture « dessinée », à meneaux, à fenestrages linéaires, à rosaces rayonnantes.

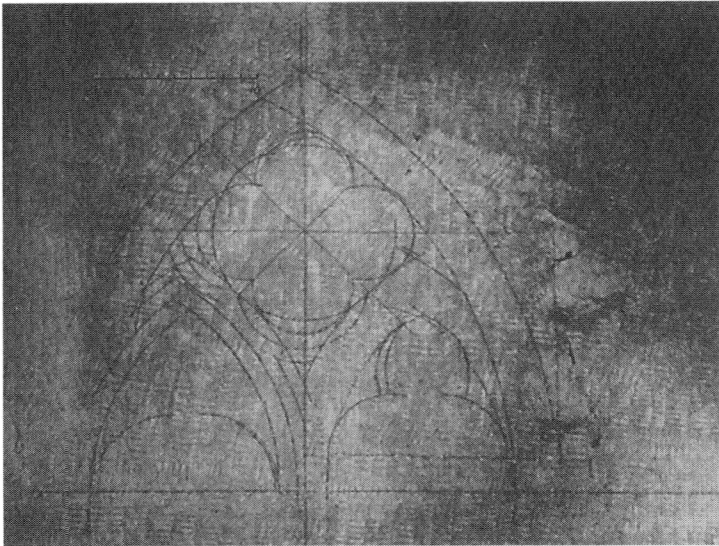

FIG. 12.—Auxerre Cathedral, graffito from chevet triforium.

NOTES

1. Cf. J. Bony, "Les premiers architectes gothiques," *Les architectes célèbres*, II, Paris, 1959, pp. 28-32.

2. G. Downey, "Byzantine Architects," *Byzantion*, XVIII, 1948, pp. 99-118, esp. pp. 114-118.

3. H. Reinhardt, *Der St. Galler Klosterplan* (92. Neujahrsblatt, Historischer Verein des Kantons St. Gallen), St. Gall, 1952.

4. W. Überwasser, "Deutsche Architekturdarstellung um das Jahr 1000," *Festschrift für Hans Jantzen*, Berlin, 1951, pp. 45-70; on the illumination showing eleventh-century Chartres, see most recently H. H. Hilberry, "The Cathedral of Chartres in 1030," *Speculum*, XXXIV, 1959, pp. 561-572.

5. For Hildesheim, see H. Beseler and H. Roggencamp, *Die Michaeliskirche in Hildesheim*, Berlin, 1954; for Cluny, K. J. Conant, "New Results in the Study of Cluny Monastery," *Journal of the Society of Architectural Historians*, XVI, pt. 3, 1957, pp. 3-11; for Fontenay, O. von Simson, *The Gothic Cathedral* (Bollingen Series, 48), New York, 1956, pp. 48-50.

5 a. Cf. the obvious absence of drawings in the construction of Romanesque cupola churches, demonstrated in R. Chapuis' " principe de la taille à la demande " (" Géométrie et structure des coupoles sur pendentifs, " *Bulletin monumental* CXX, 1962, pp. 7-39).

6. This is clearly represented in the story of Pope Leo tracing the plan of Sta. Maria Maggiore in Rome; cf. the twelfth-century visualization of the planning of Cluny (P. du Colombier, *Les chantiers des cathédrales*, Paris, 1953, fig. 6); it must be noted, however, that this does not necessarily imply a fluent working knowledge of Euclidian geometry. Models do not seem to have been used in northern Europe, with the unique and perplexing exception of St. Germain at Auxerre in the ninth century (*ibid.*, pp. 72-73; cf. L. H. Heydenreich, "Architekturmodell," in O. Schmidt, *Reallexicon zur deutschen Kunstgeschichte*, I, 1937, cc. 918-940, with an apparently incorrect reference to William of Sens).

7. R. Branner, "Jean d'Orbais and the Cathedral of Reims," *Art Bulletin*, XLIII, 1961, pp. 131-133.

8. For instance, it was said of Master Simon, during the fortification of the château of Ardres (c. 1200), "...(Symonem) cum virga sua magistrali more procedentem et hic illic jam in mente conceptum rei opus non tam virga quam in oculorum pertica geometricantem" (V. Mortet-P. Deschamps, *Recueil de textes*, Paris, 1929, p. 190); Guillaume d'Auvergne, Bishop of Paris (1218-1249), was even more categoric : "Omnium operationum prima est indubiter cogitatio, et haec est locutio intellectualis, et hoc est primum in omni artifice, qui est apud nos quicumque enim artifex qui aliquid operatur per artem exterius apud nos, primum operatus apud se ipsum, et in seipso ejusdem operis imaginem seu exemplar praecogitans ac praeordnans apud semetipsum opus quod fabricaturus est, et propter hoc pingens, in semetipso exemplar alicujus, ne oblivione a corde ejus effugiat, descriptionem ejus interdum ad quam recurrit quoties alicujus eorum, quae praecogitaverat, oblitum se videt" (*Opera omnia*, Paris, 1674, *De Universo*, XX, *Quod tribus intentionibus dicitur verbum*). The paraphrase in R. E. Swarthwout, *The Monastic Craftsman*, Cambridge, 1932, p. 103, note 1, does not seem correct in the point at issue here, for "imago" = "exemplar" and both are products of thought. "Descriptio" of course does not mean a measured drawing alone, any more than "designator", often mentioned in the eleventh and twelfth centuries, means only draftsman. P. Booz also admits that mental procedures may have been used (*Der Baumeister der Gotik* [Kunstwissenschaftliche Studien, 27], Munich, 1956, pp. 68-69). Cf. also Geoffroy de Vinsauf in E. Faral, *Les arts poétiques*, Paris, 1924, p, 198.

9. See Mortet-Deschamps, *op. cit.*, p. 291.

10. V. Mortet, *Recueil de textes*, Paris, 1911, p. 223.

11. For instance at Meaux, in 1253; see Mortet-Deschamps, *op. cit.*, pp. 283-284.

12. J. Harvey, *The Gothic World*, London, 1950, pp. 29-30; cf. L. F. Salzman, *Building in England*, Oxford, 1952, p. 21.

13. J. Knauth, "Erwin von Steinbach," *Strassburger Münsterblatt*, VI, 1912, pp. 7-52; R. Branner, "Drawings from a thirteenth-century Architect's Shop : the Reims Palimpsest," *Journal of the Society of Architectural Historians*, XVII, 1958, pp. 9-21, with bibliography, and H.R. Hahnloser in *Résumés, XIII* *Congrès international d'histoire de l'art*, Stockholm, 1933, pp. 260-262.

14. H. Reinhardt-H. Fels, "La façade de la cathédrale de Strasbourg," *Bulletin de la société des amis de la cathédrale de Strasbourg*, s. 2, III, 1935, pp. 15-27.

15. The drawings were themselves palimpsested in several cases. The windows on E were effaced and a plan, possibly for a pulpit or Sepulcher, was drawn on the parchment.

16. R. Willis, "On the Construction of the Vaults of the Middle Ages," *Transactions of the Royal Institute of British Architects*, I, pt. 2, 1842, pp. 1-69, esp. pp. 10-12; R. Branner, "Villard de Honnecourt, Archimedes and Chartres," *Journal of the Society of Architectural Historians*, XIX, 1960, pp. 91-96; E. Brunet, "La restauration de la cathédrale de Soissons," *Bulletin monumental*, LXXXVII, 1928, pp. 65-69; Mr. Carl F. Barnes II has recently found other drawings, apparently from this period, at Soissons. G. E. Street also found a capital at León Cathedral with marks on it, during the nineteenth-century restorations there (*Some Account of Gothic Architecture in Spain*, 2d ed., New York, 1914, I, p. 145), but since he did not publish a drawing of it and since no sketch is contained in those of his notebooks preserved in the R.I.B.A. Library in London, it is very likely lost (I am informed

by Professor R. J. Lambert, of Sidney Sussex College, Cambridge, that Street's other notebooks from the León period perished during the last war).

17. From about 1300 there are two windows, one on the pavement of the chamber in the pier-buttress at Bourges, and another (fragment) from the old chapel of St. John's College, Cambridge (see C. C. BABINGTON, *History of the Infirmary and Chapel of St. John the Evangelist at Cambridge*, Cambridge, 1874, pl. 9 and G. G. COULTON, "Artist Life in the Middle Ages," *Burlington Magazine*, XXI, 1912, pp. 336-344, fig. 2); cf. also R. ANDREWS, "Notice of working drawings scratched on the walls of the crypt at Roslin Chapel," *Proceedings, Society of Antiquaries of Scotland*, X, 1875, pp. 63-64 (founded in 1446; I am indebted to Professor J. F. Fitchen III for this reference), and HARVEY, *op. cit.*, pp. 30-31.

18. I am indebted to M. Robert Gauchery, sometime Architecte des Bâtiments de France at Bourges, for calling the Noirlac drawings to my attention. The St. Quentin drawing is presumably the same one published by Bénard as a rose window—although they differ so markedly as to cause some hesitation—and used by him to "prove" that Villard de Honnecourt was the architect (P. BÉNARD, *Collégiale de Saint-Quentin*, Paris, 1867); it is located on the northern wall of the northwestern radiating chapel.

FIG. 13.—Reims G 661, sheet A.

19. Pages 39-40. See H. R. HAHNLOSER, *Villard de Honnecourt*, Vienna, 1935, pp. 104-118; all references in the present article use Hahnloser's page and figure numbering. Cf. also R. BRANNER, "Three problems from the Villard de Honnecourt Manuscript," *Art Bulletin*, XXXIX, 1957, pp. 61-66 and "A Note on Gothic Architects and Scholars," *Burlington Magazine*, CXV, 1957, pp. 372, 375.

20. H. DENEUX, "Signes lapidaires et épures du XIIIᵉ siècle à la cathédrale de Reims," *Bull. mon.*, LXXXIV, 1925, pp. 99-130; F. DE VERNEILH, " Construction des monuments ogivaux. Epures de la cathédrale de Limoges, " *Annales archéologiques*, VI, 1847, pp. 139-144.

21. I am indebted to M. Gandrille, Architecte des Bâtiments de France at Royat, for his cooperation in making the photographs for the montage; it is our intention to study the Clermont drawings in greater detail.

22. For Saint Alban's, see the Royal Commission on Historical Monuments' *A Guide to Saint Alban's Cathedral*, London, 1952. p. 26.

23. J.-B. A. LASSUS, *Album de Villard de Honnecourt*, ed. A. Darcel, Paris, 1858, p. 211.

24. R. WILLIS (*A Facsimile of the Sketch-Book of Wilars de Honecort*, London, 1859, pp. 211-236) thought only page 64 was copied from a project drawing, as did H. KUNZE (*Das Fassadenproblem der französischen Früh- und Hochgotik*, Leipzig, 1912, pp. 53-55). HAHNLOSER (*op. cit.*, pp. 166 and *passim*) and UEBERWASSER ("Massgerechte Blauplanung der Gotik an Biespielen Villards de Honnecourt," *Kunstchronik*, II, 1949, pp. 200-204) were more categoric in accepting all the Reims drawings as copies of project drawings, while E. GALL (*Kunstchronik, loc. cit.*) was equally categoric in rejecting them.

25. H. DENEUX, "Des modifications apportées à la cathédrale de Reims au cours de sa construction du XIIIᵉ au XVᵉ siècle," *Bull. mon.*, CVI, 1948, pp. 121-140, esp., p. 123.

26. *Idem*, « Les 'crétiaus' de la cathédrale de Reims », *ibid.*, CIV, 1946, pp. 109-112.

27. I cannot agree with Willis's suggestion that Villard shows the vaults unfinished. The radiating chapels were certainly vaulted by the time the upper stories of the nave were undertaken, but Villard portrays both in the same way.

28. On the development of tracery forms, see WILLIS, *Facsimile*, pp. 221-223 and my « Paris and the

THE ORIGIN OF GOTHIC ARCHITECTURAL DRAWING   145

29. Lassus, *op. cit.*, p. 97.

30. Cf. Hahnloser, *op. cit.*, pp. 169-170, and Willis, *Facsimile*, p. 229. A second revision can be found in the responds at the juncture of the chevet and the transept, which were probably altered from three-quarter *piliers cantonnés* to their present form after the plinth was laid.

31. The upper figures on page 63 were also copied from templates; this is especially noticeable in 63 *a*, which has marks on it (cf. Willis, *loc. cit.*).

32. The elevations on p. 62 can only represent bays 2 through 6 of the nave (counting from the crossing). The lower portions of these bays were begun before the riots of 1233, but the upper portions were finished only after 1236 : bays 1, 2 and 3 by 1241, when the canons were installed in the new choir, and bays 4, 5 and 6 between 1241 and 1254. See R. Branner, " The Labyrinth of Reims Cathedral, " *Journal, Society of Architectural Historians*, XIX, 1962, pp. 18-25.

33. " Ves ci l'esligement del chavec me dame Sainte Marie de Canbrai, ensi com il ist de tierre. Avant en cest livre en trouverez les montees dedens et dehors, et tote le maniere des capeles et des plains pans autresi, et li maniere des ars boteres. "

34. " D'autretel maniere doivent estre celes de Canbrai s'on lor fait droit."

35. Saint Elizabeth of Hungary (d. 1231) "ayda par or et par argent à achever ledit chœur" (J. Deligne, *Sommaire des antiquités de l'église archiépiscopale de Cambrai*, cited in Lassus, *op. cit.*, p. 119. note 3) ; this perhaps provided the chaplaincy of Saints Nicholas and Catherine, in the northeastern radiating chapel, which is said to have been founded in 1230 or 1231 (A. Leglay, *Recherches sur l'église métropolitaine de Cambrai*, Paris, 1825, p. 34). The traditional date of 1227, when the *loi godefroi* temporarily established peace between the bishop, the chapter and the burgers, is therefore difficult to accept for the start of the High Gothic campaign, since three years would hardly have sufficed for the erection of the chapels. On the other hand Leglay states that Bishop Godefroi des Fontaines sent out a quest for funds that probably inaugurated the work, and this could not have been before his election late in 1218 (*op. cit.*, p. 13). See also L. Serbat, "Quelques églises anciennement détruites du nord de la France", *Bull. mon.*, LXXXVIII, 1929, pp. 365-435, esp. pp. 400-418.

36. Cf. Willis, *Facsimile*, p. 223 ; apparently the oldest extant example of the form shown by Villard is at Notre-Dame in Paris.

37. It is uncertain, however, whether the plan on p. 28 is a copy of a project drawing. The oldest extant project drawing of a ground-plan is Wallraf plan B from Cologne, of c. 1280-1308 (although Vienna plan A probably represents an older stage from the same campaign; for both, cf. *Kölner Domblatt*, XI, 1956); Reims E-2 (fig. 1) unfortunately cannot be identified with monumental architecture. It is perhaps significant that all the oldest ground-plans from Cologne are of towers, one of the most frequently represented parts throughout the Middle Ages (cf. M. Velte, *Die Anwendung der Quadratur und Triangulatur* [Basler Studien zur Kunstgeschichte, 8], Basel, 1951, pp. 29-64). The plan devised by Villard and Pierre de Corbie (p. 29 *a*) and the Cistercian plan (p. 28 *b*) are the only "theoretical" drawings known from the thirteenth century, but both are sketches without scale or details and can hardly be taken to represent project drawings. All the other ground-plans for nearly a century, including Strasbourg "21", of about 1315-1320, were copied from buildings (O. Kletzl, "Ein Werkriss des Frauenhauses von Strassburg", *Marburger Jahrbuch für Kunstwissenschaft*, IX, 1939, pp. 103-158). Project elevations, on the other hand, were common (e.g., in the Reims palimpsest, at Strasbourg, at Cologne and perhaps also at Dijon; see P. Quarré, "Saint-Bénigne de Dijon d'après la tombe de l'abbé Hugues d'Arc", *Bull. mon.*, CIII, 1945, pp. 231-242). This suggests that ground-plans

FIG. 14.—Reims G 661, sheet B.

GAZETTE DES BEAUX-ARTS

other than those of towers may still have been devised in the old geometrical manner long after elevations were regularly composed on parchment. Villard's sketches, however, prove that the ground-plan, with vaulting patterns, occupied a distinct place in early thirteenth-century architectural thought.

38. H. EICHLER, "Ein frühgotischer Grundriss des Liebfrauenkirche in Trier", *Trierer Zeitschrift*, XXII, 1953, pp. 145-166.

39. For Dommartin, see C. ENLART, *L'architecture romane et de transition dans la région picarde* (Mémoires de la société des antiquaires de Picardie), Amiens-Paris, 1895, p. 108, note 2; the drawing is now indecipherable. It was not deeply engraved into the surface and probably was not made before the end of the thirteenth century. See also G. C. COULTON, *Art and the Reformation*, 2d ed., Cambridge. 1953, pp. 178-179 and 530, and H. HUTH, "Die romanische Basilika zu Bechtheim bei Worms", *Der*

Wormsgau, IV, 1959-1960, pp. 5-97 m esp. pp. 76-79, with a list of other graffiti on pp. 78-79. I am indebted to Dr. Eichler for the last reference.

40. The tools used in drawing on parchment were probably at first the same ones that were used to trace profiles, such as the small compass and the small ruler. Between 1100 and 1200 these were refined considerably, if the development of profiles of ribs and imposts is a criterion. Further refinement must naturally have been necessary to permit their use on parchment, and this is already evident in the Reims palimpsest, as well as occasionally in Villard's manuscript.

41. For Beauvais, see R. BRANNER, « Le maitre de la cathédrale de Beauvais ». *Art de France*, II, 1962, pp. 77-92.

42. This is also apparent in the tracery of Reims E.-1.

5

The tracing floor of York Minster

J.H. Harvey

York Minster is famous for its stained glass windows, deliberately saved in 1644. Owing to destruction at other cathedrals, this splendid display is unique in Britain. The Minster deserves to be famous also for its preservation of another rare survival: the tracing-floor of plaster-of-Paris upon which details were set out geometrically by the master masons.

Such floors were in normal use during the Gothic period, on the Continent and in Britain[1] but so far as is known only two have survived in this country: one in the upper room above the North Porch of Wells Cathedral[2], the other here in York, where it occupies a large part of the L-shaped room above the Chapter House Vestibule[3]. This room may well be the only surviving mediaeval drawing office in the world, and requires more detailed study. It provided accommodation for the Minster masons to work in comfort, having a large fireplace in the south wall of the eastern arm, and a garderobe entered by a passage from the east side of the southern arm. The plaster floor now extends over the whole of the southern arm and may originally have reached to the north wall, providing an area more than 40 feet long by over 16 feet wide. Such a space was needed in order to set out accurately the shapes of large windows, the curves of vaults, and other major details of the building

Why were such floors necessary in the Middle Ages? The reason is that from the Norman Conquest to the reign of Henry VIII the greater churches were built upon a scale almost without parallel. Apart from the King's Palace of Westminster there were virtually no civil buildings in England comparing in size and complexity with the larger churches. Since the Reformation it has rarely been necessary to employ full-size geometry to determine the exact curves for tracery or vault-ribs, for these forms practically disappeared along with the Gothic style. Such setting-out as was needed was carried out on any large floor that happened to be available. Modern stonemasons carrying out structural work of large size usually have some space, such as the boarded floor of a loft, on which setting-out can be done[4]. Since the great age of church-building ended in the mid-sixteenth century, the maintenance of large plaster floors was not justified.

Mediaeval master masons discharged several functions. They were architects, in that they were responsible for the plan and design of new buildings and for the precise forms to be taken by details; they had the supervision of work as it proceeded; and they

2

were commonly responsible for accurate measurement of contractors' work, and the certifying of accounts. Unlike modern
architects, the master masons were experts in the cutting of the
individual stones of which the work was made, and in the practical
geometry requisite to see that they fitted together accurately. It
was this science of cutting materials to their true shape, or stereotomy, that made it possible to produce the exquisitely proportioned
details of Gothic architecture; and it was the fact that the greatest
of the mediaeval masons were so highly skilled in this field that
endowed their works with a vitality that has rarely, if ever, been
equalled since their day[5].

Design in the Middle Ages included the drawing out of plan
and section on large sheets of parchment, and the making of
templates showing the profile of mouldings. The templates were
normally cut out of thin wooden boards, but might also be of
thin sheets of metal. Continuity of use of the York Minster
Masons' Lodge and Tracing House (i.e. Drawing Office) is suggested by the large numbers of wooden, iron and zinc templates
(probably of 19th-century date) hanging in bundles from the walls.
(See Plate I). Each bundle will correspond to the details of a particular job, and may eventually be shown to correspond to surviving
drawings[6].

Besides the plans and sections, and the templates, there was need
for drawings to indicate the precise arrangement of stones. Where
each stone had to fit snugly to its neighbours it was necessary to
make a basic 'cartoon' of the design to full size. These full-size
details, from which the complete assembly of stones forming the
tracery of a window or the system of ribs of a vault could be determined, were drawn out upon the plaster floor.

A plaster floor was convenient because mere brushing over
would obscure earlier drawings, while the freshly scratched lines
of the current detail would show up white and sharp. Thus the
surface of the floor in the course of time came to be covered with
patterns of straight and curving lines having no connection with
one another. To make sense of any part of the conflicting patterns
one must trace all surviving lines and follow out those groups of
lines which appear to be interrelated. To do this on the floor itself
would be excessively difficult; it is much simpler to use a set of
overlapping photographs taken at a uniform distance from the
floor and mounted together[7]. Even after selective tracing there will
remain a mesh of lines not easy to interpret.

Preliminary inspection of the drawings on the floor shows that
they are of the Perpendicular period, belonging to the last two

6

centuries of work on the Minster prior to the Reformation. This agrees with the stylistic evidence of the room itself and of its timber roof[8]. The Chapter House is certainly of the end of the thirteenth century; against it, rather later, was built the Vestibule; while the upper room used as the Tracing House was a still later addition, about 1350. Fifty years after the Reformation the room was still known as 'the tracinge hows' when, in 1581, work was done on the leads above it[9]. There is a strong presumption of continuous use of the room by the Minster masons, in spite of the apparent post-Reformation break in the series of Master Masons to York Minster[10].

By good fortune at least one important drawing can be identified: the setting-out for the tracery of the aisle windows of the Lady Chapel (Fig. 1, above A). This part of the Minster—the four eastern bays—was built for Archbishop John de Thoresby between 1 August 1361, when he laid the first stone, and 6 November 1373, when he died[11]. The design for the tracery of the aisles must have been prepared fairly early, presumably by 1365. This proves that the Tracing House had by then been roofed, and was in actual use.

By whose hand was the drawing made? The master mason, with his giant compasses and square, prepared the designs from which subordinates shaped the stones. Thus it was probably the master himself who drew the incised lines upon the plaster. At that period the master was either William de Hoton (junior), granted a fee of £10 a year on 1 October 1351; or Robert de Patryngton, granted a like fee on 5 January 1368/9[12]. In both cases the grants included the dwellings in the Close formerly inhabited by the master masons of the Minster and both refer to good service already performed for the Chapter upon the fabric. Patryngton's grant shows that William de Hoton was already dead; the date of Hoton's death and of Patryngton's succession as master is not quite certain, but may well be put at 1368.

This change during the building of the Lady Chapel is confirmed by the evidence of the fabric. There is a marked difference between the tracery and details of the aisles and those of the clerestory. The work of the aisles, at a lower level, belongs to an earlier date than the clerestory, and the former can be safely attributed to William de Hoton, who was presumably the designer, and probably also the draughtsman, of the tracery identified on the floor. Apart from this we know nothing of Hoton beyond the fact that he was the son of William de Hoton (senior), master mason to the Minster before him.

The other details which can be isolated from the mesh of lines have not so far been identified with certainty, though to the north (Fig. 1, B) there is a pair of arches (upside down, compared to A) with a circular spandrel and mouldings of late-14th or early-15th century type; and further on (Fig. 1, C) a group of tracery details of mid-14th century date similar in style to the windows of the Vestries and the Zouche Chapel—probably designed by one of the two Hotons, father or son. Superimposed, upside-down, on the large detail of the Lady Chapel aisle window, is the tracery of a four-centred window, comparable in style to the work of St. Michael-le-Belfrey (by John Forman, 1525-37). It is to be hoped that more details will eventually prove recognisable.

NOTES

(1) The continental floors are discussed by Otto Kletzl, *Plan-Fragmente aus der deutschen Dombauhütte von Prag* (Veröffentlichungen des Archivs der Stadt Stuttgart, Heft 3, Stuttgart, 1939), p. 9. The plaster floor at Strasburg Cathedral was in use by 1490 and existed until 1759 when it was destroyed in a fire.

(2) See *Reports of the Friends of Wells Cathedral*, 1954, p. 7; 1957, pp. 8-11, and *The Flowering of the Middle Ages*, ed. Joan Evans (1966), p. 115 and illustration on p. 114. In another case designs have been found on wall-plaster, in the upper room known as 'Oliver Cromwell's Saddle-room' above the eastern chapels of the north transept of Christchurch Priory, Hampshire (*Gentleman's Magazine*, 3rd series, viii, 1860, p. 277; xi, 1861, p. 635).

(3) The only published reference to the room seems to be in George Benson, *The Handbook of York Minster*, 1893, p. 88: 'Above the vestibule is an apartment, with fireplace, probably the residence of former keepers of the Minster, in it are the mouldings used by the mason and joiner in the restorations, and it would prove a happy hunting ground for the architectural student in pursuit of profiles of full sized mouldings.'

(4) For mediaeval methods, see J. H. Harvey, *The Gothic World*, 1950, pp.12-38.

(5) For the problems of setting out of mediaeval vaulting, see R. Willis, 'On the Construction of the Vaults of the Middle Ages', in *Transactions* of the Royal Institute of British Architects, I, pt. ii, 1842, pp. 1-69.

(6) A large collection of drawings, including many of those used during the 19th century, has recently been transferred from the office of the Clerk of Works to the Minster Library, where it is being sorted and listed.

(7) A set of photographs of the Wells floor was taken in 1953 by Mr. L. S. Colchester; and of much of the York floor in 1967 by Mr. John Bassham of the Royal Commission on Historical Monuments (England). I have to thank the Commissioners for permission to use these photographs (nos. YM.265-YM.281) as a basis for tracing, and to reproduce the general view (no .YM.265) as Plate I. For the final drawing reproduced as Fig. 1 I wish to thank Mr. A. R. Whittaker.

(8) The roof closely resembles that formerly over the Bedern Chapel, York, built in 1348 (see sections reproduced in F. Harrison, 'The Bedern Chapel, York', in *Yorkshire Archaeological Journal*, XXVII, part 106, 1923, p. 197ff.).

8

(9) *The Fabric Rolls of York Minster*, ed. J. Raine, Surtees Society, XXXV for 1858, 1859, p. 118. For tracing-houses and the meaning of the mediaeval Latin 'trasura' see J. H. Harvey, 'The Origin of the Perpendicular Style' in *Studies in Building History*, ed. E. M. Jope, 1962, pp. 164-5.

(10) Although there is some uncertainty as to the precise dates of succession, the series of master masons to the Minster is known from the start of the Nave in 1291 under Master Simon until the death of John Forman (master from 1523) in the summer of 1558. (For details of the careers of the masters see J. H. Harvey, *English Mediaeval Architects*, 1954). The later Fabric Rolls indicate only that certain masons were at work, but do not state explicitly the name of a master or leading mason until after the Restoration. For some of the later masons see G. W. O. Addleshaw, *Four Hundred Years*, 1962; and his annotated revision in *Architectural History*, X, 1967, pp. 89-119. Mr. C. B. L. Barr of the Minster Library has kindly communicated to me a list which he has compiled from original sources of the leading masons and clerks of the works of the Minster from 1667 to the present time.

(11) W. H. Dixon & J. Raine, *Fasti Eboracenses*, I, 1863, pp. 482-491.

(12) *Fabric Rolls of York Minster*, pp. 166-7, 180-1.

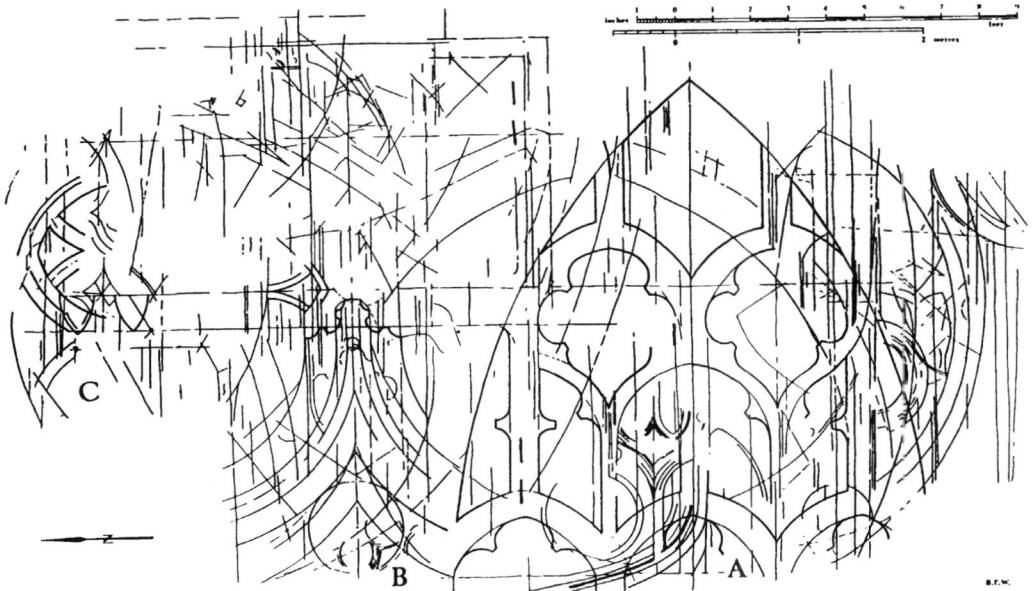

Fig. I. Drawings on the plaster tracing-floor, dating from *c.* 1360–*c.*1500
(drawing: A.R. Whittaker)

PLATE I   *Crown Copyright*

The Tracing House, looking south

The drawings are scratched on the surface of the hard plaster floor. Note the bundles of templates, showing profiles of mouldings, hanging on walls. The rear wall includes part of the earlier gable of the north transept of c. 1225-50.

Reproduced by permission of H.M. Stationery Office and the Royal Commission on Historical Monuments (England).

6

Late Gothic structural design in the 'instructions' of Lorenz Lechler

Lon R. Shelby and R. Mark

Gothic cathedrals and churches represent one of the most extraordinary periods of structural design in the history of Western architecture. Yet with all the scholarly attention which these monuments have received since the first half of the nineteenth century, there is still much to learn about the structural character of Gothic building. On the one hand, historical scholars, architects, and engineers have not reached full agreement on how the structural elements function in a Gothic church [1]. On the other hand, we know little of what medieval builders themselves thought about the structural character of their buildings. This is in large part because architects–the master masons and master carpenters–left few written or graphic records of their design techniques, at least until the latter half of the fifteenth century.

In this paper a historian and an engineer collaborate to interpret one of the few literary sources from the hand of a medieval master mason which deals with Gothic structural design, namely, the "Instructions", written in 1516 by Lorenz Lechler for his son, Moritz. Our procedure will be to summarize and analyze passages in the booklet in order to ascertain what Lechler thought about certain structural problems, and simultaneously, to provide a critique of some of his ideas from the perspective of modern structural engineering.

Lechler was a late fifteenth-century German master mason whose career developed in the Neckar valley, and particularly at Heidelberg [2]. The "Instructions" distilled much of the technical knowledge about Gothic building which Lechler accumulated during his career. Unfortunately, this small book was not put into print at the time of writing, and Lechler's original manuscript copy has disappeared. However, two late sixteenth-century manuscript copies have survived. One of these, by Jacob Feucht von Andernach, was printed in the nineteenth century [3]. The other copy, which has only very recently come to light, was made around 1600 by an anonymous copyist in the Swiss town of

---

[1] For a convenient bibliographical survey and analysis of this literature before 1960, see Paul Frankl, *The Gothic: Literary Sources and Interpretations through Eight Centuries*, Princeton, 1960. Since 1960 several engineers have applied to these questions modern engineering analysis, including 1) plastic design techniques and wind tunnel test data developed for the design of steel frame buildings; and 2) two and three-dimensional photoelastic model analysis and computer simulations, usually applied to the design of complex machine parts. For examples, see Jacques Heyman, "The Stone Skeleton," *International Journal of Solids and Structures*, II, 1966, pp. 249–279; idem, "Spires and Fan Vaults," *ibid.*, III, 1967, pp. 243–257; idem, "On the Rubber Vaults of the Middle Ages and Other Matters," *Gazette des beaux-arts*, Ser. 6, LXXI, 1968, pp. 177–188; idem, "Beauvais Cathedral," *Transactions of the Newcomen Society*, XI, 1967–68, pp. 15–35; Robert Mark and Richard A. Prentke, "Model Analysis of Gothic Structure," *Journal of the Society of Architectural Historians*, XXVII, 1968, pp. 44–48; Robert

Mark and Ronald S. Jonash, "Wind Loading on Gothic Structure," *ibid.*, XXIX, 1970, pp. 222–230; Lutz Kübler, "Computeranalyse der Statik zweier gotischen Kathedralen," *Architectura*, IV, 1974, pp. 97–111; Maury I. Wolfe and Robert Mark, "The Collapse of the Vaults of Beauvais Cathedral in 1284," *Speculum*, LI, 1976, pp. 462–476; and K. D. Alexander, Robert Mark, and John F. Abel, "The Structural Behavior of Medieval Ribbed Vaulting," *Journal of the Society of Architectural Historians*, XXXVI, 1977, pp. 241–251. For other approaches to the analysis of Gothic cathedrals, see Martin Grassnick, *Die gotischen Wölbungen des Domes zu Xanten und ihre Wiederherstellung nach 1945*, Xanten, 1963; and Jürgen Segger, *Zur Statik gotischer Kathedralen: Dargestellt am Kölner Dom und statisch verwandten Kathedralen*, Diss. Aachen, 1969.

[2] On Lechler's career, see Annaliese Seeliger-Zeiss, *Lorenz Lechler von Heidelberg und sein Umkreis*, „Heidelberger Kunstgeschichtliche Abhandlungen", N.F., Bd. 10, Heidelberg, 1967.

Bern [4]. The two copies have many parallel passages, but there are differences in terminology, word order, and technical details of design.

Apparently both copyists took liberties in transcribing Lechler's text, so that occasionally there are problems in determining what Lechler intended in his original version of the booklet. But it is evident from these copies that the original booklet was very awkwardly organized. Lechler moved back and forth from one topic to another, often without transitional sentences to let his reader know that he was changing topics. Thus we should warn our readers that the systematic presentation of Lechler's ideas in this paper does not reflect the character of the booklet itself. Furthermore, Lechler occasionally contradicted himself when he returned to a formula or rule that he had previously mentioned.

Another problem in explicating Lechler's booklet is that he made little attempt to identify the types of structures for which he was providing design rules. He used a technical vocabulary, as we shall illustrate, but this pertained only to individual elements of the building and not to classifications of building types. Lechler mentioned more than once that he intended to discuss design problems of military fortifications—and his working experiences at Heidelberg castle would have qualified him to do so. However, the surviving copies of his text do not include any discussion of fortifications; they contain only Lechler's rules for the design of churches and their interior fittings. It was natural for Lechler to have been concerned with the latter, for two of his known works are the tabernacle and choir screen which he designed and built in the Church of St. Dionysius in Esslingen during the period, 1486–89. He was also deeply involved in the design and construction of the "Mount of Olives" monument at Speyer Cathedral from 1504 to 1511 [5]. Unfortunately, he has not been identified with the design of an entire church, so that we cannot turn to one of Lechler's own buildings for assistance in explicating his rules on church design.

In spite of Lechler's lack of terminology for building types, one can determine from a careful reading of his booklet that he had in mind the hall church of the Late Gothic tradition of southern Germany. The structural system of the hall church differed significantly from that of the basilican church; in the latter, the side aisles were lower than the central vessel of nave and choir, and this permitted a clerestory of windows to be placed in the central vessel. The vaults over the central vessel of the basilican church created special buttressing problems that were solved towards the end of the twelfth century with the introduction of flying buttresses which leaped across the side aisles to carry the thrust of the central vaults to the outer buttresses. In the hall church, by contrast, the elevation of the aisle vaults was the same or close to the elevation of the vaults of the central vessel, thereby eliminating the clerestory of windows. That made it possible for the aisle vaults to buttress directly the central vaults. The aisle vaults then had to be supported at their outer extremity by buttresses placed at points of thrust concentration. However, in the particular hall church tradition within which Lechler worked, the choir was a single-vessel structure with no side aisles; hence the choir vault could be supported directly by external buttresses placed at right angles to the choir wall.

---

[3] Feucht's copy is preserved in Cologne, Historisches Archiv, Handschrift Wf. 276*. It was printed in an uncritical edition, with transliterated text, by August Reichensperger, *Vermischte Schriften über christliche Kunst*, Leipzig, 1856, pp. 133–155.

[4] The anonymous copy appeared in the manuscript marketplace in 1975 and was sold by H. P. Kraus in 1976 to the Heidelberg Universitätsbibliothek, where it is now catalogued under the signature, Hs. 3858. In this paper we shall refer to these manscripts as the Cologne and Heidelberg copies, respectively. We shall not attempt here to sort out the many technical problems pertaining to these copies and to their relationship with each other. These problems will be fully explored in a critical edition of Lechler's booklet which Shelby is preparing. For reference purposes in this paper we will first cite the page number in Reichensperger's edition and then the folio number of the Cologne copy on which this edition was based. Parallel texts, where they exist in the Heidelberg copy, will be cited by page number.

[5] For description and analysis of Lechler's work at Hei-

In Lechler's design scheme the nave did have side aisles, and he gave some attention to the structural problems of buttressing the central nave vault with the aisle vaults. But the essential character of Lechler's structural design technique can be found in details of his choir design. Hence we shall further limit this paper to a historical and critical analysis of Lechler's structural rules for designing the single-vessel choir of a German Late Gothic hall church.

Design Principles

We shall begin by summarizing the basic tenets which underlay Lechler's design rules. He mentioned each of these at some point in his booklet, although he did not systematically set them forth. First, design would be influenced by the quality of the stone to be used. If it was good stone, the dimensions of the structural elements could be reduced. If it was lower-quality stone, then dimensions had to be increased [6]. Thus one of the first steps in the design of a church was a qualitative assessment by the master mason of the stone to be used in the building.

Second, dimensions were affected by the choice of scale. For Lechler, this choice was between what he called the "Old Foot" and the "New Foot", i.e., between the larger scale of construction used in former times, and a smaller scale generally preferred in his own day. Lechler provided formulas for transposing designs from one scale to another [7]. As he pointed out, the "Old Foot" gave larger dimensions and therefore required more stone than did the use of the "New Foot". But this was still a matter of choice which the master mason had to make at the beginning of the project. As will be shown, the dimensions of individual architectural elements were affected throughout by this choice.

The third principle was that some design problems required judgmental decisions on the part of the master mason. While Lechler provided many design rules, he repeatedly remarked that the designer must not rely merely on rules, but must make certain judgmental decisions on the strength of his

overall knowledge and accumulated building wisdom. Lechler wrote to his son, "Give to this writing careful attention, just as I have written it for you. However, it is not written in such a way that you should follow it in all things. For [in] whatever seems to you that it can be better, then it is better, according to your own good thinking [8]."

The fourth and most important principle was to design a building that would maintain its structural integrity. Several times Lechler posed the issue in these terms: there is a design decision to be made; make it, and if the building does not fall, then you know you made the right decision. In the prologue to his booklet Lechler exhorted his son, "Therefore, if you give proper attention to my teaching, you can meet the needs of your building patron and yourself, and not be despised as the ignorant are, for an honorable work glorifies its master, if it stands up [9]."

Finally, it must be observed that Lechler based much of his design technique on what we have previously labelled the constructive geometry of the medieval masons [10]. Although Lechler himself did

delberg, Esslingen, and Speyer, see Seeliger-Zeiss, *Lorenz Lechler*, pp. 31–72, 87–137, and 144–157.
[6] Reichensperger, *Schriften*, pp. 133, 135; Cologne, Wf. 276*, fols. 43ᵛ, 44ᵛ; Heidelberg, Hs. 3858, p. 2.
[7] Reichensperger, *Schriften*, p. 146; Cologne, Wf. 276*, fol. 51.
[8] Reichensperger, *Schriften*, p. 137; Cologne, Wf. 276*, fol. 45: „... vnd kom disem schreiben fein Vleissig nach / wie ich dirs darvir geschriben hab / aber es ist nit darumb geschriben / daß du ihm ir allem Volgen solst / dan waß dich besser tungt / daß es besser sein khan / so bessers / nach deinem gueten gedunken ess ist einem Iedem nuz / wan er etwaß khan / Vnd waiß Zuebrauchen."
[9] Reichensperger, *Schriften*, p. 133; Cologne, Wf. 276*, fol. 43: „Darumben so Nemedt aigendtlichen Achtung / auf mein Lehr / so mögendt Ihr Eurem Bauherrn / Vnd euch Versorgen / Vnd nit Veracht werden alß die Vnwissenden / dan ein Ehrlich Werckh / lobet seinen Meister / so Es stedt." Cf. Heidelberg, Hs. 3858, p. 1: „... der maßen so gründtlich vnd eigentlich ahnzeigen vnd berichten will / daß wann Ir demselben allem fleissig nachsetzen vnd gebrauchen werdet / Ir nicht für Stümpler gehaltten werdet: Dann ein Ehrlich werck lobet seinen Meister."

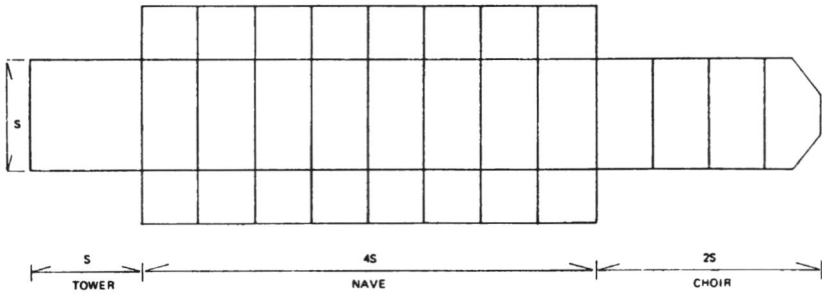

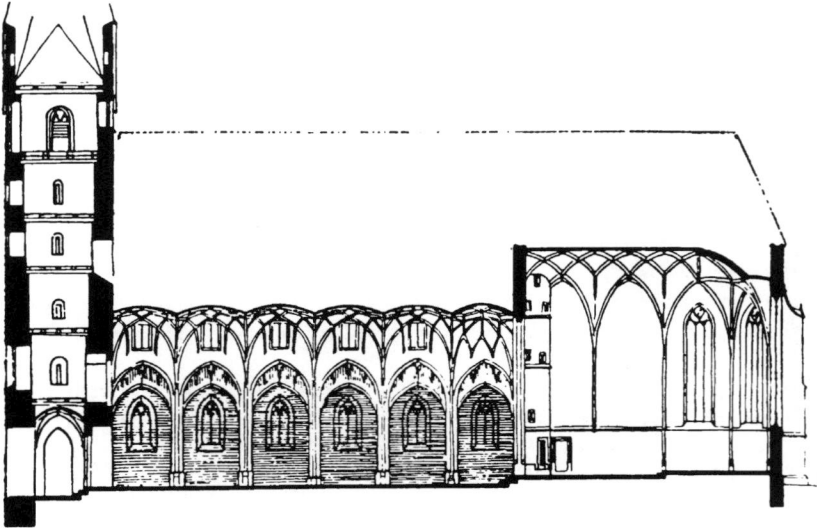

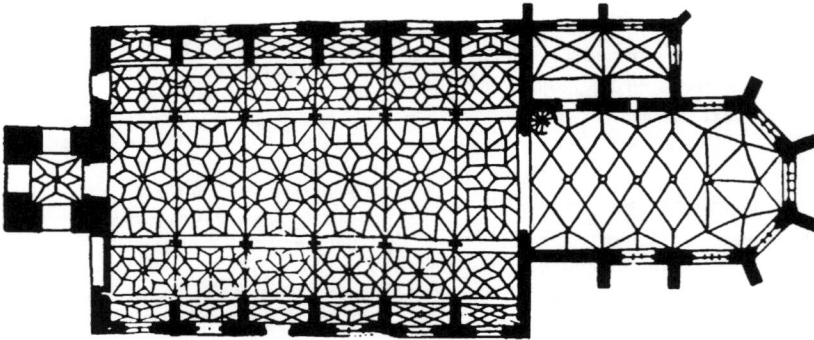

1. Schematic Plan of Lechler's Design for a Hall Church

2. Groundplan and Longitudinal Section of the Alexanderkirche, Marbach. (Hans Koepf, Die Baukunst der Spätgotik in Schwaben, Stuttgart, 1958. Abb. 21 and 22.)

116

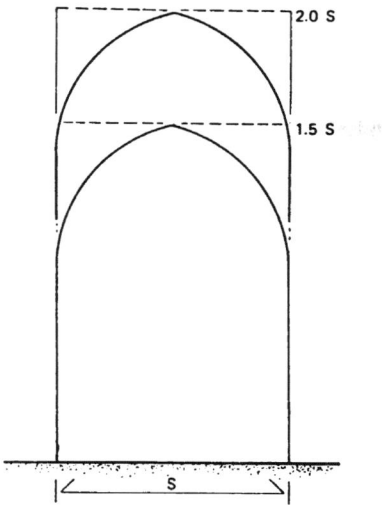

3. Schematic Cross-section of Lechler's Alternate Designs for a Hall Church Choir

not use the word geometry, his earlier contemporaries, Roriczer and Schmuttermayer, referred to the "art of geometry" used by the masons [11]. This was not Euclidean geometry; it did not require mathematical reasoning to move through the geometrical processes. Instead, it was a matter of constructing geometrical figures such as circles, squares, triangles, and octagons. These figures were then manipulated through a series of arbitrarily determined steps to produce a network of points and lines which served as a geometrical frame for the architectural elements being designed by the masons. It was in this sense that the masons used geometrical figures to "construct" architectural elements.

Building Configuration

We may now turn to the specifics of Lechler's technique for designing German Late Gothic hall church choirs. The first step consisted of setting the basic dimensions of the ground plan and elevations. The primary decision was that of determining the width of the choir, since this provided what we shall call

the macro-module of Lechler's design scheme. The choir could be set to whatever width the designer preferred, though Lechler recommended that it be either twenty or thirty feet wide [12]. Many of the architectural elements of the church were ultimately derived from this macro-module. For example, the lengths of both the choir and the nave were functions of the choir width. Lechler stated that the choir should be twice as long as it was wide, and the nave was to be twice the length of the choir [13]. The width of the nave was to be the same as that of the choir, and the nave aisles were to be half the width of the nave itself [14]. Although Lechler provided no overall ground plan in his illustrations, with these simple ratios we may easily reconstruct the basic geometric framework of his design, as shown in Fig. 1. In this diagrammatic plan we have added the outline of the west tower, which was a characteristic feature of the type of church that Lechler was describing. For example, this geometric frame can be compared with the plan of the Alexanderkirche in Marbach [15]. (Fig. 2)

[10] Lon R. Shelby, "The Geometrical Knowledge of Mediaeval Master Masons", *Speculum*, XLVII, 1972, pp. 409–410.
[11] For the texts, see *Gothic Design Techniques: The Fifteenth-Century Design Booklets of Mathes Roriczer and Hanns Schmuttermayer*, Edited, Translated, and Introduced by Lon R. Shelby, Carbondale, Ill., 1977, pp. 82, 126; for general discussion of this phrase, see pp. 61–79.
[12] Reichensperger, *Schriften*, p. 134; Cologne, Wf. 276*, fol. 43ᵛ; Heidelberg, Hs. 3858, p. 2.
[13] Reichensperger, *Schriften*, p. 153; Cologne, Wf. 276*, fols. 55, 55ᵛ. Elsewhere Lechler indicated that the choir length should be three times its width: Reichensperger, *Schriften*, p. 152; Cologne, Wf. 276*, fol. 54ᵛ. Perhaps he intended the designer to have this option for the length of the choir. However, this would have produced a very long nave by doubling a three-bayed choir. In sorting out Lechler's formulas, Paul Booz, *Der Baumeister der Gotik*, Munich, 1956, p. 45, showed Lechler's groundplan with a choir length twice its width, and with the nave twice the length of the choir.
[14] Reichensperger, *Schriften*, pp. 152–154; Cologne, Wf. 276*, fols. 54ᵛ, 55ᵛ.
[15] The groundplans of many Late Gothic churches in Swabia, the Kurpfalz, and the Middle Rhineland show

The height of the choir was also a function of its width, although it is not entirely clear what Lechler intended on this point. Early in his booklet he provided a ratio of height to width, and the Heidelberg text has his rule read in the following way: "Item, a choir that is twenty feet wide should be one and a half times as high; that is its correct height. Yet not all choirs [have this height], for sometimes the church patrons do not have this disposition, so that they make it as they please, or as it must be done in that locale. But a workable height should be twice as high as the choir is wide[16]." (Fig. 3)

The Cologne text provides the same ratios, but the wording is different and rather more confusing than in the Heidelberg copy[17]. Elsewhere in the Cologne text there is a flat statement that "as wide as the choir is, twice as long should one make it; and as wide as the choir is, twice as high should one make it[18]." In yet another passage in the Cologne copy Lechler gives quite a different rule: "Item, whoever wants to make a choir and give it the correct height should know more than one height, for there are three heights. The first height is one and a half times the width of the choir in the clear; it should be this high up to the tas-de-charge. The other height should be twice as high as the width of the choir in the clear. The third height is for the choir to be three times as high up to the tas-de-charge as its width in the clear[19]."

What are we to make of this confusion? First, one can seriously question the latter rule which would call for the height of the choir to be three times its width. Hall churches with single-vessel choirs with which Lechler would have been familiar simply did not have ratios of that magnitude. Something is wrong with the text, or else Lechler was thinking only hypothetically. Secondly, it is not clear what was Lechler's reference point for determining the height of the choir. In the last passage quoted he made it the tas-de-charge; this could

have been used, although the tas-de-charge was a structural element in the vault that could not readily be discerned once the vault was completed. More likely he was referring to the highest keystone in the vault. This was a readily identifiable reference point for the height-to-width ratio. Furthermore, a number of Late Gothic hall church choirs have a 2 : 1 or 1.5 :1 ratio, as measured to the keystone of the vault[20].

Wall and Buttress Design

The structural significance of the choir width is underscored by the fact that it determined the thickness of the choir wall. Lechler recommended that the wall thickness carry a ratio to the choir width of 1 : 10. However, he expressed this only

that Lechler's geometric frame was commonly employed in the fifteenth century. The aisleless choir with a length twice its width was particularly common, as was the practice of making the nave aisles one-half the width of the nave. The length of the nave was

more variable, although Lechler's rule for making it twice the length of the choir was not uncommon. Interestingly, these ratios were applied whether it was a hall church or a basilican church. For groundplans, see Georg Dehio and Gustav von Bezold, *Die kirchliche Baukunst des Abendlandes, Atlas*, Stuttgart, 1901, V, Taf. 446–457; Hans Koepf, „Die Baukunst der Spätgotik in Schwaben", *Zeitschrift für Württembergische Landesgeschichte*, XVII, 1958, pp. 1–144; *idem, Schwäbische Kunstgeschichte*, Bd. 2, *Baukunst der Gotik*, Constance and Stuttgart, 1961, pp. 8–27; and Friedhelm W. Fischer, *Die spätgotische Kirchenbaukunst am Mittelrhein, 1410–1520*, „Heidelberger Kunstgeschichtliche Abhandlungen", N.F., Bd. 7, Heidelberg, 1962, *passim.*

[16] Heidelberg, Hs. 3858, p. 2: „Item ein Cohr / der 20 Schuch wyt ist / sol Anderthalb mal so hoch sin / das ist sin rechte höch / doch nichs alle Cohr / dann zu zyten die kirchen Buwherren nicht gelegenheit haben / so machen sie es nach ihrem gefallen / nach dem es ouch an eim ortt gemacht muß werden / Aber ein werckliche höch einer kirchen soll Zwey mal so hoch sin / alß der Cohr wyt ist."

[17] Reichensperger, *Schriften*, p. 134; Cologne, Wf. 276*, fol. 43ᵛ: „Item ein khor der 20 Schuech weydt ist / im Liecht / der sol Anderthalben mal so hoch sein / alß weitt er ist / daß ist seine rechte höche / doch nit alle khor / dan Zu Zeiten die Pauherrn / der kirchen gelegenheit / bauen wie sie wollen / aber im Werckliche höhe / solt Zweimal so hoch sein / als weit Er ist."

[18] Reichensperger, *Schriften*, p. 153; Cologne, Wf. 276*, fol. 55: „Vnd alß weit der khor ist / Zweymal so lang

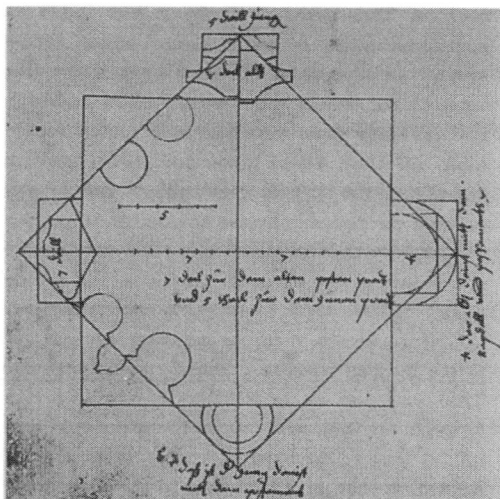

4. *Rotation of Equal Squares in Lechler's "Instructions".* (Cologne, Historisches Archiv, Hs. Wf. 276*, fol. 42.)

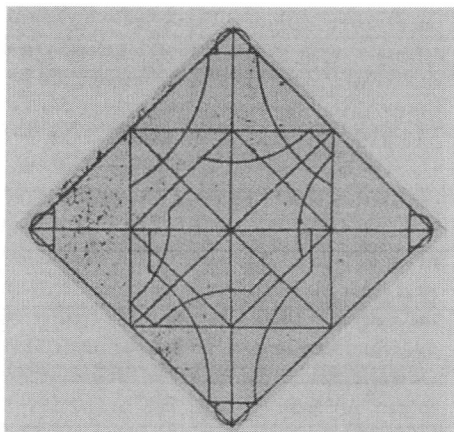

5. *Rotatiton of Halved Squares in Lechler's "Instructions". (Cologne, Historisches Archiv, Hs. Wf. 276*, fol. 42ᵛ.)*

by saying that a twenty-foot wide choir needs a wall two feet thick, and a thirty-foot wide choir should have a wall three feet thick. But within these numerical ratios, Lechler inserted still another factor which was to be based on the quality of the available building stone. With really good stone, one subtracted three inches from the width of the wall; with weak stone, one added three inches[21]. These rules for determining the choir wall thickness take on even greater importance when it is perceived that the wall width served as a "micro-module" for designing other structural elements. Lechler said to take the width of the wall for one side of a square, which he then manipulated by means of "constructive geometry" in two different ways. One was to rotate the square 45° to produce two squares which were diagonally placed over the corners of each other. (Fig. 4) The second manipulation was to inscribe successively smaller squares within the basic square. (Fig. 5) From these two geometric grids he then derived the templates that determined the forms and dimensions of the vault ribs and shafts, as well as window jambs and mullions and other molded architectural elements of the building[22].

Lechler used the micro-module of the wall width in determining the most crucial element in the structural system of the choir, namely, the external buttresses; but unfortunately his intent is not entirely clear. Since the choir had no aisles, the vaults rested directly on the choir walls, which would have had external buttresses set at right angles to provide

soll er ihn machen / Vnd auch so weit der khor ist / Zweymal so hoch soll er ihn machen."

[19] Reichensperger, *Schriften*, p. 152; Cologne, Wf. 276*, fol. 54ᵛ: „Item wer ein khor machen will / der soll in seine rechte höch geben / auch sol er wissen mehr / den einerley höch / das sindt dreyerley höch / die erste höch ist als weit der khor ist / Im liecht Anderthalb mal / so hoch / soll er sein / biß an daß haubt / die Ander höch ist / so weit der Chor ist / ihm liecht / Zweimal so hoch soll er sein / sie dritte höch ist / so weit der khor ist / Im liecht / dreymal so hoch er sein / biß auf das haubt."

[20] For elevational drawings of Late Gothic hall churches contemporary with Lechler, see the works cited in note 15.

[21] Reichensperger, *Schriften*, pp. 133–134; Cologne, Wf. 276*, fol. 43ᵛ; Heidelberg, Hs. 3858, p. 2.

[22] For a detailed analysis of these procedures, see Lon R. Shelby, "Mediaeval Masons' Templates", *Journal of the Society of Architectural Historians*, XXX, 1971, pp. 147–154.

the necessary stability. Lechler gave a great deal of attention to the design of these buttresses; however, on a critical point there is a contradiction between the two surviving texts of his booklet. The Cologne text states that for a choir with a width of thirty feet, the walls should be three feet thick; the buttress section should be three feet and two (inches?) in width above the ground table, and it should be twice as long as it is wide [23]. The Heidelberg text states that the walls should be three feet thick, and the buttresses should be two and a half feet wide and five feet long [24]. Although Lechler returned several times to the matter of designing the buttresses, and later repeated that the buttress was to be twice as long as it was wide, nowhere else in either copy of his booklet is there a specific formula for determining the width of the buttress. The Cologne version suggests a simple relationship between the buttress dimensions and the thickness of the choir wall, i.e., 1 : 1 buttress width to wall thickness; 2 : 1 buttress length to wall thickness. On the other hand, some of the actual Late Gothic single-aisled choirs of Lechler's time show buttress dimensions more consistent with the rule in the Heidelberg copy. For example, the Alexanderkirche in Marbach has a choir width of thirty feet, with buttresses that are thirty inches wide and sixty-five inches long above the ground table. Marbach is on the Neckar River not far from Esslingen, and there is good reason to think that Lechler may have had this particular church in mind when he wrote his booklet [25].

While we cannot be categorical on the matter, we favor the Heidelberg version, even though this leaves Lechler without a rule for choir walls which had widths other than three feet. Other elements of the buttress also lacked specific rules for deciding design questions which Lechler considered important. We have been discussing the dimensions of a horizontal cross-section of the buttress just above the ground table. But the entire vertical profile of the buttress is equally important in determining the mass of the buttress, and hence its ability to resist vault thrusts. It was the practice to reduce the buttress section by setbacks as the buttress rose to the

roofline. These setbacks occurred just above the horizontal tables or stringcourses which carried around the buttress at the same level where they were placed on the choir wall. Lechler indicated that normally there were four such tables: ground table, sill table, water table, and corbel table, in addition to the buttress skew table [26]. But he gave no rules for determining the amount of setback; on the contrary, he explained to his son that this required judgmental decisions on the part of the designer: "If you want to set out a choir with good carved stonework, then use measurement and give to the buttress its correct length, as beforestated. Then you may set back above the ground table however far [you wish] up to the springer or capital, as long as the buttress maintains its strength. Regarding that, give attention to the divisions of the buttress; for that which is above the springer or capital you may take whatever you think will stand up well. Thus have I heretofore described several buttresses, so that you may take that which is useful to you for your work and understanding of this technique [27]."

[23] Reichensperger, *Schriften*, p. 134; Cologne, Wf. 276*, fol. 43ᵛ: „Item Ein khor 30 Schuech weidt / 3 Schuech dickh / die mauren / aber mit den Pfeillern / den mag 3 schuech dickh 2 ob dem schreggesimbß / Vnd alß dickh der Pfeiller ist / Zweimal alß Lang sol er sein."

[24] Heidelberg, Hs. 3858, p. 2: „Item ein Chor / das dryssig Schuch wyt ist / so mach die mauren 3 schuch dick / Aber den Pfyler mach drithalben schuch dick / ob dem Schreg gesims / vnd alß dick der Pfyler ist / also mustu zwey mahlen so lang von der muren machen / das thut fünff werck schuch soll er sin."

[25] It is known from masons' marks on the piers of the nave that Caspar Lechler, a relative of Lorenz Lechler, worked at the church. See Hans Koepf, *Die Alexanderkirche in Marbach am Neckar*, Neuauflage, Marbach am Neckar, 1972, pp. 18–19; and Seeliger-Zeiss, *Lorenz Lechler*, pp. 158–159. There is no evidence that Lorenz himself worked there, but his overall design fits Marbach in several ways, particularly regarding the nave, which is a *Staffelhallkirche* of the type that Lechler describes in his discussion of the nave and aisle vaults.

[26] Reichensperger, *Schriften*, p. 134; Heidelberg, Hs. 3858, p. 3; Cologne, Wf. 276*, fol. 43ᵛ: „Item ein schlechter khor / der bedarf nicht mehr / den 4 gesimbß / schreg-

It is not surprising that Lechler had no precise means of defining the mass required in the buttress to prevent the vault thrusts from overturning it. Instead, he relied on a general rule of thumb for proportioning the buttresses relative to the width of the vault. His rule did not take into account the height of the vault, and we have seen how this might vary relative to the width. However, his failure to account for the differences in the height of the vault was not so irrational as might at first appear. Certainly as the vault height increased, the overturning forces caused by the vault thrust against the buttressing system increased in direct proportion to the height. But the mass of the buttress system (i.e., the wall and adjacent wall buttress) was also proportional to its height. Since both the overturning forces and resisting masses increased proportionately, a rule proportioning the buttressing to the vault span was a sensible way to deal with the thrust of the vault itself [28].

In determining the buttress dimensions Lechler did not provide a rule for setting the length of the vault bays, i.e., the distance between the centerpoints of the buttresses. Perhaps this was because Late Gothic builders varied bay lengths considerably in relation to the vault span. Nevertheless, a review of the plans of fifteenth-century single-aisled choirs built in the Neckar region suggests that a rule of thumb may have been operative in this area, namely, to make the length of the bay half the width of the choir. Again a good example is the Alexanderkirche in Marbach, where the length of the choir is twice its width, and the length of the vault bay is half of the choir width [29]. (Fig. 2)

Another important omission in Lechler's rules for buttress design was his lack of comment on the size of the roof, or the effect of high winds against the massive roofs that covered hall churches. Analytical studies of High Gothic cathedrals have shown that high wind forces against the roofs could be considerable [30]. To be sure, Lechler was concerned with buildings of lower height than these cathedrals, so that the effect of wind would have been less severe. Nevertheless, Lechler's failure to take into account roof size or wind forces was a curious omission in

his design process [31]. This underscores the point that his design formulas were rules of thumb based on experience of what would work, rather than analytical formulas taking into account all of the factors which affected the structural integrity of the building.

One other factor affecting the wall as a support for the vault was the size of the window openings. On this matter Lechler provided a rule that was quite precise: "Divide the space between the but-

gesimbß / Vnd kapgesimbß / Tragesimbß / vnd taggesimbß vnd Pfeiller Tagung."

[27] Reichensperger, *Schriften,* p. 137; Cologne, Wf. 276*, fol. 45ᵛ; Heidelberg, Hs. 3858, p. 6: „So du einen Chor ahnlegen wilt / von gutem gehauwen Steinwercks / so bruch du die maß / vnd laß dem Pfyler sin rechte lenge / wie vorstaht / doch so magstu Ihm abbrechen / ob dem Schregs gesims / so ferr / das der Pfyler sin Stercke behaltte / bis zu dem Anfang / oder Capenthäl / vf das / so hab merckung mit den Abkleydung der Pfyler / dann waß ob dem Anfang oder Capengesims ist / dann magst nemen / was dich dunckt vnd wol stahs / Darumb so hab ich etliche Pfyler hiefornen verzeichnet / darundter du nemen magst / der dir dienstlich ist / zu diner Arbeit / vnd wolstandt Jetziger Art."

[28] Assuming similar geometry, the vault thrust would have been proportional to the vault mass, i.e., to the third-power of the proportioned vault span dimension. The overturning movement to be resisted by the buttress was equal to the product of the vault thrust and the vault springing height; hence it was proportional to the fourth-power of the span module. Buttress resistance against overturning was also a function of the product of its mass and some fraction of the wall-buttress base dimension, and this too has a fourth-power relationship to the span module. Hence equilibrium of the vault-buttress system would not be dependent upon scale.

[29] Cf. the plan of the choir of the Leonhardskirche in Stuttgart: Koepf, „Baukunst der Spätgotik in Schwaben", p. 52.

[30] Mark and Jonash, "Wind Loading on Gothic Structure," pp. 222–230.

[31] Greater architects could get into difficulty on this score. The original Parler family design for the huge hall-church plan for the nave at Ulm Münster had to be modified into a basilican design in part, at least, because of the problems of constructing a roof that would successfully span the extra-wide nave and aisles of the proposed hall church: see Reinhard Wort-

tresses into five equal parts; give three parts to the window, and two parts to the wall on either side of the window [32]." It is interesting that Lechler expressed the window size as a fixed ratio rather than as a fixed dimension, for in this way he seemed to be assuring that whatever the bay length, or actual dimensions between the buttresses, the relative amount of solid wall working with the buttresses to support the vault would remain the same.

Summarizing Lechler's techniques for wall and buttress design, if we assume a bay-length rule (which, admittedly, he did not express) to the effect that the length of the bay was to be one-half the width of the choir, then the entire wall and buttress system of the choir can be expressed as a function of its vault span. To illustrate his scheme, we have constructed in Fig. 6 a wall segment based on a choir width of thirty feet, with a choir length of sixty feet that is divided into four bays of fifteen feet each. The dimensions of the wall and buttress mass and the window openings can be easily read from the drawing. It is interesting to note that this scheme produced a choir wall in which the wall masses and the window spaces were about equally divided into thirds. This same ratio would have been maintained with a choir width of twenty feet where all subsequent dimensions were generated out of that module.

Vault Design

While Lechler's career came at the culmination of complex designs for the vault rib patterns of German *Spätgotik*, he seems not to have had (or at least to have used) technical terms to distinguish the different types of rib patterns which modern German scholars call the star vault *(Sterngewölbe)*, net vault *(Netzgewölbe)*, winding rib vault *(Schlingrippengewölbe)*, and faceted-star vault *(Schleifensterngewölbe)*, amongst others [33]. Instead, Lechler simply used one word, *reiung*, to refer to the "ground plan" of the rib patterns in each vault bay. It was from the *reiung* that one projected the rib patterns, whatever may have been their configuration.

Although Lechler did not use special terms to refer to vault types, he did possess a fairly extensive terminology of Gothic vault components, and his terms correlate rather closely with modern usage. A review of these terms (using the spelling of the Cologne text) will give some idea of his verbal conception of the Gothic vault and how it worked. The vault *(gewelb)* rested on the wall *(mauer)* or pier *(pfeiller* or *seillen)*. A vaulting shaft *(dienst)* could begin at a corbel *(krackhstein)* and extend up the wall, or it could be attached to a pier, beginning at the base *(postament)* of the pier and extending upward to the springer *(anfang)* of the vault. At the springer, the arches or ribs *(bogen)* of the vault were bunched together and carved from one or more stones whose beds were horizontal. As the arches and ribs extended upward from the springer, they separated into their various arcs at the tas-de-charge *(haubt)*. Lechler mentioned only two particular types of ribs. By *creutzbogen* he meant cross-rib, diagonal rib, or more generally, any of the ribs within each bay of the vault. He indicated that the

mann, „Hallenplan and Basilikabau der Parler in Ulm", *600 Jahre Ulm Münster Festschrift*, eds. H. E. Specker and Reinhard Wortmann, „Forschungen zur Geschichte der Stadt Ulm", Bd. 19, Ulm, 1977, pp. 117–122.

[32] Reichensperger, *Schriften*, p. 135; Cologne, Wf. 276ᵛ, fol. 44: „Item die weite des fensters / solstu teillen / Zwischen den Pfeillern in dem khor / In funff teillen / Vnd nimb drey teill / Zum Liecht Im fenster / mit sambt den Pfosten / die Vbrige Zwey teill / Zu dem gewengen das fensters."

[33] For succinct definitions, often accompanied by illustrations, of many of the terms used in this paper, see Hans Koepf, *Bildwörterbuch der Architektur*, „Kröners Taschenausgabe", Bd. 194, Stuttgart, 1968. Heinrich Otte, *Archäologisches Wörterbuch zur Erklärung der in den Schriften über mittelalterliche Kunst vorkommenden Kunstausdrücke*, Leipzig, 1857, is still useful, particularly because Otte provided French, English, and Latin equivalents of the German terminology. For the historical development of the vault types, see Konrad W. Schulze, *Die Gewölbesysteme im spätgotischen Kirchenbau in Schwaben von 1450–1520*, Diss. Tübingen, Reutlingen, 1939; now to a large extent superseded by the more comprehensive study of Karl H. Clasen, *Deutsche Gewölbe der Spätgotik*, Berlin, 1961.

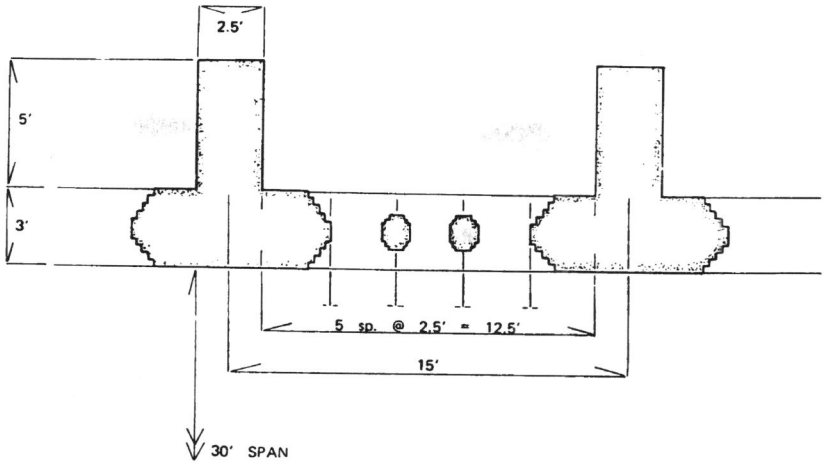

6. Choir Wall Segment Based on Lechler's "Instructions"

creutzbogen could be either curved or straight *(ge-wunden oder scheitrecht)* and thus his term included not only the diagonal, but what we today call the tierceron *(Flechtrippe)*, lierne *(Scheitelrippe)*, and ridge rib (again, *Scheitelrippe)*. By *scheitbogen* he apparently meant the transverse arch or rib that longitudinally separated one bay of the vault from another, although in modern usage *Scheidebogen* generally means the rib or pier arch that separates the nave bay from the adjacent aisle bay. Finally, the arches and ribs of the vault reached their apex in the keystone *(scloßstein)*.

Having outlined Lechler's vault terminology, we may now offer a translation from one of the longer passages in his booklet where he deals with the technique of designing ribs. Besides providing a text for comment, this will also indicate the flavor of Lechler's descriptive style: "Here I want to give you a report so that you can easily understand how you should determine the crossribs in the rib plan, whether they be curved or straight. In the first place, when you have traced out the ribs as they belong in the rib plan, then carefully draw the springers and the keystones of the ribs however large they should be. Wherever a crossrib is called for, whether it is

curving or straight, you look to the groundplan, for one rib plan is not like another regarding the cross-ribs. Where a straight crossrib is called for, you need not draw it otherwise in the rib, for [the rib plan] gives you the length of the rib itself between the springer and the keystone. However long the crossrib is between [the springer and] the keystones, just so long must you make it, and no shorter. Thus you have understood how you should make the straight crossrib. However, when you want to make a curving crossrib, notice where it comes to—that is to say, between the keystones or springers—then you must draw a template between them. You must make the curving crossrib according to this template. When you have drawn this template between them—that is to say, between the keystones or the springers—then you must terminate the same template at each end along its length. After that you must again draw a special template at the termination. Thereto you must terminate the curving cross-rib on its length, because the rib plan will be laid out with small divisions. However, when the rib plan is not laid out with small divisions, then no termination will be needed. Thus you have understood my meaning [34]."

123

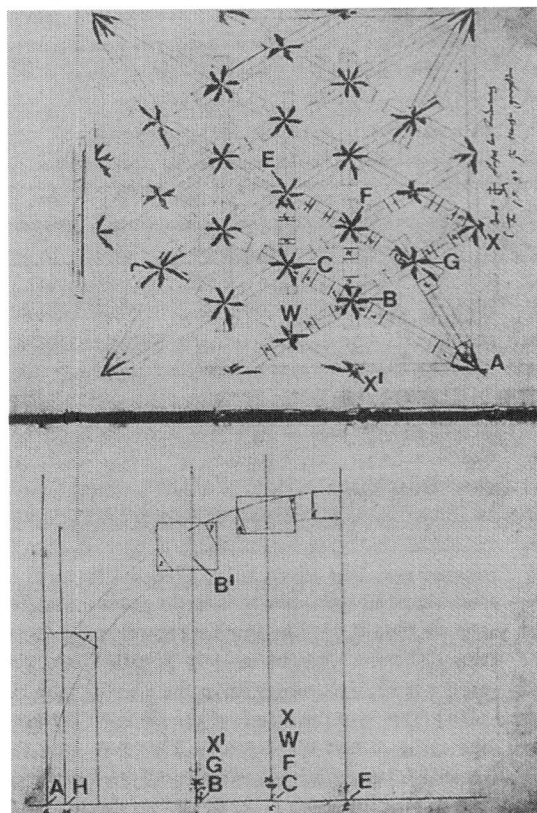

7 (above). Net-Vault Rib Plan from Jacob Feucht's Copybook. (Cologne, Historisches Archiv, Hs. Wf. 276*, fol. 11ᵛ.)
8. Bogenaustragung from Jacob Feucht's Copybook. (Cologne, Historisches Archiv, Hs. Wf. 276*, fol. 12.)

Apparently Lechler did not provide drawings in his original text to aid one's understanding of this somewhat cryptic passage. Fortunately, he was describing a technique for designing vault ribs that was commonplace in his day, and one for which numerous contemporary drawings survive[35]. In-

34 Reichensperger, Schriften, pp. 151-152; Cologne, Wf. 276*, fol. 54: „Item hie will ich dier ein bericht geben / wie du dich mit den Creüzbegen halten solt / in den Reiungen / sie sein gewunden / oder scheitrecht daß du eß leichtlich Verstehn khanst / Zum aller Ersten / wan du das gebogens / nach der Reiung aufgetragen hast / wie es gehört / so reiß den Anfang / Vnd die schloß- stein / alle fein / In daß gebogens / wie groß sie sein sollen / Darnach / wo es einen kreutzbogen / bedarfen wirdt / er sey gewunden oder scheitrecht / darnach muestu sehen / in dem grundt / Den nit ein reiung ist / wie die ander / mit den Creützbegen / den wo es einem scheitrechten Creutzbogen bedarf / so darfst du in nit anderst in daß gebogens reissen / Den eß gibt / dir die leng / selber / an dem gebogens / Zwischen den schloß- stein Vnd anfengen / Vnd also lang / der Creützbogen / Zwischen den schloßsteinen / ist / also lang muestu ihn auch machen / Vnd nit khürzen / also hastu Verstan- den / wie du den scheitrechten / Creüz bogen / machen solst / nun aber wanß du einen gewuntnen kreüzbogen machen wilst / so schau / wo er hinkombt / es sey Zwi- schen schloßstein / oder Anfeng / so muestu darzwi-

deed, Jacob Feucht himself provided examples of the technique in other parts of the manuscript which contains his copy of Lechler's text. But contrary to Lechler, who wrote an extended text without adequate illustrations, Feucht provided no verbal explanation of his drawings illustrating this particular design technique. A German chemist, Dr. Werner Müller, has patiently puzzled out the details of these and other related drawings. In a series of articles he has developed a precise explanation of the technique, which turns out to have been quite simple in its basic principles[36]. Once one has grasped the technique, Lechler's passage then becomes quite intelligible.

The historical term for the technique is *Bogenaustragung*, meaning roughly, "rib-projection." The technique was based on the idea that all the ribs of the vault could be generated out of the "principal arc" *(Principalbogen)*. First, the rib plan *(Reihung)* of the vault bay was drawn at full scale on the floor of the church, building lodge, or tracing house—wherever there was sufficient room for the task. Then it was determined what would be the highest point of the vault bay, and the distance was measured on the rib plan from one corner of the bay to that point. For example, in a quadripartite vault with semicircular diagonal ribs over a square bay, this distance would have been simply the straight line from one corner (the springer) to the center of the bay (the keystone) on the rib plan. But in the

9. *Late Gothic Rib Patterns in the Choir of the Stadtkirche, Michelstadt; designed by Moritz Lechler*

complex patterns of German *Spätgotik* the distance was determined by summing the lengths of the rib patterns from the springer to the highest keystone of the vault. For example, in Fig. 7, taken from

schen Darein Reissen / ein fürbredt / nach demselben fürbredt / muestu den gewunden Creutzbogen machen / Vnd wan du daß fürbredt / Zwischen hinein gerisen hast / es sey Zwischen den schloßstein oder Anfeng / so muestu denselben fürbredt / an einer Ietten seitten / an der leng abprechen / Vnd muest darnach / wider ein sonderlichs / fürbredt reisen / nach dem abpruch / Darumb muestu den gewundnen Creüzbogen / an der leng abprechen / dieweil die reiung mit kleinen teillen aufgetragen / wirdt / wan aber die Reiung / nicht mit kleinen teillung aufgetragen wirdt / so bedarfs / du keines abpruchs / also hastu mein meinung Verstanden."

35 The so-called Dresden Sketchbook of sixteenth-century technical drawings is entirely devoted to this particular design exercise. See François Bucher, "The Dresden Sketchbook of Vault Projection," *Actes du*

22e *congrès international d'histoire de l'art, Budapest, 1969,* Budapest, 1972, pp. 527–537.

36 Werner Müller, „Technische Bauzeichnungen der deutschen Spätgotik", *Technikgeschichte*, XL, 1973, pp. 281–300; *idem,* „Zum Problem des technologischen Stilvergleichs im deutschen Gewölbebau der Spätgotik", *Architectura*, III, 1973, pp. 1–12; *idem,* „Einflüsse der Österreichischen und der Böhmisch-Sächsischen Spätgotik in den Gewölbemustern des Jacob Facht von Andernach", *Wiener Jahrbuch für Kunstgeschichte*, XXVII, 1974, pp. 65–82; *idem,* „Die Zeichnungsvorlagen für Friedrich Hoffstadts ‚Gothisches A.B.C.-Buch' und der Nachlaß des Nürnberger Ratsbaumeisters Wolf Jacob Stromer (1561–1614)", *Wiener Jahrbuch für Kunstgeschichte*, XXVIII, 1975, pp. 39–54.

Jacob Feucht's copybook, this would have been the distance from A to G to F to E. This distance was then transferred onto a horizontal base line and became the radius of a quadrant set out from that line. This quadrant was the *Principalbogen* (Fig. 8).

The actual length and curvature of each of the ribs of the vault was then generated out of this *Principalbogen* in the following manner. First the length of each rib as shown on the *Reihung* was measured along the base line (points G, F, and E on Fig. 8). Vertical lines were then struck upward from the endpoints of each of these rib plans. The actual length of the curved ribs in the vault was determined by the intersection of these vertical lines with the *Principalbogen*. With the length and curvature of each rib set out on the *Principalbogen*, the master mason could then easily devise the templates to be used by other masons in cutting the voussoirs for the ribs, and the master carpenter could construct the temporary timber centering to be used by the masons in actually building the ribs into the vault. With this explanation of the technique one may reread the passage from Lechler quoted above and more readily grasp the basic meaning of the procedure that he was trying to describe.

It will be noted that in this passage Lechler concerned himself only with rib design; he said nothing here, or elsewhere, about the design or construction of the webbing of the vault. Indeed, he used no special term to distinguish this component, which is rather surprising, considering the importance of this structural element of the vault. Recent engineering studies have shown that the webbing was the essential load-bearing component in the configurations of High Gothic ribbed vaulting [37]. However, it is likely that the complex system of ribs typically used in German Late Gothic vaulting (Fig. 9) was considered almost a substitute for the webbing, with the actual webs being only a light infilling to close the spaces between the ribs.

We are still left with the question of how medieval masons viewed this fundamental problem of ribbed vault design. Did *they* consider the webbing to be carried by the ribs? We wish Lechler had given an opinion on this question. Since he did not even

mention webbing, obviously we will not get a direct answer. However, we may find indirect responses by looking into his technique for designing the ribs. We have seen how he determined the overall length and curvature of the ribs by means of the *Bogenaustragung*. He also determined the cross-sectional dimensions of the ribs by the same technique of modular design which he used in developing the shape and dimensions of other structural elements of the vault system. For the rib cross-sections he turned to his micro-module and stipulated that the depth of the crossribs was to be one-third the width of the choir wall. In turn, the width of the crossribs was to be one-half of their depth [38]. Since the width of the choir wall was itself a function of the width of the choir in a nominal ratio of 1:10, this meant that the cross-sectional depth of the crossrib was also a function of the width of the choir, and thereby of the span of the vault over the choir. This relationship can be reduced simply to the nominal ratio of 1:30; that is to say, a thirty-foot wide choir would have had crossribs with a depth of one foot and a width of six inches [39].

This constant ratio might suggest that Lechler viewed the rib to be acting structurally since the size of the rib was related to its span. On the other hand, in another section he told his son that the crossrib could be designed on either the large or the small scale. The first produced "the great crossrib which our forefathers used, for they had plenty of stone. However, at the present time one uses much more economy, so that you should use the small crossib [40]." He then explained that the small cross-

[37] Alexander, Mark, and Abel, "Structural Behavior", p. 251.

[38] Reichensperger, *Schriften*, pp. 147–148; Cologne, Wf. 276*, fol. 51ᵛ.

[39] We have not been able to check these figures against the actual dimensions of ribs in Late Gothic single-aisled choirs, but we did measure the ribs in the nave of the Alexanderkirche in Marbach, and they come very close to these dimensions–extending about ten inches below the web surface of the vault.

[40] Reichensperger, *Schriften*, p. 136; Heidelberg, Hs. 3858, pp. 4–5; Cologne, Wf. 276*, fol. 45: „. . . dieses ist der großkreüzbogen den Vnsere Alt Vetter haben ge-

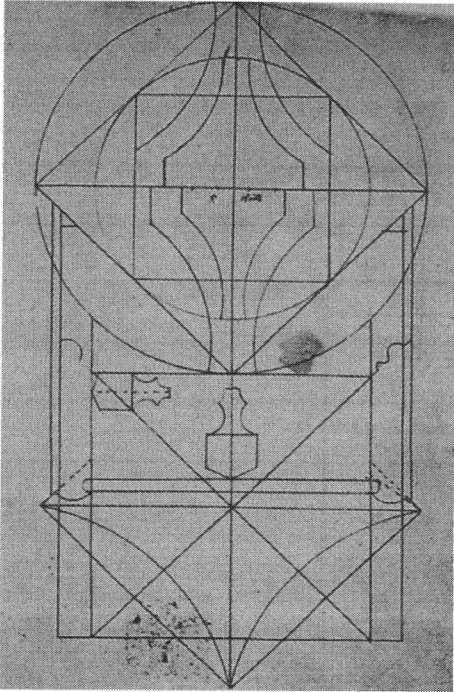

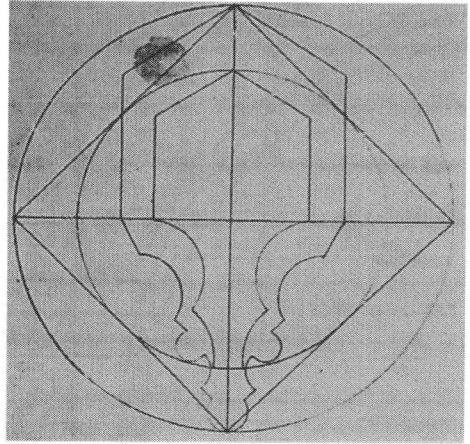

10. *Templates for Vault Ribs and Window Mullions in Lechlers "Instructions". (Cologne, Historisches Archiv, Hs. Wf. 276*, fol. 41.)*

11. *Templates for Transverse Ribs and Crossribs in Lechler's "Instructions". (Cologne, Historisches Archiv, Hs. Wf. 276*, fol. 42.)*

rib should be scaled at five-sixths of the dimensions of the large crossrib. Although there are some contradictions in the surviving texts regarding these formulas, we can feel fairly confident about what Lechler intended. The Cologne copy contains a drawing showing the depth of the large crossrib at one-third the width of a square that represents the module of the choir wall [41] (Fig. 10).

The dimensions of the transverse rib were in turn a function of the dimensions of the crossrib, for Lechler stipulated that the depth of the transverse rib was to be one-third larger than that of the crossrib of the choir [42]. Another of his illustrations shows the cross-sections of two ribs which appear to be the crossrib and the transverse rib (Fig. 11). However, they carry a ratio of 5:7 rather than 3:4. The 5:7 ratio frequently occurs in Lechler's rules and illustrations, for it is the mathematical result of his technique of rotating and inscribing squares. The successively smaller squares inevitably bear a ratio of 5:7 (actually $1/\sqrt{2}$) to the immediately larger

braucht / dan sie haben genuegsam stein gehabt / aber Zu Ieziger Zeit / braucht man gar Viel Redt darumb / Darumb so brauch du dise khleine khreüzbogen ..."

[41] In contrast to the passage cited in note 38 the Cologne copy has another section (Wf. 276*, fol. 45; Reichensperger, *Schriften*, p. 136) which states that the choir wall is to be divided into six parts and *one* of these is to be the depth of the crossrib. The copyist of Heidelberg Hs. 3858, p. 4, seems to have recognized the problem, for while he indicated that the wall should be divided into six parts, he did not indicate how many parts were to be used to determine the crossrib: „Item wann du den Crützbogen gewinnen wilt / so theill die Murdicke in 6 theil / das nim zu der brette / vnd die läng noch so lang / das ist der erste Crützbogen."

[42] Reichensperger, *Schriften*, pp. 148, 152–153; Cologne, Wf. 276*, fols. 51ᵛ ; 4ᵛ–55.

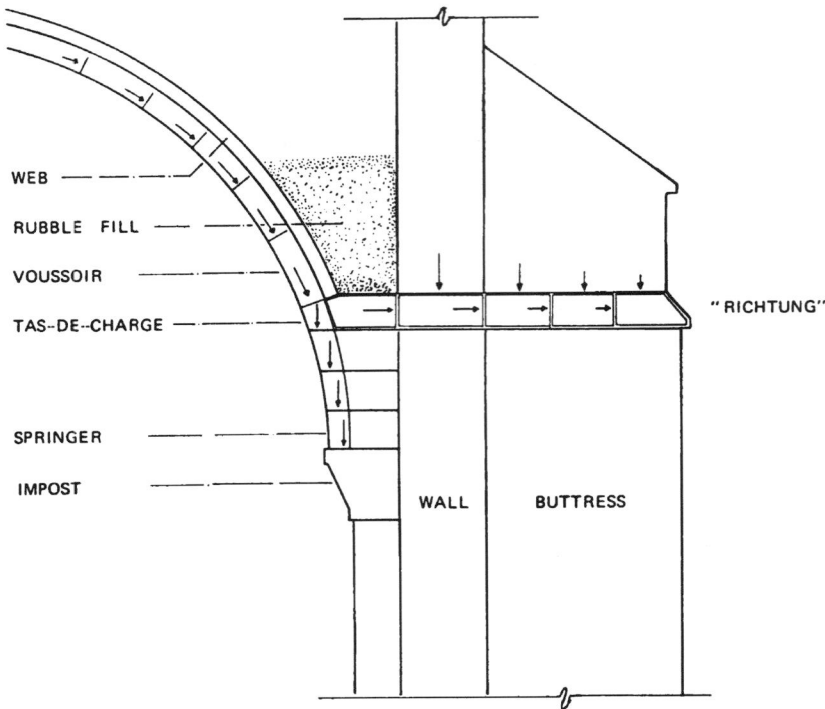

12. Schematic Cross-section of Vaults, Wall, and Buttress in Lechler's Design for a Hall Church Coir

square. This is the only explanation that we can find for this discrepancy between Lechler's text and this drawing.

While the answer is by no means definitive, it appears that Lechler did consider the ribs to be structurally significant, load-bearing elements in the vault. This is reflected in his concern to scale the cross-sectional dimensions of the crossribs to the span of the vault, and in his formula for making the transverse rib larger than the crossribs. But his silence regarding the webbing still leaves open the question of the load-bearing qualities which he might have considered the web to contribute to the vault.

That Gothic vaulting concentrates its horizontal thrust at focal points on the supporting wall was well understood by Lechler, who seems to have considered that the greatest thrust came at the level

of the tas-de-charge in the vault system. To understand his view we must review briefly the salient elements of the Gothic vault (Fig. 12). The vault rose from an impost, which could be the capital on a shaft or pier, or it could be a corbel projecting from the wall. Immediately above the impost, the vault ribs sprang out in curving arcs to define the vault. The beginning of this grouping of ribs was often carved from a single stone called the springer, the beds of which were horizontal on bottom and top. Above the springer might be two or three courses of stones which also maintained horizontal beds. These stones still had the several ribs carved in them, although the ribs were more separated from each other than they were in the springer. Eventually this separation of the ribs, as they rose in different directions, reached a point which could not be contained in a single stone. The ribs then

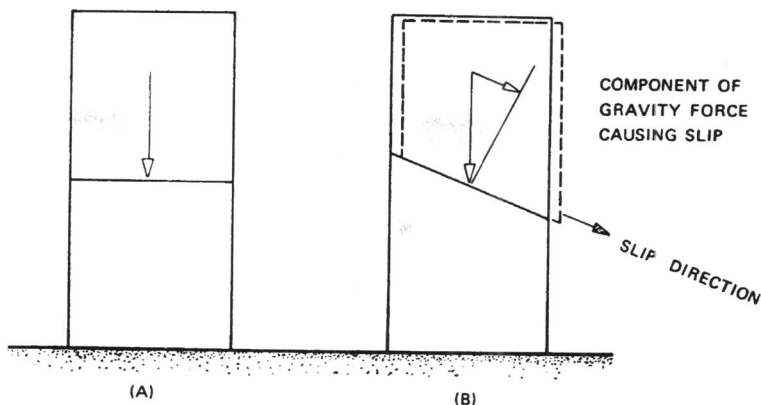

COMPONENT OF
GRAVITY FORCE
CAUSING SLIP

SLIP DIRECTION

(A)   (B)

13. Schematic Representation of Right-angle and Oblique Forces Pushing against the Faces of a Stone

had to be carved in separate stones called voussoirs, and their beds, instead of being laid horizontally, were laid at right angles to the curve of the ribs themselves. The tas-de-charge was the last stone above the springer which had a horizontal lower bed and which still had the ribs carved into it. But the top of the tas-de-charge had faces carved at an angle appropriate to meet the lower bed of the voussoirs, each of which rose in the separately curving arcs of the ribs above the tas-de-charge.

We come now to a crucial part of Lechler's understanding of the behavior of the vault: "Make *ein Richtung* and set it behind the tas-de-charge which stands on the springer. You should let the *Richtung* go out as far as the length of the buttress, so that the buttress becomes a square [43]." What was a *Richtung*? One of the basic meanings of this term is in the sense of direction, line, or course. In modern structural analysis it can have the refined meaning of "vector component of force" that develops from the outward thrust of the vault [44]. Did Lechler mean to imply this relatively sophisticated notion? It does not seem so. Throughout his booklet his language and thought are concrete and practical, not mathematical or abstract. Likewise his language here does not appear to refer to some imaginary line or mathematically determined vector of force. On the contrary, he seems to have been saying that

a *Richtung* was something which one carved and set into the masonry of the structure. Let us then translate the term as a buttress course which extended from the tas-de-charge through the wall and to the outer limit of the external buttress.

If this is the correct interpretation of Lechler's meaning, an interesting observation can be drawn from the passage. It placed considerable significance on the structural role of the tas-de-charge in the buttressing system. As we have noted, the tas-de-charge had an upper face angled to meet the voussoirs of the ribs, while its lower face was horizontally bedded. Thus the shape of the tas-de-charge transferred the vault forces into an outward thrust through the *Richtung*, and into vertical forces which passed through the springer and the wall that supported the vault from directly below. The *Richtung* was seen as the critical course of stones that resisted the outward thrust of the vault acting through the tas-de-charge. Lechler realized that a single course of horizontal stones would be insuf-

43 Reichensperger, *Schriften*, p. 146; Cologne, Wf. 276*, fol. 50ᵛ: „Vnd mach ein richtung / solstu setzen / hinder das haubt / darauf die anfang stehn / Alß vil der Pfeiller lang ist / soltu dein richtung außgehen lassen / daß der Pfeiller / in ein gefier kombt."

44 Georg Ungewitter, *Lehrbuch der gotischen Konstruktionen*, 3rd ed. Karl Mohrmann, Leipzig, 1890, I, 128–133, uses the term in this sense.

ficient to counter this outward thrust, so he completed his discussion of the *Richtung* by indicating how to carry the buttress above it to provide the mass that would prevent the *Richtung* from being pushed outward along its horizontal bed [45].

Conclusions

Let us try to summarize what has been learned from Lechler's rules regarding the structural design of Late Gothic German hall-church choirs. First, it is clear that he was not approaching the problem of structural design on the basis of mathematical or analytical theory. His rules relied on geometrical figures and numerical ratios, but there was virtually no analytical reasoning or calculation involved. Just as with the design techniques in Roriczer's and Schmuttermayer's booklets, Lechler's formulas consisted of a series of seemingly arbitrarily determined steps that one followed because that was the way Lechler said to do it, and not because he had logically demonstrated their correctness. In short, Lechler's structural design techniques were based on rules of thumb, rather than on analytical formulations. This is not to say that his rules were devoid of good sense. We may assume that they were based on his own experience and that of his contemporaries, as well as his observations of the buildings of many previous generations of masons. For Lechler, as for builders in general until the late 19th century, it was empirical evidence which constituted the best proof of the correctness of structural design. That his rules generally called for scaling up (or down) the size of building elements appears reasonable for the range of relatively small buildings with which he was concerned [46]. The variation in structural sections which was to be a function of the "quality of stone" might better be ascribed to the quality of construction which, of course, includes stone quality. His acknowledgement of such differences gives us an important insight into his thinking.

There is another point worth pondering in Lechler's design technique, namely, his concern for the design of the individual stones within the architec-

tural elements of the building. As a mason, it was natural for him to give so much attention to the design of the templates for individual stones, for that was the way medieval masons went about their business of cutting and setting the stones, one after another, as they built up the architectural elements into the whole of the building. A primary task of the mason was to insure that all forces acting against individual stones worked at right angles to the faces of the stones. Any major force working at an oblique angle to the face would tend to push the stone along that face in the direction of the oblique angle (Fig. 13).

The structural design would begin with the masons' basic unit of construction, the individual stone. In structures with complex forces at work, such as a stone vault with its buttressing system, each stone had to be cut and placed so that the forces transmitted through the stone worked at right angles to the faces of the stone. While Lechler did not specifically formulate the matter to his son in this way, he might very well have said, Design the ribs and shafts and springers and buttresses however you will; but if you make sure that the forces acting against the stones are always working at right angles to their faces, and if these forces are maintained in equilibrium, then you may be confident that the design is correct and that the building will stand up.

[45] Reichensperger, *Schriften*, pp. 153–154; Cologne, Wf. 276*, fol. 55ᵛ.
[46] This is not generally true over a large range of scaling. If all the dimensions of any building element are increased linearly, its mass will increase by the third power of its scale, while the cross-sectional area of its support increases only by the second power. Hence stresses, which are proportional to the mass divided by the support cross-section, will have higher values at a larger scale. Except for a few highly localized regions in Gothic churches, such as at the bases of piers, stress levels are generally low. Hence, structural reliability was usually determined by overall stability or by the presence or lack of tension forces in the building fabric. For an example of the collapse of a building caused by local tension, see Wolfe and Mark, "Collapse of the Vaults of Beauvais Cathedral," pp. 462–476.

Finally, it should be pointed out that in his rules for the vaulting and buttress system of a single-aisled choir, Lechler was dealing with a relatively simple design problem compared with the complexity of the structure of a basilican church choir, which might have double aisles and chapels radiating around the apse. Lechler's rules of thumb were safe, because they were based upon widespread experience with the simpler forms of Late Gothic single-aisled choirs. As we have seen, he left some judgmental decisions to be made within these rules, but those decisions did not call for great daring on the part of the designer, nor did they pertain to previously unfathomed questions. Lechler's son, Moritz, could look around and find much guidance on how to make those decisions, as indeed he did when he built the choir vault of the Stadtkirche in Michelstadt around 1540 [47]. Nevertheless, the design rules of a Late Gothic master mason like Lorenz Lechler can suggest the approach that would have been employed by the designers of High Gothic basilican cathedrals. There is no reason to suppose that late medieval master masons had lost a knowledge of calculating structural forces which earlier masons had possessed. Just as Lechler used rules of thumb which were based on geometrical manipulations and numerical ratios, but which were grounded on empirical evidence of success, so did, in all likelihood, the designers of the great cathedrals of Chartres, Reims, and Amiens. But the master masons of that earlier era experimented more with previously untried solutions to new structural questions than did Lechler. Their daring appears all the more impressive in light of the evidence from Lechler's booklet about the actual techniques used in Late Gothic structural design.

[47] On Moritz Lechler's work at Michelstadt, see Richard Lösch, *Stadtkirche Michelstadt*, 3rd ed., Michelstadt, 1975, p. 16; and Seeliger-Zeiss, *Lorenz Lechler*, pp. 160-161.

Photos: Cologne, Historisches Archiv, Figs. 4, 5, 7, 8, 10, 11. W. Kohlhammer Verlag, Stuttgart, Fig. 2. All other drawings and photographs are by the authors.

7

The ground plan of Norwich Cathedral and the square root of two

Eric Fernie

THE SEAT OF THE BISHOP OF EAST ANGLIA was officially moved from Thetford to Norwich in 1094 and the foundation stone of the new cathedral laid in 1096. Some preliminary work on the site may have been begun as early as the mid 1080s, but in any case Herbert de Losinga built the bulk of the church before his death in 1119 and his successor Eborard finished it before leaving office in 1145.[1] The Romanesque fabric is complete at ground level apart from four piers in the presbytery, one transept chapel and the axial chapel, which has however been excavated. We have, then, a virtually complete plan accurately dated to the last years of the eleventh century, laid out on a new site unmarked by an important earlier relic or church. All these factors qualify it as an excellent candidate for an attempt to establish the presence or absence of proportional systems in its plan.

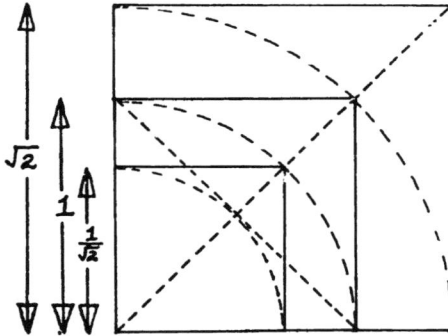

FIG. 1. The square root of two in geometrical form

The evidence shows that a single proportion used in a number of different ways underlies all the measurements of the plan, namely the relationship of the side of a square to its diagonal, which is $1 : \sqrt{2}$ or $1 : 1.4142$. That is, the side multiplied by the square root of two equals the diagonal, and divided by the same it equals half the diagonal (Fig. 1).

The interior length of the church is 132 m. (433 ft) (Fig. 2), and the average length of an aisle bay consisting of a vault and a transverse arch is 5.495 m. (18 ft 0⅓ in.).[2] 132 m. divides into twenty-four units of 5.5 m. each. The divisions coincide with the west faces of

1 The evidence for the dates will be found in H. W. Saunders, *The First Register of Norwich Cathedral* (1939), folios 1–9 (Norfolk Record Society, XI), and Bartholomew Cotton, *Historia de Episcopis*, ed. Luard (1857), 4. The best published plan of the cathedral is that by Arthur Whittingham, in the *Archaeological Journal*, CVI (1949), 86.
2 For example, the bays on the wall of the north aisle are, in metres from east to west, 5.31, 5.43, 5.29, (21.97), 5.46, 5.49, 5.50, 5.51, 5.50, 5.54, 5.53, 5.53, 5.52, 5.53, 5.54, 5.53, 5.71. 21.97 m. is the distance from the east edge of the westernmost vault in the presbytery aisle across the transept to the west edge of the easternmost vault in the nave. Divided by four it gives an average 'bay' size of 5.49 m. The same measurements in the south aisle produce an average of 5.494 m.

78   NORWICH CATHEDRAL

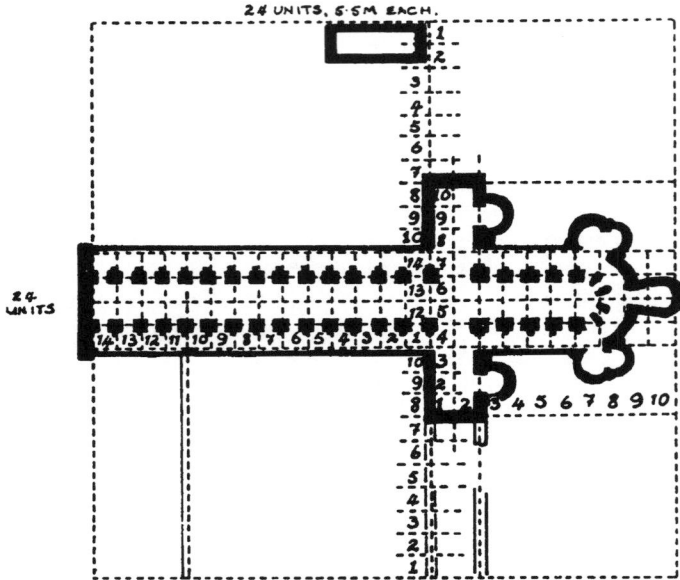

FIG. 2.   The bay system

the transverse arches in the presbytery and with their east faces in the nave. The line between presbytery and nave lies on the axis of the transept, and the half-way mark of the whole church on the entrance to the liturgical choir. On the north–south axis, the units of 5.5 m. mark the axis of the church, the centres of the arcade walls, and the interior faces of the aisle walls.[3] The eastern arm, the transept and the cloister are all ten units in size, while the nave is fourteen. Since nave and aisles are four units wide, the cloister plus the width of the church equals the length of the nave. 10:14 is a proportion of $1 : \sqrt{2}$.

In the aisle bays (Fig. 3) the diagonal of the vault equals its side plus the thickness of the arcade wall. Each transverse arch is half as thick as the arcade wall, therefore the diagonal also equals the side of the vault plus the two flanking arches.[4] The bay size is thus closely related to the square root of two since its length of 5.5 m. consists of a vault side of 4.65 m. (14 ft 11½ in.) plus half the amount generated by 4.56 × $\sqrt{2}$, namely the .94 m. (3 ft 1 in.) of the transverse arch. The aisle wall is the same thickness.[5] The relationship of

[3] For example, bay twelve from the crossing: north aisle 4.55 m. (14 ft 11½ in.), north arcade wall 1.91 m. (6 ft 3 in.), nave 9.03 m. (29 ft 7½ in.), south arcade wall 1.94 m., south aisle 4.58 m., total 22.01 m. (72 ft 2½ in.). 5.5 × 4 = 22.00.

[4] For example, the diagonals of north aisle bays seven, eight and nine are 6.53 m., 6.5 m. (21 ft 4 in.), and 6.51 m. The width of the north aisle varies from 4.5 m. to 4.6 m., averaging 4.56 m. (14 ft 11½ in.), and the thickness of the arcade wall ranges from 1.88 m. (6 ft 2 in.) to 1.94 m., averaging 1.92 m.; 4.56 + 1.92 = 6.48. The transverse arches vary from .88 m. to .99 m. and average .96 m. (3 ft 1¼ in.).

[5] The thickness of the aisle wall is measurable at very few points: door to stair off south aisle of presbytery 1.01 m.; door to bookshop in south aisle of nave .95 m. The entrances to the radiating chapels have responds of .94 m. or .95 m., but these may not represent the wall thickness.

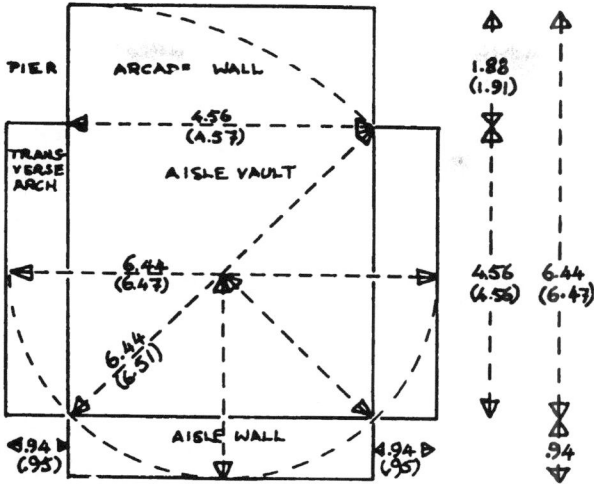

FIG. 3.   Aisle bay. Figures in brackets are the measurements of north aisle bay 9

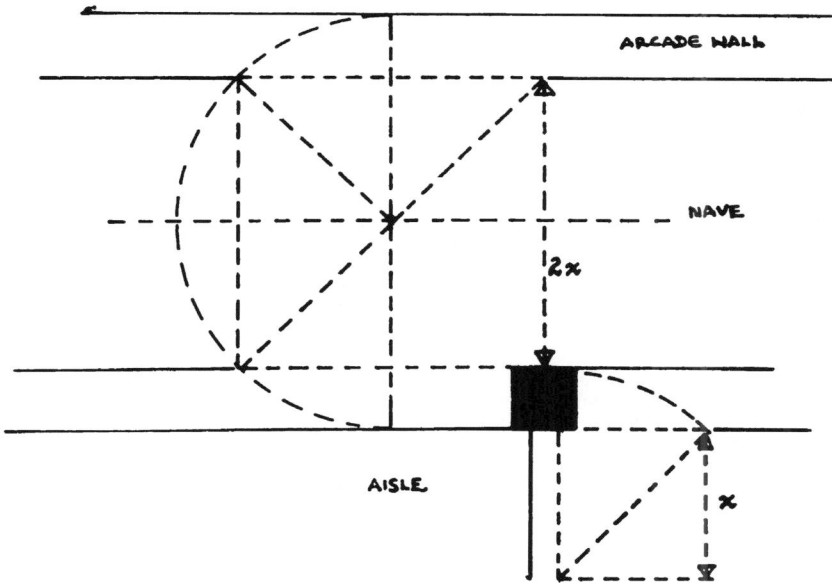

FIG. 4.   Widths of aisle, arcade walls and nave

the aisle vault to the arcade wall is exactly analogous to that of the cloister to the nave and aisles, while the elements flanking it on the other three sides, the transverse arches and the aisle wall, are like the three ranges of the cloister.[6] The nave is twice as wide as the aisle, and because the aisle width multiplied by the square root of two generates the thickness of the arcade wall, the interior width of the nave is to its exterior width as $1 : \sqrt{2}$.[7] This is the same relationship as that between the aisle vault and its two flanking arches (Fig. 4).

The piers constitute the third and smallest scale on which the proportion is used. There are four types of Norman pier, namely two kinds of minor arcade piers based on a cylinder, and the major arcade pier and the crossing pier, both based on a square. The four minor piers lying third and fifth west of the crossing are simple cylinders. Their radius (Fig. 5, i) is the distance from the centre to the point at which the transverse arch over the aisle meets the arcade wall.[8] The other minor piers are like rectangular accretions around these cylinders. The diagonal AC (Fig. 5, ii, iii) equals AF, DE and the inner order of the arch, GH. The difference between the two orders, GE, equals EI, and the diagonal GI equals

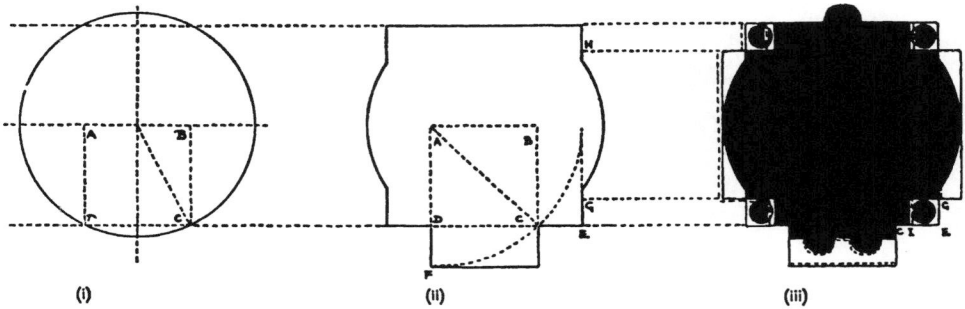

(i)   (ii)   (iii)

FIG. 5.   Minor piers. Dotted lines in (iii) represent the ninth pier of the south arcade of the nave

---

[6] Presumably for utilitarian reasons the ranges vary in width. The east is two units wide, the south slightly more, and the west one and a half.

[7] The width of the nave is 8.96 m. at bay four, 8.98 m. at bay six, 9.03 m. at bays eleven and twelve, and 9.05 m. at the crossing. 4.56 × 2 = 9.12 (29 ft 11 in.). The width of nave and aisles varies only 2 to 3 cm. from the ideal 22 m. (four units of 5.5 m.). The aisle vault size of 4.56 m. is strictly adhered to on the north–south axis to within 3 or 4 cm. The arcade walls however are consistently about 4 cm. thicker than the amount generated by 4.56 m. × √2, namely about 1.92 m. against 1.88 m., with the consequence that the nave is slightly narrower than it should be. This implies an initial laying out of the inner faces of the aisle walls, then the outer faces of the arcade walls, and finally the arcade walls. The responds supporting the transverse arches in the aisle are also thicker than the ideal, .96 m. against .94 m. (1.88 ÷ 2). This slight error in the arcade pier bases and those of the aisle responds suggests that masons tended to remove too little stone rather than too much.
      It is apposite to note here that the blind arcading on the interior of the aisle wall invades the space of the vault square. In other words the east–west length of the vault square extends on the north–south axis to the back of the blind arcading. The same feature occurs in the Norman parts of Ely and Peterborough Cathedrals, and in the late twelfth-century parts of Lincoln Cathedral. The reason for this sleight-of-hand is probably that an aisle wall which was the thickness of a transverse arch would not have sufficient room for the sort of plasticity necessary to the Norman aesthetic of the late eleventh century, while an increase in its thickness would separate it from the size of the aisle bay which generates it.

[8] The only part of the piers where √2 is not used is in establishing the radius of the cylindrical piers. The system is related to the golden section, though probably unintentionally. It may have been chosen because a diameter equal to the width of the arcade wall would leave an overhang at the arch springers, while a diameter equal to the diagonal of a square on the wall width would be very large. The compromise leaves the pier a reasonable girth while seeming to provide sufficient support for the capital and abacus. See E. Fernie, 'The Norman Piers of Norwich Cathedral', forthcoming in *Norfolk Archaeology*.

CE. The aisle respond is the same as on the major pier, described below. Figure 5, iii shows an ideal minor pier constructed in the manner just described, while the dotted lines represent the outline of pier nine in the south arcade.

In the major pier the core is a square on the width of the arcade wall (Fig. 6, i). The square ABCD is half the size of the core, and CD represents the position of the pilaster on the aisle face. The half-diagonals AE and BE (Fig. 6, ii) mark the north and south edges of the inner order and its east and west extent. The diagonal AC equals AF, F marking the extent of the base of the respond, and the half-diagonal CE equals CG, G being the centre of a halfshaft on that respond. DG equals DH (Fig. 6, iii), H being the inner edge of the order on the aisle face. I and J are the points on the nave face equivalent to D and H. IJ equals JK, K marking the extent of the nave pilaster. The square DLMF (Fig. 6, iv) has a side the depth of the aisle respond. FN equals FO, O marking the face of the pilaster, as opposed to the base, of the respond. L marks the inner edge of the left-hand shaft and GL is its radius. The shafts on the other three faces (Fig. 6, v) are arranged in the same way, allowing for the fact that those on the east and west faces are in threes: CD = PQ and RS; CE = QT and SU. Figure 6, vi is an ideal pier constructed in the manner just described, while the dotted lines represent the outline of the tenth pier in the south arcade.

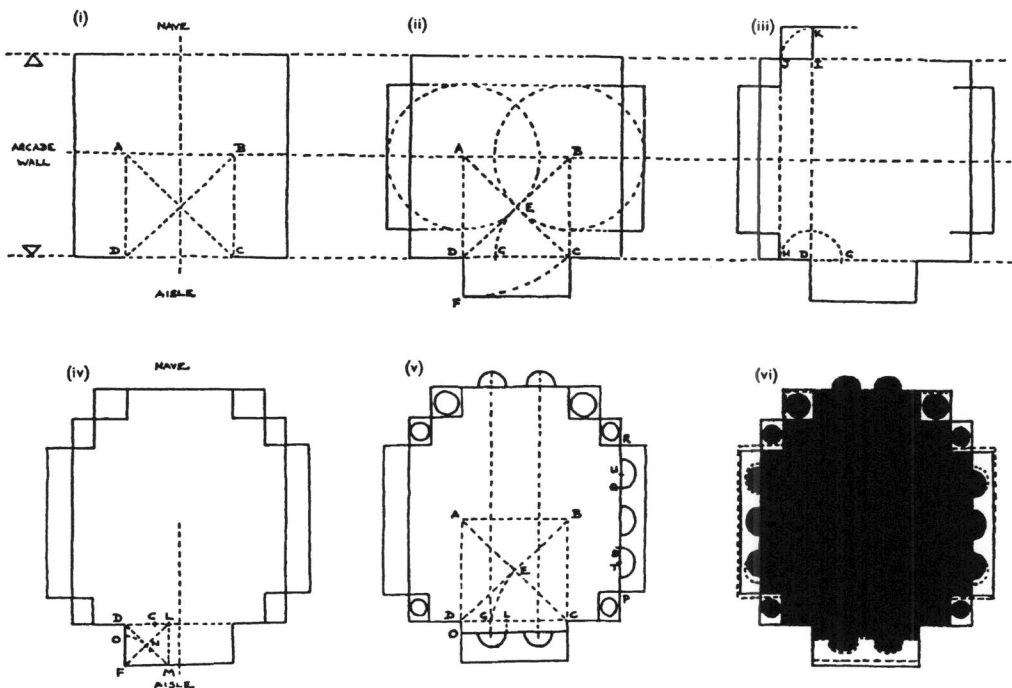

FIG. 6.   Major pier. Dotted lines in (vi) represent the tenth pier of the south arcade of the nave

The respond on the aisle wall has the same pilaster and two shafts as the aisle face of each pier. The aisle wall is as thick as the pilaster is wide (Fig. 7, AD). Half of this width, DE, equals DF. The diagonal EF equals EG and EH, while FI equals FJ, and DH equals HK. The external buttress makes up the thickness of the wall and respond to that of the arcade wall. Its width is the same as that of the interior respond, and the nook-shaft squares are the depth of the buttress divided by the square root of two.

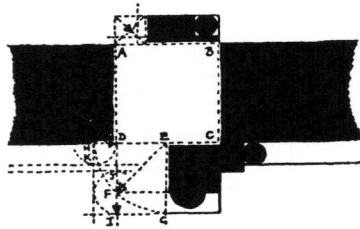

FIG. 7.   Aisle wall respond

The crossing pier is like the major pier in having a core which is a square on the arcade wall, and an inner arch support which is half the diagonal of that square. Figure 8 indicates how all the dimensions relate to the core.

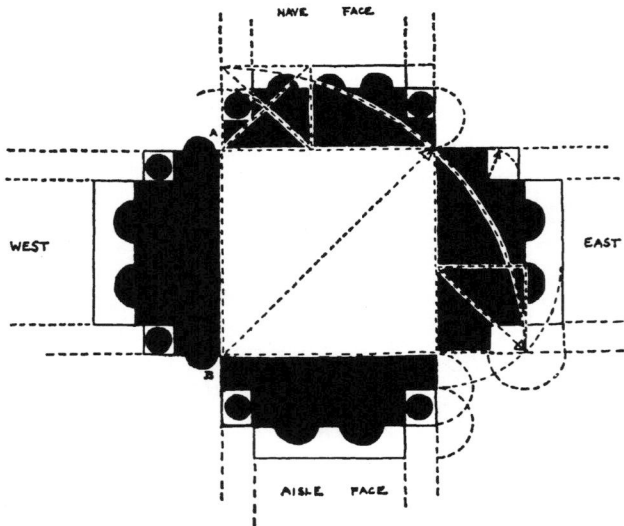

FIG. 8.   Crossing pier

The bishop's palace is also related to the square root of two. Its axis lies between 9° 35′ and 9° 55′ off the north–south axis of the cathedral (Fig. 10). If two squares relating as 1 : √2 are juxtaposed on the same base line, a line drawn through their outer corners will lie at 9° 44′ to the base line (Fig. 9).[9] The location of the palace relates to the rest of the plan via the centre of its great chamber (Fig. 10, D). This lies on the point of intersection of the centre line of the cloister (Fig. 10, DE; that is, the fifth bay of the nave from the crossing) and the north side of the triangle ABC, which is constructed as follows: its base BC lies on the line of the fourth bay unit of the nave from the west, that is the western wall of the cloister. The length of BC is the same as the twenty-four bay interior length of the church, applied north–south with its mid-point on the main axis. This length extends south to the centre of the south wall of the cloister and north to 1.2 m. beyond the north wall of the small hall in the palace, the northernmost extent of the close until the fourteenth century (Fig. 2).[10] The apex of the triangle ABC lies at the head of the main apse seventeen units from the base. The height to base ratio of 17:24 equals $1:\sqrt{2}$. Such a triangle has a base angle of 54° 44′. The axis of the palace lies at 45° to this north side and hence at 9° 44′ to the base.

FIG. 9.   Constructing an angle of 9° 44′ using the square root of two

The measurements of the palace also exhibit the proportion $1:\sqrt{2}$. The interior width of the corridor (Fig. 11, AB) multiplied by the square root of two equals the exterior width, CD, which is equal to the interior width of the great chamber, EF.[11] This interior width again multiplied by the square root of two gives the exterior width of the great chamber, GH. This is the same principle as that relating the interior and exterior widths of the nave and that relating the aisle vault to its transverse arches. The exterior width GH is the same as the interior length IJ, and since the wall thickness is dependent on the $1:\sqrt{2}$ relationship between interior and exterior widths, the exterior length follows automatically.

The radiating chapels lie at about 9° 52′ to the east–west axis of the church, and the extensions of their axes meet at a point twenty-four bay units east of the centre of the church (Fig. 10, H). Each chapel[12] is composed of two circles, the smaller forming the apse (Fig. 11). The centre of the smaller circle lies on the axis at a point where it is cut by a line KL running down from the apex of the main apse at an angle of about 9° 52′, parallel to the palace and at right angles to the axis of the chapel. This line KL cuts the exterior of the ambulatory wall at point M (Fig. 11); ML is the radius of the circle forming the exterior of the apse, while the centre of the large circle lies where the first circle cuts the

[9] In Fig. 9 BC = $\sqrt{2}+1$ and AC = $\sqrt{2}-1$; tan AB̂C = $\frac{\sqrt{2}-1}{\sqrt{2}+1}$, hence AB̂C = 9.74° = 9° 44′. I am indebted to Louis Gidney for clarifying this figure and a number of other mathematical points.
[10] See H. Harrod, 'Excavations Made in the Gardens of the Bishop's Palace, Norwich, April, 1859', *Norfolk Archaeology*, VI (1864), 27. The north edge of the square may thus have coincided with the close wall.
[11] The interior of the great chamber is .7 m. narrower than the exterior of the corridor. This appears to be due to a distortion in the laying out: the whole rectangle is offset to the east, and the east wall of the corridor gets thicker to the north, so that where it abuts the church it is in line with the eastern interior face of the great chamber.
[12] The northern chapel appears to be badly distorted.

84   NORWICH CATHEDRAL

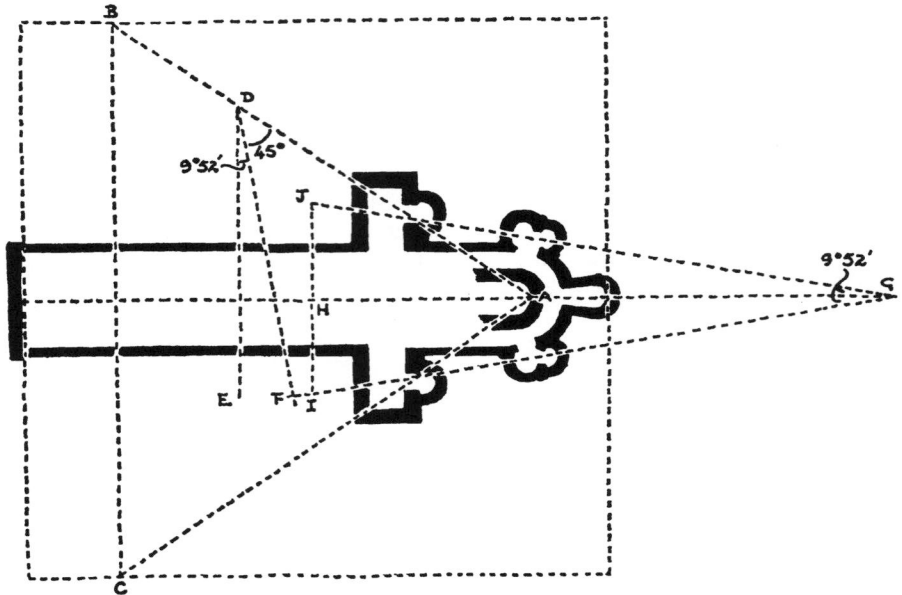

FIG. 10.   Location and axes of bishop's palace
             and radiating chapels

axis at N. Chapels and palace are thus closely related in that they lie at an angle of 9° 52′
to the east–west and north–south axes of the church respectively.

Norwich Cathedral does not stand alone among Norman buildings in using a grid of
squares and the square root of two. Grids occur at St Augustine's Canterbury, Bury St
Edmunds and York Minster.[13] The cloister is to the nave as $1 : \sqrt{2}$ at Westminster Abbey,
Bury St Edmunds, and the cathedrals of Canterbury, St Albans, Winchester, Worcester,
and Durham. In greater detail, at St Augustine's the church, excluding the west towers, is
eighteen aisle bays long. The divisions mark the east faces of the bays in the nave and the
west faces in the presbytery. From the main axis to the interior of the north wall of the
cloister is a distance of nine bays, half the length of the church. The half-way point on the
east–west axis is marked by the rood screen. The transept is eight units long and the nave
eleven, a $1 : \sqrt{2}$ relationship in round numbers. With the difference of eighteen bays to
twenty-four, this is very much like Norwich Cathedral.[14]

---

[13] For York see J. Harvey, *The Medieval Architect* (1972), 122. Investigations into other buildings are in progress.
[14] Both at St Augustine's and at Norwich some measurements bear more accurate $1 : \sqrt{2}$ relationships than those
      permitted by the grid. At Norwich the interior length of the nave is 76.36 m. (250 ft 6 in.) and that of the
      transept 54.03 m. (177 ft 3 in.). 76.36 m. ÷ $\sqrt{2}$ = 53.99 m. (The interior length of the nave equals fourteen
      bays less one transverse arch. With an ideal bay size of 5.5 m. this would be 76.06 m.). At Canterbury the
      same lengths are 58.13 m. (190 ft 8¼ in.) and 41.23 m. (135 ft 3¼ in.). 58.13 m. ÷ $\sqrt{2}$ = 41.1 m. In the transept
      at Norwich the bays are of a different size from those in the aisles, therefore it is possible that its length was
      determined not by the grid but by dividing the length of the nave by the square root of two. It is also possible
      that the length of the transept was given by the points at which the sides of the great triangle cut its west wall.
      It is not clear to what extent these follow naturally from the use of $1 : \sqrt{2}$ in the grid system, and to what extent
      they were specifically intended.

NORWICH CATHEDRAL 85

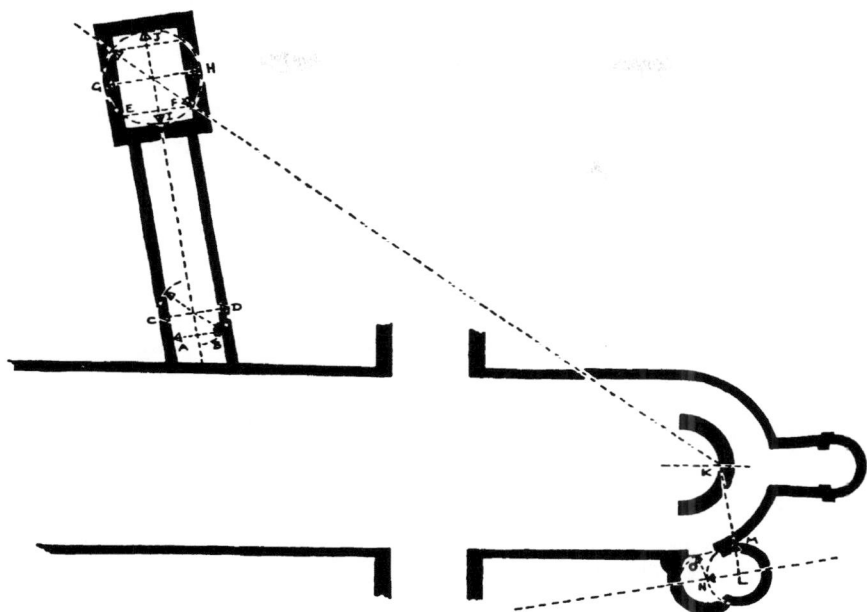

FIG. 11. Form of bishop's palace and radiating chapels

At Bury St Edmunds the church, claustral ranges and carnary are contained in a square twenty-four bays by twenty-four. The cloister plus the north and west ranges is ten bays square; the east arm of the church is ten bays long and the nave fourteen, that is $1:\sqrt{2}$. These bays are grouped into larger four-by-four squares: two for the east arm east of the crossing and a third for the extent of the liturgical choir (that is the crossing and two bays of the nave), the entrance to which thus marks the half-length of the church. There are three blocks in the western half, one of which represents the part of the nave salient from the claustral square. The aisle width generates the thickness of the arcade wall in both the crypt and the nave. All this is exactly as at Norwich.

At Peterborough Cathedral the transept is the same length as the nave. The aisle bay measurements are almost identical with those at Ely, except that the longer axis at Peterborough lies east–west. The crossing pier is a model of $\sqrt{2}$ construction.[15]

The three buildings at Ely, Bury, and Norwich are as closely related by their proportions as by more obvious features. This would suggest that no western transept was intended at Norwich, as it is the sacrifice of such a transept within the grid which permits the cloister to extend west until it is the second largest in the country. On the other hand it must be admitted that at St Augustine's and Westminster Abbey the western towers lie beyond the

---

[15] I am indebted to Peter Kidson for calling my attention to a nineteenth-century plan of this pier and to its significance as an example of proportional design.

grid and therefore beyond the $1:\sqrt{2}$ relationship of the nave to other parts of the building.

Why was the proportion used at Norwich? There are at least four possibilities. It may have been a means of (1) pegging out the building; (2) determining the disposition of the elements of the Priory; (3) establishing aesthetically satisfying relationships between parts; or (4) establishing a symbolic harmony with religious or mystical connotations.

(1) There is a lot to recommend the proportion as a simple tool for a practical purpose, not least Roriczer's use of it in his instructions for the setting out of the plan of a pinnacle.[16] The figures describing the piers at Norwich (Figs. 5–8) show that $1:\sqrt{2}$ could have enabled a mason to lay out a complex form in five or six steps without resorting to measurements in feet and inches. This however necessitates the laying out of each pier afresh, where a template could easily obviate such repetition. It would seem the device was used at design stage and not during execution. On a larger scale the proportions of the bays establish wall surfaces and not footings and are therefore of little help in setting out. The relationship between walls and spaces is the same whether the spaces are vaulted or not, therefore the use of the proportion is not a response to a problem of statics. Finally on the largest scale, there can have been no practical advantage in the particular relationship of nave to cloister, or the placing of the radiating chapels at the specific angle of 9° 52′.

(2) Despite the seeming obviousness of the second possibility it must be rejected. $1:\sqrt{2}$ only becomes a relevant factor in the design once decisions have been made to dispose parts in a particular way. The angle of the palace for example is determined by $1:\sqrt{2}$, but the decision to angle the range in the first place is unlikely to have had anything to do with the proportion. It is true that there may be a dialectic in the process, so that the decision to angle the palace could have been taken because the architect wished to introduce an element which he had used so much elsewhere. This, however, begs the question, and $1:\sqrt{2}$ must remain a designer's tool, explaining how parts of the building were designed but not why.

(3) If it is a designer's rather than a builder's tool then, given the lack of documents, the reason for its being used (as opposed to the techniques of using it) must remain unknown. The building resulting from the application of the proportion may be aesthetically satisfying, but such satisfaction cannot be presumed in the designer.

(4) The fourth possibility has on occasions lead writers into a quagmire not, as in the third possibility, because there is no written evidence, but because there is too much, all of it couched in such flexible terms that any cap can be made to fit any head.[17]

Given all this and the clear internal evidence that the designer was fascinated by the square root of two, and attempted to use it in as many different ways as possible, it seems wisest to go no further than the following: the proportion $1:\sqrt{2}$ was adopted at Norwich, at least in part, because of a desire for some form of geometrical unity in complexity.

[16] L. Shelby, 'The Geometrical Knowledge of Medieval Master Masons', *Speculum*, XLVII (1972), 395–421, fig. 6.
[17] Bearing in mind the *caveat*, there may be some symbolic intent in the linking of the bishop's great chamber, firstly with his throne along the north side of the triangle ABC in Fig. 10 (though the bishop's throne in the apse at Norwich has a later Romanesque setting, the throne itself is pre-Norman, so that there is no reason to suppose that it is not now where it was intended to be in the original building), secondly with the sanctuary in the nave along the axis of the palace, and thirdly with the cloister along the north–south line passing through the centre of both the chamber and the cloister walk. Nothing now marks this point in the fifteenth-century north walk at Norwich, however at Tintern there are remains of a thirteenth-century prior's seat in the centre of the south wall of the cloister. See O. E. Craster, *Tintern Abbey* (1968), 12.

8

The stonework planning
of the first Durham master

Jean Bony

The historical position of Durham cathedral and the very nature of its invention have given rise to so much unwitting mythmaking, and our own perception of the conditions of architectural production in that period is so uncertain, that the time seems to have come now to re-examine, in the most meticulous detail, all the material signs of work procedures and of modes of thinking that can be read from the texture itself of the fabric. For the simplest of observations may reveal, though their implications, unsuspected approaches to construction, as well as mental attitudes or cultural backgrounds, which may all ask for further probing. By making us aware of unfamiliar aspects of the process of creation, such investigations are likely to provide some of the elements needed for a better assessment of what Durham actually meant in its time and of what could, with some validity, be said to constitute its novelty and its significance.

As a first move in that direction, this essay will consider the stonework of the initial stage in the construction of the cathedral, a period of very short duration but of critical importance, since it set the tone for what was to become the testing ground for a succession of unpredictable developments.

The chronology of the works and the extent and duration of the major campaigns of construction have been established by John Bilson [1] on such solid bases that there is nothing significant to be added.[2] In the 'First Great Campaign of Construction', as Bilson has called it, which lasted from August 1093 to September 1104, were completed the whole east arm, the crossing (with enough of the two easternmost bays of the nave to ensure its stability), the south transept in its entirety and the north transept up to gallery level. That constituted an ensemble large

---

[1] This chronology was presented already in John Bilson, 'The Beginnings of Gothic Architecture, II, Norman Vaulting in England', *Journal of the Royal Institute of British Architects*, 3rd series, 6, 1899, pp. 289–319; it was further refined in, Idem.,'Durham Cathedral: the Chronology of its Vaults', *Archaeological Journal*, 79, 1922, pp. 101–60.

[2] No valid re-evaluation of the chronology of Durham cathedral has so far been presented.

enough to be put into service as the new cathedral in September 1104.[3]

Bilson was not interested in defining exactly the successive stages within that major campaign of eleven years, but at least three stages can be recognised. The initial one consisted only of the peripheral walls of the east arm together with the three pairs of detached piers in the choir area and the east walls (meaning the aisle walls) of the two transepts.[4] It did not include the east piers of the crossing or the piers which support the main eastern elevation of the transept.

The limits of the first stage in building operations can be identified from the type of plinths used inside and outside that easternmost part of the work: for the plinths change in their profile or in their level in the north-east and south-east corners of the transept, indicating at those points the beginning of a second stage of work (Plate 4a). As the main arcade and *a fortiori* the aisle vaults of the choir could not be built without the existence of the east piers of the crossing, it is obvious that this first stage of work could have been of only short duration. Given the speed at which Durham was built and the enormous means at the bishop's disposal, it is likely not to have lasted even two years. The central structure of the choir was probably taken no higher than the top of the column capitals, and the outer walls of this eastern ensemble (including the main apse) must have been raised only to about the same height, stopping short of the zone of the aisle vaults.

Although this first stage represented just a start, it involved a considerable amount of masonry work, in which all the essential features of the design were already registered; and it is therefore possible to analyse on the mere basis of this very first section the specific character of the fabric and the principles that were applied in its construction.

The Masonry Work

The stonework at Durham is perfectly Romanesque in nature, conforming to the type which had been used for more than fifty years in the great buildings of the northern half of France, of Normandy in particular, and which had been spreading in the last twenty years not only to England in

---

[3] This inauguration was marked by the translation of the remains of St Cuthbert to the apse of the new building on August 29, 1104.

[4] The liturgical choir, where the stalls of the monks constituting the cathedral priory were situated, occupied originally the crossing area and part of the bays adjoining it to the east and to the west: see Arnold W. Klukas, 'The Architectural Implications of the *Decreta Lanfranci*', *Anglo-Norman Studies, Proceedings of the Battle Conference 1983*, 6, Woodbridge, 1984, pp. 136–71 (plan p. 164); but from the second half of the 13th century the choir was transferred entirely to the east of the crossing. In view of that present position of the choir proper and for the simplicity of presentation, it has been decided in this essay to apply the term choir (or choir area) conventionally to the totality of the east arm of the Romanesque cathedral.

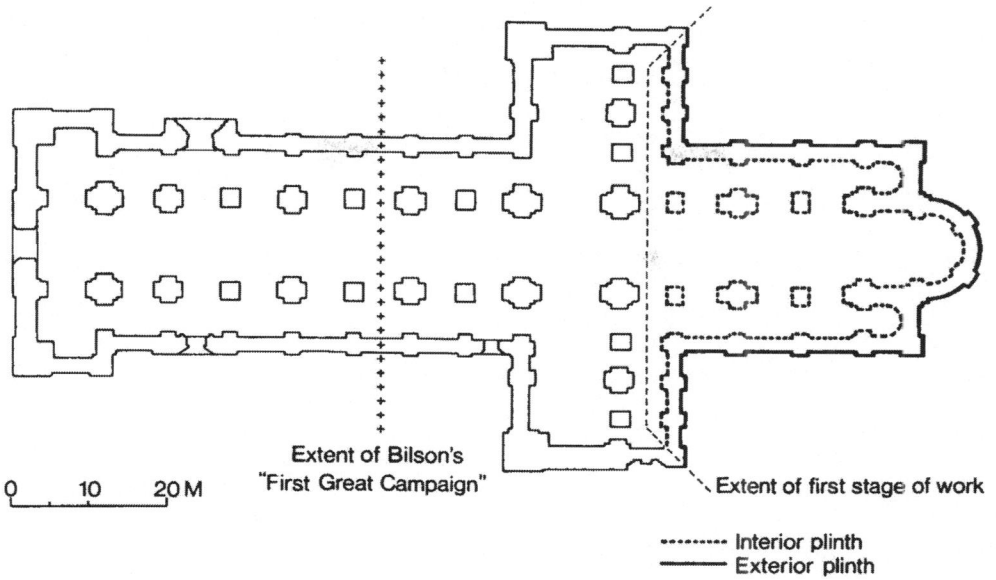

Extent of Bilson's
"First Great Campaign"

0   10   20 M

Extent of first stage of work

-------- Interior plinth
———— Exterior plinth

4a Durham Cathedral, schematic plan showing extent of first stage of construction.

4b Durham Cathedral, interior wall arcading in choir aisle bays.

5a Durham Cathedral, exterior wall arcading on south side of choir bays.

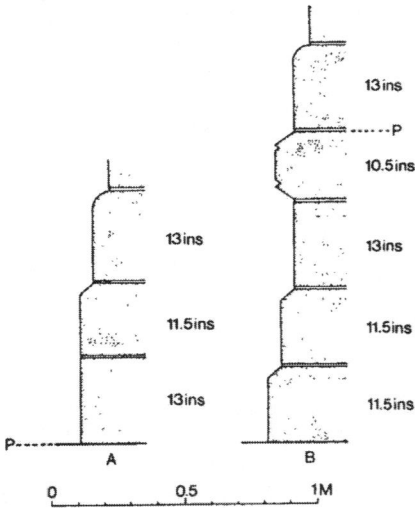

5b Durham Cathedral, plinth profiles
in first stage of work;
(i) Interior plinth; (ii) Exterior plinth.
(P: pavement level in the choir aisle)

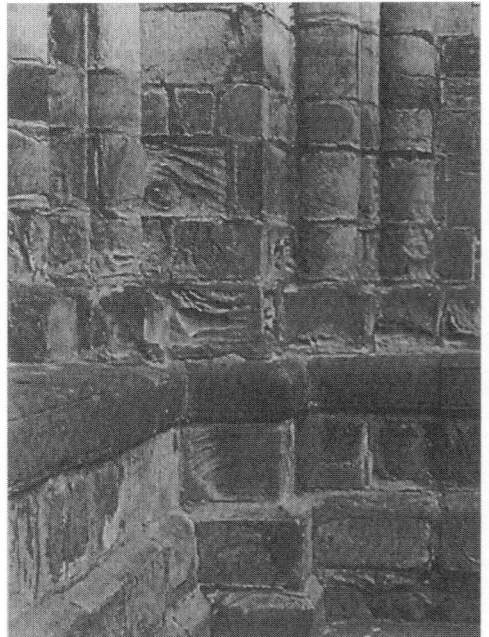

5c Durham Cathedral, exterior plinth at
contact of choir and north transept.

the wake of the Conquest, but also southwards, along the great pilgrimage roads. It is a masonry of fine sandstone ashlar, with joints of very strong mortar about 1.5 cm (.67 inches) thick. In the areas of normal walling and in all the piers of compound type, the courses vary in height between 21 and 40 cm (8 and 16 inches), the great majority of the courses ranging between 24 and 35 cm (2.5 inches on either side of 1 foot). This is very close to the norms of stone production that can be observed at Winchester cathedral, where work had started in 1079, or in the late eleventh-century buildings of Normandy.

The Durham stonework is exceptionally elaborate in the detail of its execution. All the peripheral walls of the building have been designed and built with a dado of arcading, both on the inside and on the outside (Plate 4b). Interior wall arcadings can be found at all periods since late Antiquity, and they were becoming common in the 1080s, particularly in the areas north of the Loire. On the other hand, exterior arcadings at ground level were very unusual, being found at that date only on the outside of a few apses: St-Nicolas at Caen, *c.* 1080–85 is the earliest remaining example, although the apse of St-Etienne in the same city (begun *c.* 1066–67) may well have been the prototype.[5] At Durham this exterior arcading is exceptionally deep and large, normally two arches of two orders per bay, which means that wooden centrings had to be employed already at that low level of construction all along the walls (Plate 5a). Low level arcadings also meant that courses of set height recur more frequently at given levels, being needed for the bases and for the capitals and abaci of the shafts supporting the arches; and these courses of set height are continued horizontally into the wall panels, responds and buttresses. So that changes of height in the coursing should not be viewed, at Durham, as marking interruptions in the process of construction: they indicate only that the wall was reaching a height at which a fixed horizontal level (such as an impost or a string-course) had been set in the design.

The Plinthwork

Even more demanding in terms of course height were the plinths that run at the base of the walls (Plate 5b). Inside, walls and piers have plinths of a fairly simple type, composed of three courses above the level of the pavement: first a course 33 cm (13 inches) high, now reduced to 28 cm (11 inches) through the raising of the pavement level; then a course of blocks of 29 cm (11.5 inches), with their upper edge chamfered; and above that another course of 33 cm (13 inches), out of which are carved the bases of the shafts or columns supported by the plinth. This

---

[5] If so, the original apse of Lincoln Cathedral, which is the English building closest to St-Etienne in its details of execution, would be likely to have had already a low level exterior arcading.

makes a consistent sequence of three courses of set height (two of 33 cm (13 inches), one of 29 cm (11.5 inches) at the base of all the walls and under all the piers in that first-built section of the work (Plates 4b and 6a)

Outside, the plinth is taller and more refined in its profile. It is made up of five courses of pre-set height (Plate 5c). First come, just above the foundations, two successive courses of 29 cm (11.5 inches) chamfered at the top, like the 11.5-inch course of the interior. Next comes a plain course, of 33 cm (13 inches); then a thinner course, 26.5 cm (10.5 inches) only in height, but remarkable by its 7 cm (2.75 inches) projection in a tablet-like profile, which draws a firm horizontal line at the base of the whole structure; and above that comes yet another course of 33 cm (13 inches), which serves at the base course for the outer wall arcade. In total that makes for the whole plinthwork (interior and exterior) three series of chamfered courses of 29 cm (11.5 inches), four of plain rectangular-cut courses of 33 cm (13 inches), and, for the outside plinth only, one course of 26.5 cm (10.5 inches) shaped like a powerful string-course.

As the laying down of the plinths was the first act in the construction of the cathedral above foundation level, this importance given by the designer to the plinthwork means that a considerable amount of blocks cut to a set height had had to be ordered from the quarries before construction could start and be brought on the site in advance for the masons to arrange them in the required sequences before passing to the mounting of the wall arcades and eventually of the more usual ashlar masonry.[6] To get an idea of the quantities involved, one has only to measure the overall length of the exterior and interior plinths all along the extent of this initial stage of work, i.e. from the north-east angle of the transept, around the whole east arm of the church (with its original apses as revealed by the excavations of 1895), to the south-east angle of the south transept arm.[7] The length of the outer plinth proves to have been in the order of 130 metres (426 feet 6 inches); and the interior plinth, which had to follow the deep curves of the apses, and has to be supplemented by the plinths of the three pairs of piers of the choir, can be evaluated at about 200 m (656 feet). A sampling having indicated that the blocks used in the plinthwork average in length 80 to 85 cm (31.5 to 33.5 inches), one comes to a total, for that preliminary order of stone cut to set heights, of some 850 blocks of 33 cm (13 inches), some 600 chamfered blocks of 29 cm (11.5 inches), and about 180 of the 26.5 cm (10.5 inch) type: in all, an order of well over 1500 blocks of specified height, in addition to what must also have been needed in the matter of current walling material in more variable heights.

[6] The importance given to the plinthwork is a well known characteristic of Anglo-Saxon architecture both before and after the Conquest, Repton being probably the earliest example preserved.

[7] On the excavations of 1895, see John Bilson, 'On the Recent Discoveries at the East End of the Cathedral Church of Durham', *Archaeological Journal*, 53, 1896, pp. 1–18.

The Pierwork

Impressive as it is numerically, in respect of the mass involved, the planning of the plinthwork was simple and cannot in any way be compared with the degree of sophistication of the calculations implied by the designing and execution of the two pairs of great columnar piers of the choir area (Frontispiece). These represent an extraordinary achievement in terms of stonework planning; and the procedures that had to be followed to produce them, so demanding in technical skills, were to be repeated with slight variations in the transept, built in the second stage of the works, and were to set the basic rules for the somewhat later and simpler pier variations of the nave. Two very specific and unusual elements gave its complexity to the designing of that first series of columnar-type piers: one is the spiral patterning incised on the cylindrical part of the piers, and the other the merging in those piers of two principles of shaping, for they are not fully circular in plan, one third of the circle, on the aisle side, being replaced by a respond of the compound pier type (Plate 6a).[8]

The problem of the incised patterning is solved, in the Durham choir, with the greatest mathematical elegance. Until then, patterned piers were nearly all monolithic columns, which could be carved as a single unit, in the manner of a statue, and with no particular snags once the design had been carefully set. When columns with incised patterns were built in drums, as was attempted at Christchurch, Canterbury, in part of the undercroft to the enlarged dormitory building of the 1080s, the difficulty of making the pattern incised on one drum run smoothly into the lines of carving of its neighbours became painfully evident. The Durham master, for his much larger piers, which had to be built of multiple courses of normal ashlar masonry, found the perfect solution by linking the patterning of the piers to the joint alternate pattern of the blocks [9] in the stonework and by designing a type of block diagonally incised at such an angle that, with an exact mounting, the incision would be continued automatically from one course to the next.

The actual formula that produces the spiral pattern in the piers of the Durham choir is merely the repetition, joint alternate fashion, of twenty-seven courses of rigorously identical blocks 25 cm (9.75 inches) high and 42.5 cm (16.75 inches) long, all carrying the same diagonal incision, cut

---

[8] The origin of the spiral pier and its symbolic meaning will not be discussed here. On these aspects of the question, see Eric Fernie, 'The Spiral Piers of Durham Cathedral', *Medieval Art and Architecture at Durham Cathedral*, The British Archaeological Association Conference Transactions, 1980, pp. 49–58.

[9] On this point see Jean Bony, 'Durham et la tradition saxonne', *Etudes d'art médiéval offertes à Louis Grodecki* (S.M. Crosby, A. Chastel, A. Prache et A. Chatelet, eds.), Paris, 1981, pp. 79–92.

at the same angle and to the same profile (Plate 6b).[10] There are nine such carved and incised blocks in each odd number course, these courses ending in a vertical joint at the contact with the respond part of the pier; and, in the even number courses, eight blocks of that type, plus, at both ends, two half blocks of that curved shape, which become rectangular in their second half to merge into the profile of the pilasters of the respond. The designing and execution was impeccable. Nothing of the kind had been attempted before in medieval architecture: no previous example can be found of masonry work conceived and executed as the high precision assemblage of blocks cut in advance in such a way as to be not only interchangeable but reversible (a fine piece of template designing). In fact this was only carrying to its logical extreme the type of stonework introduced into England from the continent; yet significantly it was the problem raised by the patterning of columnar piers that made the designer of Durham convert the visual surface regularity of Romanesque masonry into absolute mathematical regularity.

This first Durham master, who clearly was English, purely Saxon in sensibility, especially in his sense of plastic and linear values, but no less clearly Norman-trained and enjoying to the full the new Romanesqueness, more Norman even than any Norman in his use of exterior dado arcades, gave in those piers the most decisive demonstration that he could outdo all continental builders in the handling of ashlar masonry.[11] Producing there a virtuoso performance in rationalised stonework, he placed himself far ahead of all would-be competitors, technically at the apex of the most advanced modernity.

A well-hidden showpiece of that virtuosity was the way in which he designed the bonding blocks which, on alternate courses (the even number courses, as has just been seen), must, according to the rules of sound masonry, bind together the two parts of those columnar piers, their cylindrical section and their respond section. For these bonding blocks, which start as half-blocks of the standard curved type on slightly over 21 cm (8.75 inches), have to continue, in their second section, as blocks of plain rectangular cut, 34 cm (13.5 inches) in length and set at an exactly calculated obtuse angle in relation to their curved section, to be in the right alignment for the respond (Plate 7a). No approximation was acceptable here; and an absolutely accurate template of that unusual composite shape had to be established and strictly followed by the stone cutters for that paradoxical junction of two forms to look perfectly normal and pass unnoticed.

[10] When applied as here to blocks with a curved surface or to their templates, the term length in all this article will mean the length of the developed arc to which the block has been cut.

[11] Joint alternate coursing means that, in a masonry composed of blocks all of the same size, the vertical joints of one course are placed exactly above the middle of the blocks of the preceding course.

6a Durham Cathedral, easternmost columnar pier of choir on north side, seen from the aisle.

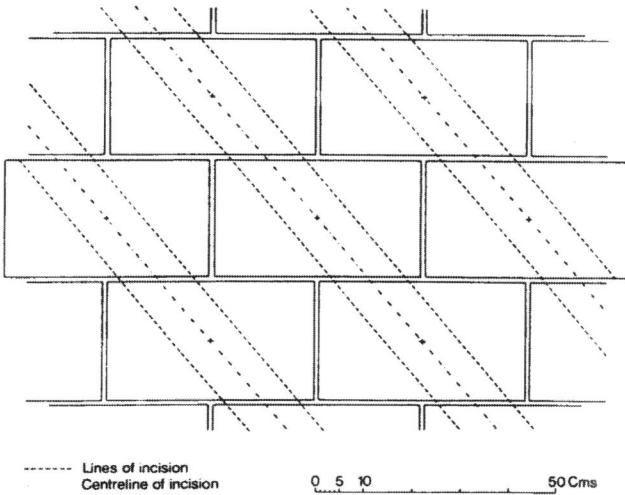

Lines of incision
Centreline of incision   0 5 10   50 Cms

6b Durham Cathedral, pattern of arrangement of incised blocks in columnar piers of choir.

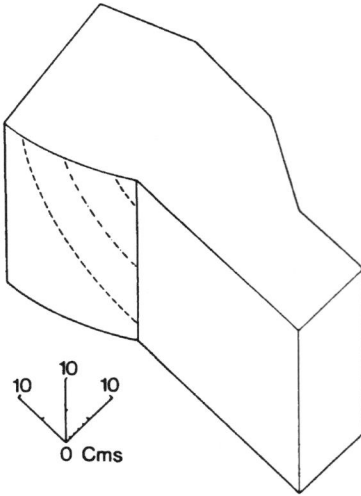

10
10 | 10
0 Cms

----- Lines of incision
- - - - Centreline of incision

7a Durham Catheral, diagram of a bonding block in columnar piers of choir.

7b Durham Cathedral, analysis of the plan of a columnar pier in choir.

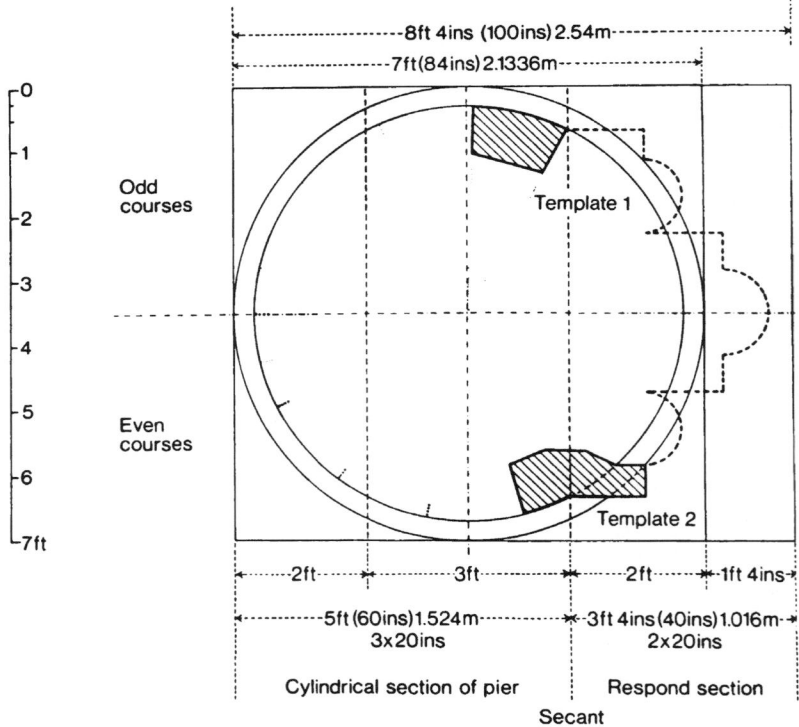

--------------8ft 4ins (100ins) 2.54m--------------
--------------7ft (84ins) 2.1336m--------------

0
1
Odd
courses
2
Template 1
3
4
5
Even
courses
6
Template 2
7ft

----2ft----|----------3ft----------|----2ft----|-1ft 4ins-
--------5ft (60ins) 1.524m--------|-3ft 4ins (40ins) 1.016m-
3x20ins   2x20ins

Cylindrical section of pier   Respond section

Secant

Each of those incised choir piers required, for the construction of its cylindrical core alone, a specified number of stone blocks cut with the highest precision to two different templates; and while, for each pier, twenty-six blocks only were needed of the complex bonding type (two for each of the thirteen even number courses), the masons had to be provided with as many as 230 of the standard 16.75 inch long curved type (14 x 9 for the odd number courses, plus 13 x 8 for the even number courses). As this first stage of work included four such piers, the quarries had to deliver within short limits of time a total of 920 blocks of the standard curved type, and 104 of the complex half-curved half-rectangular type, all impeccably sized and cut to templates.

This calculation accounts only for the rounded part of the pier. The respond part required in addition its own series of blocks to be shaped into the shafted and pilastered elements of which responds are composed. Examined on the spot (as far as the original mortar joints can be reliably recognised) this respond section seems to be composed, in all four piers, of seven blocks in the odd number courses and five only in the even number courses, in which the bonding blocks form already the outer pilasters. This would mean, for each pier, a further requirement of probably 163 blocks of various shapes, which are unlikely to have been more than roughly cut at the quarry, being carved in their finished form on the building site, but had necessarily to be all of the fixed height of 25 cm (9.75 inches) to agree with the strict coursing of the piers of which they were to become a part. The total amount of those respond type blocks for the four choir piers could then be tentatively evaluated at approximately 652 additional blocks of that semi-finished type, to be brought to the building yard at the same time as the high precision series described above, before the construction of the piers could begin.

These figures in the end become rather tiresome, although they are more or less unavoidable in any such listing of standardised units. Yet the implications of that state of affairs are not without interest. That the quarries were able to face such demanding specifications shows the high degree of competence achieved in the 1090s by the stone cutters of the Durham area.[12] More remarkable still and more significant in terms of

---

[12] In the first twenty years following the Conquest, ecclesiastical buildings of some importance had not been numerous in Northern England: at Durham itself Bishop Walcher had begun between 1072 and 1080 the construction of an ensemble of monastic buildings; but the programmes of repair and enlargement on the old sites of Jarrow and Monkwearmouth seem to have proceeded very slowly. The only new foundation of significance and requiring fine ashlar work, Tynemouth Priory, was begun only *c.* 1089. On the other hand masons and quarries must have been kept busy by the works of fortification needed for the defence of the northern frontier of the kingdom, the most important of which were the royal castles of Durham, founded in 1072, and of Newcastle, founded 1080. Richmond Castle, in northern Yorkshire, was another early stone castle. Castle building being always urgent work, the quarries must have developed the means of facing demands for the rapid production of large amounts of stone.

the mental training it pre-supposed is the amount of calculation and of advance designing carried to the minutest details of execution,which had had to be performed by the Durham master, with an unfailing decisiveness, to be able to order from the quarries those large series of perfectly shaped blocks that were so rapidly needed on the building site. It is easy for us now to observe and register on the spot the size and shape of the stone blocks and the rigour of their mounting. But how had the rules been set for that prefabrication and that assembling? Why twenty-seven courses in the piers? Why the height of 25 cm (9.75 inches)? Why that length of 42.5 cm (16.75 inches) for the standard type of curved and incised blocks? How had the templates been designed that dictated the work of the stone cutters?

This arithmetical and geometrical aspect of the designing is not so difficult to reconstruct. The height of the blocks and the number of the courses were obviously commanded in the end by the proportioning of the stories in the general design of the elevation. For the main arcade story up to tribune floor level to reach the height that had been ascribed to it, i.e. 12.471 m (40 feet 11 inches), the column-like body of the piers between base and capital had to be roughly 7 m (23 feet) tall, but slight variations were admissible and the eventual division of that height into twenty-seven courses of 25 cm (9.75 inches) resulted from the interaction of some other factors.[13]

An odd number of courses was necessary to give the proper bonding on alternate courses between the two parts of the pier (column-like core and respond); and twenty-seven courses of 25 cm (9.75 inches), plus the thickness of the joints, which averages one centimetre in the horizontal joints, comes to a total only 3.5 cm short of seven metres. The height of the blocks seems on the other hand to have been finally dependent on their length, for the solution to the problem of patterning had consisted in the creation of a type of incised block in which height and length were interdependent; and the length of the blocks was determined (as we shall see) by the geometric construction along which the plan of these dual shaped piers had been arrived at. The solution adopted for the combination in one pier of two different pier forms seems therefore to hold the key to the whole sequence of precise calculations that we find reflected in the stonework of those Durham choir piers.

Geometry of the Pier Plan

Once the architect of Durham had judged that the springers of the rib-vaults of the aisles could not be crowded onto the abacus of a column capital and that they would have to be supported on applied responds of

---

[13] The figures given here are based on Peter Kidson, *Systems of Measurement and Proportions used in Medieval Cathedrals*, Ph.D. Dissertation, London University, 1956.

the compound type, next to be decided was the point of junction between the two pier forms. A segment of the cylinder constituting the core of the pier would have to be sliced off, on the aisle side, to provide the flat vertical surface against which the wide three shaft respond could be applied. But in what precise position should the secant be placed?

Accurate plans of the columnar-type piers of the Durham choir show that the plinths on which they rest form in plan a rectangle of 2.13 by 2.54 m (7 feet by 8 feet 4 inches, or 84 by 100 inches), 2.13 m (7 feet) being their east-west dimension and 2.54 m (8 feet 4 inches) their north-south dimension.[14] They also show that the point where the two pier forms meet is situated at the distance of 1.52 m (5 feet, or 60 inches) from the choir face of the plinth. This position indicates a simple geometric construction, which must have been performed by drawing, on any tracing floor or tracing board, a full scale plan of the top surface of the plinth (Plate 7b).

First had to be drawn a 2.13 m (7 by 7 foot) square; and, inscribed within it, two concentric circles: the first one, tangent to the sides of the square and with a radius therefore of 1.06 m (3 feet 6 inches), giving the circumference of the base of the pier; and the second, with a radius of only 98 cm (3 feet 2.5 inches), shorter by 9 cm (3.5 inches) because this was the projection that had been decided for the base, giving the actual volume of the full cylindrical core of the pier. The next step was to draw a line cutting the square and its inscribed circles at the distance of 1.52 m (5 feet) from the side of the square which was to face the central space of the choir. This line cut off from the square a 61 cm (2 foot) length (a 2 to 5 relationship with the rest of the square) and at the same time it automatically also cut from the circumferences of the two inscribed circles two concentric arcs. The length of the chord subtending the arc of the smaller circle (the circle corresponding to the solid body of the round pier) gave the width available for the application of the respond on the aisle side of the pier. Then, in a third step, in order to provide a proper support for the bulk of that respond (which was to receive the three ribs of the projected aisle vaults), the 61 cm (2 feet) of plinth cut off by the secant line were extended to a length of 1.02 m (3 feet 4 inches, or 40 inches), so as to place the respond section of the plinth in a 2 to 3 relationship with the 1.52 m (5 foot) section supporting the

[14] Robert W. Billings, *Architectural Illustrations and Description of the Cathedral Church at Durham*, London, 1843, gives on plate LXA a measured plan of the northernmost pier of the south transept, which is similar to the piers of the choir, the only difference being that the plinth of that pier is reduced in its dimensions by one inch in relation to the choir piers: its width is 2.12 m (6 feet 11 inches) as against 2.13 m (7 feet) in the choir, as indicated in Billings in plate II.

32   *Medieval Architecture and its Intellectual Context*

cylindrical core of the pier.[15]

It was that geometric construction and that principle of proportioning which, by determining the precise placing of the points of junction between the two pier forms, made possible the fixing of the length and shape the blocks should have and the finalising of the design of the stonework in the piers. The establishment of the templates represented the last step in that computation and its final product. The points of contact between the two component parts of the pier now being fixed, what remained of the core became measurable and its circumference could be divided by whatever number of blocks would give a manageable size for the masons to execute the work of stone cutting, shaping to template and finally incising to pattern; and the building shows that the designer decided upon nine blocks (three sectors of three blocks), with the result that, once the width of the mortar joints had been taken into account, the length of these standard blocks turned out to be 42.5 m (16.75 inches).[16] The curve of the blocks was already given by the circle of which their length represented a sector (of about 25 degrees); and the angle at which their end surfaces were to be cut simply followed the radii of that circle. The template for the blocks could thus be cut with absolute precision in all its elements from the tracing floor.

Similarly the second template, so odd in shape, that was to be used for the bonding blocks on alternate courses, could also be copied directly from the same tracing floor: it was just a matter of following the lines inscribed, in the course of the preceding operations, on either side of the point of contact between the round pier core and the straight-sided respond pilaster. From these lines could then be read the form of the template: a first section curved, on a length of just over 21 cm (8.4 inches), i.e. half the length of the standard curved block; and a second section straight and just over 34 cm (13.5 inches) long, the measurement set in the respond design for the outer pilaster.

The whole design of this pier plan and of the templates it implied must have been settled very rapidly by so expert a designer as was clearly this first Durham master. After that, the calculations in height and

[15] It will be noted that these measurements make sense only in standard English feet and inches. This would seem to confirm that the English foot was already in use in the late eleventh century (see also E.C. Fernie, 'Anglo-Saxon Lengths: the 'Northern' System, the Perch and the Foot', *Archaeological Journal*, 142, 1985, pp. 246–254). This system of measurement, used in the small scale calculations, was combined at Durham, as shown by Peter Kidson (see n. 13 above), with another system, based on the toise of 1.42 m (4 feet 8 inches), for the large scale design of plan and elevation. The combined use of these two systems seems confirmed by the recurrence at Durham of a measurement of 2.13 m (7 feet), which is 1.5 toise and therefore the meeting point of the two systems.

[16] The vertical joints, which are not compressed by the weight of the blocks, are a little thicker than the horizontal joints: they average 1.3 cm (.5 inches), while the horizontal joints average 1 cm (.4 inches). For the meaning of the term length in this paragraph and in the following one, see note 10 above.

setting of the formula of the spiral patterning (which had to be established by tracing on the developed surface of the cylinder) could go ahead in all surety on the bases of the rigorous planning carried out at ground level.

Once this sequence of essentially simple but precise operations had been completed and fully coordinated the master could proceed to evaluate the quantities of stone required for the initial stage of work. Only then could he calculate exact numbers and place his order for two series of custom blocks (920 of the standard curved type and 108 of the composite bonding type) that had to be supplied for these first two pairs of complex columnar-type piers. The extent and accuracy of the planning required by its highly refined design placed Durham on a level of sophistication that was to remain unequalled for a long time.

There must seem to be in this essay an element of overkill, in its repetitious insistence on numbers. Yet it must be recognised that, as soon as we start examining its masonry closely, the building demonstrates that it could never had been started, not even conceived, without methods of advanced planning so unexpected in their elaboration that they force us to revise our current ideas on building yard practice in the Romanesque age. This stonework therefore becomes an essential new document on the intellectual history of the late eleventh century in that it enables us to come to some realisation of the nature of the mind that produced the design of Durham cathedral and that directed at least the early stages of its construction.

Obviously the designer who could handle with ease all these material issues was no common builder. But the date at which all this was taking place is the crucial element that gives a very special significance to what this man did and what his actions implied. His methods of serialisation and rationalisation, the evidence of his mathematical training, the urge he clearly felt for an absolute accuracy in the work of his teams as in the operations of his own mind place him as one of the identifiable fore-runners of that great mental shift of the early twelfth century. This (as we have recently been reminded) should not be viewed as a 'Renaissance' (so ambiguous and restrictive a term) but as a first Age of Enlightenment, affecting all aspects of culture in the western world.[17]

The spectacular achievements to which are attached the names of such men as Adelard of Bath in the scientific field or Pierre Abelard in matters of logic and ethics, came only some twenty five or thirty years later. But the movement had started before the end of the eleventh century, and not only with the jurists of Pavia, for the capture of Toledo and the conquest of Sicily had just given westerners access to the whole

[17] Charles M. Radding, *A World Made by Men: Cognition and Society, 400–1200*, Chapel Hill, 1985 (pp. 151 and 256 for the use of the term Enlightenment).

range of Hellenistic and Arabic science. In the generation of the 1090s, an exact contemporary of the master of Durham, like him working in England, was the other forerunner of the new science, Walcher of Malvern. In 1092 (probably the very year when the plans of Durham were elaborated) Walcher was measuring with the highest precision, by means of a Toledan astrolabe, an eclipse of the moon, to establish differences of longitude and cosmological time.[18]

Certainly Durham, situated so far to the north, seems very much out of the way. In fact it is not so surprising to find there, at that date, a remarkable manifestation, not to say manifesto, of avant-garde thinking: for the bishop who commissioned the cathedral, William of Saint-Calais, had just spent three years in exile at the ducal court of Rouen, as one of Robert Curthose's most trusted advisers, from 1088 to 1091; and there is every likelihood that his architect was part of his retinue. Arriving in Rouen in 1088 must have been an unusually stimulating experience: Toledo, reconquered three years before, had become the centre of attraction towards which scholars from all over western Europe were beginning to converge, to consult the treasure of scientific and philosophical manuscripts kept in the *armarium* of the former Great Mosque, now the cathedral. Norman scholars had doubtless been there already. Normandy was also, and more directly, involved in another no less momentous enterprise, the conquest of Sicily, by then almost complete; news was coming back of the wonderful richness of the libraries which had been found at Syracuse, captured two years before. Bishops were soon to be appointed to the new Sicilian sees and among those who were eventually chosen (he became Bishop of Syracuse in 1105 and must have been before that well known in the circles of Rouen) was that famous Norman arithmetician, Guillelmus R., who, we are told, had been the master of two royal clerks responsible, under Henry I, for the establishment of the accounting system of the English Exchequer.[19]

We can only have the most uncertain and hypothetical notion of what was happening in Rouen in those years; we cannot even say with any certainty that the designer of Durham was actually there and still less whom he could have met and how far he could have travelled outside Normandy between 1088 and 1091; but it looks very much as if the latest advances in the mathematical sciences had been directly responsible for the mastery in computation and planning that we find registered in the stonework of Durham cathedral. To read in clear this rare and precious testimony on the intellectual conjuncture of the 1090s, all that is needed is a bit of decoding.

[18] Dorothee Metlitzki, *The Matter of Araby in Medieval England*, New Haven, 1977, pp. 16–18.

[19] Michael T. Clanchy, *From Memory to Written Record: England 1066–1307*, London, 1979, pp. 108–10 and 235.

9

Archaeology and engineering: the foundations of Amiens Cathedral

S. Bonde, C. Maines and R. Mark

The significance of Gothic architecture resides in its extraordinary integration of structural and aesthetic innovation. As an architectural style, it broke decisively with Roman and Roman-derived forms to evolve new wall configurations with increased fenestration set within taller structural systems. These innovations have consistently attracted scholarly attention, of which a significant portion has been devoted to explanations of structural elements such as rib vaults and flying buttresses. However, the attention paid to the innovative technology of Gothic architecture has rarely extended to include the foundation systems which support these buildings. Despite their obvious functional importance in relation to increasingly taller buildings, and their crucial position as the earliest built phase of a new building's design, foundations have been largely ignored in architectural scholarship, and omitted from most published section drawings.

Those few studies which have treated foundations have attempted to trace the presumed formal evolution of foundation walls (H. Bernard, "Essai sur la genèse des fondations gothiques," *Cahiers archéologiques de Picardie*, 1975, pp. 85—100). However, this approach ignores the crucial relationship between any foundation and the geology of its building site. The function of a foundation is to distribute the load of the building's superstructure to an area of supporting soil or rock so as not to exceed the latter's capacity to carry that load. Buildings do not simply rest *on* the ground, but rather they interact with the soil below. Thus, analysis of medieval foundation systems must account for geological factors as well as structural principles. To take one example, analysis of the foundations beneath Amiens Cathedral reveals that medieval builders were aware of local site geology and that they used it to plan the substructure of their building.

Beneath the floors of Notre-Dame d'Amiens, a stepped, interlocking grid of below-ground walls rest on an evidently continuous raft of mortared stone. Our knowledge of Amiens comes from excavations carried out sporadically between 1850 and 1897, first under the direction of Eugène Viollet-le-Duc and then from 1874, under Just Lisch. As no report of the excavations seems ever to have been published, one must rely on two textual sources: Viollet-le-Duc's discussion of the footing below a pier buttress (*Dictionnaire*, IV, p. 175) [*Fig. 1*] and Georges Durand's plan and elevation of a section through the northern half of the choir (*Monographie*, I, pp. 202—204) [*Fig. 2*]. From these two brief descriptions, it is unclear how extensive an area of the Amiens foundations was actually revealed. At a minimum, one bay in the choir side-aisle and an area between the buttresses must have been opened, but a much larger area may have been excavated since Durand indicates that the paving was taken up in the choir (p. 202).

In spite of these limitations, there is far more information about the foundations beneath Amiens Cathedral than is available for most large Gothic structures. In the *Dictionnaire* (see *Fig. 1*), Viollet provides a description of the form, stone types, depth and soil conditions for a pier buttress of the choir at Amiens: *En A, est une couche de terre à brique de 0,40 c. d'épaisseur posée sur l'argile vierge; en B, est un lit de béton de 0,40 c. d'épaisseur; puis, de C en D, quatorze assises de 0,30 à 0,40 c. d'épaisseur chacune, en libages provenant des carrières de Blavelincourt près d'Amiens. Cette pierre est une craie remplie de silice, très-forte, que l'on exploite en grands morceaux. Au-dessus, on trouve une assise E de pierre de Croissy, puis trois assises F de grès sous le sol extérieur. Au-dessus du sol extérieur, tout l'édifice repose sur six autres assises G de grès bien parementées et d'une extrême dureté. Derrière les revêtements de la fondation est un blocage de gros fragments de silex, de pierre de Blavelincourt et de Croissy, noyés dans un mortier très-dur et bien fait. C'est sur ce roc factice que repose l'immense cathédrale.* (IV. 175—176)

Durand published a larger plan and section and provided a fuller description of the extent and depth of the excavations at Amiens: *Les fouilles faites jadis par Viollet-le-Duc à l'extérieur de l'édifice, celles qui furent faites en 1894 et 1897 lors de la réfection du dallage et surtout de la construction d'un caveau pour les évêques dans la travée 21, 23 bc, permettent de se rendre compte d'une façon à peu près complète de la disposition des fondations de la cathédrale d'Amiens. Elles sont aussi colossales que le monument lui-même....Construites à la manière d'un immense radier, elles assurent à celui-ci une assiette inébranlable et une solidité de nature à défier toutes les causes de destruction. En voici le plan et la coupe sur la travée 21, 23 du choeur.*

Le terrain sur lequel s'élève la cathédrale est remblayé sur une hauteur d'environ 7 à 9 mètres au-dessus du bon sol, remblai bien antérieur au XIIIe siècle. Dans toute l'étendue de la surface de l'édifice, ce remblai a été entièrement creusé jusqu'au sol vierge. Sur celui-ci on a étendu, aussi très probablement dans toute la surface de l'édifice, un massif de moëllonage noyé dans le mortier C D E F, plus ou moins épais suivant la déclivité du terrain, de manière à obtenir une surface horizontale à environ cinq mètres au-dessous du sol de l'église.

Sur ce massif gigantesque, véritable roc factice, sont établis de chaînages de pierre G longitudinalement sous les murs extérieurs et sous les lignes des maîtres piliers et

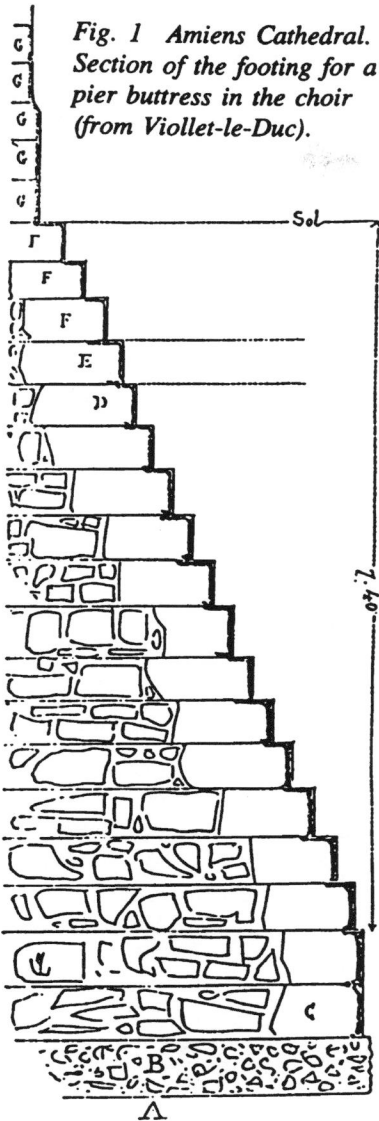

Fig. 1 Amiens Cathedral. Section of the footing for a pier buttress in the choir (from Viollet-le-Duc).

transversalement entre chaque travée, formant ainsi un vaste grillage de maçonnerie, aux intersections duquel s'élèvent tous les piliers. Ces chaînages s'élèvent presque jusqu'au sol actuel, c'est-à-dire sur une profondeur de près de cinq mètres. Epais d'environ 2m, 40 à leur sommet, ils se composent de gros moëllonages parementés de onze assises de libages de pierre que l'on croit être de Blavelincourt mais qui doit être plutôt de Belleuse, très dure et très ferme, d'environ 40 centimètres de hauteur d'assises, descendant avec des ressauts de 15 à 16 centimètres d'empattement jusqu'au massif plein. A chaque intersection, les angles sont renforcés à la partie supérieure des chaînages, en forme de goussets, de manière à donner aux piliers une assiette octogonale. Vers ·l'extérieur, les gradins de libages descendent jusqu'au bon sol, pour arrêter et soutenir le massif général, et les assises supérieures sont en grès. (I. 202—204.) The combined information of Viollet-le-Duc's and Durand's descriptions and illustrations provide the basis for a quantitative assessment of the foundations of Amiens Cathedral within their geologic context.

The stability of Amiens Cathedral may be understood by comparing the bearing pressure due to the building's weight with the corresponding bearing capacity of the supporting soil. Bearing pressure (BP) is calculated by dividing the building weight supported by a footing by the area at the base of the footing:

$$\text{BP (in metric tons/m}^2) = \frac{\text{weight in tons}}{\text{area at base (m}^2)}$$

The bearing capacity of a soil (the maximum soil pressure corresponding to acceptable foundation settlement) is today determined from on-site tests. Modern values typically range from a high of 1000 tons/m² for massive rock, to a low of 10 tons/m² for footings set on soft clays. When bearing pressures exceed the prescribed values for particular soils, the building structure is put at risk and may even be subject to catastrophic failure.

The calculations required for foundation analysis of buildings like Amiens can be simplified by using a standard engineering approach which takes advantage of the

343

modular nature of medieval construction. Because all loadings from the bays are directed onto the piers and buttresses, these can then be analyzed individually and, following engineering practice, the results can be generalized to the building as a whole. The relationship between the bearing pressure of one main arcade pier and one pier buttress of Amiens Cathedral and the bearing capacity of the soil beneath the building at those points can thus be discussed from available soil descriptions and measurements for the superstructure and foundation. In the calculations which follow, we have computed only dead weight loadings and soil pressures. Live loads, such as those caused by wind, have small enough effect on foundation behavior in buildings of this type that those loads may be neglected.

The total weight at the base of one pier of the arcade superstructure at Amiens Cathedral was taken from R. Mark and R. A. Prentke ("Model Analysis of Gothic Structure, "*JSAH*, 27, 1968, pp. 44—48). Choir weights were assumed to be equivalent to those of the nave. For the foundation beneath a choir arcade pier, the volume was estimated by multiplying the area of the footing at half of its depth by the depth of the stepped footing. [*Fig. 3*] The volume of the continuous raft below the footing was found by multiplying its area by the raft thickness. These volumes were then multiplied by an average density for limestone taken as 2400kg/m^3 and added together to arrive at the total weight:

$$
\begin{array}{ll}
\text{arcade superstructure weight} & = 0.55 \times 10^3 \text{ tons} \\
\underline{\text{arcade foundation weight}} & \underline{= 1.11 \times 10^3 \text{ tons}} \\
\text{total weight} & = 1.66 \times 10^3 \text{ tons}
\end{array}
$$

For the total weight of a pier buttress, and portions of the adjacent side-aisle wall and flying buttresses, calculations were similarly taken from Mark and Prentke. Footing weights were found following the approach used for the central vessel foundation. A portion of the footing below the pier buttress had to be calculated separately because it steps continuously downward to natural soil and does not, according to Durand, rest on the continuous platform. [*Fig. 4*] These volumes were multiplied by 2400kg/m^3 and combined to arrive at total weight.

$$
\begin{array}{ll}
\text{pier buttress superstructure weight} & = 0.84 \times 10^3 \text{ tons} \\
\underline{\text{pier buttress foundation weight}} & \underline{= 0.98 \times 10^3 \text{ tons}} \\
\text{total weight} & = 1.82 \times 10^3 \text{ tons}
\end{array}
$$

The bearing pressure under the footings of Amiens Cathedral can now be calculated by dividing the total weight by the respective area at the base of the footings:

$$
\text{BP (pier)} = \frac{\text{total weight}}{\text{area at base}} = \frac{1.66 \times 10^3 \text{ tons}}{70 \text{ m}^2} = 23.7 \text{ tons/m}^2
$$

$$
\text{BP (buttress)} = \frac{1.82 \times 10^3 \text{ tons}}{71 \text{ m}^2} = 25.6 \text{ tons/m}^2
$$

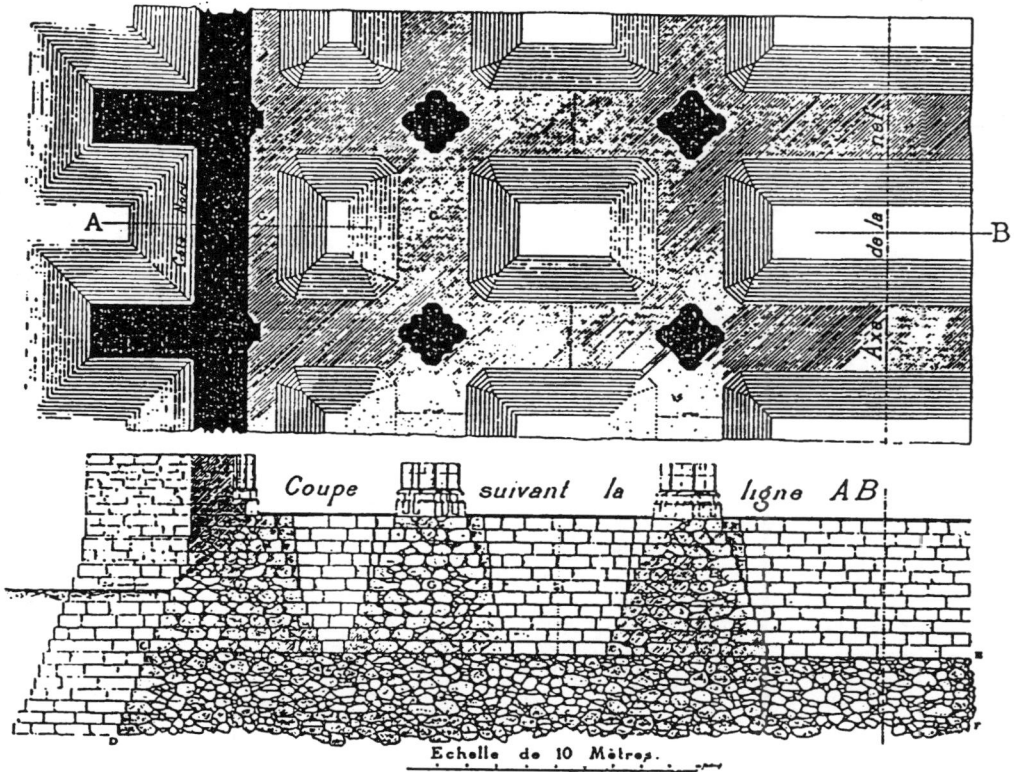

Fig. 2 Amiens Cathedral. Plan and section of the foundations in the choir (from Durand).

Viollet-le-Duc's excavation beside the pier buttress at Amiens Cathedral revealed that the soil beneath the foundation was not the bedrock one might have expected in a building of that size, but rather an „argile vierge". At a subsurface depth of nearly 8.00m, we can reasonably assume that the natural clay Viollet discovered was a Tertiary Sea deposit already pre-consolidated from the weight of its water overburden. Modern prescribed bearing capacity for hard clays, such as those deposited beneath ocean floors, is 40 metric tons/m^2. Thus the bearing pressure of Amiens Cathedral falls reasonably within the soil's capacity to carry the building.

Several important observations may be suggested from our analysis of Amiens Cathedral. First, the structure at Amiens is remarkable for its efficient use of the clay soil supporting it. By our calculations, the building uses about 60 % to 65 % of the allowable bearing capacity of the clay. Second, the bearing stress exhibited beneath the choir pier and that beneath the pier buttress are remarkably close, 23.7 and 25.6 tons/m^2 respectively. The similarity of these stresses is one of the factors which has undoubtedly minimized the risk of differential settlement in a stone building the size of

345

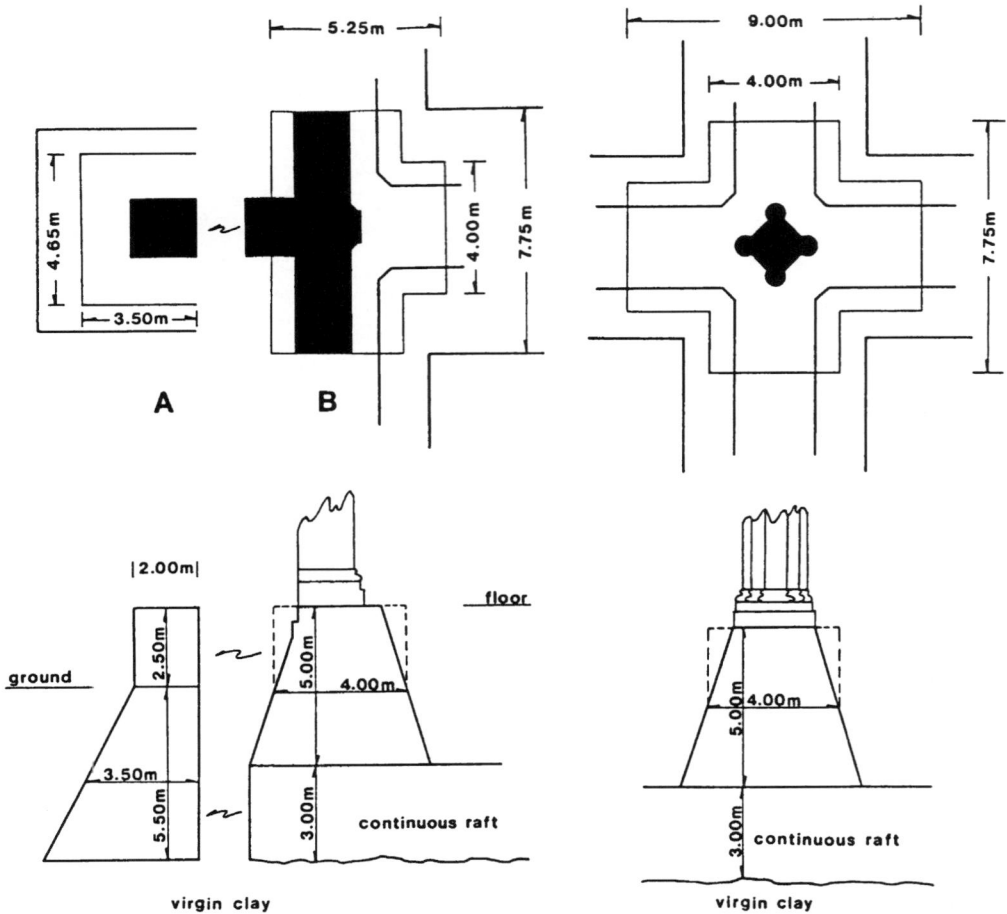

Fig. 3 Amiens Cathedral. Schematic plan and section used for estimating the volume of the footing beneath a pier of the choir arcade (on the right side).

Fig. 4 Amiens Cathedral. Schematic plan and section used for estimating the volume of the footing beneath a pier buttress (A) and the lateral wall (B) of the choir (on the left side).

Amiens. Third, although no information on the nave foundations seems to have been recovered during the 19th-century excavations at Amiens, the results presented here provide technological evidence which supports our hypothesis that they must be comparable to the foundations in the choir. Fourth, the foundations themselves are of such enormous size and weight that their construction would have revealed any problems of instability or differential settlement *before* work on the superstructure began.

A number of provisional conclusions about medieval building practice may also be suggested from this analysis. It would seem that the medieval builders had good understanding of the advantage of distributing the load of a tall building over a wide

346

Fig. 5 Amiens Cathedral. Section of the choir including foundations. This section is pastiche created from two separate section drawings originally published by Durand.

area. By opening up a huge area (presumably the entire surface area of the cathedral) to construct a broad, mortared platform raft, the builders seem to have quite deliberately distributed the building load. On a clay stratum adjacent to the river Somme, which may well have meant a high water table, this decision would seem to have been a wise one. The stepped foundation walls at Amiens were the only viable medieval solution for stone footings which needed to descend nearly 8.00m. Narrow, unstepped walls would not have spread the load widely enough over the clay soils; broad, unstepped walls would simply have been too heavy for the clay soil below. Thus, the foundations at Amiens do not represent a stylistic solution to foundation design, but rather a logical, technological response to a particular site geology. Furthermore, the development of the interlocking grid at Amiens may have evolved similarly out of a desire to stabilize large foundation walls which would otherwise have been supported laterally on the interior only by backfill.

The remarks made here on Amiens Cathedral represent the preliminary results from a larger, state of the question study which the authors now have underway on Gothic foundation systems. The methods used in the present study are extended to a series of Gothic buildings including the cathedrals of Paris, Reims and Beauvais. One of the aims of this larger study will be to publish for the first time section drawings which include foundations. [*Fig. 5*] Only by incorporating foundations into the study of Gothic architecture will a complete understanding of these magnificant buildings be possible.

This research has been funded by the National Endowment for the Humanities and the Alfred P. Sloan Foundation.

10
The collapse of 1284 at Beauvais Cathedral
S. Murray

In interpreting the factors lying behind the collapse of the upper parts of the choir of Beauvais Cathedral in 1284, certain architectural historians have attempted to go beyond the simplistic view that the limits of material and expertise had been reached in this gigantic structure, and rather to identify specific factors of weakness in its design. Such factors identified in previous studies include inadequate foundations; piers which were too widely spaced; the faulty design of certain units in the upper superstructure, notably the upper piers of the central vessel and the intermediary uprights of the flying buttresses.[2]

Recently some of the methods developed for testing design projects in modern concrete construction have been applied to problems in gothic structure: namely the system of model analysis using a two-dimensional epoxy model, charged with weights to simulate loading and wind-forces.[3] These studies have represented a break-through in our understanding of the nature of the stresses set up inside the gothic structure, and for the art historian, untrained in the technical formulae of structural engineering, are of particular value in producing visual evidence which can be read almost like a contour map.

However, none of the studies made of the Beauvais collapse have made full use of the two kinds of source which ought to provide the architectural historian with his main evidence: stylistic analysis, coupled with a review of the primary textual sources. Such sources will provide us with information as to which parts of the building collapsed, and will enable us to distinguish between campaigns of repair which followed immediately after the collapse and later work of repair and restoration. The resultant chronology of the work of repair will enable us to use the cathedral itself as a "model", with a view to establishing sources of structural weakness. Finally, comparisons with contemporaneous monuments of similar specifications will enable us to define what made Beauvais Cathedral so different from the rest.

The choir of Beauvais Cathedral was first built with three straight bays covered by quadripartite vaults, flanked by double side-aisles on each side, and terminated to the east by a seven bay hemicycle ringed by an ambulatory and seven radiating chapels of equal depth. Documentary and stylistic evidence suggests that the work was begun around 1225 and completed up to the eastern piers of the crossing by 1272 (Fig. 1).[4]

The so-called "Bucquet aux Cousteaux"[5] collection of copies made in the eighteenth century from the now-lost archives of the cathedral chapter provides the source which relates that "on Thursday November 29, 1284 at 8.00 p.m. the great vaults of the choir fell and several exterior pillars were broken; the great windows smashed; the holy chasses of St. Just, St. Germer and Ste. Eutrope were broken and the divine service ceased for forty years. Several pillars were interposed in the choir arcade in order to fortify it."[6]

The same source tells us that the disaster of 1284 was the second collapse which had occurred at the cathedral.[7] The account of the first collapse has not been taken seriously by subsequent historians, because it was said to have occurred in 1225, the date which has been assigned by modern scholarship to the commencement of work on the choir. Our text specifies that the collapse affected the straight bays of the choir; that it resulted from the over-wide spacing of the piers, and that the repairs were of a make-shift nature, involving the addition of iron ties between the piers, remains of which could still be seen by the author of the text.

The first collapse, if indeed it occurred at all, must have been a relatively minor affair, and was quickly repaired, not leaving any stylistic or archaeological evidence which would enable us to be certain as to its nature. The collapse of 1284, on the other hand, was clearly a major event and the cathedral was still not repaired in 1339. We have a copy of a text from the chapter deliberations for this year, which relates that Guillaume de Raye, master of the masonry at

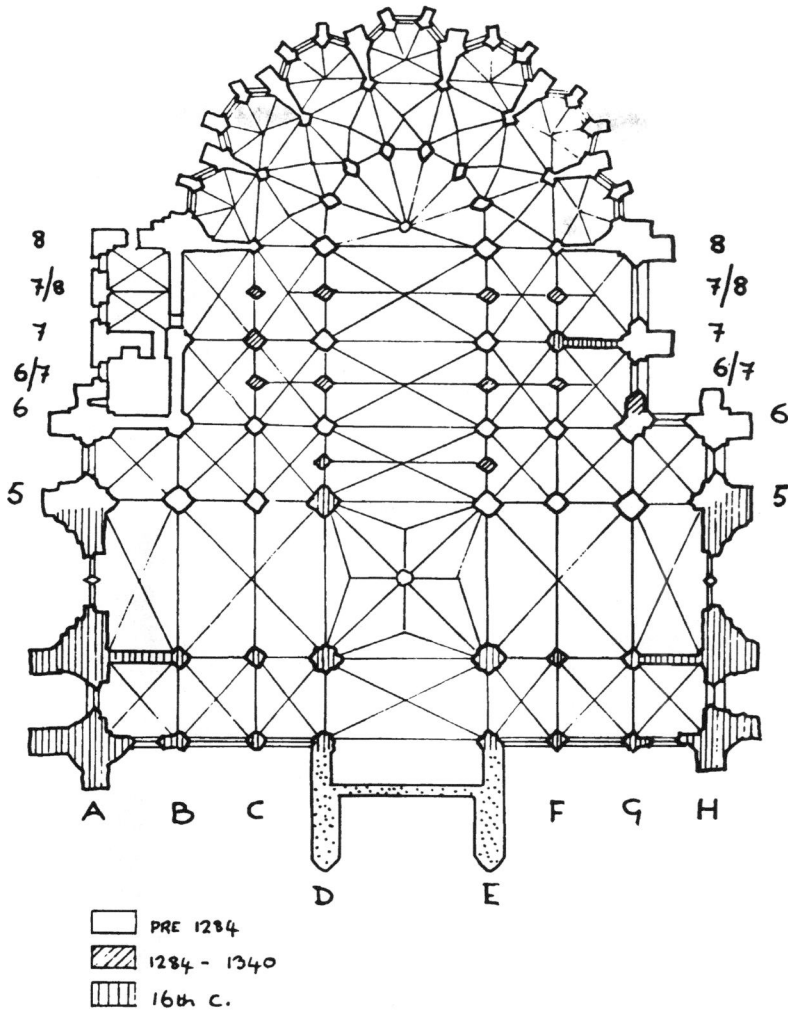

FIGURE 1.   *Plan of Beauvais Cathedral, showing repairs after the collapse.*

Beauvais Cathedral, Aubert d'Aubigny stone cutter, and Jean de Maisonchelle chaplain and master of the works had recently considered the works which were to be done at present in the cathedral, by which works the church would be all rebuttressed ("reconfortée"), all vaulted, with the scaffolding removed, and all ready for the divine service, and without which works the church

19

could not continue to stand. The author of the text then goes on to specify work on raising a great pillar ("pillier" can mean buttress, or flying buttress upright, as well as interior pier), installing the flyers, reseating the sick arches, remaking the windows and closing in the body of the church. Purchases of the materials are recorded, including 800 "pendens", or stones for the severies of vaults, 400 of which are said to be "old", or in other words, re-used, and are therefore not paid for.[8] Such re-use of material suggests that although the collapse was a serious one, some of the fallen masonry could be salvaged, and parts of the vaults may have remained intact, and could be demolished and the stones re-employed.[9]

FIGURE 2. *Detail of the triforium at bay 8 on the north side, showing the transition from pre-collapse to post-collapse tracery.*

20

A survey of the forms of the tracery, capitals and bases reveals quite clearly that everything from the sill of the triforium of the central vessel, up to the high vaults was rebuilt in the straight bays of the choir (Fig. 2). At the same time the additional piers were inserted in the main arcade, to create sexpartite vaults in the central vessel. While it is impossible to distinguish between new and re-used masonry in the vaults, it is noticeable that three of the present clerestory windows are different from the others in the tracery patterns used, raising the possibility that they might constitute units pieced together out of elements surviving the collapse (Figs. 3 and 4).[10]

The upper hemicycle (including the vault) remained intact, as did the hemicycle flying buttress system (Fig. 5). Only the exteriors of the upper hemicycle piers were rebuilt, replacing an arrangement which Viollet-le-Duc reconstructed (on paper) as openwork tabernacles, with fully detached shafts.[11] The fact that this feature of the otherwise intact hemicycle was rebuilt, presumably after the 1284 collapse (although the masonry is anonymous in terms of stylistic identifying forms) led Viollet-le-Duc to his theory that it was precisely this element in the design of the upper superstructure which caused the collapse. His account of the differential settlement of the coursed masonry, as opposed to the detached shafts has become a "classic" interpretation of the potential mechanics of collapse, and it has secured many adherents.[12] If we follow Viollet-le-Duc's thesis, then we must view the source of the weakness in the design of the upper superstructure as essentially a longitudinal one, running on the east-west axis of the building. For Viollet-le-Duc, moreover, this was a faulty detail in an otherwise well-conceived structure, the "Parthenon of French Gothic".[13]

We wish to argue, on the other hand, that the collapse occurred not because of a faulty detail which had produced a longitudinal weakness in the upper superstructure, but rather because of a critical lack of lateral buttressing at a point in the choir which is easily ascertainable; moreover that the factors causing this weakness arose from grave errors of judgement by the master planners of Beauvais Cathedral, both in the laying out of the plan, and in the nature of the superstructure.

FIGURE 3.   *Detail of re-used tracery in clerestory window.*

FIGURE 4.   *Detail of tracery made in the campaigns of repair after the collapse.*

FIGURE 5. *Exterior of the hemicycle at clerestory level.*

23

There is only one bay of the choir where we find that all of the vertical members have been rebuilt, including the piers of the central vessel (in their upper parts); the intermediary piers dividing the double aisles; the intermediary flying buttress uprights on top of these piers, and the outer massive flying buttress uprights from which the entire buttressing system is generated: namely the middle bay of the choir, marked on our plan as bay 7. This leads us to identify the stylistic features enabling us to distinguish between pre-collapse and post-collapse masonry.[14] As far as the flying buttress uprights are concerned, these criteria include above all the transition of base forms from the low "pancake"-like moulding on a simple flat-sided octagonal plinth to a much taller type with an upper rim, a flattened area below it, and a flared lower lip rather like a trumpet bell. The pre-collapse flying buttress uprights have detached en délit shafts supporting gabled arcades on their flanks, whereas their post-collapse counterparts have recessed panelling.[15]

Following these criteria we may determine that two of the main outer flying buttress uprights have been rebuilt on each side (in bays 6 and 7) and that in the straight bays of the choir only one of the intermediary uprights on each side has been rebuilt, namely at bay 7 (Figs. 7, 8, 9 and 10). The intermediary uprights at bay 6 show no sign of having been rebuilt, and employ en délit shafts and bases of the low variety (Fig. 9).

Evidence of the lateral distortion associated with the collapse of 1284 can also be established by identifying units in the interior of the aisles which were rebuilt after the collapse. The rebuild included not only the extra piers interposed in the main arcade, but also the piers dividing the double side-aisles at bay 7. Pier C 7 on the north side has a continuous moulding around it with a rounded upper rim; a flat depressed area, and a flared lower lip in the form of a trumpet bell (Fig. 11). This kind of base contrasts with the simpler forms of the adjacent piers (Fig. 12) and suggests a date well into the second half of the thirteenth century.[16] Its counterpart on the south side (F 7) has an undulating surface and hexagonal plinths for the shafts which allow it to be associated with the work of Martin Chambiges (Fig. 13). More specifically, the details of the plinth design of F 7 are similar to the piers of the north transept (1510-1518), and it seems possible that a text recording certain repairs to pillars in the choir

FIGURE 7.   *Outer upright of flying buttress, bay 7, north side.*

25

FIGURE 8.   *Outer upright of flying buttress, bay 7, south side.*

FIGURE 9.   *Intermediary flying buttress upright, bay 7, south side.*

FIGURE 10.   *General view of the south side of the choir. Cross-hatching indicates parts rebuilt after the 1284 collapse.*

completed in 1517 might refer to the reconstruction of this pier.[17] At the same time, a solid wall was built, coursed into the masonry of the pier, and extending out to the exterior wall, thus dividing two bays of the outer aisle which had been originally intended to be open.

Thus, we have seen that at bay 7 (and only in this bay) all of the vertical members of the original structure were rebuilt in some way. We suggest that there were two special causes of weakness, one of them common to all gothic chevets of this type, and the other peculiar to the design of the Beauvais choir.

Firstly, it is evident from the plan that this is the only bay where the lateral thrust of the high vaults was not countered by the heavier masonry at the east and western terminations of the choir (Fig. 1). It is of interest to find an analogous situation in the very well-documented building history of the cathedral of Troyes, where it was necessary in the 1360's and again in 1402 to consolidate the lateral buttressing at a point corresponding to this.[18] It is, of course, inherent in this kind of plan that whereas the blocks of masonry dividing the radiating chapels provide a kind of internal buttress, this kind of support is absent in the straight bays.

These inherent problems were exacerbated by two factors which are peculiar to the design of the Beauvais choir. In both plan and elevation this work has been seen as involving a marriage of elements from the Chartres-Reims-Amiens family with elements from buildings with the kind of pyramidal elevation used at Cluny III, Bourges and St. Quentin. Thus, the tall inner aisle and ambulatory with its own triforium and clerestory is distantly related to the similar arrangement at Bourges. Another point of similarity between the two buildings is the wide spacing of the piers of the main arcade in such a way as to produce aisle bays which are rectangles with their long sides running in an east-west direction, rather than approximating to squares, as at Amiens.[19] On the other hand, the plan with seven radiating chapels and a projecting transept reflects Amiens, as do the steep proportions of the central vessel, where we find that the height of the upper parts from the triforium sill to the top of the clerestory windows approximates to the height of the sill above the floor.

The bay system at Beauvais Cathedral is highly eccentric, the lateral dimension of each bay varying around 15.30m.[20] but the longitudinal dimension going from an enormous bay of 9.05m.

29

FIGURE 11.   *Pier C 7 in the north choir aisle.*

FIGURE 12.   *Pier B 7 in the north choir aisle.*

30

FIGURE 13. *Pier F 7 in the south choir aisle.*

adjacent to the hemicycle to 8.76m. and finally to a bay of 7.92m. adjacent to the crossing.[21] Even the narrowest bay at Beauvais is considerably larger than the bays used at Amiens or Reims. The greater area of each of the vaults of the central vessel would produce a unit which was heavier, and exerting a greater outward thrust. On the other hand, the relatively narrow inner aisles produced inner flyers which were quite short, and which did not therefore have the weight and inward thrust of a flyer with a wider span.[22] It is obviously important to note that the heaviest vault of the central vessel (bay 7-8) was supported on its western side by a lower superstructure which was, for the reasons defined above, significantly weaker than in the adjacent bays.

We have seen that the intermediary aisle piers at bay 7 were rebuilt at two different periods, the pier on the north side towards 1300, and the pier and chapel dividing wall on the south side around 1517. It is very significant to note that the corresponding piers at the east side of the largest vault of the central vessel (vault D E 7-8) began to fail towards the end of the nineteenth century. Photographs

31

FIGURE 14.   Drawing made for the restoration of pier F 8 by the architect
              Sauvageot in 1897 (MH. 201374).

and drawings[23] made at this time reveal precisely how they would have failed, had they been left unattended (Fig. 14). The plumb line indicated in our restoration drawing reveals that the inward buckling of the piers in the chapel mouths on either side of the choir at bay 8 occurred at the height of the springing of the vaults of the first radiating chapels and adjacent aisle bays. This buckling would probably have resulted from the inward thrust of these vaults coupled with the rotational movement produced by the tendency of the higher inner aisle vaults to push outwards (see rotational arrows sketched on the section, Fig. 15). The chapel wall at bay 8 provided the extra strength at this point which allowed the piers at C 8 and F 8 to remain solid for six centuries after the completion of the choir: we suggest that the absence of such support in bay 7 was a critical factor leading to the collapse of 1284.[24]

We must lastly examine the second idiosyncratic feature of the Beauvais choir which may have contributed to the collapse: the placing of the vertical members in the upper superstructure in such a way that their entire mass was not directly over a supporting pier, but instead projected partially over a void. This practice is generally termed *"porte-à-faux."*[25] The existence of such a lack of axial alignment in the placing of the intermediary uprights of the flying buttresses around the hemicycle has, of course, been common knowledge since the publication of Viollet-le-Duc's dictionary, in which he gave a section of the upper parts of the choir at bay 8, where the straight bays turn into the eastern hemicycle.[26] Benouville later published a full section of the choir at bay 8, including the lower parts.[27] It is particularly unfortunate that most subsequent discussions of the collapse have been based upon the evidence of these drawings, since it was precisely this part of the choir which remained solid, and which does not, therefore, embody all of the weaknesses which led to the 1284 collapse.

We are obviously caught in something of a dilemma since we cannot be certain as to the nature of the elevation at bay 7 before the rebuild, and particularly as to whether the intermediary flying buttress uprights were pushed slightly over the inner aisle, as were their counterparts in the hemicycle. Measurements carried out in the adjacent bay 6 (on the north side, where easy access is possible to the tops of the aisle vaults) have confirmed that the intermediary upright

33

A. Gosset, del.

Cathédrale de Beauvais.

Coupe du chœur.

FIGURE 15.   *Section of Beauvais choir at bay 7* (Congrès Archéologique, *1905).*

was, indeed, carried *porte-à-faux* here, and there is thus every reason to believe that the same arrangement was used throughout the straight bays.[28]

34

The *porte-à-faux* device in itself should not be considered as a direct cause of the collapse, since it was also used in the turning bays of the hemicycle without unfortunate consequences. The question is obviously a relative one: in the turning bays the weight of the central vault was less (since its area is smaller and it has more supports) and the massive outer uprights were more stable, since they are solidly based on the divisions between the chapels.

This leads us to a consideration of the nature of the outer flying buttress uprights at bay 7. The discussion of the nature of the collapse given by both Viollet-le-Duc and Heyman was based upon the premise that these outer uprights remained firm. We have seen, however, that the stylistic evidence reveals quite clearly that the units at bay 7 on each side were rebuilt, and a closer examination of the unit on the south side reveals that it cannot be considered as the massive and stable prop as supposed in earlier accounts of the collapse. The projection of the unit is quite shallow (about 3.00m. beyond the surface of the wall), and a significant proportion of the total depth of the unit (about a quarter) projects over the outer aisle (Figs. 15 and 16). We are thus dealing with *porte-à-faux* not only in the intermediary upright, but also in the heavy outer upright at this point.[29] We are not equipped with the engineering expertise which would enable us to predict the effect that such an overhang would have upon the interior transverse arch which partially supported it. This would depend to some extent upon the coursing of the masonry involved. In the intermediary uprights, the stones are of great width, some of them running across the entire width of the pillar, so that the weight of the overhanging portion could be carried by the corbel action of these wide stones, and would not bear down directly upon the arch underneath. Had smaller stones been used, allowing the weight of the vertical unit to bear directly upon the arch partially supporting, the arch would have suffered unfortunate consequences. This can be demonstrated using the cathedral itself as a model. The massive outer buttress at G 6, which was originally intended to form the corner of a transept tower, was constructed with à portion of its depth projecting over the adjacent chapel window (Figs. 10 and 17). The stones of the projecting portion of the buttress being small, it seems certain that the enclosing arch of the chapel window must have borne a certain amount of the weight of the overhanging

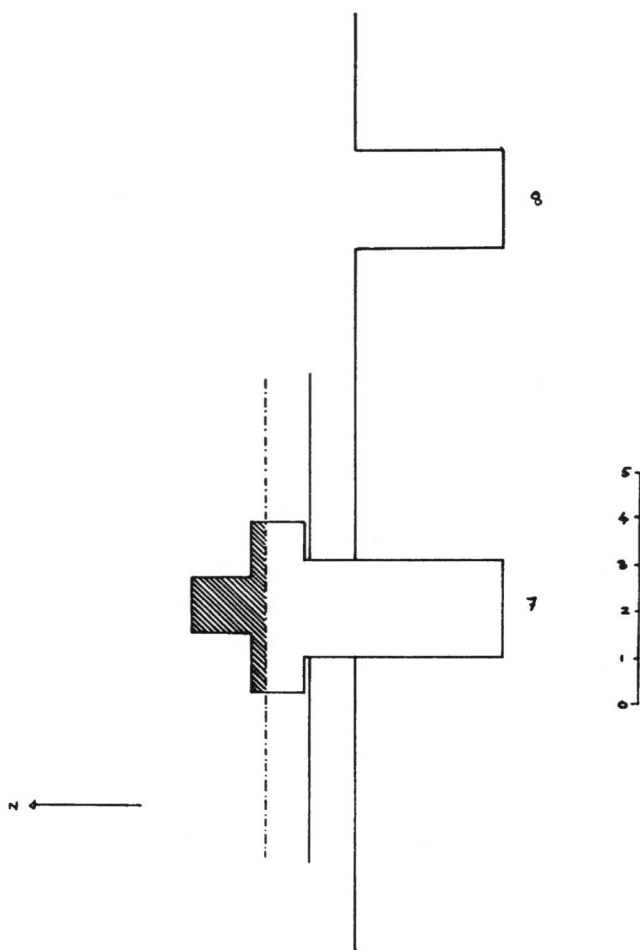

FIGURE 16.   *Section of outer flying buttress upright at G 7 (cross-hatching indicates* porte-à-faux*).*

buttress above. The thin voussoir stones of the window have failed, several stones at the crown of the arch have broken, and the geometry of the arch has been severely distorted. It was found necessary in the campaigns of repair immediately following the collapse to insert a narrow strip of masonry to eliminate the overhang.[30]

    We suggest that a similar situation of *porte-à-faux* on the part of the exterior flying buttress uprights at bay 7 caused the transverse

FIGURE 17.   *Chapel window at G 6-7.*

arches over the outer aisles to fail.[31] This would cause the intermediary aisle piers at C 7 and F 7 to be pushed towards the central vessel in a way similar to the distortion noted in the nineteenth century restoration drawing (Fig. 14).

We can best summarize and demonstrate our thesis by means of the following three propositions:

1. We know that the original flying buttress upright at G 7 was carried *porte-à-faux*, projecting out over the outer aisle. We assume that the coursing of the upright was such that its weight was carried partially by the transverse arch over the outer aisle. This unusually heavy charge has caused the intermediary aisle pier to hinge inwards, as shown in Figure 15.

2. We know that after the rebuild following the 1284 collapse (i.e. the present buttress) it was still carried *porte-à-faux*, but that it received less weight from the high vaults of the central vessel, transmitted by the flyers, since the doubling up of the piers of the central vessel reduced the size, and therefore the weight and outward thrust of each vault. Even with this diminished weight and lateral thrust, the interior transverse arch and pier F 7 failed in the early sixteenth century, necessitating the rebuilding of the pier and the addition of the interior wall to support the transverse arch: work completed around 1517.

3. If the reduced charge in the rebuilt cathedral produced a situation of failure by 1517, we may be certain that the greater stress existing before the doubling up of the piers of the central vessel would have been disastrous.

The text cited above blamed the collapse of 1284 upon certain unspecified exterior pillars. We suggest that these pillars were the flying buttress uprights at bay 7. The outer uprights had been built *porte-à-faux* over the outer aisle, and had been subjected to exceptionally high lateral thrust for reasons associated with the inherent problems in this kind of chevet design, and with the idiosyncratic bay system of the Beauvais choir. The extra weight imposed on the transverse arch over the outer aisle has caused the intermediary aisle piers C7 and F7 to hinge inwards. This tendency to hinge would in turn dislocate the lower parts of the intermediary uprights for the flying buttresses. The collapse thus resulted from the failure of the vertical members of the buttressing system at bay 7.

NOTES

1. This paper was read at the Annual Conference of the College Art Association, Chicago, 1976, and also at the conference organized by the Center for Medieval and Early Renaissance Studies at Binghamton, April 1976. I wish to express gratitude to M. Ph. Bonnet-Laborderie of Beauvais, and Professor Robert Mark with whom I spent several hours in visiting the cathedral in 1975.

2. E. Viollet-le-Duc, "Cathédrale" and "Construction", *Dictionnaire raisonné de l'architecture française*, Vols. 2 and 4 (Paris, 1854-1869); G. Desjardins, *Histoire de la cathédrale de Beauvais* (Beauvais, 1865); V. Leblond, *La cathédrale de Beauvais* (Petite Monographie, 1956); J. Heyman, "Beauvais Cathedral," *Transactions of the Newcomen Society*, Vol. XL (1967-1968), 15-36; M. Wolfe and R. Mark, "The Collapse of the Beauvais Vaults in 1284," *Speculum*, 51 (1976), 462-76. The theory of inadequate foundations has been convincingly dismissed by M. Wolfe and R. Mark ("The Collapse") and also by J. Kerisel, "Old Structures in Relation to Soil Conditions," *Géotechnique*, 25 (1975), 433-483. Kerisel drew on data assembled by the archaeologist E. Chami to show that the foundations of the gothic cathedral penetrated the level of compressible marsh material to rest upon a solid bed of chalk. Kerisel favored the idea of deformation due to unequal settlement in the upper piers, and pointed to analogous failures in modern tunnels.

3. R. Mark and R. E. Prentke, "Model Analysis of Gothic Structure," *Journal of the Society of Architectural Historians*, XXVII (1968), 44-48, and many other articles including specific studies of Chartres, Bourges, Amiens and Cologne Cathedrals.

4. R. Branner, "Le maître de la cathédrale de Beauvais," *Art de France*, II (1962), 77-92.

5. A manuscript collection of notes and copies made by Jean-Baptiste Bucquet in the mid-eighteenth century amounting to ninety-five hand-written volumes. A printed inventory of the collection was made by V. Leblond, *Inventaire sommaire de la Collection Bucquet aux Cousteaux* (Paris and Beauvais, 1908).

6   Bucquet aux Cousteaux Collection, Vol. XXIII, p. 1, "Le vendredi 29
    novembre 1284 a 8 heures du soir les grandes voutes du choeur tomberent
    et quelques pilliers en dehors rompurent les grandes vitres brisees, les
    saintes chasses de St. Just, St. Germer et Ste. Eutrope en furent . . . le divin
    service cessast durant 40 ans et on fit quelques pilliers entre le arcade du
    milieu du choeur pour le fortifier."

7. *Ibid.*, p. 5, "Les successeurs de Roger de Champagne continuerent ce
    superbe ouvrage avec le Chapitre autant qu'ils le püvent mais ce ne fut pas
    sans quelques interruptions a cause de plusieurs accidens contre lesquels les
    ouvriers ne s'etoient point assez precautionnés. La trop grande distance
    quils avoient laissée entre les pilliers du choeur qui suivoient les huict du
    sanctuaire devoit leur faire connoistre que l'elevation des voutes ou la
    pesanteur d'un si grand amas de pierre ne pouvoient subsister longtemps,
    de sorte que n'ayant point en d'appuy suffisant elles tomberent par leur
    propre poids aubout dun certain nombres dannées. Ce fut en vain qu'on en
    fit la reparation en 1225. On crut que les tirandes accrochées dans de fort
    gros crampons de fer qui paroissent encore aujourdhuy dans quelques
    pilliers, mettroient en seureté les nouvelles voutes, et sur sur (sic) cette foible
    apparence, les chanoines resolurent de commencer l'office divin dans ce
    nouveau choeur. Quelque temps apres que toute cette reparation fut
    achevée ils le firent en effet en 1272 ou lundy veille de la feste de tous les
    saints: on verra cy apres que cet ouvrage ne subsista quenviron 12 ans quil
    noferit qu'a attirer une plus grande ruine."

    See also *ibid.*, p. 10, "Les voutes du choeur de ceste eglise tomberent pour
    la seconde fois de son temps en lan 1284 quelques uns des pilliers du dehors
    furent rompus et toutes les grandes vitrees brisees." The writer who copied
    our first text above was clearly confused as to the date of the "first
    collapse" since he first says that the repair was made in 1225, and then
    implies that it was not completed until 1272. It is not stated whether the
    aisle vaults or the high vaults were involved.

8. Bucquet aux Cousteaux Collection, Vol. XXVII, p. 34 (note: the
    handwriting of this text is barely legible in places, and there are several
    problematic words), "Veu et consideré du commandement de chapitre par
    Maistre Guillaume de Raye à che temps maistre de la machonnerie de
    l'Eglise de Beauves, par Aubert d'Aubigny apparilleur à che temps de ladite
    Eglise et par Jean de Maisonchelle chapelain de ladite Eglise et Conseilleur
    de la fabrique tous les ouvrages qui sont à faire à present en ladite Eglise,
    et par lesquels ouvrages ledite Eglise sera toute reconfortée toute voutée et
    en tel point que on pourra tous les escafoux oster, et toute preste pour
    faire le serviche divin et sans lesquels ouvrages ledite Eglise ne se peut
    porter ne soustenir, mes devroit on compter et reputer aussi qu'il
    pourvoient tous cous et mises que en ledite Eglise ont esté mis et depensés

depuis 40 ans en encha (?) lesquels cous et despens montent bien selonc le tesmoing des anchiens et les comptes de ledite fabrique a 80 (= 80,000 ?) livres ou plus. Primes pour le grant pilier huiché les arsboutens asseoir et tenir les maunes (?) ars, refourmir les fourmes (windows), asseoir, clorre (close in) le corps de l'Eglise tant comme le breque ("breche" = trou, hole) contient estassavoir pour deux machons et six baardeurs (carpenters) 50s. le semaine 750 lb. Pour plonc 16 lb. Item mortier cloies et forge 20 lb. Item pour les carriets et voitures 20 lb. Item pour lavantpas (= "avantpis," a balustrade?) et pour la moitie de 7 piliers 15 lb. Somme 120 lb. Item pour faire les vitres et pour eguis ("ogives" = ribs?) et doubliaux (transverse arches) pour (300 piés 2s. 6d. le pié valent 36 lb. par. Item pour 700 de pendans avec quatre qui sont vieus chacun de 6 lb. 10s. — this phrase deleted) 300 pieds 2s. 6d. pour le pié valent 38 lb. par. Item pour 800 de pendens avec 4 qui sont vieus chacun 6 lb. 10s. valent 15 lb. 10s. Item pour tailler liens 10 de pendans 20 lb. Item pour les 2 devines (?) 15 lb. Somme 1103 lb. Item pour faire une vis depuis les basses voyes jusques à la haute charpenterie lequelle vis aura 160 pieds de haut 157 lb. Somme toute 500 lbs."

9.   Such re-use of existing masonry was, of course, a very common economy measure in medieval construction. We have found an analogous situation in the workshop of Troyes Cathedral, when, in 1362 a visiting expert named Pierre Faisant recommended that certain flyers be rebuilt at a lower level using the old masonry, see S. Murray, "The Completion of the Nave of Troyes Cathedral," *Journal of the Society of Architectural Historians*, XXXIV (1975), 125.

10.   The three windows involved are in the south side bays 6/7-7 and 7-7/8 and the north side 6/7-7. The differences, once noticed, are glaring. The three windows have a main oculus which is larger in relation to the rest of the tracery than in the case of the other windows. The sticks of tracery seem flatter and with less projection; the central unit of the three lancets lacks the upper trilobed element used in the other windows, and the other two side lancets do not have "spherical triangles." Finally, the enclosing arches of the three windows are composed of accurately defined geometric curves, whereas in all of the other clerestory windows a more awkward curve is apparent. Of course, these differences could have arisen from the existence of a subdivision within the workshop, with two autonomous groups of masons working simultaneously on different windows, using different plans, but the re-use of pre-collapse material provides a much more likely explanation. If this were, indeed, the case, the tracery elements re-used would provide us with the possibility of reconstructing the pattern used in the original choir as six-unit windows with two groups of three lancets.

41

11. E. Viollet-le-Duc, *Dictionnaire*, Vol. 4, pp. 178-181.

12. Including J. Heyman and J. Kerisel, in the works cited in note 2.

13. E. Viollet-le-Duc, *Dictionnaire*, Vol. 1, p. 71.

14. Of course a unit built after the collapse, re-using old masonry might look very similar to pre-collapse work. We have found no evidence to suggest that such a situation might exist in either the piers or the flying buttresses, however.

15. The development of the flared "trumpet moulding" in the second half of the thirteenth century might be illustrated by many well-known examples — we have chosen the securely dated example of the new south transept of Sens Cathedral, where such mouldings are used at a date towards 1300 (Fig. 6). See C. Porée, "Les architectes et la construction de la cathédrale de Sens," *Congrès Archéologique*, LXXIV (Avallon, 1907), 559-598. The use of panelling rather than *en délit* shafts supporting arcade work is well in line with the tendency of architects of the middle and later thirteenth century to abandon the detached shaft in favor of more streamlined effects. The uprights of the flying buttresses of the choir of Amiens Cathedral, from mid-century employ such recessed panels.

FIGURE 6. *Sens Cathedral, pier base in the south transept.*

16. Compare the pier from the Sens Cathedral south transept, Fig. 6.

17. Bucquet aux Cousteaux Collection, Vol. XXVIII, p. 248, "Martin Cambiche fait son rapport de certains gros pilliers du choeur qui netoient pas encore acheves et dit quil falloit les reparer pour prevenir plus grand danger." It is also possible that the "great pillars" might be exterior members in the buttressing system — for example the pier D 8 was consolidated in this period.

18. S. Murray, "The Completion," p. 127.

19. Dimensions taken from G. Durand, *Monographie de l'église Notre Dame cathédrale d'Amiens* (Amiens, 1901).

20. As opposed to 14.60m. and 14.65m. at Amiens and Reims Cathedrals.

21. As opposed to 7.40m. at Amiens.

22. The width of the inner aisle at Beauvais Cathedral varies considerably from bay to bay, but approximates to 7.00m. As will be pointed out later, the upright member supporting the inner flyers is pushed towards the central vessel, thus reducing the length of the flyer to around 4.10m., as opposed to about 8.00m. in the nave of Reims Cathedral and 6.01m. in the choir of Amiens (dimensions derived from Viollet-le-Duc). We have thus a situation where each of the central vaults of Beauvais Cathedral was considerably larger than their counterparts at Reims or Amiens, whereas the span and weight of the inner flying buttresses was considerably less. Moreover the flyers of Beauvais are not particularly steep, and thus lack the extra buttressing force of the steeply pitched units at Bourges Cathedral.

23. Monuments Historiques, Service Photographique No. MH 46055 and MH 192866 are photographs of the inner aisles, showing beams placed as braces to prevent inward buckling of the intermediary piers at bays 6, 7 and 8 on the north side, and bay 8 on the south side. No. MH 201374 is a restoration drawing made by the architect Sauvageot in 1897 for the restoration of pier F 8 (Fig. 14).

24. Several choir plans of the following generation of gothic remedied this weakness by placing solid walls between the bays of the outer aisle, for example Clermont Ferrand and Limoges.

25. Such overhang is a fairly common feature in gothic design. The use of *porte-à-faux* at Beauvais and in the flyers of Notre Dame at Dijon excited the interest of Viollet-le-Duc because he considered that the practice exemplified the use of the principle of "elasticity", which he considered as a key to gothic structural design.

43

26. E. Viollet-le-Duc, *Dictionnaire*, Vol. 1, p. 70; Vol. 4, p. 178.

27. L. Benouville, "Etude sur la cathédrale de Beauvais," *Encyclopédie d'Architecture*, ser. 4, IV (Paris, 1891).

28. The overhang cannot be seen, nor can it be directly measured, but must be calculated on the basis of the difference between the spacing between the given units in their lower and upper parts.

29. Having established this fact by measurements taken above the vaults at G 7, I later noticed that the same discovery had already been made by the architect who made the section of the choir at bay 7, given in the article by Abbé Marsaux, *Congrès Archéologique*, LXXII (Beauvais, 1905), here reproduced as Fig. 15.

30. This narrow strip of masonry is datable to the post-collapse period because of the characteristic base type with the "trumpet bell" flared lower rim.

31. These transverse arches are considerably narrower than the window enclosing arch just examined.

11

Beauvais Cathedral

J. Heyman

It is fashionable to deny that Gothic became decadent in the second half of the thirteenth century, and to assert that the cathedrals of Amiens, of Beauvais, and, perhaps, of Cologne, were not the supreme achievements of the discipline. As a matter of aesthetic opinion, it is possible to argue endlessly about the relative merits of this or that structure, built in this or that century. As a matter of structural fact there is almost no argument possible. The decay sensed by the eye after about 1250 stems from a slow relaxation of the firm structural grasp that had been acquired during the preceding hundred years. Certainly, one hundred years later, by the end of the fourteenth century, the Milan expertises demonstrated that the Italians, at least, did not know how to build a cathedral. After the first stoppage, in which the argument ad quadratum/ad triangulum was settled (in favour of the triangle) on advice by a mathematician, Stornaloco, work proceeded fairly smoothly for a time. Only a few years later, however, it was again found necessary to consult foreign experts.

Jean Mignot, on his arrival from Paris, drew up a list of fifty-four points in which the unfinished cathedral was defective. Some of these criticisms are curiously irrelevant and pedantic; Frankl[1] finds Magnot a sort of Beckmesser, which seems an apt description. Mixed up with structural criticisms (e.g. insufficient buttressing) are purely aesthetic comments (e.g. canopies placed too high above figures) and it is clear that neither the Italians, nor Mignot himself, can distinguish very clearly between the structural and the aesthetic function of various portions of the fabric. Mignot, in fact, criticised according to a set of rules that he knew to be "correct"; indeed, at one point he demanded the calling of further experts from Germany, or from France, or England, knowing that he would be supported by any other architect trained in a well-known Lodge. And, of course, Mignot won the day; the Italians answered on the basis of inferior schooling or, worse, tried to invent rational answers *ab initio*.[2] But Mignot's victory, and subsequent appointment as architect at Milan, did not mean that he himself *understood* the rules to which he worked; rules were, to Mignot, something out of a book, which had been tested by time and practice ("ars sine scientia nihil est"), but whose meanings and, indeed, whose very reasons for existence, were becoming more and more dim with the passage of time.[3]

The structural principles of Milan are simple, and had existed for at least 200 years; there are rib

---

[1] Paul Frankl, *The Gothic; Literary Sources and Interpretations through Eight Centuries*, Princeton University Press (1960). Frankl makes very clear the meanings of ad quadratum and ad triangulum, and illuminates the whole of the Milan controversy. For a fuller account, see James S. Ackerman, "Ars sine scientia nihil est," Gothic theory of architecture at the Cathedral of Milan. *Art Bulletin* **31** (1949), 84–111.

[2] For example, '. . . archi spiguti non dant impulzam contrafortibus." The Italian counterargument is a nice example of a tripartite legal defence. To Mignot's criticism that the buttresses were not strong enough, the reply was: First, they are strong enough. Second, without prejudice to this first defence, if they are not strong enough they are in any case not necessary, since pointed arches do not thrust. Third, without prejudice to these defences, if pointed arches do thrust, iron ties are in any case provided between the arches to absorb the thrust. (These ties may be seen today, and contribute to the general oppressiveness of the Duomo.)

[3] It is probably no accident that Mignot's famous phrase is an almost exact paraphrase of the second sentence of Chapter 1 of Book 1 of Vitruvius: "ea (the architect's knowledge) nascitur ex fabrica et ratiocinatione"; *cf.* Frankl (*op. cit.*), p. 89f.

BEAUVAIS CATHEDRAL

vaults, whose thrusts are resisted by flying buttresses. This is the structural essence of Gothic, developed and refined and slowly codified in the twelfth and thirteenth centuries.[1] There is no structural development, either explicit or implicit, between say Cologne, 1248, and Milan, 1386; indeed, Cologne, despite some flaccidity and repetitiveness, is a finer structure than Milan. Thus rules, adequate for Cologne, would be adequate for Milan, and would have needed no modification in the century that separates the two. The rules, formulated and tested, lay guarded in the lodges, dead; a fourteenth-century architect would have forgotten the long struggle necessary to achieve the balance of forces in the real structure, reflected in the final statements of proportional rules.

On the basis of twentieth-century engineering technique, it seems that numerical rules of proportion are precisely those required for masonry construction[2]; the stress level in most of the fabric of a Gothic cathedral is so low that strength of the material is of only secondary importance. But such numerical rules could have been constructed in the twelfth and thirteenth centuries only empirically, by trial and error, and by taking note of structural successes, and most importantly, of structural failures. Once this experience was no longer within living memory, there was no way of reconstructing the rules by any process of thought or intellectual argument. If the rules were followed, a safe structure probably resulted; if they were tampered with (say for "aesthetic" reasons) a safe structure probably still resulted. Any structural disaster after the middle of the thirteenth century, it is almost safe to say, can be attributed to ignorance of the rules, to gross tampering with the rules, to workmanship so poor as to be fraudulent, or to a simple Act of God.

The rules finally developed were very good. As Ackerman points out, for example, "the entire plan (*for Milan*) was conceived for a cathedral considerably different from that which was erected; buttresses and piers rose toward an unknown objective, and finally the primaty and basic structural problem, that of the vaults, was the last to be solved."[3] Nevertheless Milan continues to stand as a tribute to the efficiency of the structural elements of which it is composed; evidently a flying buttress, for example, can resist a wide range of thrust, and can thus support almost any type of vault.[4] Milan, in its final "ad triangulum" form (and low triangles at that), poses no great structural problems; almost any reasonable rules of construction would have ensured stability of the cathedral. Much better rules would have been needed a century and a half earlier for the construction of Amiens, 1220, Beauvais, 1247, and Cologne, 1248. It is not surprising that rules developed for such cathedrals as these could easily suffice for Milan.

The observation that structural elements could be assembled almost at random and yet result in a safe structure tempts the conclusion, in Ackerman's words, ". . . that structure plays a secondary role in the process of creation." This conclusion may be true for Milan; once the conclusion is believed to be true for *any* cathedral, then the slide has started from Gothic into Renaissance. This situation has been discussed by Harvey: "The Gothic rules were so complicated that no one who had not

---

[1] The development of one of these "building codes" can be seen in the manuscript of Villard, *c.* 1235; see H. R. Hahnloser, *Villard de Honnecourt*, Anton Schroll (1935). Not every Gothic structure has flying buttresses, of course; the Sainte Chapelle, Paris (1243) has none, nor does King's College Chapel. Indeed the latter does not have a rib vault either, yet remains undeniably Gothic; a stone vault presses down and out, and requires massive external buttressing.

[2] A discussion of the stability of masonry construction, and, in particular, a justification of the use of models, is given in J. Heyman, "On the rubber vaults of the Middle Ages, and other matters", *Gazette des Beaux-Arts*, 6th Period, 67 (1966). An earlier paper, J. Heyman, "The Stone Skeleton", *Int J. Solids Structures*, 2 (1966), establishes that limit theorems, originally developed for steel framed structures (see e.g. Sir John Baker, M. R. Horne, J. Heyman, *The Steel Skeleton*, vol. 2: *Plastic Behaviour and Design*, Cambridge (1956), or J. Heyman, *Beams and Framed Structures*, Pergamon Press (1964) can be adapted to the analysis of masonry. The work is extended to other masonry elements in J. Heyman, "On Shell Solutions for Masonry Domes," and in Heyman: "Spires and Fan Vaults," *Int. J. Solids Structures*, 3 (1967).

[3] Ackerman, *op. cit.*

[4] A properly designed flying buttress can, indeed, resist an enormous range of thrust, e.g. from 3 ton to 1,000 ton at Lichfield (see Heyman, "The Stone Skeleton").

16

BEAUVAIS CATHEDRAL

served a long apprenticeship and spent years of practice could master them; whereas the rules of Vitruvius were so easy to grasp that even bishops could understand them, and princes could try their hand at design on their own."[1]

BEAUVAIS CATHEDRAL: COLLAPSE AND REBUILDING

Amiens (despite some trouble with flying buttresses) and Cologne have been satisfactory structures; by contrast, Beauvais seems to have been particularly unfortunate. The apse and choir were started in 1247, and finished in 1272. On 29 November 1284 the vault fell, according to Desjardins and Pihan because the choir piers were too widely spaced, and according to Leblond because of consequent failure of some external buttresses.[2] Viollet-le-Duc has a different explanation which will be given later. Whatever the actual reason, it was certainly believed at the time that the pier spacing was too large, and the repairs over the next 50 years included the intercalation of piers between these originally built for the choir, so that the bays were halved from about 9 m. to about 4·5 m. The choir had been rebuilt by about 1337 but work was interrupted for the next 150 years by the Hundred Years War and by the English occupation. It was not until 1500 that a start was made on the transept, the *maître de l'oeuvre* being Jean Vast the elder with Chambiges as *maître architecte*. Jean Vast died in 1524, and was succeeded by his son Jean Vast; Chambiges died in 1532, and seems to have been succeeded as *principal architecte* by Jean Vast, François Mareschal eventually becoming *maître de l'oeuvre*.

The transept being well under way, the Bishop and Chapter called an expertise in 1544 on the subject of a tower over the crossing, at which masons and carpenters were to decide whether the tower should be of stone or timber. Models were examined in 1547, but it was not until 1558 that a decision was made in favour of a masonry tower, and it was 1564 before Jean Vast started the construction. This tower, completed in 1569 (Plate JHI), was immense, rising 153 m. from the ground, and alarmed the Chapter from the first.

Several examinations were made, and the detailed report of two King's masons, Giles de Harlay and Nicolas Tiersault, about 2 years after completion of the tower, found that the four main crossing piers were beginning to lean. Those on the choir side, out of plumb by 4 in. (*pouces*) and 2 in., were not thought to be dangerous, since the piers were considered to be well-constructed throughout their thicknesses[3]; the principal danger lay in the two other piers "tirant au vide", i.e. next to the non-existent nave, which were out of plumb by 5 or 6 and by 11 in., for lack of "contreboutement."

The masons proposed as remedy the immediate erection of two nave bays, and the strengthening of the pier foundations. In the meantime, temporary walls were recommended between the crossing

---

[1] John Harvey, "Mediaeval design," *Transactions of the Ancient Monuments Society*, New Series, 6 (1958), 55–72. See also Harvey, *The Gothic World*, London (1950). Frankl (*op. cit.*) gives an account of the organisation of the lodges (of course different in different places) and traces the typical progress of a boy from apprentice to journeyman to foreman to what might be called the career grade of master. A very few outstanding men might then become students again in the drawing office, finally reaching the status of architect; every architect had at one time been a master. This contrasts sharply with modern practice, which is based on the Renaissance concept of the "gentleman" architect; indeed, the twentieth-century engineer is, perhaps, Gothic man, and the architect Renaissance man.

[2] G. Desjardins, *Histoire de la Cathédrale de Beauvais*, Beauvais (1865), p. 8 f., "Le vendredi 29 Novembre 1284, veille de la Saint-Andrè, à l'heure du couvre-feu, les voûtes trop écartées s'affaissèrent, brisant tout au-dessous d'elles." Desjardins gives as authority for this statement P. Louvet, *Histoire et antiquitez du diocèse de Beauvais*, 2 (1635), 474. L. Pihan, *Beauvais*, Beauvais (1885), p. 10, expands the explanation to: "L'écartement trop considérable et l'élévation extraordinaire des piliers déterminèrent la chute d'une partie de la grande voûte le 29 Novembre 1284, à l'heure du couvre-feu." V. Leblond, *La Cathédrale de Beauvais*, Paris (1926), 15, states ". . . en 1284, quelqués contreforts extérieurs se rompent, une portion des grandes voûtes tombe . . . Cette catastrophe était due á l'écartment excessif des piles. . . ."

[3] Desjardins (*op. cit.*) quotes (p. 93): ". . . les dicts pilliers sont masonnez de cartiers par les dedans des corps d'iceux, il n'en peult venir faulte si promptement."

17

BEAUVAIS CATHEDRAL

piers. The Chapter was pusillanimous, sought further advice, and only two years later finally decided, on 17 April 1573, to put the work in hand. Thirteen days later, on Ascension Day, 30 April, the tower fell; Desjardins, Pihan, and Leblond all say that the two "open" crossing piers failed first. The clergy and people had just left the cathedral in procession; only three people were left inside, and all three escaped. The Chapter decided, in 1577, to celebrate annually on 30 April the signal protection that the faithful of Beauvais had been afforded.[1] Otherwise, however, the Chapter lost heart at this stage. By 1578 all necessary repairs had been made (but the tower had not been replaced); equally, all the money set aside for the nave had been spent. There were sporadic attempts to complete the cathedral, but in 1605 the decision was taken to consolidate the existing work, and Beauvais became what it is today, a choir and transept without a nave. "Lé temps n'était plus à bâtir des cathédrales. Les écoles d'architectes, de sculpteurs, de verriers, de peintres, que leur construction avait fait surgir, se mouraient de toutes parts."[2]

*   *   *   *   *   *

Branner has pointed out[3] that, despite all that has been written about the colossal dimensions of Beauvais, they are not much greater in fact than those of the great cathedrals of the first half of the thirteenth century. The centre-line width between main piers of the choir of Beauvais is 15·0 m., almost exactly that of Bourges, Chartres, Amiens and Cologne, and slightly more than Reims; the total width of the choir (about 42 m.) is about the same as Bourges and less than all the others, so that the width of the side aisles is, significantly, less than the others. Only the height of the vault, 48 m., is greater than the others, and Cologne has a height of 46 m. As for the spacing of the piers in the axial direction, the three original bays of the choir at Beauvais varied from about 8 m. to 9 m.,[4] slightly smaller than the largest bay at Amiens, and almost exactly the same as at Reims and Cologne.

Whatever reasons can be given for the fall of 1284, therefore, they cannot be tied to any unusual daring on the part of the designer. The question to be asked is, rather: if Amiens and Cologne stood, then why not Beauvais? Cologne took even longer to build than Beauvais, the choir being finished only in 1322 and the nave not yet started. Amiens was constructed between 1220 and 1288, spanning in time the first phase of Beauvais. Thus the evidence is that Gothic technique was efficient at least until the end of the thirteenth century; or, at any rate, that a building *designed* in the first half of that century could be built in the second half.

Benouville, reporting on the structure of Beauvais in 1891, finds that there is a difference in the quality of the work above and below the triforium; below "l'appareil . . . est très soigné, très regulier," but above, less so. Fifty years is, of course, too long for a single man to have been in charge, and Benouville concludes that the final phase was directed by someone less skilled than a *maître d'oeuvre*, but working to existing drawings.[5] Branner has made a detailed study of the chronology, and finds

---

[1] Legends have grown up about this spectacular collapse. For example, no one would undertake the dangerous job of clearing the rubble in the partially destroyed building. Finally, four months later, a condemned criminal was offered his life if he would demolish the ruins. He had only just started when his footing gave way, and he fell, but managed to catch hold of a rope hanging from the roof beams, and so climbed to safety. "La corde qui devait être le supplice de ce misérable fut son salut." (Desjardins, *op. cit.*, p. 99).

[2] Desjardins (*op. cit.*), p. 110.

[3] Robert Branner, "Le maître de la Cathédrale de Beauvais," *Art de France*, II, Paris (1962), 77–92.

[4] E. E. Viollet-le-Duc, in his *Dictionnaire raisonné de l'architecture Francaise du XIᵉ au XVIᵉ siècle*, 10 vol., Paris (1858–68), 7, 551 ff. (article Proportion), comments on the bay spacing; to reduce the lateral thrust on the crossing piers, the adjacent bay was reduced in width. "L'architecte . . . sent que les grandes archivoltes . . . vont exercer une poussée active sur la première pile . . . du choeur, qui n'est plus etrésillonée à la hauteur de ces archivoltes. D'abord il augmente la section de cette pile, puis il diminue l'écartement de la première travée. . ."

[5] L. Benouville, "Étude sur la Cathédrale de Beauvais," *Encyclopédie d'architecture*, 4th series, 4, Paris (1891–2), 52–54, 60–62, 68–70. ". . . nous en conclurons que, lorsqu'on a achevé Beauvais, le chantier n'était pas dirigé par celui qui l'avait commencé. Les plans existaient; un nouveau maître d'oeuvre ne fut pas appelé, ce fut un sousordre qui fut chargé de terminer les travaux."

18

BEAUVAIS CATHEDRAL

that the work was taken up to triforium level between 1225 and 1245 under the first *maître*. There was an interim period, 1245–50, when a small amount was done under a second *maître*, and from 1250 on the work was carried on under the direction of a third *maître*. Whether this third master was an actual *maître d'oeuvre* or not, Branner believes that he constructed the high vaults higher than was intended by the first master, and created the famous intermediate buttresses in "porte-à-faux," although there seems to be no evidence that this construction was not intended by the first master. However this may be, in 1272 there existed at Beauvais vaults standing 48 m. from the ground; in 1284 these vaults fell. Twelve years is a long time for a masonry structure to tremble, trying to decide whether its main piers are too widely spaced (too wide for what?); similarly if, as Leblond says, the external buttresses failed, why did they not fail immediately? (Benouville believes[1] that the crossing tower was imprudently started late in the thirteenth century, but this seems to be a conflation with the story of the sixteenth-century collapse.)

A Gothic cathedral stands by virtue of a more or less delicate balance of forces; the vaults thrust out, and these thrusts are taken out and down through the flying buttresses to the main buttresses, and so to the ground. Certainly Beauvais is high, and the balance of forces must therefore be more rather than less critical; if, however, a system of thrusts can be found which indicates that, at least originally, the structure was safe, then the lower bound theorem of limit analysis[2] states that, under normal circumstances, the structure will continue to be safe.

Indeed, the structural analysis given below is almost not needed. The fact that Beauvais did stand for 12 years is ample experimental evidence that it was possible to achieve equilibrium between thrust and counter-thrust. Further, it can be shown, by the use of the same theorem of limit analysis, that any small shifts in the structure (for example, the settlement of a main pier) cannot of themselves promote collapse of the structure, providing the overall geometry is not significantly changed. Thus the fact that mediaeval mortar was slow drying, and liable to shrinkage over years or decades, would be significant only if the structure were in such delicate equilibrium that it would have collapsed in any case under any slight live loading, e.g. wind, or if such shrinkage could cause some secondary failure of an important structural component, and hence trigger off a catastrophic collapse. In fact the analysis given below indicates that Beauvais in its original state was comfortably in equilibrium. As a means of determining the cause of the vault collapse of 1284, therefore, the overall structural analysis is *a posteriori* irrelevant. However, the positive conclusion may be drawn from the analysis that collapse must be attributed either to some essentially trivial, but far-reaching cause, or to an unforeseen event (e.g. an earthquake). In the absence of any record of such a natural catastrophe, the reason for the collapse must be sought in some detail of the construction rather than in a major fault of the design.

Structural Analysis and Comparison

The plan in Fig. 1 is Viollet-le-Duc's reconstruction[3] of the original design of Beauvais, together with his idea of how the nave might have been built. Even if the number of nave bays is indeterminate, Viollet-de-Duc's plan shows something very close to what the first master must have had in mind; five typical bays are drawn, and each of these would have had the same structure. In a sense, these typical bays contain the structural (as well as the visual) essence of Gothic; the other portions of the complete structure, the chevet, the transept, even the towers are developed from the structure of the nave bays. In fact, of course, an actual cathedral, built under several masters, often altered in design

---

[1] *Idem.*, "A la fin du treizième siècle on éleva imprudemment sur la croisée du transept, et avant d'avoir monté la nef, le clocher central."
[2] See note 2 on p. 16.
[3] Viollet-le-Duc, *op. cit.*, 2, 334, Fig. 22 (article Cathédrale.)

BEAUVAIS CATHEDRAL

as it was built, so that the nave and choir are different in structure; in the case of Beauvais, only one *typical* bay was ever completed, i.e. the second bay east of the crossing (B in Fig. 1). The next bay east (A) is already part of the chevet; the previous bay (C) is part of the transept. In fact this single bay is structurally the most critical of the choir, receiving no additional buttressing from either the chevet or the transept, and it is this bay that Benouville analysed, and which will be discussed here. Corroyer's cross-section[1] of the "typical" bay is shown in Fig. 2, and Benouvilles' "coupe restaurée" in Fig. 3. Comparing these two figures, it will be seen that the present structure has two extra flying buttresses, added in the sixteenth century; these will be removed from the analysis, which will deal, as far as possible, with the completed choir of 1272.

Benouville's analysis consists in the drawing of force polygons for a dozen sections of his cross-section, Fig. 3, and he evidently considers this exercise to be so straightforward (as do the Editors of

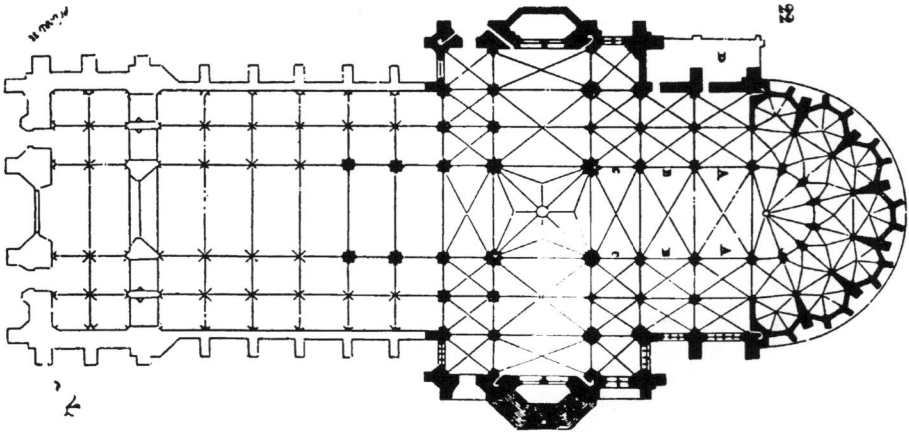

Fig. 1. Beauvais Cathedral: Reconstruction of the original plan by *Viollet-le-Duc.*

the Encyclopédie) that no explanations are given. As an example, Fig. 4 shows a redrawing of the conditions at the tas-de-charge, section CC in Benouville's analysis. On the tas-de-charge act: First of all, an inclined "poussée totale" from the rib vault, of magnitude 42,500 kg., i.e. 42·5 tonnes; secondly, a slightly inclined force of 76·2 tonnes[2] due to the weight of the material above the formeret, the parapet, and the great timber roof, combined with a small horizontal force contributed by the upper flying buttress, this total force of 76·2 tonnes having been determined from previous analysis (not given by Benouville) of a higher section; and, thirdly, two thrusts, of magnitudes 3 and 5 tonnes, contributed by the lower flying buttress. All these forces are summed in the force polygon of Fig. 4(b), and give a resultant of 111 tonnes transmitted on to the next cross-section to be considered, and eventually, of course, to the main nave pier and so to the foundations.

Benouville gives no account of how he determined the forces of 3 and 5 tonnes contributed by the lower flying buttress (nor, indeed, of why he splits up a single force into these two components,

---

[1] E. Corroyer, *L'architecture Gothique*, Paris (1891), p. 73.
[2] 1 tonne = 0.98 ton.

BEAUVAIS CATHEDRAL

although, as will be seen below, his train of thought on this matter is in fact quite clear). However, according to the principles of limit analysis applied to masonry construction, referred to above, there is no need for him to account for the values of these forces. What Benouville has done, in effect, is to calculate lines of thrust for the complete cross-section of Beauvais, for which equilibrium is satisfied everywhere, as exemplified by the force polygons. Further, these lines of thrust lie completely within the masonry (as they must); indeed, Benouville's solution involves thrusts lying in all cases very close to the centre lines of the members. No further test of stability is necessary. The safe theorem of limit analysis states that, if a thrust line *can be found* lying wholly within the masonry, then the structure is stable, and *there is no need to calculate the actual thrust line* (not that this could, in any case, be done with any confidence). Thus there is no need to determine the actual forces in

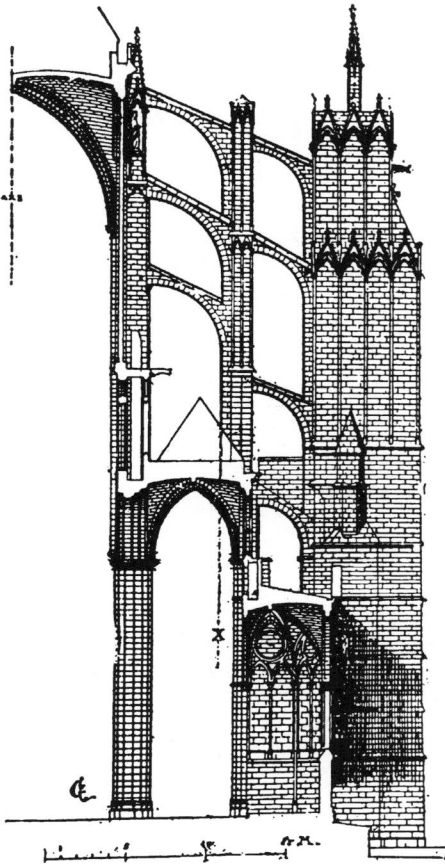

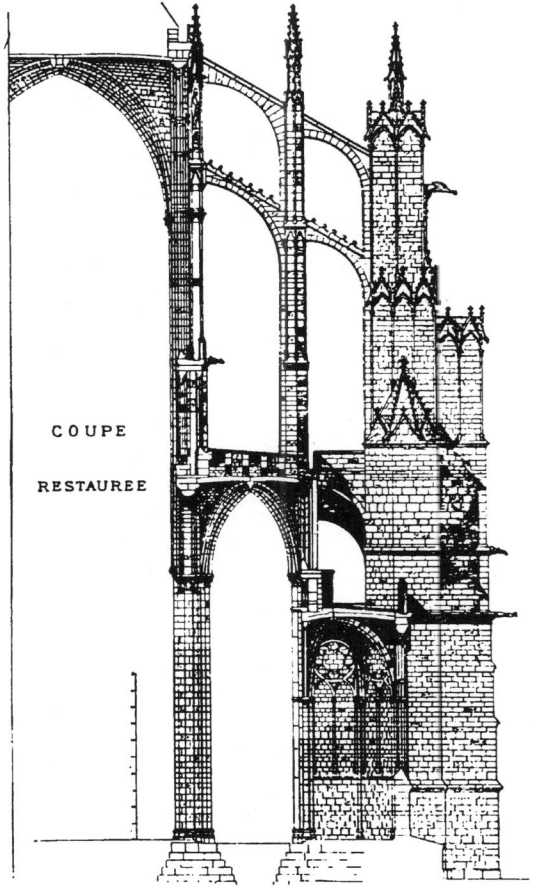

Fig. 2. Typical bay by *Corroyer*. Fig. 3. *Coupe restorée* by *Benouville*.
Beauvais Cathedral: part cross-sections of the choir.

BEAUVAIS CATHEDRAL

the lower flying buttress; if Benouville can demonstrate (as he has) that forces of 3 and 5 tonnes will assure overall stability of the structure, then the structure is indeed stable under whatever forces actually occur in the flying buttress.

Benouville's solution astonished him by the magnitudes of the stresses that he found; the largest stress in the masonry he calculated as 13 kg./cm.² [1] This value may be compared with the crushing strength of a medium sandstone of from 150–400 kg./cm.² (Ungewitter[2]), and confirms the generally low state of stress normally found in Gothic constructions. While Benouville cannot have had the

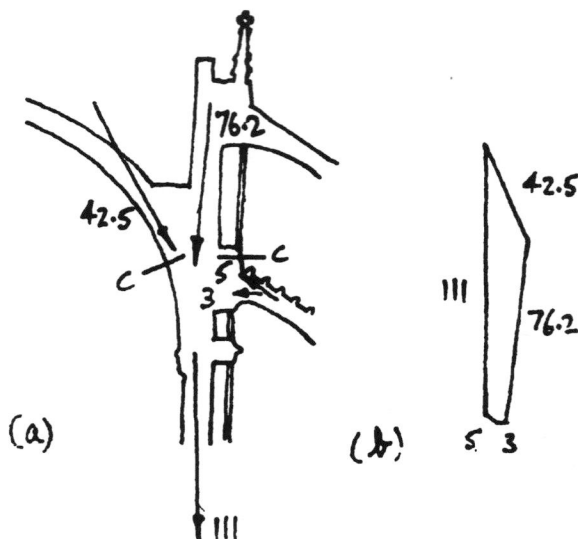

Fig. 4.
(a) Forces at the tas-de-charge of the main vault.
(b) Force polygon for these forces.
Beauvais Cathedral (*Benouville*).

comfortable assurance given by the limit theorems of structures that his analysis was correct, nevertheless his techniques had been used earlier for masonry, for example by Poleni in the analysis of the dome of St. Peter's,[3] or by Yvon Villarceau in bridge design.[4] In a sense, the engineer has always determined, as best he could, a "reasonable" set of forces in a structure on which to base his design; the limit theorems have now made respectable this pragmatic guesswork.

---

[1] Benouville, *op. cit.*, ". . . or (nous-mêmes avons été surpris de ce résultat) sait-on à combien travaille la pierre la plus chargée de l'édifice? *A treize kilogrammes* par centimètre carré."
[2] G. G. Ungewitter, *Lehrbuch der Gotischen Konstruktionen*, 2 vol., Tauchnitz (1901), 142.
[3] G. Poleni, *Memorie istoriche della Gran Cupola del Tempio Vaticano*, Padua (1748).
[4] Yvon Villarceau, L'établissement des arches de pont, c.r. Acad. Sci., Paris, *Mémoires présentés par divers savants*, **12**, 503 (1854). Yvon Villarceau's inverse design method for masonry arches consists in the assumption of a thrust line, to which the whole structure is then designed; he states quite clearly that elastic theory, leading to "correct" solutions, is not the tool for masonry arch design. For further discussion of this and of the dome of St. Peter's, Rome, see Heyman, "The Stone Skeleton."

BEAUVAIS CATHEDRAL

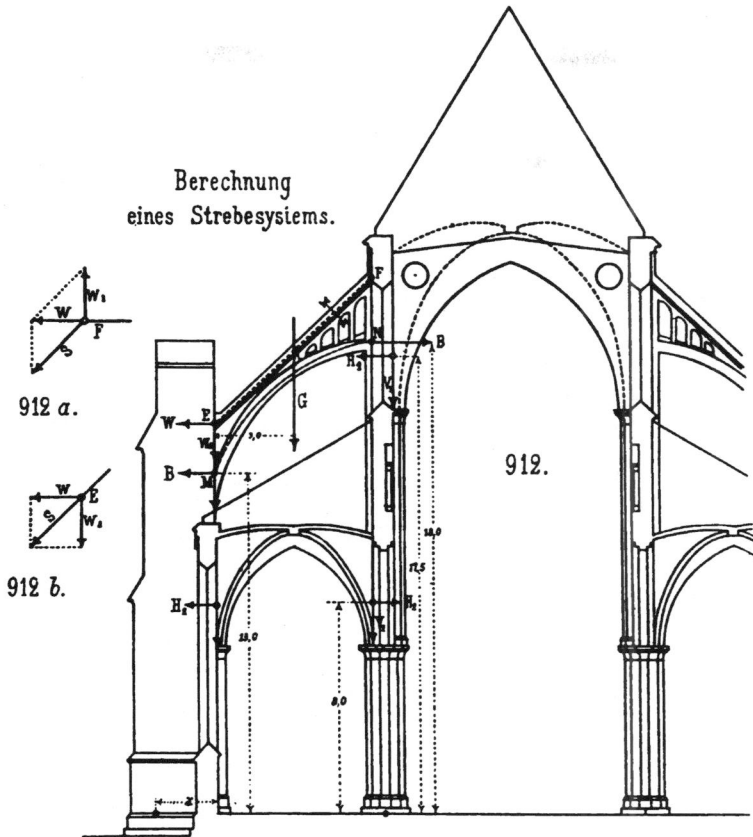

Fig. 5. Forces acting at a typical cross-section of a large Gothic church
(*Ungewitter*).

Ungewitter quite frankly uses a final desired force distribution to calculate the magnitude of the buttress force. In Fig. 5, reproducing Ungewitter's Fig. 912, Plate 87, are shown the main forces acting on a typical cross-section. Considering first the main pier, Ungewitter determines the high vault thrust H_1 and vertical reaction V_1, and similarly the aisle vault thrust H_2 and reaction V_2. Making allowance for weights of walls, and so on, carried eccentrically by the pier, the question is then asked as to the magnitude of the buttress thrust B necessary for the total thrust line to pass precisely through the mid-point of the pier at the base.[1] If moments of the forces are taken about this midpoint, the value of B is then determined immediately (in Ungewitter's example, for which $H_1 = 3.24$ tonnes, the value of B is found to be 3.02 tonnes).

---

[1] Ungewitter, *op. cit.*, p. 406. "Es soll zunächst berechnet werden, wie gross der Gegenschub des in 18 m Höhe anfallenden Strebebogens sein muss unter der *Voraussetzung*, dass der Druck unten durch den Mittelpunkt der Grundfläche des Mittelpfeilers geht." (Italics added.)

23

BEAUVAIS CATHEDRAL

It will be seen in Fig. 5 that the flying buttress consists of two ribs, of which the lower is curved, both conducting the calculated thrust B to the main external buttress and also supporting the upper straight rib. Ungewitter assumes that this upper rib is normally free of load, but acts as a wind brace, necessary for the support of the upper part of the structure. The load W shown acting on this

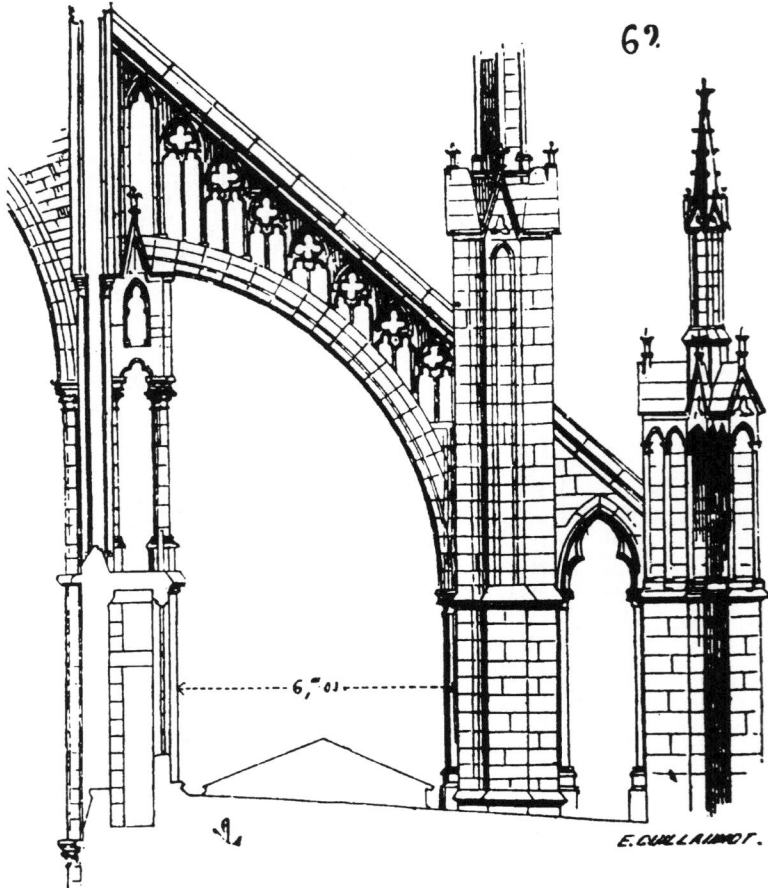

Fig. 6. Amiens Cathedral; original flying buttresses
(*Viollet-le-Duc*).

rib is computed from the wind pressure acting on the roof and part of the wall. The two ribs are only partially separated for the relatively low cathedral considered by Ungewitter; a complete separation was made at Amiens, Fig. 6,[1] necessitated by the taller construction. The solution at Amiens was not satisfactory, almost certainly because of the tracery connection between the two ribs,[2]

[1] Viollet-le-Duc, *op. cit.*, I, 72, Fig. 62 (article Arc-boutant).
[2] The probable buckling mode for the flying buttresses at Amiens is given in "The Stone Skeleton."

BEAUVAIS CATHEDRAL

and buttresses of this type now survive only at the chevet; the nave buttresses, originally of similar design, buckled, and were replaced in the fifteenth century.

Although the lower buttress at Beauvais is completely solid, Benouville evidently continued to think of it as a curved rib supporting a straight brace, and, to the horizontal force of 3 tonnes contributed by the curved rib he finds it necessary, to obtain the desired thrust distribution in the structure as a whole, to add an inclined thrust of 5 tonnes in the upper rib. As pointed out above, this is a completely legitimate procedure. While accepting the validity of Benouville's analysis, the examination below of the magnitudes of the forces acting on the fabric of Beauvais can help in understanding the particular form given to the structure, and specifically to the flying buttresses.

The main design requirement of a flying buttress system is to provide support both against the static vault thrust and also against the dynamic wind load, which has a resultant acting much higher up. An apparently completely satisfactory design was achieved by the use of two flying buttresses, as to the nave at Reims, Fig. 7[1]; Fitchen[2] has pointed out that when such a double system is used, the lower buttress absorbs the vault thrust, and the upper acts as the wind brace. The separation of the two buttresses is dictated by the height of the parapet above the tas-de-charge; for Ungewitter's example, Fig. 5, for which the total height of the cathedral is about half that of Beauvais, the separation is small.

The height of the parapet is, in turn, related to the amount of doming given to the vaults by the designer. The parapet will be high if the vaults are virtually cylindrical, with level soffits, as at Beauvais, Fig. 3, or Reims, Fig. 7. If the vaults are strongly domed, as at Notre Dame, Paris, Fig. 8,[3] then it may be possible to use a single flying buttress to counteract both the vault thrust and the wind forces.

Now the great flying buttresses at Paris span over *two* side aisles, and a buttress of these dimensions is unusual. It was more common to use an intermediate pier, as, indeed, was the case originally at Paris; the late twelfth century design is shown in Fig. 9.[4] This design was destroyed, and rebuilt in its present form, et about the middle of the thirteenth century.[5] Similarly, the *choir* at Reims, which has two side aisles on each side, uses intermediate piers in the buttressing system. And, of course, Beauvais has the intermediate buttresses in porte-à-faux.

The typical cross-section of Beauvais will be assumed to "look after" an axial length 9 m. of the structure, i.e. the axial pier spacing will be taken as 9 m. Thus with a choir width of 15 m., the plan area of a typical bay is 135 m²., and the weight of a half-bay of vaulting may be estimated from Ungewitter[6] as 36 tonnes. Ungewitter's Table also gives lines of action of the forces on the vault, and these are entered in the sketch of Fig. 10. The line of action of the horizontal thrust H, 6·5 m. below the crown of the vault, coincides exactly with the placing of the tas-de-charge at Beauvais. For equilibrium ($H \times 6 \cdot 5 = 36 \times 3 \cdot 6$), from which H is determined as 20 tonnes. The inclined reaction $R = \sqrt{(20)^2 + (36)^2} = 41 \cdot 2$ tonnes agrees well enough with Benouville's "poussée totale" of 42·5; the angle of inclination of the thrust, $\tan^{-1} 36/20 = 61°$, also agrees). Thus a horizontal thrust of 20 tonnes acts on the tas-de-charge; as Benouville demonstrated, not all of this need be transmitted by the flying buttress.

The outline of the lower flying buttress, spanning 4·5 m. between the tas-de-charge and the intermediate pier, and weighing about 5·0 tonnes, is shown in Fig. 11. The dotted line is the trace of the

---

[1] Viollet-le-Duc, *op. cit.*, 2, 318, Fig. 14 (article "Cathédrale").
[2] J. Fitchen, A comment on the function of the upper flying buttress in French Gothic Architecture, *Gaz. Beaux-Arts*, **45**, 69 (1955).
[3] Viollet-le Duc, *op. cit.*, 1, 68, Fig. 59 (article "Arc-boutant").
[4] *Idem*, 2, 289, Fig. 2 (article "Cathédrale").
[5] *Idem*, p. 288 ff.
[6] Ungewitter, *op. cit.*, p. 139, Tabelle 1. Class IVc gives a unit weight of 530 kg/m², from which the weight of the half bay is determined as ($\frac{1}{2} \times 530 \times 135$) = 35,800 kg.

BEAUVAIS CATHEDRAL

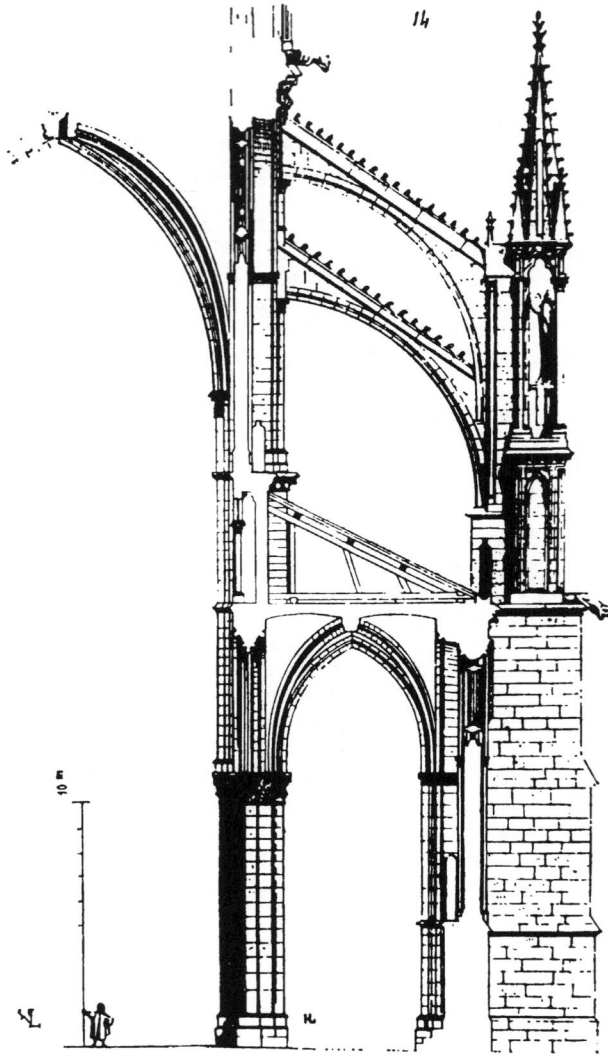

Fig. 7.   Reims Cathedral; flying buttresses
(*Viollet-le-Duc*).

26

BEAUVAIS CATHEDRAL

passive line of thrust,[1] corresponding to the minimum possible horizontal thrust necessary for stability of the flying buttress (i.e. in the absence of a vault, the buttress would, in any case, thrust against the tas-de-charge with a force of minimum value, in this example, of 1·1 tonnes). Thus the upper flying buttress, normally unloaded, also pushes against the fabric with a minimum thrust of approximately the same magnitude.

The function of the intermediate buttress now becomes clearer. Had it been omitted, each flying

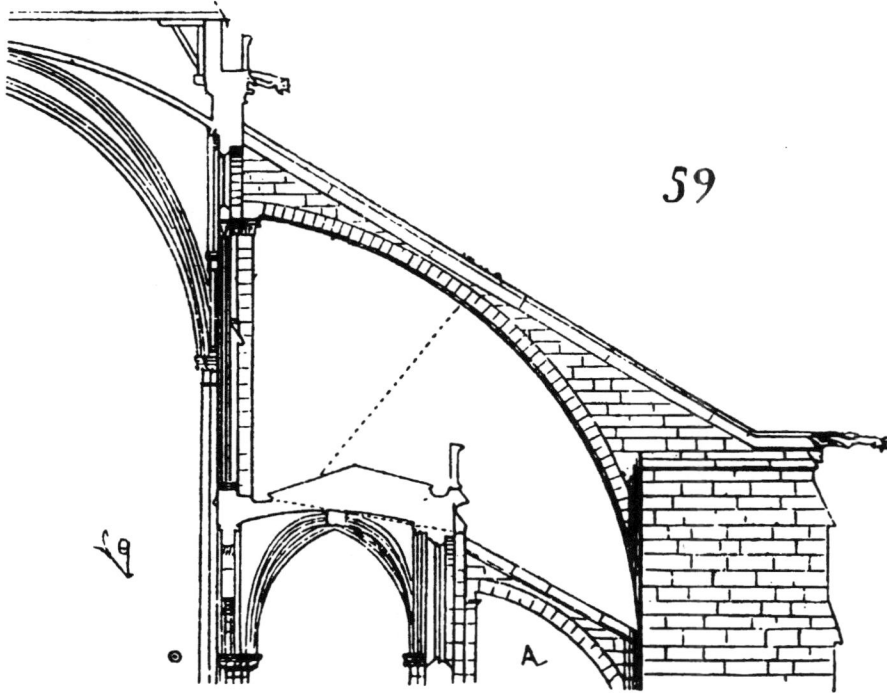

Fig. 8. Notre Dame, Paris; flying buttresses
(*Viollet-le-Duc*).

buttress would have had to span 9·5 m. instead of the 4·5 m. of Fig. 11. Rules of design were numerical rules; the buttress of 9·5 span, would, almost certainly, have been geometrically similar to that of 4·5 m. span, i.e. all its proportions would have been increased in the same ratio, 9·5/4·5. Scaling up Fig. 11 in this *linear* ratio, all the forces should be increased by the factor $(9·5/4·5)^1$, i.e. the larger flying buttress would weigh some 47 tonnes, and the minimum passive buttress thrust becomes 10·3 tonnes. The lower buttress would still be satisfactory, since it would transmit some proportion, between 10·3 and 20 tonnes, of the full vault thrust of 20 tonnes.

The upper flying buttress, however, would lean against the flat wall of the parapet, also with a mini-

---

[1] For the calculation of such a line, see "The Stone Skeleton."

27

BEAUVAIS CATHEDRAL

Fig. 9. Notre Dame, Paris; late
twelfth-century design
(*Viollet-le-Duc*).

BEAUVAIS CATHEDRAL

mum force of about 10 tonnes, and some distress could be caused to the masonry in this region.[1] The introduction of the intermediate pier causes a dramatic reduction in the values of the passive thrusts of the flying buttresses, which are turned into elegant light props, hardly pressing against the fabric when at rest, but capable of enormous thrusts if called upon to resist wind.

The relative narrowness of the side aisles at Beauvais helps to lead to this elegance. The flying buttresses over the *single* nave aisle at Reims, Fig. 7, have a span of 7·5 m., and the passive thrust of each is therefore about $(1·1)(7·5/4·5)^3$, or about 5 tonnes. The masonry as a whole at Reims is, of

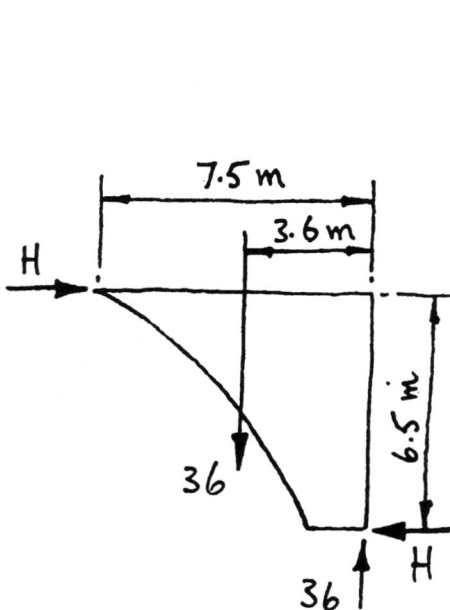

Fig. 10.   Forces acting on tas-de-charge (in tonnes).

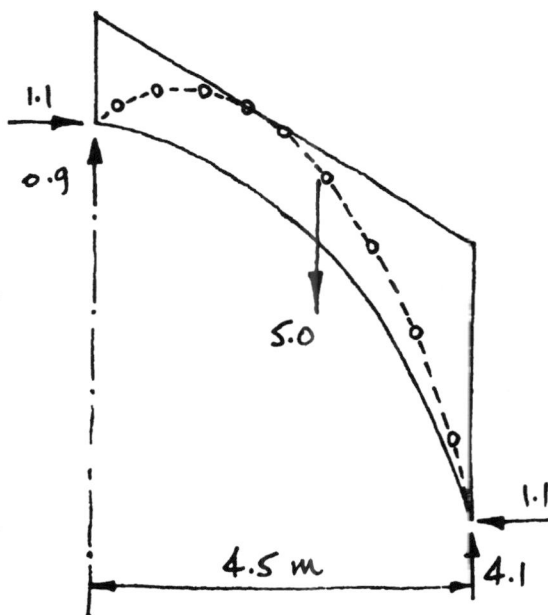

Fig. 11.   Forces acting on the lower flying buttress between the tas-de-charge and the intermediate pier.

Beauvais Cathedral.

course, very heavy; the vaults are about 60 cm. thick, compared with the more usual 20 cm., and the parapets are massive. The single great buttresses at Paris span about 11 m.; similar rules of proportions would give a passive thrust of 16 tonnes, which agrees well with the value given by a direct calculation.[2]

It is of interest to calculate wind loads. Fitchen[3] estimates that the upper flying buttress at Reims is subjected to a maximum wind load of 15 tons. At Beauvais, if the great roof presents an area (per masonry bay) of $12 \times 9 = 108$ m.² to the wind, and the unit wind pressure is say 150 kg./m.²,

---

[1] No distress would be caused, however, at the chevet, since the buttress no longer thrusts against a flat surface; the curvature of the plan allows a "triangulation" of the forces.
[2] 12 to 15 tons is given in "The Stone Skeleton."
[3] Fitchen, *op. cit.*

BEAUVAIS CATHEDRAL

then the wind force on the roof is about 16 tonnes. In addition, both flying buttresses must resist the pressure of the wind on the vertical wall of the choir.

The "porte-à-faux" of the intermediate buttresses has not yet been discussed. Although they are not placed vertically over the supporting piers, there is in fact no sense in which they are falsely carried. Had they been placed farther *inwards*, say to over the centre of the aisle, Viollet-le-Duc would have agreed that, with proper design, the aisle arch could have carried them. Indeed, the question should be inverted: Can any possible mechanism of collapse be thought of for the intermediate buttress? The answer, just as for the flying buttresses of Beauvais, is no; only collapse of the outer main buttresses will permit consequential collapse of the intermediate buttresses.

Thus the fabric of Beauvais in 1272 seems to have been, in the large, designed almost perfectly to fulfil its function. No mention has been made of the main external buttresses, but Benouville's analysis shows them to be very stable indeed. Similarly, no attention has been paid to the design of individual vault severies; it is extremely unlikely that individual panels would have fallen, and that was not the sort of catastrophe that evidently occurred in 1284. The possibility therefore cannot be overlooked that the collapse began with some trivial accident and spread thence to the whole of the fabric. Viollet-le-Duc indeed gives such an explanation.

His cross-section is shown in Fig. 12[1]; this cross-section is not of the typical bay, but is taken at the chevet, where the ground plan, Fig. 1, permits the main external buttresses to be placed closer in. A perspective sketch is given in Fig. 13.[2]

Viollet-le-Duc considers that the slender twin columns A (Fig. 13) failed.[3] The mortar, slowly drying in the adjacent pier B, shrank (perhaps because the work was too hastily done, as Benouville believes), and more and more load was thrown onto the twin columns until they eventually fractured. As a consequence, the lintel L broke, and the great block M, the tas-de-charge, loaded by the gigantic statue N, was no longer supported. Viollet-le-Duc then considers that gross deformation occurred, which is plausible enough. He suggests that the block M slid out. It is more likely, however, that the block M would tilt outwards and so drive the line of thrust (shown broken in Fig. 11) outside the section of the flying buttress which would then collapse. There would be nothing to counteract the vault thrust; the vault would then collapse in that bay, almost certainly completely across the choir, so that at least one complete bay of vaulting would fall. The collapse would be likely to spread, since each bay of vaulting buttresses the next in an axial direction.

The typical Gothic structure is, in fact, an example of an assemblage of structural elements acting one on another to assure complete equilibrium of the whole. The structure can accomodate a wide range of forces, but, take away one portion, and all the rest is likely to fall. Without trivial accidents, however, and assuming no Acts of God, the complete structure was so stable that it would remain for centuries.

* * * * * *

Guessing the weight of Jean Vast's tower as 2,000 tonnes, and taking each crossing pier as of area 3 m.[2] the tower would have imposed an extra stress of some 16 kg./cm.[2] on these piers. This is very small, and again there is no question of the *strength* of the material governing the behaviour.

[1] Viollet-le Duc, *op. cit.*, 4, 178, Fig. 101 (article "Construction").
[2] *Idem.* p. 181, Fig. 101 *ter*.
[3] *Idem*, p. 180 f. "Il est certain cependant que cet énorme édifice aurait conservé une parfait stabilité, si l'architecte eût posé les colonnettes jumelles au-dessus du triforium plus fortes et plus résistantes, s'il eût pu les faire de fonte, par exemple. Les désordres qui se sont manifestés dans la construction sont venus tous de là; ces colonnettes, trop grêles, se sont brisées, car elles ne pouvaient résister à la charge qui se reporta sur elles lorsque les piles intérieures vinrent à tasser par suite de la dessiccation des mortiers. Se brisant, les linteaux L cassèrent (fig. 101); les gros blocs M, en bascule, s'appuyèrent trop fortement sur la tête du premier arc-boutant, celui-ci se déforma, et la voûte suivant le mouvement, la pression sur ces arcs-boutants fut telle qu'ils se chantournèrent presque tous; leur action devint nulle, par suite les arcs-boutants supérieurs lâchèrent un peu, puisque la voûte ne pressait plus sur eux. L'équilibre était rompu."

BEAUVAIS CATHEDRAL

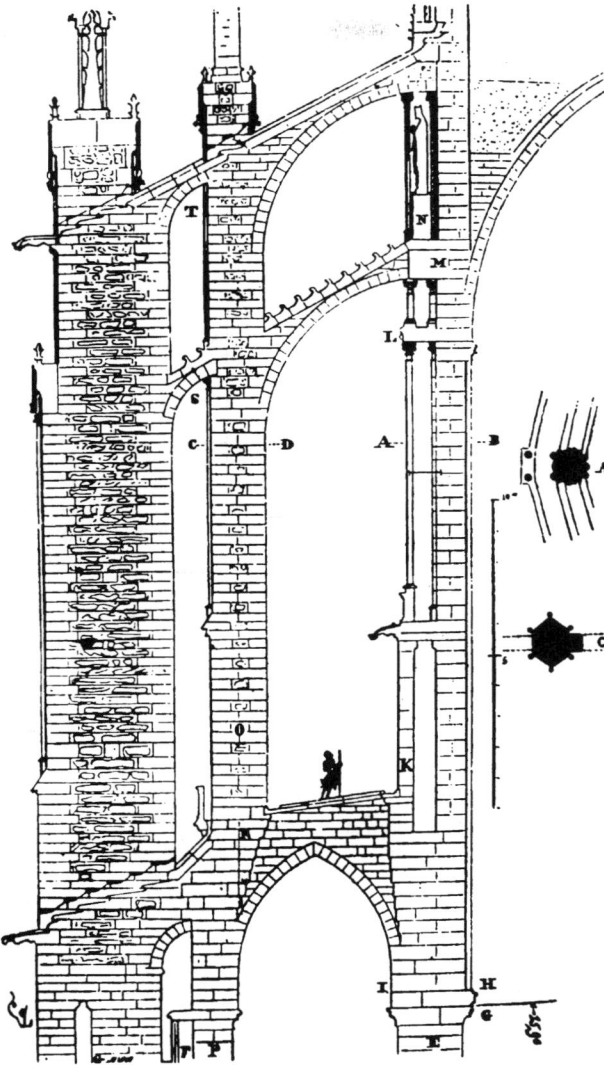

Fig. 12. Part cross-section at the chevet.

Fig. 13. Perspective sketch of
the upper part of the chevet pier
shown in Fig. 12.

Beauvais Cathedral
(*Viollet-le-Duc*).

31

BEAUVAIS CATHEDRAL

There seems little doubt, however, that had the advice of the King's masons been taken in 1571, and the crossing piers braced, the tower might be standing today.   It seems that the structure, from 1569, when the tower was completed, until 1573, when it fell, was never truly in equilibrium.   Desjardins reports[1] numerous small movements, and Leblond[2] reports fractures occuring, during these 4 years.[3]

The structural system of a massive tower supported by four unbraced piers would be liable to "drift", the movement restrained by tensile and shearing stresses developed in the mortar, and by possible interlocking of stones.   Eventually, however, the columns would have been pushed so far out of true as to be useless.   The Chapter was right to call a halt in 1605; "le temps n'était plus à bâtir des cathédrales."

DISCUSSION

In reply to a comment that, when the erection of additional columns doubled the number of arches, no buttresses were built to the new intermediate piers, Dr HEYMAN replied that 'perhaps light could be thrown on this by referring to Paris.   Great flying buttresses now occur at every pier in Paris, but there was a celebrated dispute just before the war—never really settled—as to whether the Cathedral had been built with a flying buttress at every bay or only every other bay—quite clearly the main thrust occurred at every other bay.   It is possible, Dr Heyman thought from the evidence at Beauvais, that Paris could have been built with alternate buttresses omitted and that they were added later.   It is practically certain that the architect at Beauvais would have seen the Paris construction.

Mr. R. J. MAINSTONE said that Dr. Heyman had set alongside the recorded fact, that the high vaults of Beauvais had stood for some 12 years, a demonstration that equilibrium between thrust and counterthrust was possible (barring gross deformations) and had concluded from the demonstration that the subsequent collapse must be attributed to some detail of construction rather than to a major fault of design.   Mr. Mainstone accepted the demonstration (granted that Benouville's reconstruction of the original cross-section was correct), but he questioned the conclusion.

He accepted also Dr. Heyman's assertion that Gothic rules of design were numerical ones (taking this to include also geometrical ones), but questioned the interpretation placed on these rules. It seemed a great over-simplification to suggest that they resulted simply in a direct geometrical scaling of similar elements from one structure to another.   Indeed both the present paper and the earlier one on "The Stone Skeleton" contained ample evidence that this was not the case.   The comparative heaviness of the masonry at Reims had, for instance, been commented on, and the proportions at Beauvais seemed, on the contrary, to be significantly more slender than usual elsewhere.   One almost had the impression that Reims was a lower structure than it should have been but that Beauvais was a higher one than originally intended.

The relevance of this to the possible cause of the collapse lay in the behaviour of the structure as the mortar of the piers slowly dried out.   Though mean stress levels were low, non-axiality of the thrust would lead to progressive deformations, and the more slender the piers the greater the likelihood that the deformations would increase to the point where equilibrium was no longer ensured. An element of doubt about the precise form of the vaults and supporting structure above triforium level prior to the collapse made it undesirable to be dogmatic.   The spread at the springings of the reconstructed vault was, however, measured in 1903 by William Goodyear.   He found it to be about one metre and to exceed that at any other major cathedral.   It is reasonable to assume that, immediately prior to the collapse of the original vault, it was at least similar to this and such a gross deformation

---

[1] Desjardins, *op. cit.*, p. 92 f.
[2] Leblond, *op. cit.*, p. 30.
[3] A very similar problem was encountered at Wells in the early fourteenth century.   The designer, William Joy, successfully braced the crossing piers with the famous "strainer arches."

BEAUVAIS CATHEDRAL

seems, by itself, to be sufficient to bring the slender upper piers of Benouville's cross section to the point of bucking and to lead to the collapse of the vault with or without the actual prior buckling of the piers.   The manner of reconstruction, with its emphasis on stiffening the piers and intercalating others is, moreover, entirely consistent with this interpretation.

Mr. Mainstone therefore preferred to attribute the collapse primarily to an excessive slenderness of the upper parts of the main piers in relation to their manner of construction and to consequent excessive deformations under eccentric load.   He strongly suspected, though, that a close examination of the present fabric to determine the precise extent and manner of the reconstruction would provide the necessary basis for a more definitive conclusion and regretted that the paper was silent about this.   Could Dr. Heyman say what evidence he had been able to find?

Professor A. W. SKEMPTON said he remembered very clearly an expedition to various French cathedrals during which he received an overwhelming impression that the proportions of Beauvais were markedly different from the other structures.   If the major failure at Beauvais began as a trivial accident why did the people concerned decide to double the number of piers when rebuilding?   From the manner of the reconstruction it would seem evident that originally something must have been wrong with the piers.   Moreover, Professor Skempton did not entirely agree with the idea that the stresses were very low.   The piers were probably built with a good masonry casing having a rubble core.   Recently he had seen the remedial works in progress on the piers at Winchester.   No doubt here the average stresses would also be apparently quite small, but nevertheless many of the piers of the Middle Gothic reconstruction were severely cracked, with fissures several inches wide running almost from top to bottom.   Given the extraordinary proportions at Beauvais, and taking into account the fact that the piers were doubled in number in the rebuilding, there could be little doubt that overstressing of the piers was a principal cause of the disaster.

Professor Skempton was intrigued by Dr. Heyman's reference to the evidence for using large models.   One knows, of course, that models were built, the equivalent of present-day architectural models, to show the client what was intended, but he had never been convinced that structural models were used in medieval times.

Dr. HEYMAN said that there was no clear account of how the collapse of 1284 did occur; there are conflicting accounts—none contemporary.   One report says that external buttresses failed, but this is not supported by others.   There is no evidence to put against Mr. Mainstone's view that the vaults fell without bringing down the external buttresses.   As to the rubble filling, at least in the sixteenth century they were very much alive to the stress on the main crossing piers and the King's masons made the specific remark that these piers were solid all the way through.   The evidence for the use of models was quite strong, and several examples are given by Frankl in *The Gothic*.

Mr. GRANT stated that at Chartres the original flying buttresses had consisted of two slightly curved struts held apart by a series of pillars rather like the spokes of a wheel.   The expertise of 1313 recommended the addition of a third strut above the other two.   At Amiens in 1497 an expertise recommended a third strut below the existing openwork buttress.   At the same time an iron band was fixed round the whole cathedral to stop the spread at the head of the columns at the crossing which was leaning outwards.

Dr. Heyman replied that the slow distortion and spread of the great timber roof might have much to do with the eventual requirement for a third buttress at parapet level.

Mr. MAINSTONE said that he found it very difficult to visualise how the tas-de-charge M could slide out.   According to Viollet-le-Duc's drawing it was built substantially into the main pier.   Also it was under compression from both sides; the vault pushing it out from one side and the buttress arch pushing it in from the other.   Even it if cracked it was in no different situation from that of any

BEAUVAIS CATHEDRAL

other voussoir of the buttress arch except that, being weighted from above, it could not so easily ride up the pier as this inclined outward.

Dr. HEYMAN commented that he thought the tas-de-charge could have crushed or cracked and dropped out rather than slid out with very little deformation on the main buttresses. He had no evidence either way about the original structure having been altered between 1250 and 1350; and there is no evidence that the structure was built other than in the form we see now; apart from the intercalated piers and reconstruction of the vaults themselves.

Mr. R. J. M. SUTHERLAND asked if it was known for certain that the choir which collapsed in 1284 was designed exactly as it is now; was the amount of buttressing as great as now; or was it rather like the Lincoln Chapter House where flying buttresses were added nearly a hundred years after the Chapter House was built. He would like to support Professor Skempton in the point about stresses. The stresses could become heavily concentrated and cause splitting as in the anchorage zones of pre-stressed beams. The strength of an individual brick or stone gives little indication of the strength of the wall in such circumstances. On the question of how the failure of the choir went: if the tas-de-charge did slide out then the buttresses would have had to move outwards in order to have let the tas-de-charge out, and if that happened Mr. Sutherland imagined that the buttresses could have remained standing even after the arcade had fallen. Alternatively it could have been one of those concentrated-load failures in which a piece crushed, split and came out in bits.

Mr. H. CLAUSEN said he had filmed most of the stained glass of the French cathedrals and he would like to support Professor Skempton's impression that Beauvais presented completely different proportions compared with the other great French cathedrals. He commented that the builders of Beauvais were always short of money. They got permission from the Pope to sell indulgences, and did so, to raise the money to build the tower. Why did they not go for the nave which would have provided the buttressing for the crossing? Isn't this the only case where the tower was built before the nave? In most cathedrals the nave and the choir and the transept were built first, and then the tower added after the base structure was more or less complete.

Mr. IAN DAVIDSON also emphasized the lack of buttresses or other means to resist longitudinal thrust. The obvious remedy was to put in interpolated piers thus halving the span and reducing the unbalanced longitudinal thrust.

Dr. HEYMAN agreed that the unbalanced longitudinal thrust, although small, could well have contributed to the collapse of the tower in 1573, but did not think that this effect explained the main collapse of the high vaults in 1284.

Mr. M. H. L. STANDEN asked whether the iron ties and staves at the top, had they been there for hundreds of years, would not have rotted away, or grown very much larger by corrosion so lifting and disrupting the masonry.

Dr. HEYMAN said that the iron ties that could be seen were said to be fourteenth century; he thought they were not there originally. He had not been up and did not know the state of the stonework at that level. In good building iron would be sealed with lead; this would reduce the risk of rusting.

Mr. E. W. H. GIFFORD stated that at Salisbury the masonry tower contains a great deal of iron (from about 1360 and later) some of it enclosed in lead, some not. The stone, as wet inside as out and more continuously so, gives it no protection. In some places it has remained in very good order; in others it is giving trouble.

In proposing the vote of thanks Dr. S. B. Hamilton said that Dr. Heyman had given a most interesting address, on a subject of great interest to some, but one on which the Society did not hear much at its meetings. The cause of failure at Beauvais may have been composite. Dr. Hamilton thought

34

BEAUVAIS CATHEDRAL

that the possibility of foundation movement could not be ruled out altogether. In a building of that size there is almost certain to have been some differential settlement. Allowance was apparently made for wind but Dr. Hamilton wondered if they realised what the movement from wind on a building of that size would be. Dr. Hamilton said there would be both plastic and elastic distortion. Some of the piers were heavily loaded on one side and, as we know, they do bend; for instance in Westminster Abbey there is a difference of some inches between the distance apart of the main columns of the crossing at capital level and at floor level which is quite obvious without plumb lines or instruments. This type of distortion can be seen in nearly all tall Gothic buildings. There is also the end thrust of the arcade down the length of the building, as mentioned by Mr. Davidson. We know that this caused collapse at Hereford Cathedral, Malvern Abbey and several other well-known buildings, and that it could quite well have contributed to the trouble at Beauvais. One could imagine the building settling, thrusting out at the ends, spreading crosswise and then, twelve years after building just being unable to stand any more movement, coming down with a rush. If Dr. Heyman hadn't cleared up all the mystery he had given the audience a great deal to think about that should help us to realise the cumulative movement that goes on in many of these important and most interesting Gothic structures.

PLATE V

Beauvais Cathedral as completed in 1569.

J. M. Fugère sculp.   Frontispiece from Desjardins

12
The structural behaviour
of medieval ribbed vaulting
K.D. Alexander, R. Mark and J.F. Abel

As EARLY AS 1845, Robert Willis published perhaps the first and certainly one of the most comprehensive studies on the construction of medieval vaulting.[1] Willis observed that the Gothic ribbed vault "consists, as is well-known, of a framework of ribs or stone arches, upon which the real vaults or actual coverings of the apartment rest. . . . The ribs are the principal features, and the surface of the vaults subordinate."[2] The implication that the Gothic rib plays the predominant structural role has since been widely accepted. However, this view is based more on faith in the validity of visual impressions than on comprehensive understanding of the vault's structural action. A different interpretation of ribbed vault behavior is developed in this paper, drawing conclusions from studies conducted with numerical computer "modeling," a modern engineering tool used to aid designers in understanding structural behavior. Computer analysis not only permits clarification of the complex behavior of a particular form, but its application to a variety of alternative configurations also provides some of the missing empirical "evidence" that might have been available to the

medieval builder. Various structural theories may be tested against this generated bank of information for a more complete interpretation of the role of the rib in Gothic church construction.

The Debate over the Function of the Rib

The lively controversy in the architectural-historical literature regarding form and function is responsible for focusing perhaps too much attention on the Gothic rib. In attempts to answer complex technical questions, the vaulting has usually been considered as nothing more than a simple assemblage of arches. Frankl reviewed much of the (pre-1960) literature on this subject and concluded that even "physicists" do not agree among themselves as to the role of the Gothic rib, and furthermore, "the historian of art who knows enough mechanics and mathematics to read this branch of literature intelligently is nevertheless not qualified to decide whether the rib carries or sometimes carries and when."[3] It is possible, however, to discern how certain structural phenomena have been misinterpreted by following the literature and identifying elements of confusion; the use of the new computer models should resolve much of the controversy.

The major attack on the rationalist's view of the indispensable structural function of the vault rib was mounted in 1934 by Pol Abraham who noted that the comments of the foremost rationalist, Viollet-le-Duc, often refute rather than support the structural approach.[4] Indeed, Abraham uncovered many errors in Viollet's reasoning; nevertheless, some of Abraham's own structural arguments are as confusing and ambiguous as Viollet's.

For example, Abraham observed that much of the understanding of vault behavior relies upon the concept of hori-

The work reported in this paper was performed as part of a National Endowment for the Humanities-supported interinstitutional program, Architecture and the Scientific Revolution. The authors acknowledge with gratitude the support of the Endowment as well as help received at various stages of the investigation from Dombaumeister Arnold Wolff of Cologne Cathedral and from the late Robert Branner.

1. R. Willis, "On the Construction of the Vaults of the Middle Ages," *Transactions of the Royal Institute of British Architects of London*, I, Part ii, 1842, 1–69. Willis's study, as he tells us himself in its summary, actually centered on vault geometry and construction, leaving out questions of behavior: "Thus, I have said nothing respecting mechanical principles, and have confined myself to form and management. But it appears to me from examination of the works of the Middle Age architects, that the latter considerations had an infinitely greater influence upon their structures than the relations of pressure, then very little understood, and about which they made manifest and sometimes fatal errors."

2. Willis, "On the Construction of the Vaults," 3, 24. For a discussion of Willis's technical background and his approach to the structure of historic buildings, see R. Mark, "Robert Willis, Viollet-le-Duc and the Structural Approach to Gothic Architecture," *Architectura*, VII, No. 1, 1977, 52–64.

3. P. Frankl, *The Gothic: Literary Sources and Interpretations Through Eight Centuries*, Princeton, 1960, 810.

4. P. Abraham, *Viollet-le-Duc et le Rationalisme Medieval*, Paris, 1935.

242

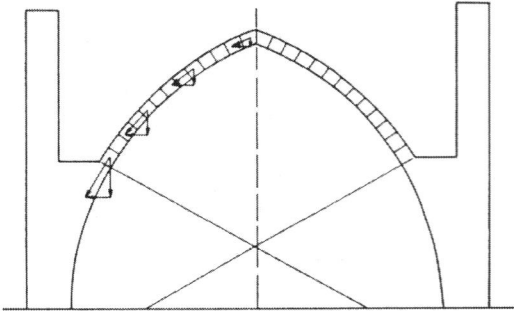

Fig. 1. Force polygons for an arch (after Abraham).

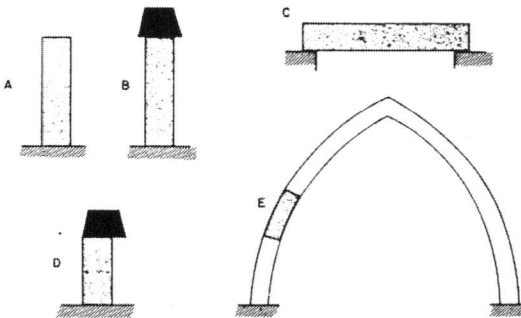

Fig. 2. Structural modes: (A) Stone element undergoing axial *compression* from its own weight. (B) Stone element undergoing axial compression from both its own weight and a concentrically applied external load or weight. (C) Stone element undergoing *bending* from its own weight; + denotes region in tension, – denotes region of compression. (D) Stone element undergoing not only axial compression from its own weight and an external load but also bending from the eccentric application of the external load; if the external load and eccentricity are sufficiently large, this stone will experience tension in the region indicated by the + symbol. (E) Voussoir acting essentially in compression as in cases A and B, but with a small amount of bending both from its own weight as in case C and from eccentricity of the forces acting on it from the adjacent voussoirs as in case D.

zontal thrust[5] and that this can best be apprehended by first considering a simple arch, within a single plane in space, instead of the three-dimensional vault form. Since the arch consists of material inclined at various angles with respect to the horizontal, he contends that the major forces will also be carried at these inclined angles. The simple graphical device of the force polygon resolves the oblique "resultant of the pressures," as Abraham calls it, into components—one horizontal and one vertical (Fig 1). At any point in the

5. In all cases, Abraham uses the word *thrust* to refer to the horizontal force: "thrust always signifies horizontal thrust": Abraham, *Viollet-le-Duc*, 10.

arch, the magnitude of the vertical component is the sum of the weight of all the material between the point and the crown. If the oblique resultant acts at exactly the angle of inclination of the arch as Abraham assumes, there is a unique relationship between the three vectors of the triangle at the point, and it follows that the horizontal thrust is directly related to the vault weight. Abraham goes on to conclude that neither the type of stone (given equal density) nor the strength of mortar binding the individual stones in any vault arch will change this relationship between vertical and horizontal thrusts.

While this line of reasoning gives a first approximation to the structural behavior of this relatively simple system, it overlooks the existence of bending which can be crucial to the integrity of the arch. Specifically, the assumption that the resultant forces always follow the inclination of the arch violates the fundamental precept that all forces acting on a structure must balance. Just as the vertical components must balance the weight, the horizontal components acting at the two ends of any voussoir must offset each other; therefore, the horizontal component in an arch loaded only by (vertical) weight must be constant throughout the span. The horizontal and vertical components which satisfy this equilibrium do not necessarily combine to give a resultant force vector aligned with the inclination of the arch. When they do not, bending must occur because the resultant tends to wander from the centerline of the rib, that is, the force becomes eccentric (Fig. 2). In short, the specific form of the arch in relation to the form of the loading has a great effect on the amount of bending, an action not accounted for in Abraham's interpretation. Bending can become a particularly serious problem when the eccentricity becomes large enough to compress one side of a member while stretching the other. Even a structure with good compressive strength will not be able to withstand a large amount of bending if it does not also have utilizable tensile capacity, and masonry construction is notorious for its poor tensile strength.

Abraham attempted to explicate the difference between the behavior of a planar arch and a three-dimensional vault by considering a long rectangular space covered with a series of parallel arches. Viollet, he pointed out, conceived of a Gothic vault as two such series, set at right angles to each other, supported by arches along the diagonal intersections. Based on this concept, Viollet thought each of the sections of vault webbing could move independently should some sort of settling occur. Abraham pointed out that such independence would imply "breaks" or cracks at the intersections of the web sections since the diagonal arches and webbing meeting them would be affected simultaneously. Indeed, Viollet's theory that the vaulting acts as independent series of arches supported by the diagonal ribs is also refuted by the separation often observed between the diagonal ribs and the vault webbing: the ribs separate from the webbing and can no longer support the vault, while the webbing remains intact.

Abraham then offered the famous analogy that a marble released along the top ridge of the vault, rolling in the "direction of the greatest curvature" as along A–B in Figure 3 directly to the support, or rolling first toward the groin and then down the groin toward the support (path C–D–E), indicates the direction taken by the forces within the vault.[6] Later in discussing vaults of double curvature (having principal curvature in two orthoganal directions), he analyzes the effect of the additional curvature but concludes in the end that the path taken by the ball "will not be essentially different from that described by a cross-sectional cut perpendicular to the axis of the vault."[7] Doubly curved vaults, he judges, behave in approximately the same manner as vaults of single curvature.

In all of this, Abraham is discounting Viollet's belief that the webbing leans or presses against the vault ribs and thus produces a thrust on the exterior walls along the window arch as shown in Figure 4. (Viollet might even have been interpreting the webbing as a series of beams between the diagonal arches.) On the other hand, the rolling ball analogy leads to the conclusion that the curvature of the webbing directs the vault forces to the groin. The thickened webbing along the groin, Abraham states, then carries the forces down to the pier where they meet it on a diagonal;[8] the webbing does not thrust against the exterior walls, but simply at the top of the piers. Interestingly enough, although Abraham notes the presence of these diagonal forces, he does not choose to distinguish their horizontal components, including the horizontal thrust at the piers which requires counteracting flying buttresses. Was this overlooked, or did he see this as the underpinning of a functionalist position?

Abraham supports his argument further with minor observations ranging from citing the lightness of the vaults to raising the question of why the Gothic master made the diagonal ribs nearly half as large in cross-section as the transverse arches, which span a shorter distance, if he in-

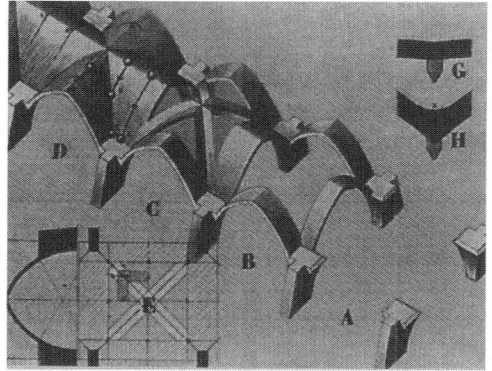

Fig. 3. Vault behavior (from Abraham).

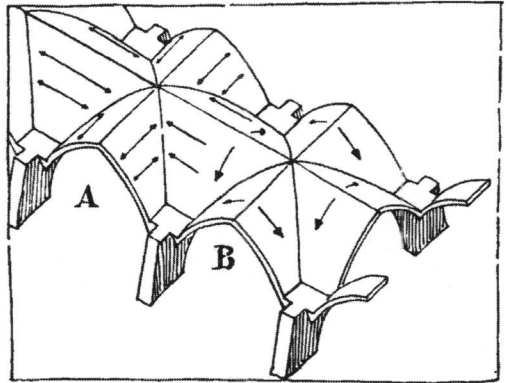

Fig. 4. Vault behavior (from Abraham, after Viollet-le-Duc).

6. Frankl translates this portion of Abraham's passage as follows: "[the marble] does not roll to the wall, of course, but in accordance with the theory that every section parallel to the wall results in a *barrel vault* which carries itself, along the curve of the section. The ball always rolls in the direction of strongest curvature and, reaching the hollow line above the groin, rolls on in this channel to the pier." Frankl, *The Gothic*, 807.

7. Abraham, *Viollet-le-Duc*, 34: "ne sera pas très sensiblement différent de la section droite."

8. This is to prove the uselessness of the diagonal ribs themselves. Abraham realized that the cross-section of the vault along the groin indicates an *arêtier* or a thickening of the vault which in itself (see Fig. 3) is sufficient to carry the diagonal forces resulting from the loads transmitted to the groin.

tended them to bear the major portion of the load. To further discredit Viollet's view, Abraham points to the use of ribs in many types of vaults as purely decorative elements; he notes that entire sections of ogival ribs have been blown out by bombs without precipitating the collapse of the vaults above. Abraham also comments on the Gothic builders' approach to construction. He notes that they probably started from simplified assumptions for the intersections of the different sections of a bay and traced them out in plan. Then he shows that they could have adjusted the intrados to match the lines of intersection with a compass. In this sense, it would be most natural to use the rib primarily as permanent centering and to take advantage of its ability to cover up unsightly gaps and difficult stone work along the lines of intersection.

The real purpose of Abraham's argument is to give the

244

rib a larger role than the purely structural one. The basic statics and behavior, he argues, do not change between a vault with ribs and a vault without ribs. The ogives do not lift any weight off any part of the vault proper; they do not alter the direction of vault forces in order to channel them directly into the rib.

Erwin Panofsky felt it necessary to respond to the debate as a kind of mediator between the extremes, although he himself was certainly capable of carrying the functional argument to the extreme.[9] Nevertheless, Panofsky's real contribution is his awareness of the context in which Gothic structure must be interpreted. To study Gothic architecture, according to Panofsky, is to face the "visual logic" or the "self-explication of reason" that permeated the entire era.[10] He argues, "that Gothic vaults have been known to survive when the ribs were blasted away by artillery fire in World War I does not prove that they would have survived had they been deprived of their ribs after seven weeks, instead of after seven centuries. . . ."[11] He observes that the ribs of Caen and Durham "began by saying something before being able to do it" while the buttresses of these two churches "began by doing something before being able to say it," and ultimately, "the flying buttress learned to talk, the rib learned to work and both learned to proclaim what they were doing in language more circumstantial, explicit, and ornate than was necessary for mere efficiency. . . ."[12] For Panofsky, and for any serious scholar, the form versus function debate did not in itself present the entire picture.

A more technical insight can be gained from the recent discussions of vaults by Professor Jacques Heyman and by Luis Carlos Curcio, both engineers. Heyman theorizes that a crease in a shell surface (i.e., the groin) marks a line of weakness, as well as a location of stress concentration; hence, ribs were required along these creases in order to provide necessary reinforcement. Furthermore, he believes that the transverse and wall ribs are less important because there are no creases at these locations; these ribs merely carry their own weight. Heyman also lists the additional rib "functions": their covering of unsightly gaps and their role as permanent centering.[13]

Curcio sheds some further light on the behavior of ribbed vaulting in his *Study and Reflection on Medieval Structures and the Equilibrium of the Gothic Cathedral of Reims*.[14] In discussing the legitimacy of the "positivists" (rationalists) of the 19th century he conveys the complexity of vault action: "Gothic architecture is not just a simple mass at rest, but the expression of an animated interlacement of forces, a process in perpetual motion—vibrant in every nerve of the building."[15] In these vaults, he notes, there exists strong three-dimensional interplay of thrusts. The fact that an arch is a structural element has led to the belief that vaults ribbed by arches must therefore be rational structures. However, the equilibrium of the *shell section* of a vault must remain essentially the same with or without the ribs. He investigates, in detail, what he calls the "virtual arch" or the thickened section of the Gothic vault along the groin—a thickening, as already noted by Abraham, which results primarily from the nature of the intersecting vault geometry. The true effect of the groin is that the internal forces in the webbing are diffused about the groin region; thus they cannot be simply analyzed with a planar, two-dimensional force-polygon or arch simplification. Curcio cites the engineers Torroja and Saboret to corroborate the complex nature of the vault stress flow and the improbability that the forces run toward the groin and then instantly change direction.[16] Instead, he suggests that the geometric nature of the vault intersection along the groin itself establishes vault equilibrium. If this is the case, how, he rightly asks, is it possible that ribs along this groin could appreciably alter the overall effect? A very stiff rib might attract some additional forces, yet a section cut across the groin reveals much greater depth for the virtual arch than for the real rib (Fig. 3H). He then estimates the strength of the virtual arch section and shows that the diagonal ribs have even far less influence than proposed by Panofsky, whom he interprets as suggesting that the ribs exist for duplication of strength in a vault which can very well carry its own weight.

9. This is demonstrated by his discussion of the cross-section plan of a Gothic pier. He argues that the more rational pier is the one which expresses the entire structure in its plan by carefully distinguishing the diagonal and transverse ribs. E. Panofsky, *Gothic Architecture and Scholasticism*, New York, 1957, 51-52.

10. Panofsky, *Gothic Architecture and Scholasticism*, 58, 59.

11. Panofsky, *Gothic Architecture and Scholasticism*, 54.

12. Panofsky, *Gothic Architecture and Scholasticism*, 57-58.

13. J. Heyman, "On the Rubber Vaults of the Middle Ages and Other Matters," *Gazette des Beaux Arts*, LXXI, 1968, 177-188. Heyman discusses the problem of stability as that of a structure's tendency to overturn or rotate to destruction. He describes the concept of hinging (which usually is accompanied by cracking) and the extent of hinging required to produce a

"mechanism" of collapse. Problems of stability are considered the primary danger by Heyman but he believes the Gothic masons assured stability by proper proportioning. He reasons that if they had had an understanding of statics, they might have developed better rules.

14. L. C. Curcio, *Estudio y Reflectiones Sobre Estructuras Medievales y Equilibrio de la Catedral de Reims*, Buenos Aires, 1967. Curcio's insight into the difficulties of interpreting Gothic structure leads him to present some very useful structural information in a way that applies directly to the Gothic work and is easily understandable to the general reader.

15. Curcio, *Estudio y Reflectiones*, Exordium.

16. Curcio gives the following references: E. Torroja, *Razon y ser de los tipos estructurales*, Madrid, 1960, 113-116; and V. Saboret, P. Abraham, and H. Focillon, *Storia Sociale dell'Arte*, 1957, 370 (sic). In the prologue to his book Curcio cites the series of lectures given by Torroja at the University of Buenos Aires in 1952 which led to his subsequent correspondence with Torroja that stimulated his writing an "ensayo historico-filosofico."

245

However penetrating these observations may be, none of them is based on quantitative structural analysis of a *complete* vault system, and thus many questions are left unanswered. Such investigations are reported in the following sections.

Structural Modeling

The first comprehensive three-dimensional structural study of quadripartite vaulting, carried out by Mark, Abel, and O'Neill, was based on a combination of small-scale photoelastic and numerical computer modeling.[17] Two bays of the 13th-century choir vaults at Cologne Cathedral (Fig. 5) were constructed at 1/50- scale from stress-free, cast epoxy plastic (Fig. 6). The actual vault form was altered only slightly; each vault segment was taken as part of a surface of a shell of revolution as shown in Figure 7, allowing the model to be assembled from four standard components, right- and left-hand longitudinal and transverse webbing.[18]

17. R. Mark, J. F. Abel, and K. O'Neill, "Photoelastic and Finite-element Analysis of a Quadripartite Vault," *Experimental Mechanics*, XIII, 1973, 322–329.

18. Other assumptions and simplifications made for the first model tests were: 1) the vaults were considered uncracked, 2) the vault thickness was kept constant, corresponding to a uniform thickness of 13 in (the actual vault thickness varies from about 10 in at the crown to 20 in near the piers), 3) the *loading* of the rubble fill at the haunches of the vault was taken into account, but any structural interaction (i.e., increased support or stiffening of the vaults) by the fill was neglected, 4) the buttress system supporting the vaults was considered immovable (i.e., no translation or rotation was permitted at the vault springing), and 5) the model was constructed without ribs.

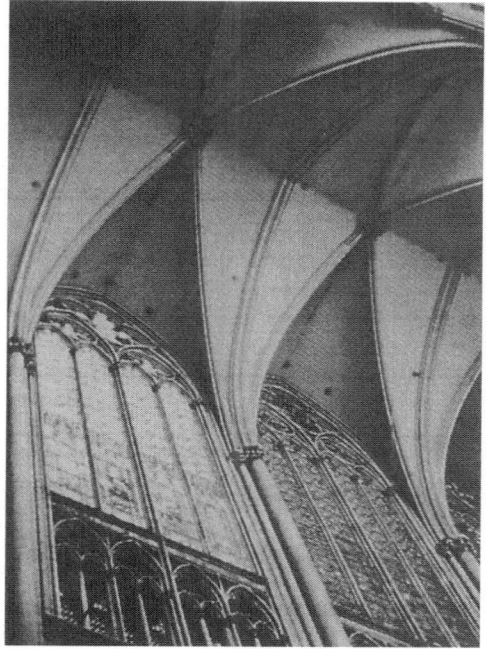

Fig. 5. Cologne Cathedral. Choir quadripartite vaulting (photo: R. Mark).

Fig. 6. Photoelastic model of Cologne vaulting.

246

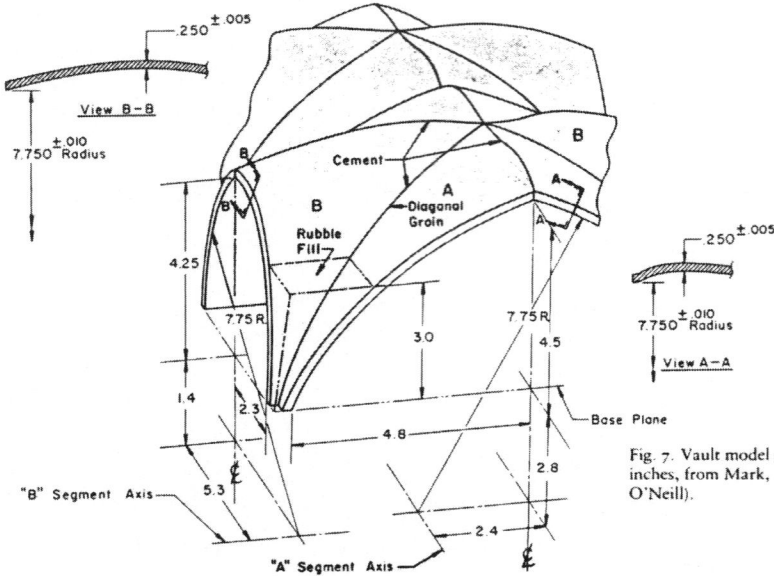

Fig. 7. Vault model geometry (in inches, from Mark, Abel, and O'Neill).

The finished model was then loaded in an oven as normally prescribed for a "stress-freezing," photoelastic test.[19] Following this cycle, the model was observed in a polariscope to study "locked-in" photoelastic patterns, and then slices were taken from it to reveal internal stress distributions. The structural action of the model, including magnitudes, distributions, and directions of structural forces as well as support thrusts, was quantitatively determined from these data and used with scaling theory to predict the structural action of the full-scale vault system.

Although three-dimensional photoelastic modeling is a very powerful analytic tool, it has the limitation that the model must be reconstructed following a first testing and slicing if variations in loading and geometry (e.g., a model with ribs vs. one without ribs) are also to be studied. It is only within the past several years that numerical computer "modeling" techniques have been developed that can feasibly handle a geometry so complex as groined vaulting. Although these programs require much time-consuming, patient effort to prepare, once the programming is accomplished, the computer models offer the flexibility to study easily any number of load and geometric modifications. In a sense, numerical modeling involves the same procedure as used with a physical model. However, the form is now described by a series of coordinates taken at discrete intervals over the structure's surface. These coordinates define a "mesh" which becomes the geometric model for the computer. A series of equations relating the loading conditions and the material properties are then used to calculate the displacements at all the mesh points (almost 750 points for these vault models), so that the displacement pattern for the entire structure is obtained. The pattern is then related to the properties of the material in the real structure through additional material-behavior equations to obtain the same type of overall structural force-distribution information that is derived from a photoelastic model test.

It is advantageous to be able to compare information from both types of models because if agreement can be obtained it provides verification of the computer solution. After demonstrating good agreement, the numerical model can be used with a great degree of confidence for studying modified configurations. This confidence is all the more important because cathedral vault geometries themselves introduce so many variables onto the study. No *detailed* drawings exist for the vaults of any of the major medieval churches (although ongoing photogrammetric studies may soon remedy this situation). The only accurate information usually available is that which is readily obtainable by tape measure and plumb line, such as vault crown to floor heights. For Cologne, three elevations were known along the top of the nave vaults as were the elevations of the top of the clerestory windows. This information allowed the

19. See R. Mark and R. S. Jonash, "Wind Loading on Gothic Structures," *JSAH*, xxix, 1970, 222-230.

247

er surfaces of the vaults to be approximated by a mathe-
ical expression giving the radius of vault curvature at
point (typical sections of these surfaces are shown as
and B–B in Fig. 7). Both the photoelastic model and
initial computer model geometries followed this ap-
ximate formulation.

dripartite Vault Behavior I

ults from the *unribbed* photoelastic model of the Co-
ie choir vaulting were shown to agree well with those
in the equivalent numerical model. Hence, the following
mary of observations from the first series of tests could
argely derived from either modeling approach. While
ie results were based on an unribbed model, it was the-
ed that taking the ribs into account would have little
itional effect because of the local thickening along the
ins as discussed above. But of course, this view was not
definitive.

The only regions of high compressive stress observed
those where the vaults join the piers. Here, the ribs may
ie been employed as a device to reduce local vault
sses, although the maximum highly localized stresses in
unribbed (13-in thick) vaults of 400 pounds per square
i do not exceed maximum values found in other regions
ypical Gothic cathedral sections.

Bending moments throughout the vault are low so that
ile stresses caused by dead weight of the vault and rubble
ire almost nonexistent. While the rubble fill may per-
i a function in helping to transmit thrust more uni-
nly to the supporting buttressing system, its weight has
iignificant effect on the overall vault performance.

The horizontal component of thrust of an (unribbed)
ogne high vault acting at the pier top was found to be
co pounds as compared to a vertical force of 170,000
nds. These calculations are based on a total vertical
it loading of each full bay equal to 340,000 pounds,
uding the rubble fill.

The major in-plane compressions are directed toward
vault supports as shown by the heavy dashed lines of
ire 8. They do *not* follow the trajectory of a rolling ball
iggested by Abraham.

his body of information was the starting point of the
ent study which set out to determine the structural
ts of adding ribs to both quadripartite and sexpartite
ts. In addition, the influences of variations in vault ge-
try were studied.

dripartite Vault Behavior II

original mathematical formulation of the Cologne
netry used in the first study was found not to be easily
table to other modifications. A simpler and more uni-

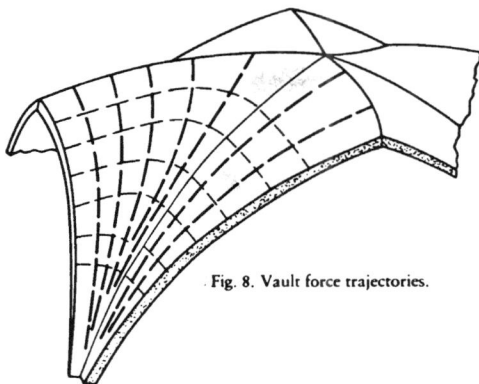

Fig. 8. Vault force trajectories.

versal analytic form was then derived based on the observation that the surface generated by a circular arc spanning
the diagonal of a bay, intersecting with circular arcs of
varying radii originating along a single line located at approximately one-third of the bay span and parallel to the
longitudinal vault axis, has a geometry very similar to that
of Figure 7. In fact, the computer was used to draw perspective projections and to aid in reaching the final form.
Once the geometry is stored in the computer, it may be re-
called as seen by an observer at any vantage point; views of
the computer-drawn vaulting are illustrated in Figure 9. This
technique represents a modern application of the very approach to understanding medieval vault geometry suggested by Willis in his pioneering study.[20] Since this simplified geometric model is applicable to a plan of any aspect
ratio and to any vault height, when the model was combined with the computer structural analysis program it became possible to investigate how the changing of dimensions affects overall vault behavior.

Force distributions within the modified vault configurations were studied: first, with the fill but without the ribs
(as in the photoelastic model tests), second, without the
ribs and the rubble fill (Figure 7), third, with the ribs but
without the fill, and fourth, with both the fill and the ribs.
Corresponding supporting forces as predicted by the computer for these four configurations are compared with the
earlier predictions in the table of Figure 10. In all four
modified-geometry cases, the stresses everywhere within the
vaults except in the regions of the *tas-de-charge* are low, of
the order of ten pounds per square inch compression. Near
the *tas-de-charge*, the springing point of the vault, the maxi-

20. See figs. 9a and 10 of Willis, "On the Construction of the Vaults,"
10, 19.

248

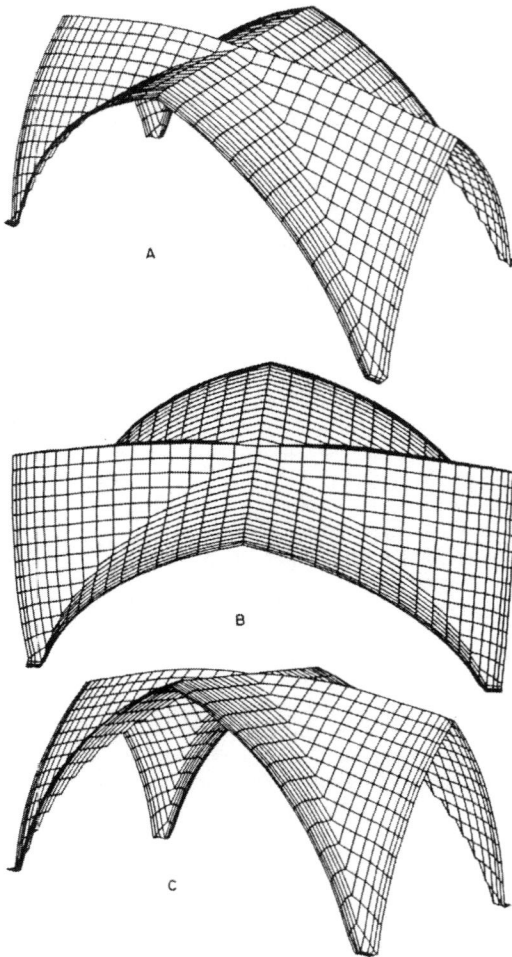

A

B

C

Fig. 9. A, B, and C. Computer-drawn quadripartite vaulting.

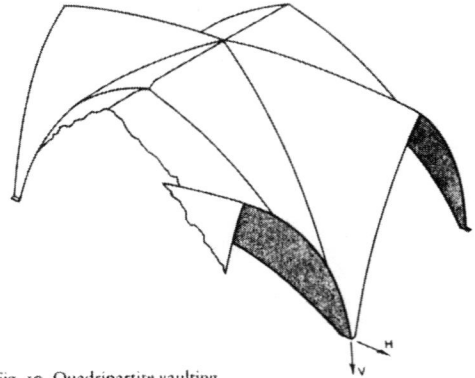

Fig. 10. Quadripartite vaulting.
Forces at the supports.

Case	Geometry	Ribs	Rubble Fill	Forces, thousands of pounds V	H	H/V
1	model, Fig. 7	No	Yes	170	57	0.34
2	modified, Fig. 9	No	Yes	172	62	0.36
3	" "	No	No	117	55	0.47
4	" "	Yes	No	145	63	0.43
5	" "	Yes	Yes	200	69	0.35
6	singly curved, Fig. 11	Yes	Yes	205	69	0.34

directions of the major compressive forces were again seen to flow along the shortest paths through the shell toward the support at the pier in the fashion shown in Figure 8. This primarily compressive behavior implies that the small cracks often seen in vaults are relatively unimportant to the overall vault stability since the weakness that they introduce would only affect a structure undergoing tension.

In addition to the vertical and horizontal support forces given in Figure 10, there is a third, longitudinal horizontal force between bays. (However, this force does not act on the pier as it is balanced by an opposite force of the same magnitude from the adjacent bay.) For case five, the prototypical Cologne vault, this force was found to be 20,000 pounds.

Computer results for axial and bending stresses in the diagonal and transverse ribs indicate that in almost all sections, the average compressive stress carried by the rib is far greater than any tensile stress resulting from bending alone. In the few sections where bending stress combined with average axial compressive stress does produce a small amount of tension in the ribs, it is only three or four pounds per square inch, and nothing of the disruptive order suggested by Curcio.[22]

mum local compressive stress of 250 pounds per square inch for case two is reduced to 70 pounds per square inch when the loading is shared by the ribs in cases four and five.[21] The

—————

21. The computer program used in this analysis was SAP IV, written by E. L. Wilson and K. Bathe at the University of California at Berkeley. The program would have allowed the handling of varying thicknesses of both the vault webbing and the arch ribs (see fn. 18), but this would have required a fair amount of additional effort. The assumption of uniform thickness does not greatly affect the overall distribution of forces; however, it does change local magnitudes of stress. The program actually produced a higher

value for stress in the unribbed 32-cm-thick webs near the *tas-de-charge* than indicated in the text; the cited value includes compensation for the actual web thickness of this region.

22. Curcio, *Estudio y Reflectiones*, 103–104.

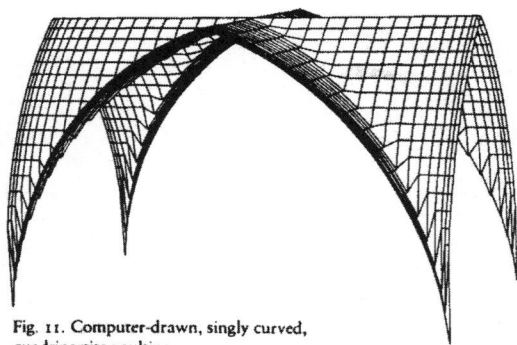

Fig. 11. Computer-drawn, singly curved, quadripartite vaulting.

It was also perceived that small changes in vault geometry have little effect on overall patterns and magnitudes of vault forces; for example, the results from the two different representations of the Cologne geometry are similar (Fig. 10, cases 1 and 2). This is an important observation in that relatively small differences in vault and bay dimensions from similar buildings or from one bay to another within a single building can be expected to produce only relatively small changes with respect to total loadings or thrusts from a given general vault geometry. This also helps to explain how a similar vault form might be copied and used successfully at another location. In addition, it indicates that approximations to actual geometry used in physical or computer modeling do not inhibit the drawing of general conclusions regarding structural behavior.

A final modification of the Cologne geometry was made to test the structural effect of the doubly curved webbing. Comparison of the behavior of a computer-produced singly curved quadripartite surface including both fill and ribs (Fig. 11) with the doubly curved vault with fill and ribs indicates similarity in all respects—particularly in the directions taken by the web forces. The support forces for this last geometric modification, shown as case six in Figure 10, are also similar. Hence, it may be reckoned that characteristic English vaulting, which approaches single curvature, behaves in very much the same fashion as more typical French doubly curved vaulting. Low sensitivity of vault behavior to double curvature also provides the basis for study of a sexpartite vault. It was concluded that one could indeed determine the general structural characteristics of doubly curved sexpartite vaulting using a simpler singly curved vault model.

Sexpartite Vault Behavior

The approach taken in generating the more complicated sexpartite geometry is essentially the same as that described

Fig. 12. Bourges Cathedral. Nave sexpartite vaulting (photo: R. Mark).

above for the quadripartite geometry; however, there is much less information available for describing the shape of a typical sexpartite surface (Fig. 12). Branner's drawing of the Bourges choir[23] is chosen as a starting point. Portions of the vault are taken initially to be of constant radius. Again, the entire surface is made continuous after successive alterations of the unknown dimensions with the aid of the computer-drawn projections (Fig. 13). Thus a doubly curved surface is not generated for the sexpartite vault geometry, making it similar to Pol Abraham's simplest case, which most "clearly" sent forces to the groin. Support conditions are the same as for the quadripartite studies and, due to the similarity of the rib sizes and shell web thicknesses between Cologne and Bourges, the same dimensions are used here as well. This is done not only for simplicity but also to facilitate comparisons between the quadripartite and sexpartite vaults by minimizing the variables between the two types of systems.

23. R. Branner, *La Cathedral de Bourges et sa Place dans l'Architecture Gothique*, Paris/Bourges, 1962, drawing in pocket.

250

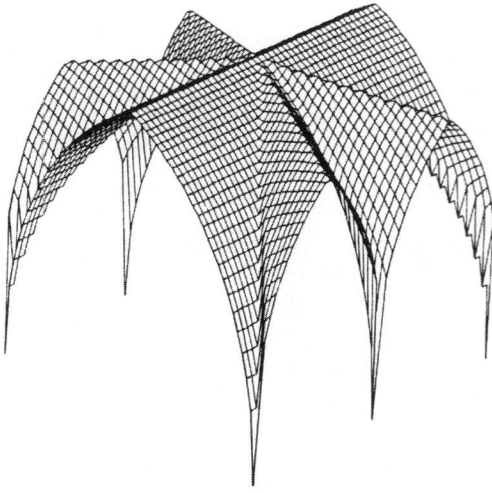

Fig. 13. Computer-drawn sexpartite vaulting.

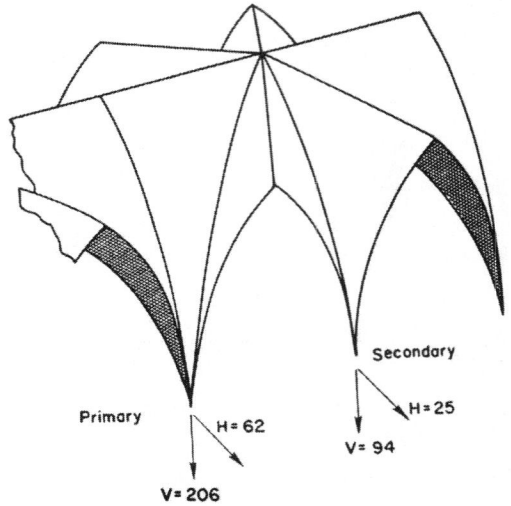

Fig. 14. Sexpartite vaulting. Forces at the supports (in thousands of pounds).

At first glance, it would seem that the radically different quadripartite and sexpartite geometries would cause the vaults to exhibit substantially different behavior; however, this is not found to be the case. The principal compressive forces are again seen to be directed through the shell in the shortest path toward the nearest pier. Hence, the ribs do not take on any more structural importance in sexpartite vaulting than they do in the less complicated quadripartite system. This is further borne out by Figure 14 showing that the ratio of main to intermediate pier forces bears no relationship to the number of ribs joining the piers at either of these locations as would be expected if the ribs, in fact, supported a major portion of the vault load. At the primary pier there are three main ribs joining the *tas-de-charge* (not counting the window ribs) and at the intermediate pier there is only one rib. This would indicate a ratio of 3 to 1 for the thrust values between the two piers; the computer results indicate that the actual ratio is 2.5 to 1.

Another significant comparison between the two systems of vaulting concerns the relationship of weight and thrust. The Bourges and Cologne vault bays are of similar dimensions; i.e., transverse and longitudinal pier spacing in the two buildings is similar (43.3′ × 21.7′ for Bourges, 45.0′ × 22.5′ for Cologne), and the rise or height of the vaults is about the same (23.1′ vs. 24.4′). The total system weights including ribs and fill are 600,000 pounds for the Bourges (singly curved) sexpartite vault vs. 820,000 pounds for the equivalent two bays of the Cologne (singly curved) quadripartite vault, and the horizontal thrusts on the primary piers

are 62,000 pounds and 69,000 pounds, respectively. On the other hand, although no direct comparison can be made for thrust of the sexpartite vault at the secondary pier (the quadripartite system having only primary piers), the lower thrust value of 25,000 pounds is indicative of its inherent potential for covering a large area with a single visual unit, lighter construction, and less total horizontal thrust. The rib weights are a major factor in making the sexpartite configuration a lighter structure. In comparing the two system geometries, it will be noted that there are in fact two fewer groin ribs to contend with in the equivalent expanse of sexpartite vaulting; these weigh nearly 8,000 pounds each and their absence helps explain the lower thrust. One may speculate that this was indistinctly but nevertheless sufficiently well understood by the master who made the repairs to the Beauvais choir vaults after the failure of the original quadripartite vaults in 1284.

Conclusion

These analyses are dependent on certain initial assumptions. Two of the most important concern the supposition that the supporting piers are held rigid by the buttressing system and the implication, inherent in this type of analysis, that gravity forces only begin to act after the vault construction has been completed. Usually neither of these conditions is strictly satisfied for actual vaults. However, where rigid centering was used to form the vault, its removal following the completion of construction would suddenly "turn on" the gravity forces because these forces affect the vault only when it is allowed to deform. Spreading of the upper portion of piers is commonly observed, often with characteristic cracks in the vault along the top of the clerestory (Fig. 15). The spreading of the piers might appear to alter the vault force distributions drastically, but if subsequent cracking occurs along a trajectory (flow) line of principal compressive force (e.g., as in Fig. 15), it should not greatly change the general pattern of vault support. Thus, these suppositions are the usual kind of engineering assumptions that make meaningful calculations possible when certain desirable factual information is unavailable.

The most important revelation from these studies concerns the distinctive structural behavior of the quadripartite and sexpartite thin vault webbing: that it clearly acts as a three-dimensional structure and that its standard treatment in the art-historical literature, as a series of almost parallel arches, can be quite misleading. Much of the problem in comprehending the structural role of the rib can be traced to this misunderstanding, as can the erroneous "rolling ball" analogy wherein the ball is imagined to roll down the vault surface in a path suggesting the presence of an *individual* arch. The fact that the vault-supporting forces are distributed throughout the webs points instead to the minimal effect that the ribs could have, since they are located only in certain discrete regions. Pol Abraham was essentially correct in his arguments against Viollet-le-Duc's analysis of Gothic ribs and was particularly astute in observing the importance of the *arêtier* or thickening of the vault along the groin. It is a less obvious conclusion that the entire webbing, even at the scale of Gothic construction, can be structurally self-supporting as has been indicated in this paper. In this light, the *constructional* and the *aesthetic* aspects of the rib take on greater importance than the *structural* function.

Fig. 15. S. Etienne, abbaye aux Hommes. Crack in webbing at the wall (photo: R. Mark).

The observation regarding similarity in behavior among bays of similar form suggests that vaults could be copied in rough detail and used in other buildings. This helps to explain the success of medieval masons with these complex structures in a period where design was largely a matter of "trial and error."

The rib has been shown to be lacking structural function in the completed vault system; however, there remains the question of the medieval builder's view of its role. Perhaps Paul Frankl has pointed to part of the answer by placing the beginning of the Late Gothic at a time (ca. 1300) when the "function of the rib is ignored."[24] From this, it can be inferred that the rib had originally been assumed to strengthen the vaulting and that through further experimentation in building, by 1300, its true, nonstructural nature was well understood.

24. P. Frankl, *Gothic Architecture*, London, 1962, 146.

13

Fan vaulting

Walter C. Leedy, Jr.

One of the central problems in architecture is how to enclose and articulate interior space. The problem becomes more critical as the size of the space increases, and it was particularly acute in a medieval cathedral or abbey. There the upward sweep of the walls strongly emphasized the ceiling, which was consequently expected to have a striking design. Furthermore, the ceiling was generally made of stone and was very heavy. The solution adopted in the most representative Gothic churches was to construct the ceiling as a series of vaults in which planar sheets made up of stone blocks ran between sharply pointed arches. The arches served as ribs bearing at least part of the load.

As medieval masons gained experience with ribbed vaults they found that construction could be made easier and the design more versatile by including additional ribs in the ceiling structure. Eventually the vaulting came to be a complicated lacework of ribs, which could be given many shapes. In England in the 14th century the proliferation of ribs led to the invention of an entirely new structure: the fan vault. In the fan vault the planar stone sheets of the ribbed vault were replaced by a rounded structure called a conoid that resembles a cone cut in half along its axis and erected with the pointed end down. Four conoids were joined to span each modular unit of the vaulting. On the curved surface of each conoid numerous ribs radiate upward and outward toward the peak of the ceiling. It is from this fanlike pattern of ribs that the style takes its name. Between the ribs is intricately carved stone tracery.

In a ribbed vault the arches were expected to bear the weight, but the conoid of a fan vault can act as a shell structure, in which the stresses are distributed fairly evenly throughout the stone fabric. The medieval builder had no way to analyze the stresses in a shell. Nevertheless, modern engineers have shown that the conoid of a fan vault is a very stable structure. Between about 1350 and 1540 English designers exploited this quality to construct more than 100 fan-vaulted ceilings spanning increasingly large spaces. Among the most famous are the ceiling of the chapel of King's College at the University of Cambridge and that of the Chapel of Henry VII in Westminster Abbey.

The elaborate ceiling of a fan-vaulted building provides a powerful visual experience. The rounded conoids flow into one another along the planes of the wall and give an impression of gracefully and continuously molded volumes of space, in contrast to the angularity of the ribbed vault. The intricate carving on the surface of the conoid emphasizes its volume and contributes to the impression of unity. The visual effect was probably foremost in the minds of the masons who built the fan vaults. It should not be forgotten, however, that underlying the aesthetic experience is the solution to a fundamental architectural problem arrived at by designers relying on practical experience in the absence of highly refined theoretical skills.

The development of the fan vault is closely connected to the physical properties of stone and the aesthetic preferences of the English masons and their patrons. Stone can withstand large compressive forces, but it has little tensile strength: it cannot readily withstand being pulled or stretched. The downward pull of gravity can cause substantial tensile stress in a horizontal beam supported at both ends. In a stone ceiling the tensile stress must somehow be converted into a compressive stress. In the pointed arch of the ribbed vault the conversion is accomplished by raising the arch stones above the horizontal plane defined by the points from which the arch springs. The force of gravity is thereby transformed into outward and downward compressive forces that are transmitted through the stones of the arch toward the walls.

The modular unit of a vaulted ceiling is usually a rectangular compartment called a bay, whose long axis is perpendicular to the major axis of the hall. In a ribbed vault the bay generally includes three pairs of arches. The transverse arches cross the hall. The wall arches run along the wall at the short sides of the rectangle, most often with large clerestory windows in the wall under the arches. The opposite corners of the bay are connected by diagonal arches that intersect in the center of the rectangle. Tall piers placed at the springing, the point at the corner of the bay where the lower segments of three arches meet, help to support the vaulting high above the floor. The thrust from the arches is conveyed to the ground by the piers and by flying buttresses that meet the exterior of the wall near the springing.

In the construction of a rib-vaulted ceiling the arches were erected first, utilizing a complex wood scaffolding called centering as a form on which to lay the stones. When the six arches in each bay had been erected, they served as the framework for the web: the sheets of stone laid between the ribs.

Ribbed vaulting, such as that in the cathedrals of Amiens and Rouen, has a strongly linear appearance, which is the basis of Gothic artistry. The effect was achieved by emphasizing the vaulting ribs as a series of frames with the web filling the space between them. Furthermore, the large transverse arches draw attention to the division of the ceiling into bays. In contrast, although the fan vault is still composed of linear elements, the unity of the interior space is emphasized. The viewer's gaze is drawn along the ceiling by the repetitive and graceful form of the conoids; the distinction between bays generally has little visual importance.

Although the aesthetic principles of the two kinds of vaulting are quite different, the historical transition between them was not abrupt. On the contrary, the fan vault developed gradually out of the construction of ribbed vaults. Ribbed vaulting was common in England, France and other countries of continental Europe, whereas fan vaulting developed only in England; the question naturally arises of why the new

KING'S COLLEGE CHAPEL at the University of Cambridge has one of the most beautiful examples of fan vaulting. The main vault of the chapel, shown here, was completed in 1515. The ceiling has the longest span of any known fan vaulting: the distance across the hall is 12.7 meters. The rounded intersecting surfaces are the main weight-bearing structures of the fan vault. Called conoids, they act as shells in which stresses are distributed fairly evenly. The carving, derived from the tracery of stained-glass windows, serves to unify the building's interior.

Fan Vaulting   3

4   Walter C. Leedy

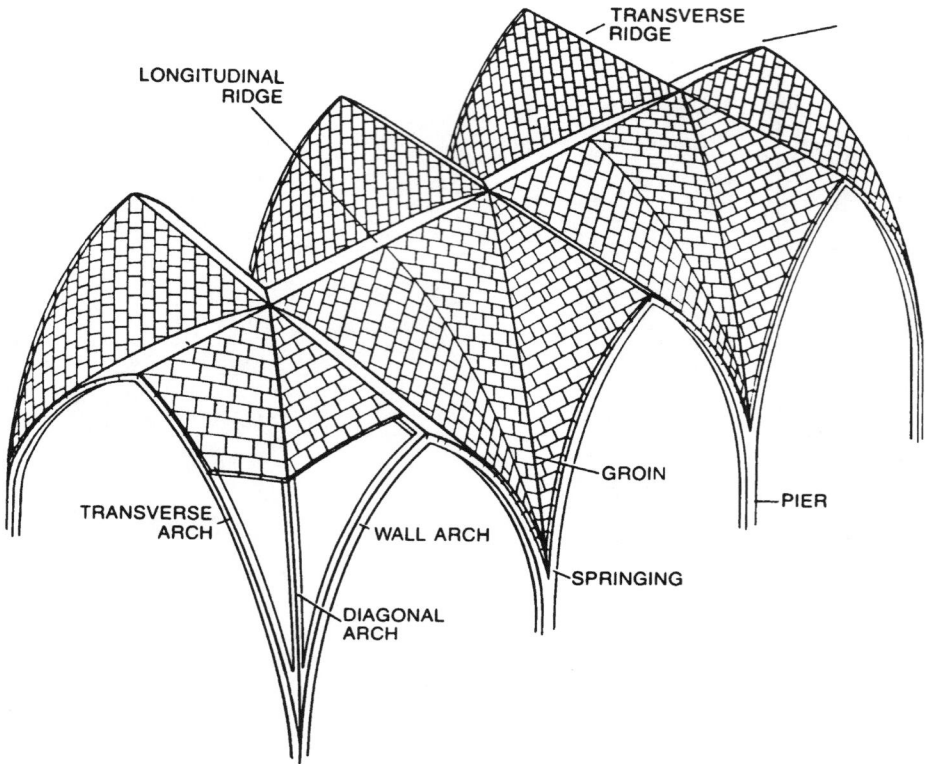

RIBBED VAULT, the classic Gothic ceiling design, was conceived by medieval masons as a frame structure in which pointed arches serve as weight-bearing ribs. Each of the modular units of the vaulting, called a bay, includes six arches arranged as a rectangle with intersecting diagonals. The load of the ceiling is converted into outward and downward thrusts that are conveyed in part through the arches to the piers and walls. The surface of the vault between the arches is filled with stone sheets called webs. In England the stones of the web were laid at an angle of approximately 45 degrees from the longitudinal or transverse ridge to the groin.

form emerged only there.

The answer lies in the aesthetic values and building practices that prevailed in England in the Middle Ages. The comparison with France is instructive. In France the courses, or rows of stone, in the web were laid parallel to the sides of the rectangle covered by the vault, so that the courses run in horizontal rows from the bottom of the vault to the top. Each quadrant of the bay resembles a peaked roof; the ridge at the peak of the quadrant is called a transverse or a longitudinal ridge depending on the orientation of the quadrant. Along the groin,

where the quadrants meet, the courses of the web were laid parallel to both ridges, forming right angles to the sides of the bay. For a while English masons followed the French precedent. Soon, however, they began to lay the courses from the ridge of the quadrant to the groin at an angle of about 45 degrees to both ridges. This style of coursing led in part to the conoid's being conceived of as a rounded structure.

English practice came to differ in other ways as well. In the French vault the web courses were laid over the top

Fan Vaulting   5

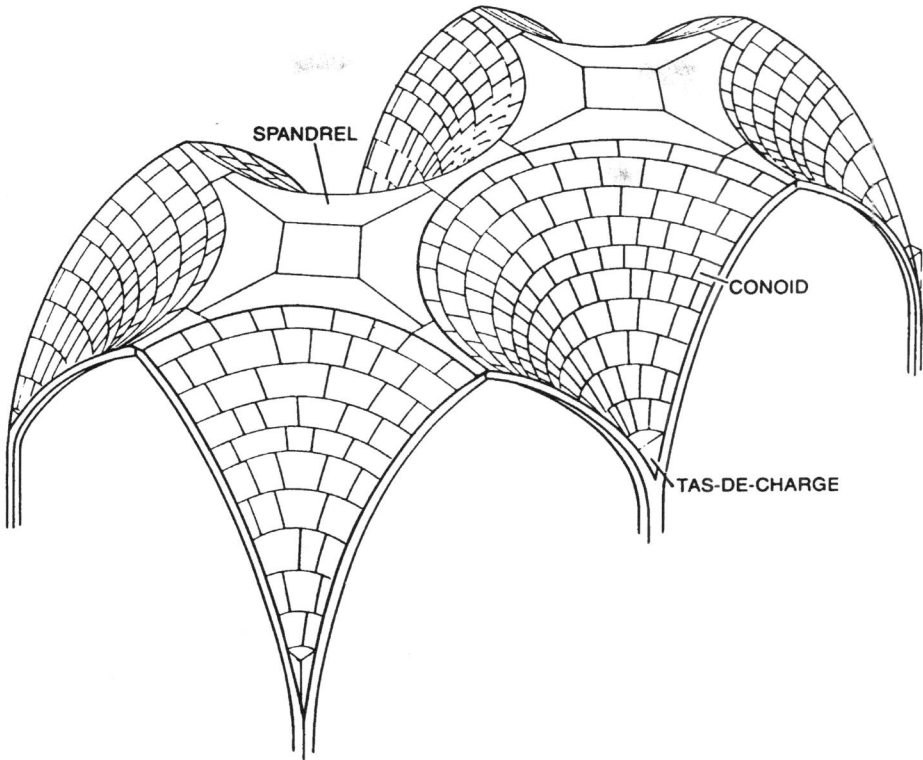

**FAN VAULT developed from the ribbed vault in the 1350's. The fan-vault conoid is a sur-
face of rotation, created by rotating a curve about a vertical axis. Viewed from the center of a
bay, each conoid has a convex horizontal section and a concave vertical section. This form allows the conoid to act as a set of vertical and horizontal arches that convey thrust to the wall.**

of the rib; the rib and the web were connected only by mortar. In England the upper edges of the rib were deeply notched and the web stones were set into the notches before the stones were cemented together. This practice, called rebating (or in modern English rabbeting), tended to make the rib and web a single assembly. Here is architectural evidence that the English mason saw the rib as a weight-bearing member.

Even the design of the arches in England differed from that in France. In France height, illumination and openness of structure were the most highly prized architectural qualities. To increase the area of the clerestory windows the wall arches were often stilted by raising the bottom of the arch above the springing on vertical wall shafts. Stilting had several significant consequences. The web near the window had to be warped to accommodate the shape of the arch. In addition the area of the wall between the clerestory windows was much reduced and the outward thrust from the vaulting was concentrated on a narrow strip of wall. As a result the flying buttresses had to be placed near their optimal position or the wall would collapse.

In England there was less concern with height and openness. The walls were generally lower and thicker. The clerestory windows were smaller and hence the wall arches did not have to be

Walter C. Leedy

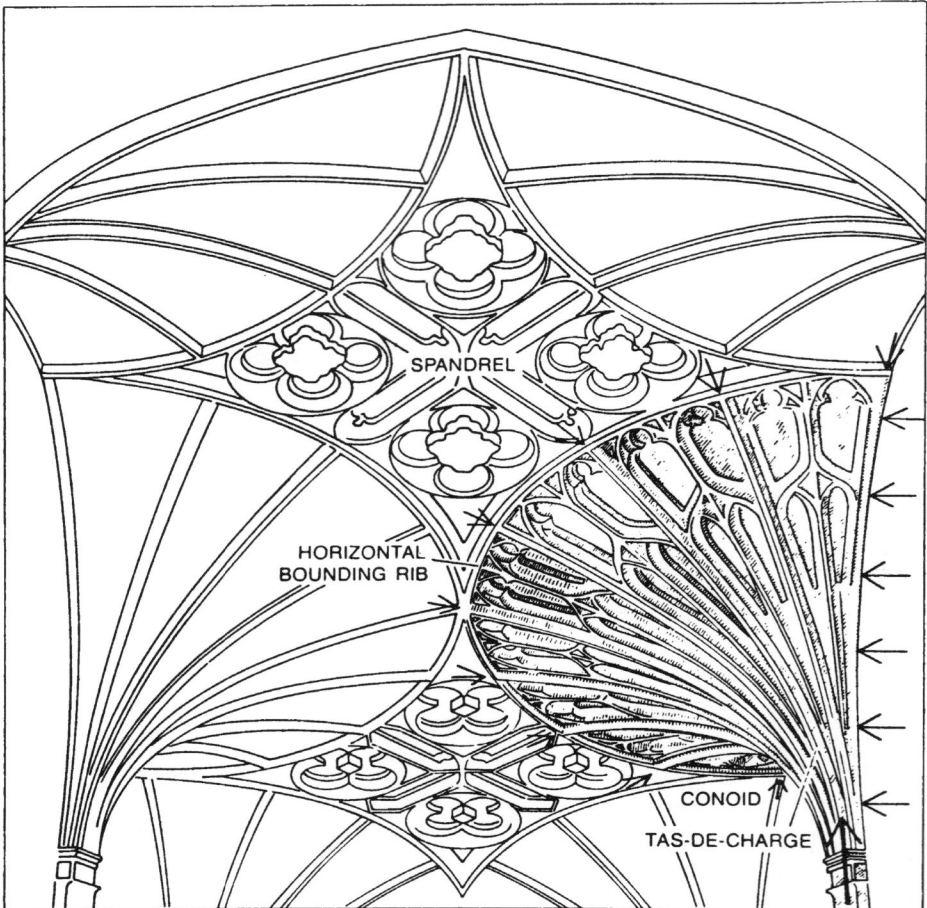

CONOID OF A FAN VAULT must be compressed along all its edges to be in equilibrium. The drawing shows the vaulting in the Church of St. Mary at North Leigh, Oxfordshire, which was finished in about 1440. The heavy spandrel panel between the conoids provides a compressive force along the horizontal bounding rib and serves as the keystone for the entire vault. The tas-de-charge built out from the wall at the corner of the bay supports the bottom of the conoid. The walls provide inward thrust to oppose the thrust transmitted outward through the conoids. Ribs carved on the surface are primarily aesthetic rather than structural elements.

stilted. This increased the area of the wall across which the vaulting thrust was distributed and reduced the need for buttressing.

The ribbed vault spread through much of Europe in the 12th century. Medieval masons soon learned that the addition of two further types of ribs could increase the flexibility of the vaulting design. The first type, the tierceron rib, ran from the springing to the ridge of the vault, roughly bisecting the angle between the diagonal rib and either the transverse rib or the wall rib. The lierne rib was a shorter structural member inserted at an angle between the longer ribs.

English masons built numerous tierceron vaults. The tierceron design was well suited to the English style, in part because it made possible great flexibility in adjusting the curvature of the ribs. In contrast to architectural practice in France, where rib curvature was constrained by the striving for height and illumination, in England the tierceron ribs radiating from the springing could be formed into an appealing geometric pattern.

In the high Gothic ribbed vault the volume defined by the webs intersecting along the groin in the corner of the bay was quite irregular and the shape was determined by other design considerations. As more ribs were added to the tierceron vault the volume became more regular and began to appear as a satisfying form worth designing for its own sake. At the same time the English coursing pattern began to yield an intriguing result. As I have mentioned, in England the web courses were laid at an angle from the longitudinal or transverse ridge to the groin. In tierceron vaulting the web courses in the corner of the bay began to approximate the form of regular, horizontal polygons. The polygons were concentric on the springing of the vault and increased in size toward the peak of the ceiling.

By the early part of the 14th century the main preconditions for the fan vault had been created in tierceron vaulting with concentric courses and rebated ribs. Two further technical advances were necessary for the fan vault to come into being. The first was the practice of carving a segment of the rib and the adjoining portion of the web from the same block of stone. The carved stones were then fitted together to form the vaulting much as blocks of ice are fitted to form an igloo. This kind of construction, which is termed jointed masonry, was employed in much fan vaulting.

The use of jointed masonry by English masons does not seem to have been motivated primarily by engineering considerations. On the contrary, it seems to have been largely the result of aesthetic concerns. English builders valued a high degree of carved articulation of the vaulting surface. Such articulation is much easier to achieve with large, closely fitted blocks than with ribs and small panels of stone inserted between them. As the size of the blocks was increased to provide a carving surface, the blocks tended naturally to assume a structural function.

The introduction of jointed masonry led to another innovation: the major axis of the rectangular cross section of all the ribs was made perpendicular to the vaulting surface. In a ribbed vault the major axis was generally perpendicular to the floor. When jointed masonry is used in a fan vault, however, the joint between the portions of a rib on two different blocks can be made perfect only if the ribs are perpendicular to the vaulting surface. The proliferation of ribs and hence of intersections between ribs made this an increasingly important aesthetic consideration.

By the middle of the 14th century the major elements of the fan vault had evolved in England. According to most modern investigators, the defining characteristics of a fan vault are as follows: vaulting conoids with a regular geometric form; regularly spaced ribs, all with the same curvature; a distinct spandrel, or central ceiling panel; ribs perpendicular to the surface of the vaulting, and patterned surface carving. This set of properties seems to have appeared first in small square canopies in the tombs of great nobles. (The origin of the style may never be identified with certainty because of the destruction of many important sites during the reign of Henry VIII.) Because the tomb vaults are small, they are often referred to as toy, or decorative, fan vaults. The first structural fan vaulting could well be that in the Trinity Chapel of Tewkesbury Abbey in Gloucestershire; the vaulting was completed in about 1380. The first large structural fan vaulting and one of the most significant early examples of the form is the vaulting in the cloister of

8   Walter C. Leedy

CONSTRUCTION OF VAULTING in King's College Chapel was accomplished in four stages. The stone blocks were carved in workshops next to the building site. It seems likely they were put in place with their decorative surfaces only partially finished. Wood center-ing, or scaffolding, was utilized as a form for the vaulting during construction. After the walls and roof were up the large transverse arches that divide the ceiling into bays were erected (*upper left*). The conoids were then laid, proceeding inward from the transverse arch-

Fan Vaulting 9

es toward the center of the bay (*lower left*). The conoids were assembled one horizontal level at a time. When the horizontal arches of adjoining conoids met in the center of the bay, the large transverse ridge stone that serves as the keystone for both arches was put in position. The stones of the longitudinal ridge were then laid (*upper right*). Finally, the large, heavy boss at the center of each spandrel was dropped in place from above (*lower right*). After the centering was taken down any final carving of the ribs and tracery was done.

Gloucester Cathedral, which was begun in the second half of the 14th century.

An essential feature of the fan vault is that the vaulting conoid is a surface of rotation, that is, a form produced by rotating a curved line about a vertical axis. The axis is the corner of the bay and the curved line is the rib. One consequence of this form is that the conoid has a horizontal section that is a circle or a portion of a circle at every vertical level. The circle is convex with respect to the center of the bay. The rib, however, is concave with respect to the center of the bay. Thus at every point on the surface of the conoid a convex horizontal curve intersects a concave vertical curve.

Since the conoid extending from the corner of the bay is circular, fan vaulting is best suited to a square bay. (Fan vaults were also built in rectangular modules, but ingenuity was required to adapt the form to the bay.) In a square bay the conoid has a horizontal section that is a quarter of a circle. At the top of the ceiling is an approximately diamond-shaped space formed by the intersection of the four circular conoids. This space is filled by the spandrel. At the curved boundary between the spandrel and the conoid the builder often inserted a horizontal bounding rib to divide the two structures visually.

The medieval mason was no formalist. He felt free to alter the elements of the fan vault in order to accommodate the design requirements or his own taste. For example, in the chapel of King's College the vertical ribs appear to be equally spaced, but actually there are small discrepancies resulting from the fact that the decision to fan-vault the ceiling was made after the building had been partially erected. Sherborne Abbey in Dorsetshire has beautiful examples of fan vaulting that were quite influential among 15th-century designers. The conoids in the chancel (the area that includes the altar) have horizontal sections that are polygonal rather than circular. Moreover, the conoids are constructed of separate ribs and small flat panels rather than jointed

masonry. Nevertheless, in both King's College and Sherborne Abbey the ceilings must be described as fan vaulting because of their overall appearance, in spite of the deviations from the definition of an idealized fan vault.

The construction of fan vaults can be divided into three periods beginning with the construction of toy fan vaults in tomb canopies. The first period ended with the completion of the cloister of Gloucester Cathedral in 1412. Between 1412 and 1430 no major fan vaults were built. This was a time of labor shortage, high taxes and economic depression; fan vaulting was costly and it is likely that no individual or institution could muster the capital needed to undertake construction. The design of the Sherborne Abbey chancel vaulting in the late 1430's marks the beginning of the second period, which lasted until about 1475. In the third period, from 1475 to about 1540, many of the largest and most important fan-vaulted ceilings were constructed.

The overwhelming majority of fan vaults were built in ecclesiastical buildings, and a considerable number of these were constructed in chantry chapels. A chantry was a fund established to pay for masses for the soul of the founder; the chantry chapel was where the masses were said. Most of the early chantry chapels were endowed by noblemen and were intended as much to call attention to the greatness of the founder as to ensure the safety of his soul. The fan vault with its striking visual form and intricately carved tracery was well suited to the purpose.

In the second half of the 15th century the taste for fan vaulting spread to the middle class and to the King. St. George's Chapel at Windsor Castle, which was a royal institution, has fan vaults dating from the 1480's. Shortly after 1500 work began on the most magnificent of all fan vaulting: the ceilings in the Chapel of Henry VII at Westminster Abbey, which Henry conceived as a huge chantry chapel for the Tudor dy-

Fan Vaulting 11

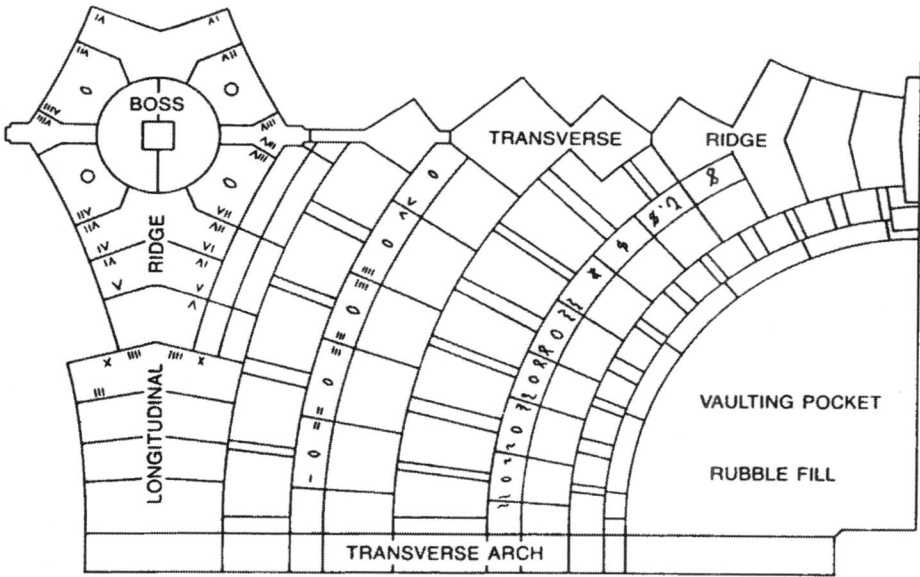

PLACEMENT MARKS on the stones for the vaulting of King's College Chapel enabled the stone setters on the building site to complete the ceiling by merely following the marks. The symbols were carved in the workshops where the vaulting was prefabricated. The large circular mark at the center of the stone indicates the quadrant of the bay in which the stone was placed. The marks at the ends of the stone next to the joints are the Arabic numerals 3 and 4. They indicate the horizontal level in the conoid where the stone was to go, the position in that level and the timing of insertion. The placement marks can still be seen on the upper surface of the vaulting stones, as is indicated by the plan view of one quadrant of a bay in the main vault.

nasty. Hence the church, the nobility, the middle class and the crown all contributed to the diffusion of the new architectural form.

The spread of fan vaulting was accompanied by an increase in the scale of the projects. The vaulting completed about 1380 in Trinity Chapel at Tewkesbury Abbey spans 1.7 meters across the hall. The main vault of the King's College chapel, which was completed in 1515, spans 12.7 meters. This is the longest span of any fan vaulting. One reason it took English builders more than 150 years to increase the span of the vaulting to its maximum is that they had little conception of how a fan vault works.

Twentieth-century engineers utilizing advanced mathematical tools have shown that for the conoid of a fan vault to be in equilibrium it must be supported along all its edges. A substantial weight must apply a compressive load at the upper edge along the horizontal bounding rib that separates the conoid from the spandrel. The load is provided by the spandrel itself, which is a very heavy stone plate. The large bosses, or raised decorative stones, that constitute the center of the spandrels in the King's College chapel weigh some 1,400 kilograms each.

The spandrel serves as the keystone of the vertical arches in the conoid. As in the ribbed arch, the thrusts are directed outward and downward. The downward thrusts are conveyed to the bottom of the conoid, where the structure rests against the wall. At this point the conoid is supported by the tas-de-charge, a projection built out from the wall in the corner of the bay. As noted above, the conoid is composed of concentric courses radiating upward from the springing, with each course having the form of an arch whose alignment is approximately horizontal. The outward thrust is conveyed through the horizontal arches to the wall, which provides an opposing thrust. Thus the conoid is compressed between the spandrel, the walls and the tas-de-charge.

The stresses in the conoid tend to be distributed fairly evenly rather than concentrated in the ribs. The specific way the stresses are distributed has important implications for the stability of the conoid, which, as I have noted, is a shell structure. An architectural shell is defined as a stressed structure much thinner than it is broad. The thrust acting on each point of the shell can be represented by a vector with components directed outward and downward; the set of all such vectors forms an imaginary surface called the thrust surface. Mathematical analysis shows that for the conoid to be in equilibrium the thrust surface must lie within the physical surfaces of the conoid shell. If the thrust surface emerges from the surfaces of the conoid, the structure is likely to collapse.

The builders of fan vaults had no way of knowing this fundamental fact. The analysis of stresses in a shell was far beyond their theoretical capacities. On the basis of long experience, however, the masons developed several techniques that greatly increased the stability of the vaulting. The narrow volume at the base of the conoid above the tas-de-charge is termed the vaulting pocket. In many fan-vaulted ceilings the pocket was filled to a height of about a meter with a solid rubble made up of stone chips and cement. In the chapel of King's College the rubble reaches precisely the same height in each conoid, indicating that it was a planned element of the structure rather than a casual afterthought.

The presence of the rubble fill has three significant consequences. First, the thrust surface extends into the rubble fill, which conveys the thrust to the wall. In this way the thrust is distributed over a much larger area of the wall than it would be if the shell alone transmitted it to the wall. Second, the rubble acts as a weight cantilevered over the floor. The weight opposes the outward thrust of the conoid reaching the walls in the corner of the bay and thereby further reduces the stress put on the walls.

Most significant for the stability of the conoid, the rubble fill reduces the span of the vaulting that functions as a shell structure. The rubble cannot act as a shell because it is a substantial three-dimensional form rather than a thin sheet. Since the vaulting thrust passes into the

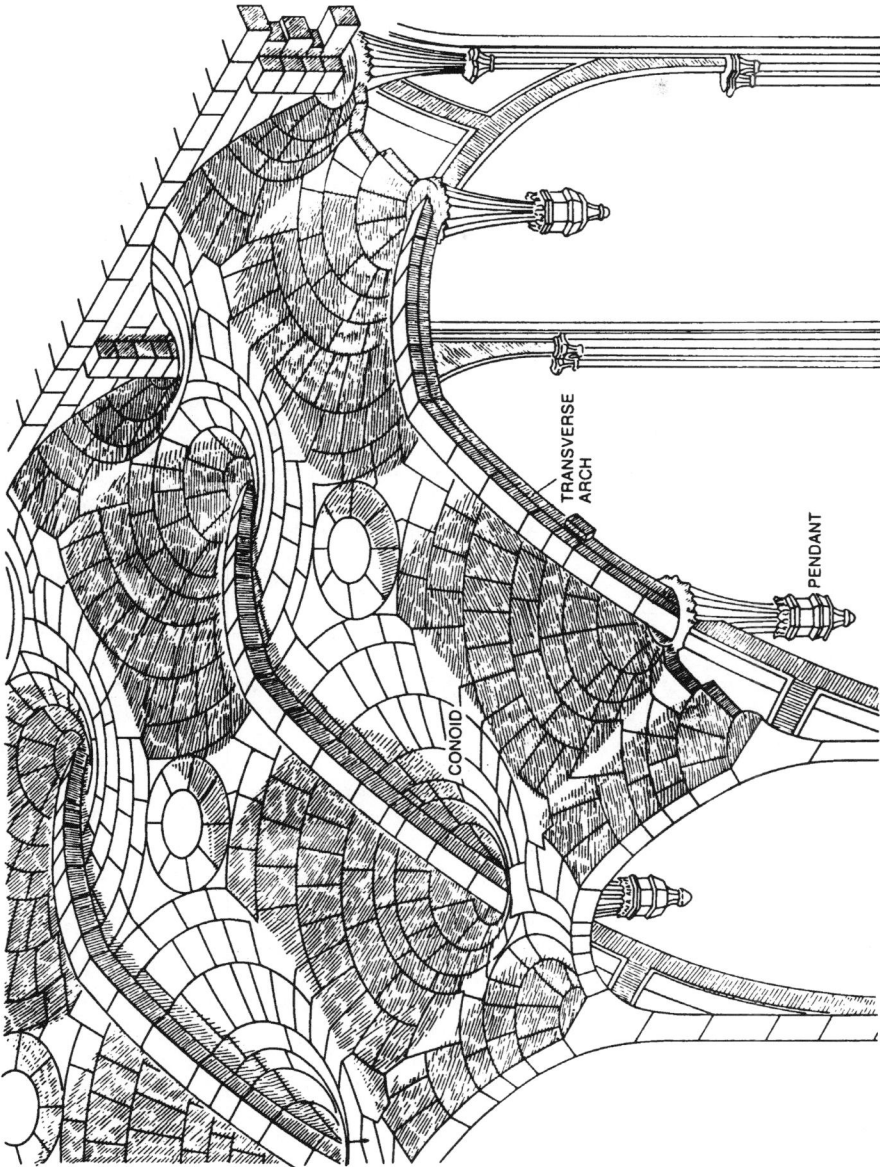

CONOID

TRANSVERSE ARCH

PENDANT

CHAPEL OF HENRY VII in Westminster Abbey has the most magnificent of all fan vaulting. Work on the vaulting was begun in about 1500. The ingenious design of the ceiling combines frame structures and shell structures. Substantial transverse arches support large pendants near the wall of the bay. The conoids are built over the pen-

dants. At the pendants the transverse arches run up through the vaulting and along its upper surface, so that the arches are not visible in the central part of the ceiling. Most of the weight is taken up by the arch where it passes through the conoid at the pendant. The arch functions as a frame member and the conoid as a shell structure.

rubble, the lower part of the conoid does not act as a shell. Jacques Heyman of the University of Cambridge has shown that the part of the vaulting that functions as a shell is the part between the top of the rubble fill and the lower edge of the spandrel. This distance can be quite short in relation to the entire span of the vaulting. Reducing the length of the shell increases the ratio of thickness to length. The shorter and thicker the shell, the greater the probability that the thrust surface lies between the surfaces of the conoid.

Thus without significant theoretical knowledge the mason managed to make the fan vault quite stable. If stability can be ensured, the only other major risk to be considered is compressive failure, that is, the crushing of the building material. Since stone has tremendous compressive strength, however, it is not likely that the vaulting will fail in compression no matter how large the vaults. Therefore large fan vaults can be constructed on the same principles as smaller ones. Nevertheless, it took many decades for builders to understand that the stability of the fan vault is not compromised by its size. By the end of the 15th century this fact had been grasped and quite audacious ceiling designs were being attempted.

The construction of the largest fan vaults entailed some intriguing building techniques. The stone for the vaulting in the King's College chapel was taken by boat up the river Cam or by wagon to the building site. The rough stones from the quarries were in large blocks; a typical rough stone was probably a cube about two-thirds of a meter on a side. Temporary workshops were erected next to the building site and the stones were partially carved there.

Indeed, in the workshops the chapel vaulting was prefabricated. The rough stones were carved to the correct size and shape and the surfaces where the stones were to join were precisely finished. The rib and the tracery on the face of the block that was to be on the ex-

posed surface of the conoid, however, needed only to be roughly carved. The stone was marked with a symbol indicating the quadrant of the bay it was to go into. Arabic or Roman numerals were carved at the end of the stone near the joint to designate the horizontal level of the conoid the stone was intended for, the exact location within that level and the timing of the insertion of the stone in the construction process.

The stone setter on the building site had only to observe the markings to put the stone in its correct place. The prefabrication of the vaulting stones may have required more skill than the actual construction. From a study of the construction markings visible on the upper surface of the conoids in the King's College chapel I have concluded that the bays in the main vault were built as follows. First the walls were erected and the roof was installed, following the common medieval building practice. Then the large transverse arches between bays were erected. All the stone setting probably required elaborate centering. The next step was the construction of the conoids, beginning at the long sides of the bays along the transverse arches and proceeding inward. One horizontal level at a time was constructed in each conoid. When the arches met in the center of the bay, the large stone that constitutes the keystone for both horizontal arches was placed. When the conoids were complete, the longitudinal ridge stones on each side of the central bosses were put in position. The final step was the insertion of the heavy central bosses.

The bosses were dropped in place from above. Because of their great weight they induced compression in the vaulting, which slightly changed its shape and allowed the centering to be readily taken down. When the centering had been removed, the final carving of the surface of the conoid was probably done. After any cracks that might have appeared were mortared the conoids were sometimes painted. The vaulting in the chapel of King's College was originally to have been painted and gilded.

Although the main vault at King's College has the longest span of any fan vaulting, the most splendid fan vaulting is that in the Chapel of Henry VII. The span of the chapel, 10.6 meters, is only slightly less than that of King's

CEILING TRACERY in the Chapel of Henry VII is among the most intricate examples of the patterns carved on the surface of fan vaults. The chapel was meant to glorify the Tudor dynasty; the carving is the material analogue of ornate praise for the Tudors. The vaulting is constructed entirely of jointed masonry, in which closely fitted blocks of stone compose the vaulting surface. Jointed masonry affords the best surface on which to carve elaborate designs.

16   Walter C. Leedy

College and the overall design is awesome, combining a simplicity of three-dimensional form with a very complex and regular surface pattern. Because of the high degree of surface articulation, which was considered fitting for a monarch, the vaulting was constructed entirely of jointed masonry.

The structural solution in the Chapel of Henry VII is ingenious. Large, mostly hidden transverse arches support pendants placed near the wall. Conoids are built upward and outward from the pendants. At the pendants the transverse arch rises through the vaulting conoid and continues along the upper surface of the vaulting, disappearing from view in the central area of the bay. Practically the entire structural load is transmitted to the transverse arch at the pendant. The arch carries the thrust to the walls and buttresses. Hence the ceiling combines shell structures (the conoids) and frame structures (the transverse arches).

Unlike the conoids of many fan vaults, where each horizontal cross section is only a segment of a circle, the central conoids in the Chapel of Henry VII are fully circular in horizontal section. This is symbolically appropriate because the circle, representing the revolution of the heavens and the disk of the sun, was an important symbol in Tudor political iconography. In 1500, when Henry's chapel was designed, the Tudors were a young and self-conscious dynasty; Henry had seized power only

15 years before on the basis of a slim claim to the crown. The magnificently carved ceiling of the chapel with its multiple circular forms was intended as elaborate praise for a monarch anxious to secure a place for himself in the cosmic and historical orders. The ceiling tells us more about the chapel's patron and his society than it does about the personality of the designer, whose identity we can only guess at.

The ceiling of the Chapel of Henry VII represents the culmination of fan vaulting. In the 1540's Henry's son Henry VIII drastically reduced the authority and wealth of the monastic orders in whose buildings much fan vaulting had been included. The construction of fan vaulting came to a virtual halt for almost 100 years, until it was revived by the Church of England at the University of Oxford, which was then a stronghold of Anglicanism.

The construction of fan vaults has never ceased entirely. Some modern cathedrals have vaulting in imitation of the earlier styles. The later examples, however, lack the charm and interest of fan vaults built between 1350 and 1540. This interest comes not only from the beauty of the ceilings themselves, which is considerable, but also from the way the form was worked out by the builders, who were guided solely by aesthetic impulses and empirical results but who were nonetheless dramatically successful engineers.

14

'Ars mechanica': Gothic structure in Italy

Elizabeth Bradford Smith

Introduction[†]

It has often been said, with regard to the fortunes of French Gothic architecture in the Italian peninsula, that the Italians never really accepted it except as a decorative system which they applied in a superficial manner to windows, facades, or to whatever elements they wished to bring in line with current fashion—and that even then, they tended to alter the borrowed forms and to blend them with classical elements in an eclectic mixture all their own which was not truly Gothic. Furthermore, this reluctance to adopt the French Gothic structural system has been largely perceived as born of incomprehension and has been linked to the earlier failure of Lombard masons to grasp the possibilities inherent in the pointed arch and the rib vault, of which they were making ample use in the Romanesque period, and to exploit these features to develop the same structural system as developed in Northern France. In short, the Italians are seen as having 'missed the boat'.[1]

Certainly it is true that the Italians did not invent the French Gothic structural system and that French Gothic architecture in Italy is definitely an import. Earlier scholars credited the French monks of the Cistercian Order with the introduction of Gothic architectural features in Italy.[2] But whereas in some the Gothic style with elements of their own native traditions, tending all the time to make it less Gothic. It retarded without being retrogressive. It never tried to revert to the Romanesque or to pure classicism, but with complete impartiality, borrowed elements from both these styles and from the Gothic'. See also J. White, *Art and Architecture in Italy 1250-1400*, Harmondsworth, 2nd ed., 1987, p. 41: 'A carefree and at times uncomprehending attitude to Gothic structure is the one recurring feature of Italian architecture'. Regarding the Lombard masons in particular, see A. K. Porter, *Lombard Architecture*, New Haven, 1917, pp. 115-122; and M. Aubert, 'Les plus anciennes croisées d'ogives; leur rôle dans la construction', *Bulletin Monumental*, 1934, pp. 5-67; 137-237, esp. p. 12: 'Les architectes lombards, qui avaient su utiliser sous la voûte d'arêtes ces arcs maçonnés apportant quelque renfort à la voûte dont ils cachent les arêtes toujours difficile à monter, n'en ont pas, en perfectionnant ce procédé, tiré ses conséquences logiques'.

[2] The literature on the influence of French Cistercian architecture in Italy is extensive. A basic treatment is in C. Enlart, *Origines françaises de l'architecture gothique en Italie*, Paris, 1894. See also R. Wagner-Rieger, *Die italienische Baukunst zu Beginn der Gotik*, Graz/Köln, 1956-1957, vol. I, pp. 29-102; vol.

[†]   Author's original English text: Elizabeth B Smith, ' "Ars Mechanica". Problemi di struttura gotica in Italia', in *Il Gotico europeo in Italia* (ed. Valentino Pace & Martina Bagnoli), Electa Napoli, 1994, 57-70.

[1] Thus, for example, P. Frankl, *Gothic Architecture*, Harmondsworth, 1962, p. 144: 'With the spread of the Gothic style to Tuscany, architecture there entered a stage where it was no longer Romanesque, still not classical in the sense of the Renaissance, but not pure Gothic either. It was a superficial application of Gothic elements to traditional local forms'. And again, p. 170: 'The Italians amalgamated elements of

2

areas, such as the Abruzzi or Piedmont, the Cistercian buildings are relatively French in character and do seem to have had some impact on later structures in the surrounding area, in most regions they either absorb the characteristics of the local school, blending into the landscape, as in Lombardy, or else, like Fossanova and Casamari, they remain as isolated examples of French architecture, exotic and foreign in their Italian settings.[3] Similarly, when we see a non-Cistercian building that looks particularly French, such as the Upper Church of San Francesco in Assisi or the choir of San Lorenzo Maggiore in Naples, there is usually evidence that imported workmanship played some part in its creation.[4] Even into the fourteenth century many churches, even those most richly endowed, continued to adhere to the timber-roofed basilical form, trad-

---

II, pp. 10-75; 222-239; L. Fraccaro de Longhi, 'Elementi francesi ed elementi lombardi in alcune chiese cistercensi', *Palladio*, I-II, 1952, pp. 48-58; and by the same author, *L'architettura delle chiese cistercensi in Italia*, Milan, 1958; A. M. Romanini, 'Monachesimo medievale e architettura monastica. Introduzione', *Dall'eremo al cenobio. La civiltà monastica in Italia dalle origini all'età di Dante*, G.P. Caratelli, ed. Milan, 1987, pp. 456ff.

[3]   Fraccaro, 'Elementi...', p. 50, and *L'architettura...*, pp. 21, 31-33. Reflections of Fossanova can be seen, piecemeal, in the work done in the churches of neighboring towns during the first half of the thirteenth century.

[4]   On San Francesco in Assisi see W. Krönig, 'Hallenkirchen in Mittelitalien', *Kunstgeschichtliches Jarhbuch der Biblioteca Hertziana*, II, 1938, pp. 1ff., esp. pp. 36ff; and by the same author more recently, 'Caratteri dell'architettura degli ordini mendicanti in Umbria', *Atti del VI Convegno di Studi Umbri*, Perugia, 1971, pp. 165-98, esp. 167-70. A great number of architectural renderings can be found in G. Rocchi, *La Basilica di San Francesco ad Assisi*, Florence, 1982. An extensive review of the bibliography on San Francesco appears in A. Cadel, 'Studi sulla basilica di san Francesco ad Assisi. Architettura', *Arte Medievale*, 1988, 2 #1, pp. 79-103; 1989, #1, pp. 117-36. On San Lorenzo Maggiore see J. Krüger, *San Lorenzo Maggiore in Neapel: Eine Franziskanerkirche zwischen Ordensideal und Herrshaftsarchitektur*, (Franziskanische Forschungen, 31), Werl/Westfalen, 1986. For a current opinion, see the article by C. Bruzelius in *Il Gotico europeo in Italia*.

itional in Italy since Early Christian times, as, for example, at Santa Croce in Florence (begun in 1294).[5] Thus, French Gothic architecture did not take root in Italy; the Italians never really did adopt the aesthetic/structural package of French Gothic. Leaving aside for the moment the reasons why the Italians might have rejected the French High Gothic system, and assuming that some, at least, of the buildings erected in Italy during the Gothic era can qualify as 'Gothic', let us ask the following questions: What kinds of structural approaches were used by the masons and architects of Gothic buildings in Italy, and how do they compare with the ones used in France?

Such a comparative study must obviously begin with the structural system used to erect the great Gothic cathedrals of Northern France. This inquiry will therefore be restricted to religious architecture, and it will focus on two major elements intrinsic to the French Gothic structural system—the vaulting over the main nave and its abutment. Although the two elements are interdependent, we will at first examine them singly. We will use the cathedral of Amiens (begun 1220) as a typical example of French High Gothic, and we will compare its structure to that of those contemporary buildings which best serve to illustrate the types of solutions adopted in Italy. Discussion will be limited to large-scale structures, since they provide the best proving-ground for the success or failure of a specific structural solution to a given architectural problem. Most of the examples will be drawn from Central Italy, where the economic expansion of the communes and the religious ferment of the Mendicant Orders combined to produce a number of important buildings in the 13th and 14th centuries.

Vaulting

In his *Dictionnaire raisonné* of medieval architecture in France, under 'Symmetry', Eugène Emmanuel Viollet le Duc says the following: 'For the master of the Middle Ages, it is the thing to be carried that is the

---

[5]   On Santa Croce, often attributed to Arnolfo di Cambio insofar as the original design, see F. Moisè, *Santa Croce di Firenze. Illustrazione storico-artistico*, Firenze, 1845; W. & E. Paatz, *Die Kirchen von Florenz*, Frankfurt am Main, 1952-55, I, pp. 497-701; A.M. Romanini, *Arnolfo di Cambio e lo 'stil nuovo' del gotico italiano*, Milano, 1969, pp. 196-221; for the timber-roofed churches of Central Italy, see Krönig, 'Hallenkirchen...'; and K. Biebrach, *Die holzgedeckten Franziskaner-und Dominikanerkirchen in Umbrien und Toskana*, Berlin, 1908.

3

1) Amiens Cathedral, ground plan (Dehio, *Die Kirchliche Baukunst des Abendlandes*, Stuttgart, 1884-1901, taf. 363)

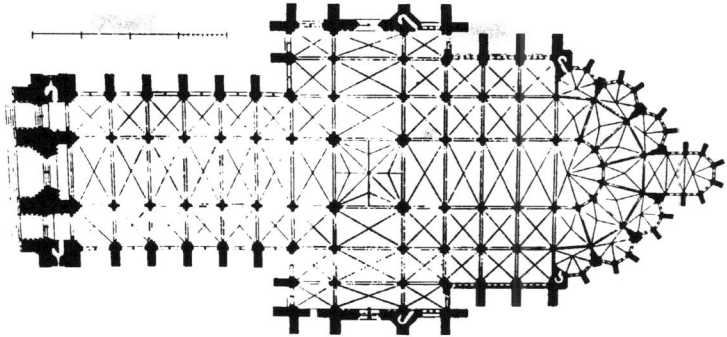

2) San Francesco, Lodi, ground plan (Romanini, *L'Architetttura gotica in Lombardia*, Milan, 1964, vol. I, fig. 27)

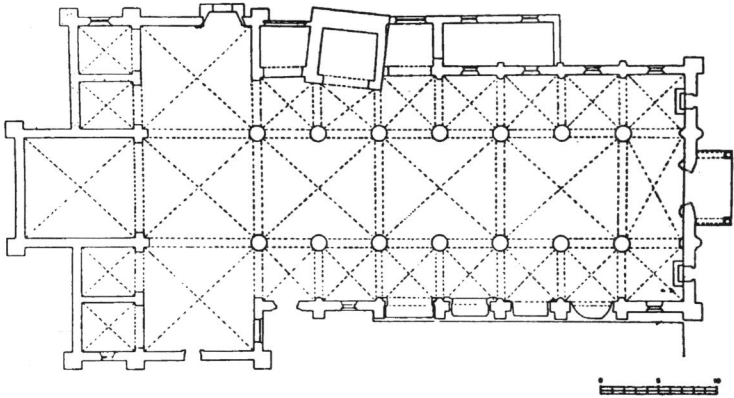

3) Santa Maria Novella, Florence, ground plan (Dehio, taf. 534)

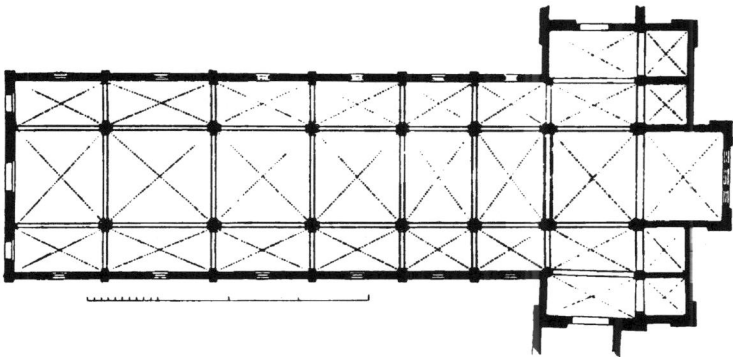

4

main objective, it is this vault that he must support and buttress. It is the vault, therefore, which governs the symmetry of all the elements'.[6] The French High Gothic vaulting system consists of a quadripartite rib vault erected over a nave bay which is rectangular in plan, wider than it is deep, and which corresponds to one aisle bay approximately square in plan (figure 1). In contrast, the most common Italian solution to the problem of vaulting the nave was to erect a quadripartite vault over a square bay. This square nave bay could correspond either to two aisle bays, also square, as at San Francesco at Lodi (begun c. 1280) (figure 2), or to one aisle bay, rectangular in plan and deeper than it is wide, as at Santa Maria Novella in Florence (begun c. 1279) (figure 3).[7] The 2:1 aisle:nave ratio of San Francesco at Lodi, often used in Lombardy, derives from the local Romanesque tradition, visible at Sant'Ambrogio in Milan. The disposition of Santa Maria Novella, apparently a solution of the Gothic era, occurs more frequently in Central Italy.[8]

The type of quadripartite vault most often used in Italy does not differ in plan alone from its French counterpart; it is also different in elevation. A comparison of the nave of Amiens with the nave of Santa Maria Novella (figures 4 and 5) makes clear that whereas the vaults of Amiens have even-level crowns,

4. Amiens Cathedral, interior, nave (Photo Hirmer)

---

[6]   E. E. Viollet-le-Duc, *Dictionnaire raisonné de l'architecture française du xiè au xviè siècle*, 10 vols, Paris, 1854-68, VIII, p. 517.

[7]   On S. Francesco in Lodi, see A.M. Romanini, *L'architecttura gotica in Lombardia*, I-II, Milano, 1964, pp. 110-114; L. Motta and A. Novasconi, *Il tempio di San Francesco a Lodi*, Lodi/Milano, 1958. On S. Maria Novella, see J.W. Brown, *The Dominican Church of Santa Maria Novella of Florence*, Edinburgh, 1902; W. Paatz, *Werden und Wesen der Trecento Architektur in Toscana*, Burg B.M., 1937, pp. 7-22; Paatz, *Die Kirchen...*, III, pp. 663-845; G. Villetti, 'Descrizione delle fasi costruttive e dell'assetto architettonico interno della chiesa di Santa Maria Novella in Firenze nei secoli XIII e XIV', *Bolletino della Faccoltà di Architettura dell'Università degli studi di Roma*, XXVIII, 1981, pp. 5-20.

[8]   This type of disposition is also found, for example, in the Florence cathedral and in Venice, at SS Giovanni e Paolo. Occasionally, as at S. Francesco, Bologna, the square bay with a 2:1 aisle:nave ratio is covered with a sexpartite vault, but this is exceptional in Italy, as are the flying buttresses and radiating chapels in the choir of the same building.

in which the keystones of the transverse arches, the wall arches, and the vaults themselves are all at the same height, the crowns of the vaults of Santa Maria Novella rise higher than the transverse and wall arches, and so qualify as domical vaults. Like the 2:1 aisle:nave ratio, such a vaulting system, going back to 12th century Lombardy, can be found in a wide range of buildings in many parts of Italy in the Gothic era, from the 13th through the 15th centuries. Visually, it creates an effect profoundly different from that engendered by the French Gothic even-crown vault, dividing the space of the nave into distinct canopy-covered cells, instead of uniting it into an articulated whole.

The domical rib vault over a square bay developed from the domed groin vault, which in turn had been developed by Lombard masons of the Romanesque era probably as an improvement on the even-crown vault.[9] By raising the crown of the vault

---

[9]   The construction and action of domical vaults are discussed in the following: Viollet-le-Duc, *Dict-*

cutting as occurred in a regular groin vault at the intersection of the two vault webs, since the domical vault
could be built in concentric coursing, like a dome. It
could also be erected with less centering, because of its
steepness, thus economizing on wood. Finally, the
domical vault brought the masons a new freedom,
since the conformation of the vault surfaces was now
determined by the shape of the boundary arcs, including the diagonals, instead of resulting from the projection of some generating arch, and thus different shapes
could be given to the side and diagonal arches, and the
vault curvature could be modified at will. For example, the webbing of some domical vaults flares up in
each separate panel. On the downside, however, the
domical vault was heavier than an even crown vault
and required heavy and powerful support.

Abutment

Although the domical vaults produced approximately
20% less outward thrust than even-crown vaults, they
nevertheless required some lateral abutment.[10] What
means did the Italians generally adopt for the abutment
of domical vaults? What did they use in those
instances where they erected even-crown vaulting
instead?

As anyone who has travelled through Italy
will know, the flying buttress is rarely to be seen.
Although Camille Enlart's remark that in the entire
Italian peninsula there might not be more than seven
churches provided with flying buttresses may be an
exaggeration, it is certain that this device was not
much used.[11] In fact, visible salient buttressing of any
kind is rare in medieval Italian structures. Arthur
Kingsley Porter, eloquent as always, suggests that with
regard to buttressing, the builders of medieval Italy
resembled those of ancient Rome, who 'never perceived in the buttress any possibilities of architectural
or decorative development, studiously avoided it wherever walls could be made to stand by any other means,
and...only employed [it] in places where decoration
was entirely subordinated to utilitarian considerations.
In short, it was considered merely a mean and ugly
makeshift, much as we should regard today an
unsightly prop applied to a stone building.' Similarly,
the medieval builders of Lombardy 'always adopted it
reluctantly, and abandoned it wherever it was possible

5. Santa Maria Novella, Florence, interior, nave
(Photo Marburg)

above the keystones of the semicircular boundary
arches, the contour of the diagonal groin was changed
from a semi-ellipse to a semicircle, in other words,
from a weaker, flatter form producing stronger
outward thrusts, to a stronger arch requiring less
abutment (figure 6). When, in the Gothic era, the
transverse arches acquired a pointed profile, the vaults
became steeper, thus reducing thrusts even further. A
second bonus of the domical vault from the mason's
point of view, was that there was no difficult stone

ionnaire..., vol. IV, 'Construction', pp. 108ff; C.
Ward, _Medieval Church Vaulting_, Princeton, 1915,
pp. 44-58; Porter, _Lombard Architecture_, vol. I, pp.
109-110; Aubert, 'Les plus anciennes...',pp. 8-12; J.
Fitchen, _The Construction of Gothic Cathedrals_,
Oxford, 1961, pp. 55ff; J.H. Acland, _Medieval
Structure: the Gothic Vault_, Toronto, 1972, pp. 76ff.
For an explanation of the forces in even-crown quad-
ripartite vaulting, see R. Mark, _Experiments in Gothic
Structure_, Cambridge, Mass., 1982, pp. 102-117.

[10]   J. Bony, _French Gothic Architecture in the Twelfth
and Thirteenth Centuries_, Berkeley, 1983, p. 18.
Bony's statement is based on calculations by R. Mark.
[11]   Enlart, 'Origines...', p. 5.

6

to do so, and even in some cases where subsequent experience proved that it could not be abandoned with safety.[12]

In order to understand the Italian approach to abutment, let us once again compare the transverse sections through the nave of our French model, the cathedral of Amiens (figure 7), with Italian examples. Santa Maria Novella provides an example of one of the ways of abutting domical vaulting (figure 8). In place of the double range of flying buttresses of Amiens, Santa Maria Novella substitutes two other architectural members: 1) the vaults of the side aisles, much higher in relation to the nave vault than those of the side aisles at Amiens; and 2) transverse buttresses rising over the aisles from the pier extensions to the outer walls. At Santa Maria Novella the vaults of the side aisles, also domical, rise up to the level of the springing of the nave vaults, and so help to counteract their lateral thrust. Further support is provided by the transverse buttresses which transmit the horizontal thrust of the high vaults to the exterior walls of the aisles. In the case of the transverse buttresses, the height of the aisles also made an aesthetic contribution, for it allowed the medieval Italian architect to conceal the buttresses—those 'clumsy accessories', to use Porter's words—under the roof of the aisles.

The way in which the architect of Santa Maria Novella solved the problem of supporting the high vaults of the nave resulted in high and wide arcades, creating an interpenetration of the space in the aisles with that of the main nave. This elegant and airy interior has been much admired for the clarity and simplicity of its beauty, and has been recognized as a model for a number of later Mendicant churches, both in Italy and throughout Europe.[13] It most likely would not, however, have served so successfully as a formula if it were not also a model of elegance and clarity from a structural point of view. Corroboration of this is provided not only by the repetition of this structural formula in other churches of comparable size, as for example the cathedral of Arezzo (begun 1277/78) (figure 10), but also by the fact that the builders of the cathedral of Santa Maria del Fiore in Florence elected to use a similar system in the gigantic nave which went up over the course of the 14th century (figure 9).[14]

Since any flaws in a structure will be compounded in a larger building, the adoption in the Florence cathedral of the square bays, the domical vaults, the high side-aisles, and the transverse buttresses of Santa Maria Novella, clearly reveals the builders' confidence that this would indeed do the job on such a large scale.

An Italian Gothic System?

What seems apparent from the analysis of the structure of Santa Maria Novella and the cathedral in Florence is that the Italians, as a result of their long acquaintance with the domical vault, eventually evolved a successful way to use it as a covering over vast spaces and at great heights. In sum, out of their constructional practices they had developed a Gothic system of their own that would respond to their needs.[15] This Italian Gothic system is not strictly limited to the Santa Maria Novella type; there are variations. The three-aisled hall church with domical vaults, such as the cathedral of Perugia (begun 1437), can be considered one of them, based on a similar principle of mutual

---

[12] Porter, *Lombard Architecture*, I, p. 127.

[13] Bony, *French Gothic...*, p. 456.

[14] On the cathedral of Arezzo, see A. del Vita, *Il Duomo di Arezzo*, Milan, n.d.; M. Salmi, 'La Cattedrale d'Arezzo', *L'Arte*, XVIII, 1915, pp.373-391;

Paatz, *Werden und Wesen...*, pp. 19-22. The literature on the cathedral in Florence is too vast to be cited in the context of this article. A recent study of the structure of the nave is by G. Rocchi et al,, *S. Maria del Fiore. Il Corpo Basilicale. Rilievi, documenti, indagini strumentali. Interpretazione*, Milan, 1988. Rocchi notes (p. 12) that the nave of Santa Maria Novella is a meter lower than the side aisles of the cathedral—26 as opposed to 27 meters—while the nave of the cathedral, at 42.59m., is only slightly lower than the nave of Amiens.

[15] H. Hartung, *Ziele und Ergebnisse der Italienishen Gotik*, Berlin, 1912, pp. 14-16; 24-25, sets forth an Italian Gothic system, in many ways similar to the one proposed here, but does not relate it to the use of the domical vault, which he sees as a shortcoming. W. Gross, *Die Abendländische Architektur um 1300*, Stuttgart, 1948, pp. 171-184, analyses precisely the aesthetics of the naves of Santa Maria Novella and of Arezzo and argues convincingly that they are indeed Gothic though of a particular Italian brand, but does not relate this to the domical vault; for him the more important factor is the relationship between the interior space and the wall. The view of early examples of Lombard Gothic (*e.g.* Morimondo) as experiments leading to the Santa Maria Novella solution, has still to be confirmed. Cf. Romanini, *L'Architettura*, I, pp. 32-35. for Morimondo; pp. 11ff for her definition of Lombard Gothic architecture, in which the wall is the 'protagonista'.

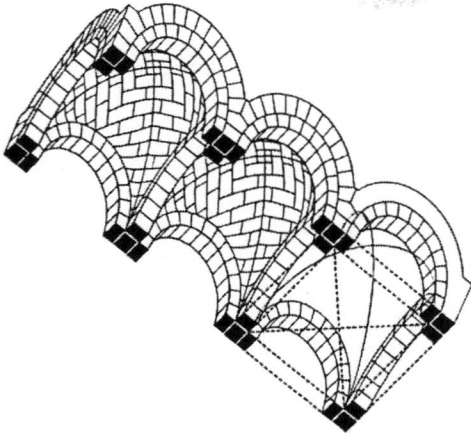

6. Diagram of domical vaults (Fitchen, *The Construction of Gothic Cathedrals*, Oxford, 1961, p. 60)

7. Amiens Cathedral, transverse section, nave (Dehio, *Die Kirchliche Baukunst des Abendlandes*, Stuttgart, 1884-1901, taf. 599)

8. Santa Maria Novella, Florence, transverse section, exterior elevation; interior elevation (Hartung, *Ziele und Ergebnisse der Italienischen Gotik*, Berlin, 1912, fig. 44)

9. Florence Cathedral, transverse section (Dehio, nave, taf. 542)

8

10. Arezzo Cathedral, interior, nave (Alinari, 9682)

buttressing among the vaults of the three aisles (figure 11).[16] Another variation occurs when, instead of transverse buttresses, the structure is given additional support by means of lateral chapels along the aisles. Aside from any liturgical use they may have had, these lateral chapels seem to have served as a buttressing device. They are found in a wide variety of buildings, from the domical-vaulted basilicas of Lombardy and Emilia such as Santa Maria del Carmine Pavia,[17] designed c.1370 (figure 12), to the Kingdom of Naples, where they are used in conjunction with open

---

[16] Krönig, 'Hallenkirchen...', p.91-115, suggests that Perugia cathedral is a copy of San Domenico in Perugia (begun 1304), which he considers the truly original contribution of Italy to the Gothic hall church as an architectural genre. The inner fabric of San Domenico was destroyed in the 17th century and subsequently rebuilt.

[17] On S Maria del Carmine, Pavia, see Romanini, L'Architettura, I, pp. 415-425.

9

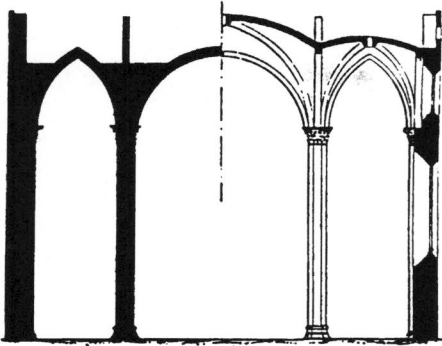

11. Perugia Cathedral, transverse section (Dehio, *Die Kirchliche Baukunst des Abendlandes*, taf. 601)

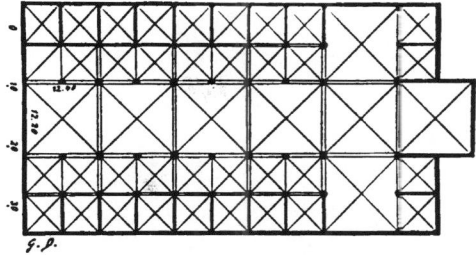

12. Santa Maria del Carmine, Pavia, ground plan
(Dehio, p. 514)

timber roofing over the nave at San Francesco in Messina. Like the transverse buttresses hidden under the roof of the aisles, the lateral chapels also responded to an aesthetic requirement—they helped to maintain the flat wall surface of the exterior, apparently intrinsic to the Italian concept of architectural beauty.

The Italian builders were less successful in dealing with the problem of abutting even-crown vaulting, most likely because they were less familiar with its action on the structure. At San Francesco in Assisi, for example, the original semi-cylindrical buttresses of the Upper Church were reinforced by flying buttresses in the 14th century. And in the Franciscan church of San Fortunato of Todi (begun 1292), a three-aisled hall church, the increased horizontal thrusts of the even-crown vaulting, erected here in place of the more usual domical vaults, necessitated the insertion of masonry struts across the side aisles (figures 13 and 14). These, with the lateral chapels, allowed the ext-erior to retain the desired flatness, so that, as John White describes it, 'it squats like a huge hanger on a hill'. White's comment on the interior, that 'such audacity flies in the face of structural reality', is also worth noting in this context.[18] Yet it seems likely that the builders of San Fortunato were coping with the problem of even-crown vaults using the means at their

disposal in the way they saw fit. If they had used the more stable domical vaults—like the architects of San Domenico and the cathedral in Perugia, or those of Saint-Serge in Angers, one of the West French hall churches which Wolfgang Krönig suggests as models—they probably would not have needrd the stone struts.

The fact that the men who built San Fortunato apparently did not mind the look of what White refers to as 'lumpish elements' brings us to the discussion of another way in which Italian builders of the Middle Ages strengthened their vaulted structures: by the use of wooden tie beams and iron chains and tie rods.[19] Both wood and iron reinforcements in walls and across the internal spaces of buildings were common through-out the medieval period in Italy.[20] Because of the relative scarcity and great expense of iron, how-ever, it was at first used for clamps and dowels and was not used with great regularity for tie rods until the 12th century, when iron production increased. Wood was the usual material for cross-bracing in the earlier Middle Ages. Both wood and iron were used in masonry constructions for the same reason—because of

---

[18] White, *Art and Architecture*, pp. 40-41; Recent publications on S. Fortunato in Todi include C. Calano, 'San Fortunato a Todi: una chiesa 'a sala' gotica', *Quaderni dell'Istituto di Storia dell' Architettura*, XXIV, 1977/8, pp. 113-128; G.de Angelis d'Ossat et al., *Il Tempio di San Fortunato a Todi*, Milan, 1982.

[19] On the uses of wood and iron reinforcements in medieval architecture, see Viollet - le - Duc, *Dictionnaire...*, II, 'Chainage', pp.396-404; Fitchen, *The Construction...*, pp. 275-278; R.P. Wilcox, *Timber and Iron Reinforcement in Early Buildings*, London, 1981; W. Haas, 'Hölzerne und eiserne Anker an mittelalterlichen Kirchenbauten', *Architectura*, 13.2, 1983, pp. 136-151.

[20] They were also used, to a lesser extent, in antiquity. A. Hoffmann, in an as yet unpublished paper delivered at a symposium on the Roman villa at the University of Pennsylvania in April 1990, discussed the way in which the architects of the Villa Hadriana exploited the capabilities of iron in vaulted structures.

10

13. San Fortunato, Todi, interior (Anderson 31306)

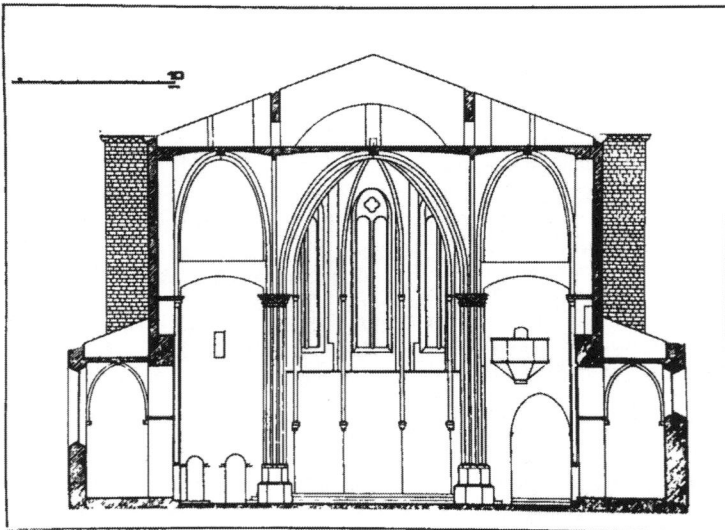

14. San Fortunato, Todi, transverse section (By Calano, in G. De Angelis d'Ossat, *Il tempio di San Fortunato a Todi*, Milan, 1982, p. 55, fig. 39)

15. SS Giovanni e Paolo, Venice, interior, nave
(Anderson 22707)

their tensile strength—although the medieval mason would not have been able to express it in those terms. He would have thought of them simply as material for ties. As suggested by Walter Haas, it is likely that at first there was a vague feeling that the brittleness of masonry required the elasticity of wood for its completion, in order to become more capable of resisting deformation. Eventually, out of a manner of construction developed for specific situations—earthquakes, common in Italy, or unreliable ground conditions, as in the mud lagoon of Venice—artisanal habits were formed, the basis for which was no longer placed in question.[21]

As a result, whereas the architects of the French cathedrals relied on wood and iron reinforcement to ensure stability during construction only, and removed them when the building was complete and the mortar dry, the Italians intended them as permanent features.[22] The grid of beams spanning the nave and

aisles of SS Giovanni e Paolo in Venice (begun c. 1333) (figure 15) are an integral part of its structure and should be considered as such in any architectural analysis of this building, as should the ties in the multitude of other buildings similarly constructed.[23] It may be that these elements actually contribute little to the strength or stability of the structures of which they are part, but it important that the architects who used them—to paraphrase Ruskin—believed them to be useful and knew them to be, if not beautiful, at least not something to be hidden. It may strike us as odd that the Italians avoided visible external buttressing but did not object to visible tie rods inside their buildings, but Viollet-le-Duc, champion of iron, was adamant in his support of the Italians' attitude to its use in masonry buildings. According to him, the ties in Italian structures 'make no pretension whatever to be a decorative feature', since 'the proper function of iron in masonry vaulting is that of a tie, whenever we wish to avoid having recourse to the expensive contrivance of buttresses and abutments'.[24] Certainly the men

---

[21] Haas, 'Hölzerne...', p. 147.

[22] Viollet-le-Duc, Dictionnaire..., II, 'Chainage', p. 403, illlustrates this from a pier in Amiens cathedral.

[23] On SS Giovanni e Paolo see P. Rambaldi, La Chiesa dei SS Giovanni e Paolo, Venice, 1913; G. Fogolari, I Frari e i SS Giovanni e Paolo, Milan, 1932; H. Dellwing, Studien zur Baukunst der Bettelorden im Veneto, München/Berlin, 1970, pp. 98-117, and Die Kirchenbaukunst des späten Mittelalters in Venetien, Worms, 1990, pp 95-100. Dellwing considers the tie-beams to be an integral part of the aesthetics of the nave of SS Giovanni e Paolo, as well as of its structure. It should be noted that the vaults of SS Giovanni e Paolo are built of wood, with only the transverse arches and ribs in stone. Both the wooden vaults and the tie beams were probably used here because of the instability of the soil and a desire to limit the total weight of the structure.

[24] Viollet le Duc, Discourses on Architecture, 2 vols, Boston, 1875, II, p. 67. Viollet-le-Duc's thoughts on the subject are worth quoting at greater length: 'How is it that while in France we object to the appearance of interior ties beneath our masonry vaulting, our sight is not offended by the presence of those which are so profusely employed in Italian buildings? I shall not attempt to explain this inconsistency; I merely remark that the architects who sketch the Italian buildings of the Middle Ages and the Renaissance suppress these iron ties in the edifices built in imitation of them, which leads one to suppose that they regard them as offensive on this side of the Alps; why then should they have no objection to them on the other side? I will add that the ties across the springing of the Italian

12

responsible for the construction of the cathedrals of Florence and Milan would have agreed with Viollet le Duc. Documents from 1365 and 1366 attest to the fact that the Florentines were counting on iron chains and tie rods, still in place above the capitals of the piers of the main arcades, to help contain the thrusts of the high vaulting in the middle bays of the nave.[25] As for Milan, the records of the famous Expertise of the last decade of the 14th century, in which the Milanese respond to the criticisms of the Parisian Jean Mignot, demonstrate their faith in the power of iron tie rods—*'strictores ferri magnos'*, whose very name in Latin conveys the type of job they were meant for—to strengthen the vaults they intended to construct.[26]

The Expertise of Milan

Having invoked the Expertise, we should perhaps turn to it at this point. Consisting of records of consultations between the Milanese Council responsible for the construction of the cathedral and various outside experts called in for advice on the vaults they were about to construct, it is a rich source of contemporary opinion as to how a Gothic building should best be erected and cannot easily be ignored in any survey of Gothic structure in Italy. Especially interesting in terms of the structure is the exchange with Mignot. To

---

vaulting make no pretension whatever to be a decorative feature; they are simply iron bars. It is fortunate, however that it has not occurred to the Italian clergy to have those bars cut away in their churches, as our French curés have done with the tie-beams of all the timber ceilings; for had they done so, many an edifice which now excites the admiration of travellers would have fallen'.

[25] The documents referred to are given in C. Guasti, *Santa Maria del Fiore, la costruzione della chiesa e del campanile secondi i documenti*, Firenze, 1887, pp. 163, 173. See also Rocchi, *Santa Maria del Fiore*, p. 96, on this matter.

[26] Major analyses of the 'Expertise' are by J. S. Ackerman, 'Ars sine Scientia nihil est: Gothic Theory of Architecture at the Cathedral of Milan', *Art Bulletin*, XXXI, 1949, pp. 84-111; and P. Frankl, *The Gothic*, Princeton, 1961, pp. 62-83. For the text of the 'Expertise', see C. Cantu, *Annali della fabbrica del duomo di Milano*, 9 vols, Milan, 1877. On the cathedral see also M. L. Gatti Perer, *Il Duomo di Milano*, (Atti del congresso internazionale, 1968), Milan, 1969; and Romanini, *L'Architettura...*, I, pp. 351-414.

date, the Milan Expertise has served to reinforce a negative view of the Italians' understanding of Gothic structure. Certainly James Ackerman, in his study of the Expertise, considers them lacking in this respect. For example, when the Milanese defend the strength of their piers and abutments with the apparent assertion that pointed arches do not exert a thrust on buttresses (*'archi spiguti non dant impulzam controfortibus'*), Ackerman takes it as 'a sign, not of superior confidence but exceptional ignorance'. He mocks the Italians: 'In a single sentence the Milanese have cast off what we consider to be the major structural problem in Gothic architecture', and accuses the Council of fabricating 'strange theories to suit its own ends'.[27] Similarly, Paul Frankl takes the Milanese statement at face value, noting that it is 'historical testimony as to what was thought in Milan in the year 1400 about the statics of pointed arches'.[28] Only White inserts a note of caution, reminding us that the Expertise is a written condensation of long verbal discussions taken down by a notary in 'very crude and occasionally uncomprehending Latin'. Regarding the claim that pointed arches have no thrust, White suggests that what the Milanese actually meant was that those arches they planned to erect would not exert a lateral thrust too great for the existing buttresses to withstand. Moreover, Mignot seems to have understood the Milanese in this way, says White, because he doesn't accuse them of saying that pointed arches don't exert any thrust, and he surely would not have missed such a 'god-sent opportunity to demolish their arguments completely'. Instead, in his response Mignot accuses them of saying that pointed arches have less weight (*i.e.* thrust) than round arches, whereas any arch, no matter how pointed, exerts a thrust which needs proper abutment, failing which it will collapse.[29]

If this last is indeed what the Milanese were saying—that pointed arches have less thrust—then they were right in principle, not only about the general properties of arches but also of vaults, regardless of whether they were right about the particular arches and vaults planned for their cathedral. Part of the problem may have been mutual incomprehension. The usual Lombard vaults, with which the Council would have been most familiar, were domical, like those of Santa Maria del Carmine, Pavia (figure 16). If domical

---

[27] Ackerman, 'Ars...', p. 97.

[28] Frankl, *The Gothic*, p. 72.

[29] See White, *Art and Architecture*, pp. 517-528, for his discussion of the Milan Expertise.

13

16. Santa Maria del Carmine, Pavia, interior, nave
(Alinari 39781)

17. Milan Cathedral, interior, nave (Alinari 14193)

vaults are made steeper, as they are here, the lateral thrust they engender becomes proportionately less than the vertical, and so the Milanese remark about pointed arches applies also to domical vaults. Even-crown vaults, however, exert a strong horizontal thrust no matter how pointed their arches may be, and as we have seen, Italian masons and architects seem to have had some difficulty in dealing with this thrust. Mignot, however, as a Parisian, would have known mainly, or even exclusively, even-crown vaulting. Thus Mignot is right, too, in his statement that such vaults, no matter how pointed, require sufficient butt-ressing. What we may be witnessing here is a 'dialogue de sourds', born of the Milanese decision, apparently taken in mid-stream, to build a northern Gothic cathedral according to Italian structural tra-ditions. Curiously, the vaults eventually built over the nave are a sort of hybrid, even-crown along the length of the nave and domical along the sides, and in fact did not receive flying buttresses until the 19th century (figure 17).

It might be worthwhile to look at the Milan Expertise once more, from a structural/constructional point of view. Similarly, other instances cited as examples of a structural incomprehension of Gothic architecture on the part of Italian builders, such as Maitani's buttressing of the transept and apse of the cathedral of Orvieto, should probably be reexamined. The way in which the Florentines and Milanese dis-cussed problems arising during the course of construct-ion, sometimes calling in outside experts as advisors, is most likely not at all unusual. What is unusual is that we have a written record of it. Such discussions must have occurred at most building sites when some-thing developed unexpectedly, or when a critical point in construction had been reached. As Robert Mark reminds us, the builders of the Middle Ages, lacking any theoretical knowledge, worked empirically, depending mainly on two resources: 1) previous exp-erience from structures they had built or knew about, both successful and unsuccessful; and 2) the building

14

under construction.[30] From the first they knew what had and had not worked. When and if trouble appeared in a structure as it was going up, they used the knowledge gained from past experience to remedy it. The second resource was especially useful if the builders were attempting something new. The structure under way then became a laboratory for their experiments. Thus, if we had more complete documentation for 12th and early 13th century France when the High Gothic system was being perfected, we would no doubt see similar meetings called, to debate how best to respond to threatening developments in the structure. Both the Florentines and the Milanese were undertaking buildings of unprecedented size. This factor alone ensured that they would encounter a host of unforseen problems, and it is not surprising that they engaged in lengthy discussions. These are not proof of incompetency but rather serve as examples of medieval building practices.

In light of the above, it seems safe to suggest that rather than incomprehension of Gothic on the part of Italian architects, we may be dealing with an alternative Gothic structural system which, while not French, is nonetheless a reasoned system and one which patently worked. Although the Italian architects chose to hide the structural system of their buildings under roofs and inside smooth exterior surfaces, it does not necessarily follow that they thought exclusively in terms of walls, nor that their approach was either Romanesque and retardataire or proto-Renaissance and forward-looking, as has also been said.[31] Both the buildings and their creators were in and of their own time, and participated in the 'new modernity' of Gothic culture which enveloped all of western Europe.[32] Nearly one hundred years ago, Enlart insisted that Italian architecture of the Gothic era should indeed be called Gothic. He did not see Gothic as rigid, limited to a single system, but rather as supple and capable of modification depending on the physical conditions and tastes of different countries.[33]

The modifications made by the Italians to the French Gothic system reflected a variety of conditions. Some were physical, such as the recurrence of earthquakes, which accounts for the use of tie beams and tie rods; or climatic, such as the abundance of sunlight, which explains why a high clerestory with large windows was not a major requirement of their churches. Other conditions were aesthetic, such as the reluctance to display structural elements on the exterior; or religious, such as the need for spaciousness and visibility in the Mendicant churches. Still others were based on constructional materials or traditional techniques, such as brick, open-timber roofs, or domical vaulting. In time, given all the factors—local conditions, previous knowledge and practice, aspirations—and incorporating some features imported from France, the masons and architects of Lombardy and Central Italy developed what could be called an Italian Gothic structural system. This system, as outlined above at Santa Maria Novella, was based to a large extent on the domical vault.

Further evidence that the Italians were indeed thinking in terms of a structural system comes to us not from any written documents but from Florentine wall paintings of the Trecento contemporary with the construction of Florence's cathedral. Both Taddeo Gaddi and Matteo de Pacino depicted the cathedral in their frescoes at Santa Croce.[34] Pacino's painting, in the Cappella Guidalotti-Rinuccini (figure 18), dates from the late 1360s when the critical phase of the nave was under way, and resembles the cathedral more closely. Pacino shows the building as a skeletal framework, including piers, vaults, and transverse buttresses, but omitting the wall on the viewer's side. Admittedly, leaving out the wall on one side is a painter's device for letting us see what is happening inside the building. But it does show the structure clearly, and it suggests that the painter may have seen a similar model of the building, meant to demonstrate the structural system. What is more important, however, is that even if the painting does not result from knowledge of an architectural model it implies that the man who painted it was capable of thinking of the cathedral in this way, as a structural skeleton and not solely in terms of the

---

[30] Mark, *Experiments...*, pp.10-11.

[31] Regarding the tendency among historians to view Italian architects of the 13th and 14th centuries in this way, see Rocchi, *S. Maria del Fiore*, p.11.

[32] Bony, *French Gothic...*, p.421.

[33] Enlart, in 'Origines...', views the Italians as generally on the receiving end as far as Gothic architecture is concerned; nevertheless, on p.5, he says 'nulle dénomination autre que celle de gothique ne convient à l'architecture qui fait l'objet de cette étude. ...Essent-

iellement souple, savante et raisonnée, cette architecture se modifie nécessairement selon les conditions matérielles et les tendances du goût des divers pays'.

[34] Taddeo Gaddi included the cathedral in his fresco of Scenes from the Life of the Virgin, in the Baroncelli Chapel.

18. Matteo de Pacino, Presentation of the Virgin, Cappella Guidalotti-Rinuccini, Santa Croce, Florence (Alinari 3953).

wall. If a painter could conceive of architecture in this way, surely the architect could and did as well.

In conclusion, let us return to the Expertise of Milan. At one point in the argument with Mignot, the Milanese respond to Mignot's dictum, 'ars sine scientia nihil est' (art without science is nothing), with the reverse, 'scientia sine arte nihil est' (science without art is nothing). They then go on to defend their 'art'. Ackerman, while agreeing with the Milanese that their 'ars' may be of superior quality, translates the word as prowess in masonry, as an art of stone carving, but not as an art of structure. Thus, when in opposition to Viollet le Duc's notion of Gothic architectural forms as deriving from the structure, Ackerman proposes instead that Gothic architecture is primarily a union of theory and art, he is omitting structure from the equation. To structure he assigns a secondary role in the process of creation.[35] If, however, we accept the more positive image proposed here of the Italians as the creators and masters of their own system of Gothic structure, then we can perhaps broaden Ackerman's narrow interpretation of the word 'ars' to include 'ars mechanica', the art of engineering, and thereby remove the stigma of incomprehension from the builders of Italian Gothic architecture.

———————————

[35] Ackerman, 'Ars...', p. 107. Cf. also Frankl, The Gothic, pp. 108-109, for a discussion of the meaning of 'ars' and the place of architecture within the hierarchy of arts and sciences in the Middle Ages.

15

Brunelleschi's dome of S. Maria del Fiore and some related structures

Rowland J. Mainstone

The completion, between A.D. 125 and 128, of the dome of the Roman Pantheon[1] marked the main culmination of a period of intensive development of wide-spanning concrete vaults that had begun some 60 years previously. This dome was emulated but never challenged in the subsequent years of the Western Empire. In the East, in Justinian's church of St. Sophia in Constantinople,[2] something that was in several ways more daring was attempted between 532 and 537, but the central dome which there crowned a complex system of billowing semidomes had a diameter little more than two thirds of that of the Pantheon. Only with the completion, in 1434, of the dome of Santa Maria del Fiore in Florence—the new cathedral that replaced the earlier church of Santa Reparata—was this pre-eminence of the Pantheon's dome lost. Brunelleschi's Florentine dome is both slightly greater in mean span than that of the Pantheon and, being carried at a considerable height on a relatively slender drum, called for a greater refinement in its structural design. Today it still dominates the city skyline in a manner unequalled anywhere else and it remains the greatest masonry dome in existence (Plate XII).

Unlike the domes of both the Pantheon and St. Sophia, it was built on piers commenced long previously by other men. Its span was fixed, its basic form largely predetermined and even its drum had been completed before detailed design began in 1417. Brunelleschi's achievement was, therefore, primarily a technical one with little scope for spatial or formal invention and it was thus that it was recognised and acclaimed by his contemporaries. What chiefly caught their imagination was the fact that the small (but increasingly important) city of Florence had at last demonstrated to the whole world that, in vaulting the vast space of the crossing of the new cathedral, it could equal and even outdo the greatest achievements of the ancient Roman Empire. This feat had been all the more remarkable because none of the massive temporary supports normally considered necessary at the time had been used and because both the unsightly external buttresses of northern gothic and the visible internal ties generally preferred in Italy had been dispensed with.

In the records which have come down to us, the acclamation is best summed up in the dedication to Brunelleschi of the 1436 manuscript of the Vernacular version of Alberti's *De Pictura*. "Who is so stubborn or so jealous as not to praise Pippo the architect when he sees here such a great structure, rising above the heavens, broad enough to cover with its shade all the people of Tuscany, made without any aid of falsework or other quantity of timber (*facta sanza alcuno ajuto di travamenti o di copia di legnami*), an achievement which, if I am not mistaken, was believed to be impossible in our time even as, among the ancients, it may well have been unknown and unheard of."[3] It was given

[1] The best account of the Pantheon is now Kjeld de Fine Licht, *The Rotunda in Rome,* Jutland Archeological Society, Copenhagen, 1968.
[2] See R. J. Mainstone, "The structure of the church of St. Sophia, Istanbul," *Transactions of the Newcomen Society,* XXXVIII (1965–66), 23–49, and idem, "Justinian's church of St. Sophia, Istanbul: Recent studies of its design, construction and first partial reconstruction," *Architectural History,* 12 (1969), 39–49.
[3] Quoted here from the edition of Luigi Malle, *Leon Battista Alberti: Della Pittura,* Florence, 1950, 54.

BRUNELLESCHI'S DOME OF S. MARIA DEL FIORE

more formal expression in the cathedral records and in the epitaph to *"Philippo Brunellesco antiquae architecturae instauratori"* set up in the cathedral after his death in 1446 and was echoed finally, at much greater length by his biographers Manetti[1] and Vasari.[2] Both refer to the more purely architectural achievements of the other building commissions, culminating in the design for S. Spirito, as well as to the cathedral dome, but treat the latter as unquestionably the central and major achievement. Both also emphasise again the feeling, which must have been almost overwhelming at the time, of having matched up to the Romans; of having, in Manetti's words, brought back to life *"questo modo de'muramenti, che si dicono alla Romana ed all'antica."*[3]

I shall concentrate in this paper on some structural aspects of the design and construction of the dome, largely passing over other almost equally important aspects of the total technical achievement such as the organisation of the work (including the supply of materials) and the design of new equipment for handling and lifting.[4] Nothing will be said of the vexed question of the roles of Ghiberti and other collaborators. I do not think that anything significant would be added to our understanding of the nature of the technical achievement if it could be definitely established that certain contributions were made by others, so it seems sufficient, for the present purpose, to note that, as early as 1423, Brunelleschi was officially recognised (in a prize award of 100 gold florins) as *"inventor et ghubernator maiori Cupule"*[5] and that he almost certainly took the major initiative from the start. Subsequent references to Brunelleschi reflect this judgement but should not be read as necessarily excluding collaborators.

Today the dome still stands virtually untouched save for the renewal of the circumferential timber tie (which had rotted badly by the seventeenth century) and extensive reconstruction more than once of the crowning lantern. Thus it is still possible to inspect at close quarters nearly all details of the original construction that are visible on the surface. This I have done on several occasions spread over the past 17 years, most recently and most fully in the summer of 1967 at the invitation and in the company of Professor Howard Saalman of Carnegie-Mellon University, to whose stimulus and interest this paper largely owes its origin. Much, however, remains hidden beneath the surface, so that considerable reliance must also be placed on documentary evidence. Fortunately this is far more comprehensive than for almost any other comparable structure, though, for various reasons, it contains serious ambiguities that have given rise to much conflicting speculation. To the extent that these ambiguities arise from the difficulties of giving clear expression in the legal Latin or Vernacular of the time to new structural concepts, any translation into the more precise vocabulary of Modern English necessarily involves a particular narrowing of the meaning and makes it essential to return to the original if a new understanding is sought. This I have endeavoured always to do,

---

[1] Antonio Manetti, *Vita di Filippo di Ser Brunellesco,* probably written soon after Brunelleschi's death but unpublished until 1812. I have used the editions of Carl Frey, *Le Vite di Filippo Brunelleschi Scultore e Architetto Fiorentino scritta da Giorgio Vasari e da Anonimo Autore,* Berlin (1887), 59–118, and Elena Toesca, *Antonio Manetti: Vita. . .,* Rome, 1927. These are cited below as Frey and Toesca. Since the paper was written a new edition has appeared edited by Howard Saalman, *The life of Brunelleschi by Antonio di Tuccio Manetti,* Pennsylvania State University Press, 1970. This has an English translation of the Italian text by Catherine Enggass.

[2] Giorgio Vasari, *Le Vite de piu eccellenti Architetti, Pittori et Scultori Italiani. . .,* Florence, 1550. English translation by William Gaunt, Everyman's Library, London, 1927.

[3] Frey, p. 61; Toesca, p. 2.

[4] For these aspects see especially Piero Sanpaolesi, *La cupola di Santa Maria del Fiore,* Florence, 1941 (cited below as Sanpaolesi); F. D. Prager, "Brunelleschi's inventions and the 'Renewal of Roman masonry work',", *Osiris,* IX (1950), 457–554 (cited below as Prager) and Gustina Scaglia, "Drawings of Brunelleschi's mechanical inventions for the construction of the cupola," *Marsyas,* X (1961), 45–68. Revised versions of the latter two papers with some additional related material have now appeared in a single volume: Frank D. Prager and Gustina Scaglia, *Brunelleschi: Studies of his technology and inventions,* MIT Press, 1970.

[5] Cesare Guasti, *La cupola di Santa Maria del Fiore illustrata con i documenti dell'archivio dell'Opera Secolare,* Florence, 1857, doc. 177, p. 71. This basic source is cited below as Guasti, *Cupola.*

BRUNELLESCHI'S DOME OF S. MARIA DEL FIORE

though it will be apparent to those who have made similar attempts that my debt to previous scholars is considerable.[1]

THE CONSTRUCTION BELOW THE DOME

Construction of the new cathedral began at the west end as far back as 1294.[2] The overall width of the nave was then established, but the bays were initially much narrower than at present, suggesting that a timber roof may have been intended as at S. Croce and the cathedral of Pisa. Beyond this the intentions are obscure, though it is probable that some sort of dome over the crossing was envisaged from the start, such as already existed at both Pisa and Siena. Progress was slow and almost came to a halt when energies were turned to the campanile in the second quarter of the next century.

The building, as it now exists (Plate XIII) dates effectively from 1355–57, when the present almost square nave bays were decided upon and the first explicit mention of a dome over the crossing appears in the official records. Even this plan was not definitive though. Renewed intensive discussion and debate in 1365–67 led to fresh proposals and a significant enlargement of the whole project. A fourth bay was added to the nave, the central octagon over which the dome was to rise was increased in width to the full width of nave plus aisles (72 braccia = 42 m.), and it was decided to raise the entire dome above the rest of the church by interposing an octagonal drum between it and the supporting arches. This design was embodied in a substantial brick model erected alongside the church and, after it had been formally approved in 1367, all those responsible for work on the church were required, up to 1421, to swear to adhere to it.

The model no longer exists, though there are two oblique reflections of the discussions that led up to it in a fresco in the Spanish chapel of S. Maria Novella and in a large tabernacle in the church of Or San Michele, both of which show a pointed octagonal dome, the first with a lantern but no drum and the second with a drum pierced by round arches. Other evidence suggests that, without indicating any details of construction the model did at least establish all the major dimensions including the basic curvature of the dome.[3] The problem eventually to be faced in constructing the dome was thus closely defined in its main essentials without being solved, indeed without any real assurance that it was soluble. A hope was expressed, nevertheless, that there would be no need for the visible tie bars just decided upon in the nave as a result of ominous cracks in the recently completed vaults there. More generally, the character of the structure now envisaged was no longer that of the purely gothic west end of 70 years earlier, nor even that of the somewhat later gothic of Milan cathedral. In the centralised planning of the octagonal crossing and three identical radiating arms to south, east and north, and in the insistence on a structure of maximum formal clarity, unencumbered either externally or internally by obtrusive props or ties, a character later to be recognised as Renaissance was already emerging. This cannot have been without influence on Brunelleschi, whose first

---

[1] Only the most directly relevant recent studies are cited in these notes. Fuller references are given in Howard Saalman, "Giovanni di Gherardo da Prato's designs concerning the cupola of Santa Maria del Fiore in Florence," *Journal of the Society of Architectural Historians*, XVIII (1959), 11–20; idem, "Santa Maria del Fiore: 1294–1418," *Art Bulletin*, XLVI (1964), 471–500 (respectively cited below as Saalman, *Cupola* and Saalman, *S.M.F.*), and in Sanpaolesi and Prager. There is no recent survey as comprehensive as that by G. B. Nelli in S. B. Sgrilli, *Descrizione e studi dell' insigne fabbrica di S. Maria del Fiore*, Florence 1733. For the dome itself, however, some more accurate data may be found in *Rilievi e studi sulla cupola del Brunelleschi eseguiti dalla Commissione nominata il 12 Gennaio 1934*, Florence, 1939 (cited below as *Rilievi e studi*) and further measurements are now in progress. Piero Sanpaolesi, *La cupola del Brunelleschi* (in the series *Forma e colore*), Florence, 1965, is an excellent album of recent photographs.

[2] The basic source for the period prior to the construction of the dome is Cesare Guasti, *Santa Maria del Fiore : La costruzione della chiesa e del campanile secondo i documenti tratti dall'archivio dell'Opera Secolare*, Florence 1887 (cited below as Guasti, *S.M.F.*). The present brief summary is based largely on the analysis in Saalman, *S.M.F.*

[3] Saalman, *S.M.F.*, pp. 491; idem, *Cupola*, p. 14.

BRUNELLESCHI'S DOME OF S. MARIA DEL FIORE

direct involvement with the construction was to advise, with other artists, on a departure from the 1367 model that seemed to jeopardise this character.[1]

The drum was complete by 1413 and, as work on the side arms continued, the problem of the detailed design of the dome must have come increasingly to the fore. Brunelleschi was certainly involved by 1417.[2] In 1418 an open competition was announced[3] and a period of protracted discussions and debates ensued. It ended in 1420 with the appointment of Brunelleschi, together with Ghiberti and Battista d'Antonio, to superintend the construction, and the approval of the design to be followed.[4]

SPECIFICATION FOR THE DOME

The design had already been embodied in a substantial masonry model, but this time the model was supplemented by a written specification.[5] It is a document of outstanding importance since, with modifications agreed in 1422 and 1426, it is the primary source for following the development of the design and for interpreting those features of it whose nature or intention is not clear from inspection of the structure today. Its twelve clauses are too long to quote here in full but the main provisions may be summarised as follows:

(1) to (3) define the basic form of the dome. It is to have two octagonal shells, the inner being initially $3\frac{3}{4}$ braccia thick and the outer $1\frac{1}{4}$ braccia, separated by a void initially 2 braccia wide. (1 braccia = 0·584 m.) Both shells are to taper uniformly[6] to about two-thirds of their initial thicknesses at a central eye (later to be covered by the lantern). The curve of the inner shell at the internal corners is to be "*quinto acuto*"—i.e. it is to be a circular arc of radius one fifth less than the diameter across the corners. The outer shell is to keep the inner one dry and, at the same time, to make the dome more imposing (*piu magnifica e gonfiante*) and to provide covered access within the intermediate void.

(4) calls for 24 internal ribs to bind the two shells together (*e legano insieme le decte due volte*). Eight of these, in the angles, are to be 7 braccia wide at the foot and the remaining 16, two to each side, 4 braccia wide at the foot. These ribs also are to taper uniformly[6] to the eye.

(5) and (6) indicate that the separate shells and ribs are to commence at a height of $5\frac{1}{4}$ braccia, up to which the construction is to be solid, but they are mostly concerned with the specification of 6 rings of large sandstone blocks well clamped together and having, in addition, iron chains set above them. Successive pairs are to be 2, $1\frac{1}{2}$ and 1 braccia high and at the foot there are also to be long transverse blocks of sandstone to provide a base for both shells.

(7) calls for further interconnections between the shells in the form of small barrel vaults between adjacent ribs at 12 braccia intervals to carry ambulatories and, below these barrel vaults, of oak and iron chains to tie the ribs together (*che legano i decti sproni*) and encircle the inner shell

(8) specifies the materials to be used—briefly stone up to a height of 24 braccia and, thereafter, a lighter material such as brick or tufa.

(9) to (11) are concerned with matters of limited structural importance.

---

[1] Guasti, S.M.F., doc. 425, pp. 299, 300. It should be pointed out, however, that there is no direct reference to the architectural character in the document. This refers only to a departure from the model in the construction of one of the triangular buttresses at the lowest level visible in Plate 1. The buttress had been built too high and was to be lowered.

[2] Guasti, *Cupola*, doc. 16, p. 17.

[3] Guasti, *Cupola*, doc. 11, p. 16.

[4] Guasti, *Cupola*, doc. 71, pp. 35–7.

[5] Two copies of this exist, one published by Guasti, *Cupola*, doc. 51, pp. 28–30 and the other by Alfred Doren, "Zum bau der Florentiner Domkuppel," *Repertorium fur Kunstwissenschaft*, XXI (1898), 249–62. I am quoting from Doren's definitive version.

[6] More accurately "like a pyramid (*piramidalmente* or *piramidalmente . . . per iguale proporzione*)."

BRUNELLESCHI'S DOME OF S. MARIA DEL FIORE

(12) deals with the method of erection. "The shells are to be built as specified without any supporting centering (*sanza alcuna armadura*) up to at least 30 braccia, but with such working platforms as may be decided by those in charge of construction; and thereafter as considered advisable, since in building practice teaches what should be done (*perche nel murare la praticha insegnera quello che ss'ara a seguire*)."

The amendment of 1422[1] introduced two modifications, both for the purpose of reducing weight. The intermediate ribs were reduced in thickness to 3 braccia and the height of the change from stone was reduced to about 12 braccia, above which brick was to be used.

That of 1426[2] was wider in scope. It called first for small openings in the faces to permit access to the interior, etc. Then, in place of the small barrel vaults between the ribs, arch-like bands of brickwork were to be constructed above the second ambulatory, 1 braccia high, "for the perfection of the circle of the outer shell so that the live arch shall be complete and not broken (*per perfectione del cerchio che gira intorno la cupola di fuori, accio che detto archo vivo sia intero e non rotto*)." If judged to be unsightly or an obstruction, these additions could be removed upon completion of construction, their primary purpose being to allow the construction to proceed with greater safety. Revised details of the final pair of "stone chains" (*catene di macigno*) and superimposed iron chains follow. Finally comes a call for the manufacture of special large bricks to be laid in a herringbone (*spinapescie*) bond as determined by the master in charge, and a confirmation of the earlier decision to dispense with a supporting centering. In the absence of this, a device with three cords (*gualandrino con tre corde*) is to be used to control the curvatures.

As usual, some discretion over details is left to those directly in charge of construction. The formal latin authority (to Brunelleschi and his associates) that accompanies this last amendment specifically confers a wider discretion though,[3] and this has naturally led to speculation about the extent to which the specification, as amended, was subsequently followed. There are three principal matters of interest: the ring of horizontal arches; the "stone chains" and the oak chains and barrel vaults previously called for but passed over in silence in 1426; and the manner of construction.

CONSTRUCTION OF THE DOME

It is immediately apparent from an inspection of the structure today (Fig. 1) that not merely one but no less than nine rings of horizontal arches of about the size specified were built in the annular void between the two shells. They were never removed and, since they seem to be bonded into the outer shell, it was possibly decided at an early stage that they would be left.

The omission of the small barrel vaults previously called for followed inevitably from the decision to construct these other additions and all the oak chains were likewise omitted except the first—which had already been constructed in 1423 of chestnut. The ostensible purpose of the barrel vaults had been to carry ambulatories. These ambulatories are, in fact, carried instead by regularly spaced heavy sandstone blocks spanning between the two shells and strongly reminiscent of the interconnecting blocks between the first pair of stone chains called for in 1420. It is difficult to escape the conclusion that they are indeed performing a similar function in relation to the second and third pairs of stone chains. If so, it seems certain that both pairs were constructed, as further documents suggest. Professor Saalman and I found final confirmation of this in 1967 in a detail which has not previously been noticed in this connection. We believe that, though it has hitherto been assumed that all six chains lie wholly within the thickness of the masonry, a length of almost 4 m. of the second outer chain is exposed where the outer shell is cut back to allow the internal

[1] Guasti, *Cupola*, doc. 52, pp. 30–1.
[2] Guasti, *Cupola*, doc. 75, pp. 38–41.
[3] Guasti, *Cupola*, doc. 75, p. 40. Authority is given to vary the design "*in addendo, minuendo ac disponendo, plus et minus . . . et prout et sicut dictis operariis . . . videbitur et placebit, non obstante dicto raporto.*"

BRUNELLESCHI'S DOME OF S. MARIA DEL FIORE

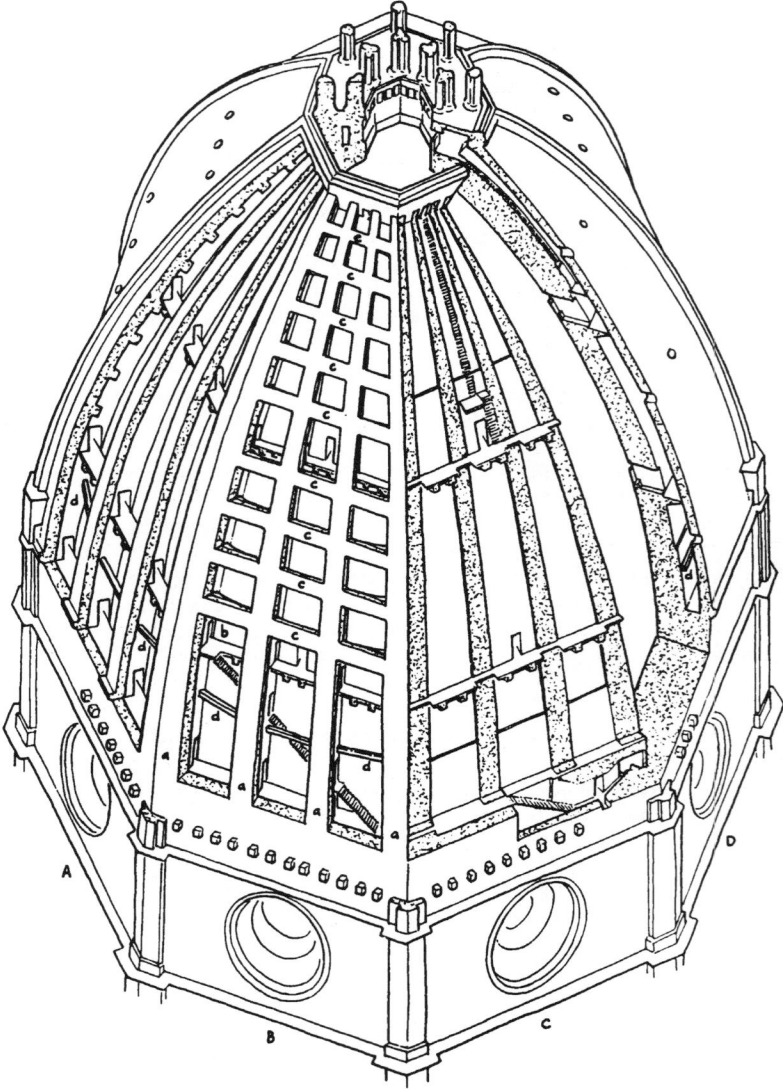

Fig. 1. S. Maria del Fiore, Florence. View of the dome partly cut away. The timber chain is marked *d*.

Shown above A and B: the outer shell cut away between radial ribs, *a*.
A: the circumferential arches *c* also cut away.
C: only the inner shell and the ambulatories.
D: a complete cross-section through both shells.

Drawn by the Author. Partial sources—an unpublished drawing by Siebenhuener, a drawing by Sanpaolesi, the survey published by Sgrilli.

112

BRUNELLESCHI'S DOME OF S. MARIA DEL FIORE

stairway to ascend to the second ambulatory.[1] (Plate XIV (c)) It is reasonable to assume that the superimposed iron chains were also constructed in all cases, but there is no means of telling what form they took without exposing them.

The documentary evidence and a continuous literary tradition are unanimous in stating that the construction of the dome was completed without *armadura*. Alberti's contemporary testimony has already been quoted. Alongside it may be placed that of the official cathedral records on Brunelleschi's death 10 years later. There he is described as having completed the dome "*sine aliqua armatura . . . ut clare asseruerunt dicti operarii, viderunt et notum est in civitate predicta*" (as clearly stated by the operarii and seen and known in the city).[2] This apparently simple fact has, nevertheless, been repeatedly questioned in more recent times because, among other reasons, there are references in the documents to the making of centers (*cientine*).[3] I believe that the description of the method of construction given by Nelli,[4] a later architect to the cathedral, is correct in describing the herringbone bond of the brickwork as the feature which made it possible to dispense with supporting centering, and that the centers referred to were used only to assist in keeping to the specified profile.

The herringbone bond, which is today such a characteristic feature of the construction above the second ambulatory (Plate XIV (a)) involves essentially the setting of a brick vertically on edge at regular intervals in each otherwise horizontal course. The special large bricks that were called for in 1426 are only about 50 mm. thick, so that each vertical brick rises through four or five horizontal courses. Being set with one face against the most exposed face of the corresponding brick of the previous course, it forms part of a diagonally ascending band. In a circular dome this would be quite straightforward and the ascending bands, being set radially in plan, would automatically converge towards the top (Fig. 2).[5] Here the same convergence is seen but nothing was straightforward on account of the octagonal plan, the consequent elliptical profile of each side of the dome and the introduction of ribs built integrally with the shells. In part the problems were overcome by the manufacture of a number of special non-rectangular bricks and by the cutting in-situ of others and in part by the introduction in places of special levelling courses of tapered bricks. But they must have remained a troublesome preoccupation to the end.[6] There was also a further drawback which is unlikely to have been overlooked. This is that the inclined bands of vertical bricks would inevitably constitute planes of weakness in resisting tensile hoop stresses. A much greater tensile strength could have been achieved in the brickwork by laying all the bricks horizontally with all vertical joints lapped and the tensions transmitted from brick to brick by friction and bond shear at the horizontal joints.

It is inconceivable that these difficulties and drawbacks were accepted right to the end unless they were indeed directly linked with (and justified by) the avoidance of the expense, difficulties, and risks of constructing a heavy vault-spanning supporting centering. More positively, the herringbone bond would have made it possible to proceed without such support. Since every complete course

---

[1] Parts of two sandstone blocks, 470 mm. high, are visible towards the upper left of the photograph, notched over one of the transverse beams that has been cut to allow the stairway to pass. The other (inner) end of this transverse beam is seen projecting at the right to carry the ambulatory, here cantilevered out from the inner shell.

[2] Guasti, *Cupola*, doc. 120, pp. 56–7.

[3] Guasti, *Cupola*, docs. 170–2, p. 70.

[4] G. B. Nelli, *Ragionamento sopra la maniera di voltar le cupole senza adoperavi le centine* in *Discorsi di Architettura del Senatore Giovan Batista Nelli*, Florence, 1753, 53–74. This account is very discursive, the relevant passage appearing on pp. 65–7. The word *spinapescie* never appears, but Nelli gives what seems to be a fairly detailed description of the bond actually adopted in the upper part of the dome.

[5] Piero Sanpaolesi, "Le cupole e gli edifici a cupola del Brunelleschi e la loro derivazione da edifici Romani," *Atti del I Congresso nazionale di storia dell'architettura*, Florence, 1936, 37–41, first drew attention to this drawing which illustrates this simple application very well.

[6] Sanpaolesi (pp. 25–9) gives a detailed account of the practical problems with illustrations of some of the expedients adopted. His Plate Vb attempts to show the internal details of the bond as they would have been seen during construction, but I do not find it wholly convincing.

BRUNELLESCHI'S DOME OF S. MARIA DEL FIORE

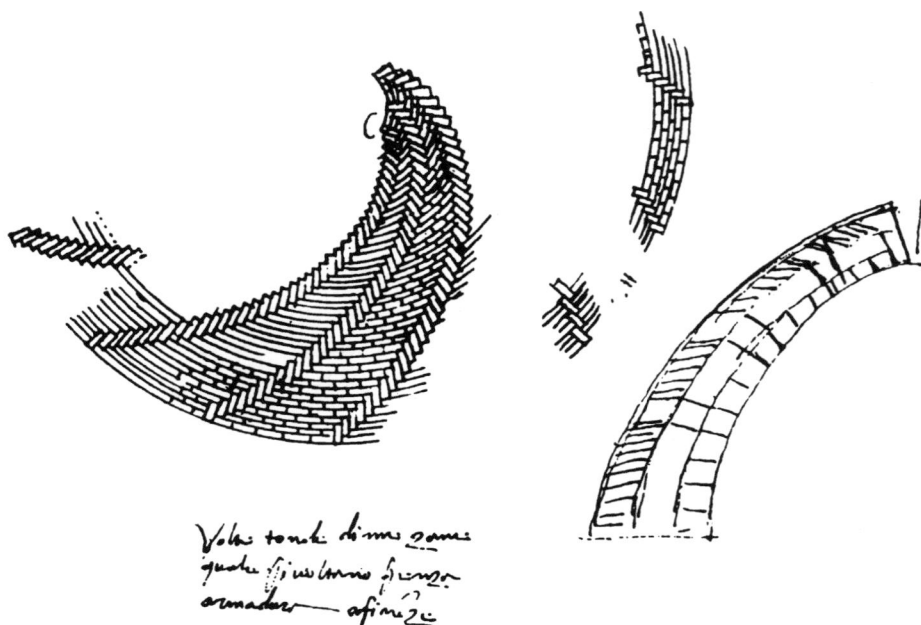

Fig. 2.   Detail of a drawing by Antonio da Sangallo the Younger.
(*Uffigi 900—courtesy of the Soprintendenza alle Gallerie*)

of a dome is, in effect, a closed horizontal arch, it has no need of support other than that of the structure below.   Only an incomplete course needs, temporarily, some further support.   The inclined bands of vertical bricks do not eliminate this need, but they subdivide each course into a large number of short sections never more than about five bricks long and reducing to single horizontal brick at the eye.   Each of these sections, when completed, would constitute a short flat arch which would be self-supporting as soon as the mortar had hardened sufficiently.   It may be assumed that, working together, teams of two or three bricklayers spaced around the dome on light working platforms would each be responsible for one section and would complete this from the intrados outwards, the only difficulty being to keep the bricks on the intrados in place until the short flat arch was complete and self-supporting.   There are several ways in which this might have been done, depending partly on the nature of the mortar used.   The mortar beds, as well as the bricks, are thin and they are inclined to the horizontal at an angle never more than 60°.   Friction alone would keep them in place up to an angle of, say, 30° and even at the eye the shear on the joint would be only 0·001 N/mm².   This may be compared with a bond shear of at least two orders of magnitude greater for a modern cement-lime-sand mortar after 28 days.   One possibility, therefore, was simply for assistants to hold the bricks in place for a time in much the same way as they still do when constructing vaults freehand in the Middle East today.   Another simple possibility suggests itself however.   This is the use of timber boards just long enough to span from one band of upright bricks to the next and either held against these bricks as each course was laid or tied back to them.   Since the average rate of construction was less than one course a week, there would be ample time for each course to become self-

114

BRUNELLESCHI'S DOME OF S. MARIA DEL FIORE

supporting before the next was added and thus to make this procedure practicable. Possible confirmation of its actual use may be found in a passage in Alberti's *"De re aedificatoria"*.[1] Its precise range of possible meaning turns on the words *amenta* and *ansas*, the second of which could be read as referring to something like the bricks that project vertically. Leoni's translation here seems as good as any: "Yet it will be necessary, when you have laid one or two rows of stone, to make little light stays, or catchers jutting out (*amenta et ansas*), on which, when those rows have settled, you may set just framework enough to support the courses next above . . . ."[2] The context is a general discussion of the construction of polygonal domes without centering and there is no explicit reference to Brunelleschi's dome, but the procedure described could have been suggested to Alberti by what has been proposed here.[3]

Control of the form without first replicating it in a complete system of self-supporting centering called for some system of measurement. The *gualandrino* with three cords specified in 1426 can be envisaged thus: two of the cords (or chains) have one end of each fixed at opposite ends of a horizontal line, drawn at right angles to a corner diagonal of the octagon and passing through the centre of the circular arc that traces, in the vertical plane of that diagonal, one corner of the intrados of the inner shell. The other two ends are joined together so that they would sweep out this arc and define it uniquely in space. The fixed ends would be attached to eyes on the ambulatory inside the dome at its springing. The third cord was presumably associated with the control of thicknesses and of the profile of the outer shell. This device would be cumbersome and was presumably used only at intervals in each corner in turn.[4] Day-to-day control would be the role of the *cientine*. There was apparently one for each corner and their low cost[5] indicated that they cannot have been more

---

[1] L. B. Alberti, *De re aedificatoria*, Florence, 1485, Book III, chapter XIV (but not so subdivided until late editions and with the pages unnumbered): "*Tamen conferet ductis iam una atque alteris lapideis et duratis coroni levia illic subillaqueare amenta et ansas; quibus tantum armamenti committas; quantum sat sit ferendis coronis hi quae aliquot inde in pedes superastruantur quoad siccescant. . . .*" Prager has since made the interesting suggestion to me that the passage should have read "*Tamen conferes . . . levia sub illaque [testudine] ar[m]amenta et ansas*"

[2] James Leoni, *Ten books on architecture by Leone Battista Alberti*, 1726, 59. (Facsimile edition edited by Joseph Rykwert, London 1955). Cosimo Bartoli in the Italian translation on which Leoni's English one is based, *L'architettura di Leonbatista Alberti*, Florence, 1550, 90–1, renders the words "*amenta et ansas*" as "*spranghe o perni non gravi.*" Portoghesi has more recently translated them as "*leggeri lacci e grappe.*"

[3] There is one difference between the procedure I have suggested and those described by Nelli and Alberti. I have assumed that one course at a time would be laid and left to harden. Nelli (p. 65) says that three courses are constructed at a time and Alberti (*loc. cit.,*) even speaks of a few feet. The former would be equally reasonable, but the latter would not unless the "*ansas*" were something other than the projecting vertical bricks of the herringbone. Another possibility is that Alberti had in mind the built-in iron rings set in horizontal rows about 2·2m, apart in the inner surface of the inner shell that are described and illustrated by Sanpaolesi (pp. 16–18). This spacing corresponds, roughly, to a year's work, so I think that Sanpaolesi is correct in interpreting them as supports for working platforms for the following year rather than for any kind of formwork. Since Alberti is only giving general advice, it is quite likely that he ran together two distinct aspects of Brunelleschi's method which he may not even have fully understood.

[4] This system of control of the curvature in the corners would virtually preclude the use of any large working platform such as those suggested to Sanpaolesi (Plate VI*b*) and Prager (Fig. 11, p. 508) by the drawing published by C. von Stegmann and H. von Geymüller in *Die Architektur der Renaissance in Toskana* (Munich, 1885, Fig. 12, p. 48), and, as an engraving, in Nelli (*op. cit*). This drawing was said to be by Brunelleschi, but is clearly much later and is inaccurate in important respects. I take it to be a later record of a structure associated with the construction of the lantern and perhaps with the final stage of work on the dome at the level of the eye. In relation to the suggested points of attachment on the ambulatory, it should also be pointed out that recent surveys show the centres of curvature of the circular arcs that best fit the present internal profiles in the corners to lie below the level of the ambulatory. This would seem, *prima facie*, to call for lower points of attachment. As yet, however, the surveys have not been properly correlated with one another. (The only one yet published has even claimed that the present profiles are elliptical arcs with small and widely varying eccentricities.) Also there is insufficient further data to permit the necessary allowances to be made for the movements (and consequent changes in profile) that have occurred since construction—let alone the divergences likely to have occurred during construction.

[5] Professor Saalman has pointed out to me that the recorded payment for the first eight amounts to less than the average daily wage of the lowest paid labourer for each and was only slightly higher for the remaining two (presumably replacements).

BRUNELLESCHI'S DOME OF S. MARIA DEL FIORE

than light templates formed to the appropriate curve and used in exactly the same way as the similar templates employed in 1909 in constructing the large dome of the cathedral of St. John the Divine in New York.[1] The most effective control of the straightness of the sides would have been simply by sighting from one corner to the next along each course. If the procedure suggested above is correct, this would have been done when setting each vertical brick. The horizontal planks, or short lengths of string would have given the line between the vertical bricks.

The dome was completed up to the eye in 1434. Brunelleschi then made models and other preparations for the construction of the lantern, but work on this had barely started when he died in 1446.

STRUCTURAL ASSESSMENT

Before turning to a discussion of the likely origins and development of the design, it is worth considering the probable structural actions, both during construction and afterwards, of the structure as built.

The proportions of the dome are very massive, even in relation to its span and rise. Wind loading is therefore relatively unimportant. Since earthquakes were not a serious threat either, it will be sufficient to consider only gravity loading. Under such loading the actual cross sections, even of the shells considered independently between a pair of ribs, are quite thick enough for the entire dome to stand as two rings of independent vertical arches, each ring with a shared keystone at the central eye—providing that the supporting structure does not yield too much under the thrusts generated. But this thickness does not, in itself, preclude the alternative possiblity of a more shell-like action in which the outward thrusts developed in the upper parts of the shells and ribs are contained lower down by internal horizontal tensions. Evidence of cracking referred to below shows that the present behaviour of the dome is closer to the first of these possibilities than to the second, but it still leaves open the question whether this was also the case right from the completion of construction.

The construction of possible thrust lines for action of the shells as rings of independent arches merely provides a theoretical demonstration of the first possibility, with reference to the dome only and not its supports—a demonstration that is hardly necessary in view of the fact that it does still stand despite some radial cracking.[2] A more revealing insight—and one that is more relevant to the design problem that was posed after completion of the drum in 1413—is given by considering the second in terms of a simple thin-shell analogue. In plan, even at the base, the thickness of the inner shell exceeds by about 25 per cent the difference between the internal radii measured at the corners and at the centres of the sides. It is therefore possible to inscribe within the thickness two circles spaced apart a distance equal to this excess, and to consider the actual octagonal shell as containing within itself another shell of the same profile but circular in plan and only one-fifth of the thickness. Such a shell is statically determinate, so that it may be analysed without calling for unknown effective moduli of elasticity of the materials. It is, of course, a pointed shell of revolution open at the top throughout construction, but eventually carrying a relatively light lantern—i.e. a distributed more-or-less vertical load—on the uppermost ring of a central eye.

The radial symmetry of the gravity loading, as of the shell's geometry, means that it will be subject only to direct tensions and compressions in the meridional and horizontal circumferential directions. Adopting Flügge's notation (Figures 3 and 4) the equations of equilibrium for an element of the shell are:[3]

---

[1] "Erecting a large dome without falsework," *Engineering Record*, 60 (1909), 508–10.
[2] W. B. Parsons, *Engineers and Engineering in the Renaissance* (Baltimore (1939), reprinted, M. I. T. Press (1968), 594–8) gives two such analyses and some related caclulations.
[3] Wilhelm Flügge, *Stresses in shells*, Berlin, 1960, chapter 2, "Direct stresses in shells of revolution," pp. 18–24. For an analysis of an alternative analogue of the actual dome, see Guido Guerra, *La statica delle cupole a padiglione in muratura*, Rome, 1951 (reprinted from *Giornale del Genio Civile*, Sept. 1951) and idem, *Statica e tecnica costruttiva delle cupole antiche e moderne*, Naples (1958) 48–9.

BRUNELLESCHI'S DOME OF S. MARIA DEL FIORE

$$\frac{d}{d\phi}(rN_\phi) - r_1 N_\theta \cos\phi = -p_\phi rr_1, \tag{1}$$

and

$$\frac{N_\phi}{r_1} + \frac{N_\theta}{r_2} = p_r \tag{2}$$

Solving for N_ϕ and integrating:

$$N_\phi = \frac{1}{r_2 \sin^2\phi}\left[\int r_1 r_2(p_r\cos\phi - p_\phi\sin\phi)\sin\phi\, d\phi + C\right] \tag{3}$$

N_θ is then given by equation (2). The constant of integration is given by the known value of N_ϕ at the open top ring at any intermediate stage of construction or at the uppermost ring of the central eye on completion of construction. Thus, at any intermediate stage of construction, $N_\phi = 0$ for $\phi = \phi_c$, where ϕ_c is the value of ϕ at the then uppermost ring. Similarly on completion of the dome to the central eye but before construction of the lantern, $N_\phi = 0$ for $\phi = \phi_e$, where ϕ_e is the value of ϕ at the eye; and, on completion of the lantern, $N_\phi = -P/\sin\phi_e$, where P is the weight per unit length (measured around the eye) of the lantern.

Since the springings may be assumed to be vertical, the meridional stresses there will, of course, also be vertical, their outward components higher up being wholly contained by horizontal circumferential or hoop tensions. These hoop tensions are the really interesting stresses, since they are the only ones likely to approach the ultimate strengths of the materials used.

Fig. 5 shows a number of plots of the hoop stresses, $\sigma_\theta = N_\theta/t$, both tensile and compressive, expressed non-dimensionally as multiples of ρr_1 where ρ is the density, assumed uniform, and r_1 is the radius of curvature of the generating arc as in Fig. 3. Of the two curves passing through the origin, the upper one represents the progressive increase in tension at the base as construction proceeds,

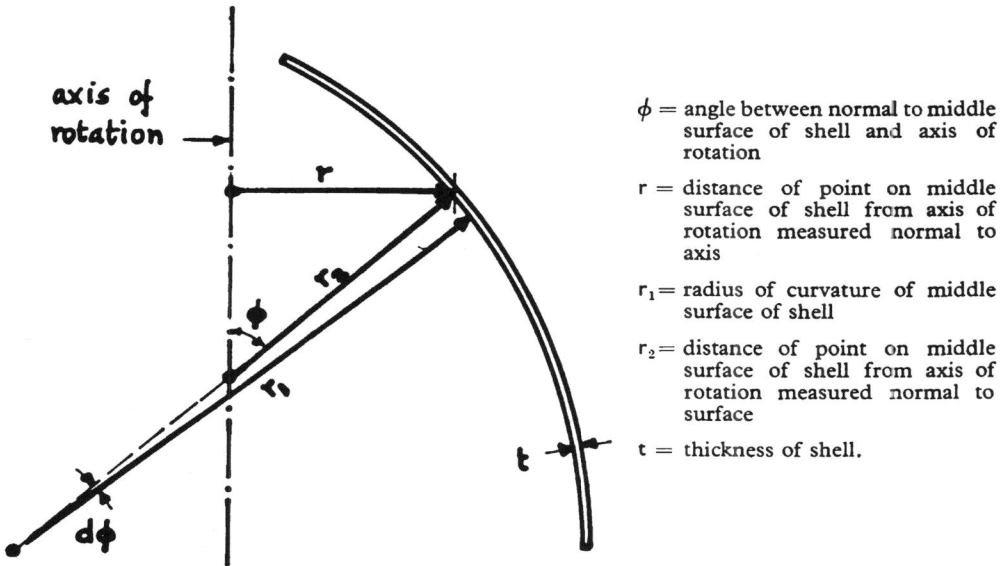

ϕ = angle between normal to middle surface of shell and axis of rotation

r = distance of point on middle surface of shell from axis of rotation measured normal to axis

r_1 = radius of curvature of middle surface of shell

r_2 = distance of point on middle surface of shell from axis of rotation measured normal to surface

t = thickness of shell.

Fig. 3.

117

BRUNELLESCHI'S DOME OF S. MARIA DEL FIORE

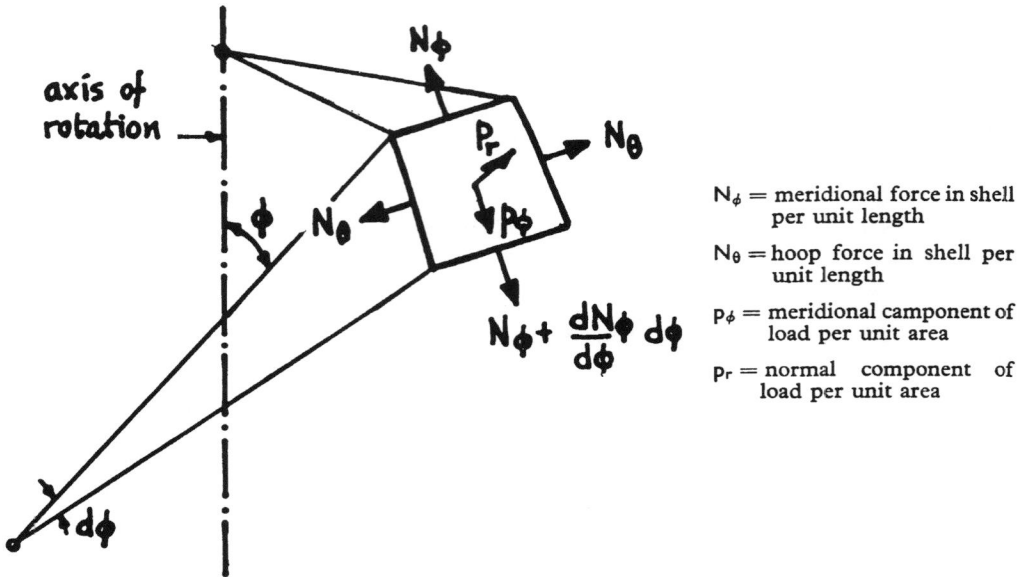

N_ϕ = meridional force in shell per unit length

N_θ = hoop force in shell per unit length

p_ϕ = meridional camponent of load per unit area

p_r = normal component of load per unit area

Fig. 4.

and the lower one the initial compression at the open top ring at any stage of construction. The family of curves roughly parallel to the first similarly represent the increases in tension (appearing for a time as reductions in the initial compressions) at various higher levels corresponding to the change from stone to brick, the timber tie, and the second ambulatory. These levels are denoted, in terms of ϕ, by the abscissae of the lowest points on the curves. The other family of curves, shown in broken line, represent the variations in stress with height throughout the dome when construction has just reached the level denoted again by the abscissa of the lowest point on the curve. The additional stresses due to construction of the lantern cannot be shown in the same way since they depend on the thickness of the shell, but, in the dome as built, they are spread over such a thickness near the base as to be negligible there in relation to the stresses plotted.

From the lower curve passing through the origin it will be seen that the hoop stress in the ring that at any stage terminates the construction will always be compressive, being virtually zero for the first ring. It will increase with the height of this ring to a maximum of $0.33\,pr_1$, when the structure is about three-quarters complete. Thereafter it will reduce again to about half this value. But at any particular level this initial compression will continually decrease as construction advances higher, the rate of decrease being roughly the same at all levels. Up to a height rather more than midway between the second and third ambulatories the stress will, at some stage, change to tension: indeed at the very foot it will be tension almost from the start. The final value of the tension at the foot will be $0.48\,pr_1$, which may be compared with $0.99pr_1$ for the corresponding hemispherical shell. At the level of the change to brick it will be $0.25pr_1$, at that of the timber tie $0.21pr_1$ and at that of the second ambulatory, where the herringbone starts, $0.12pr_1$. Taking pr_1 as roughly 0.7 N/mm², these maximum tensions are, respectively, about 0.34, 0.18, 0.15 and 0.08 N/mm² and, if construction proceeds at a uniform rate and takes 14 years, they will develop from the initial compressions by

BRUNELLESCHI'S DOME OF S. MARIA DEL FIORE

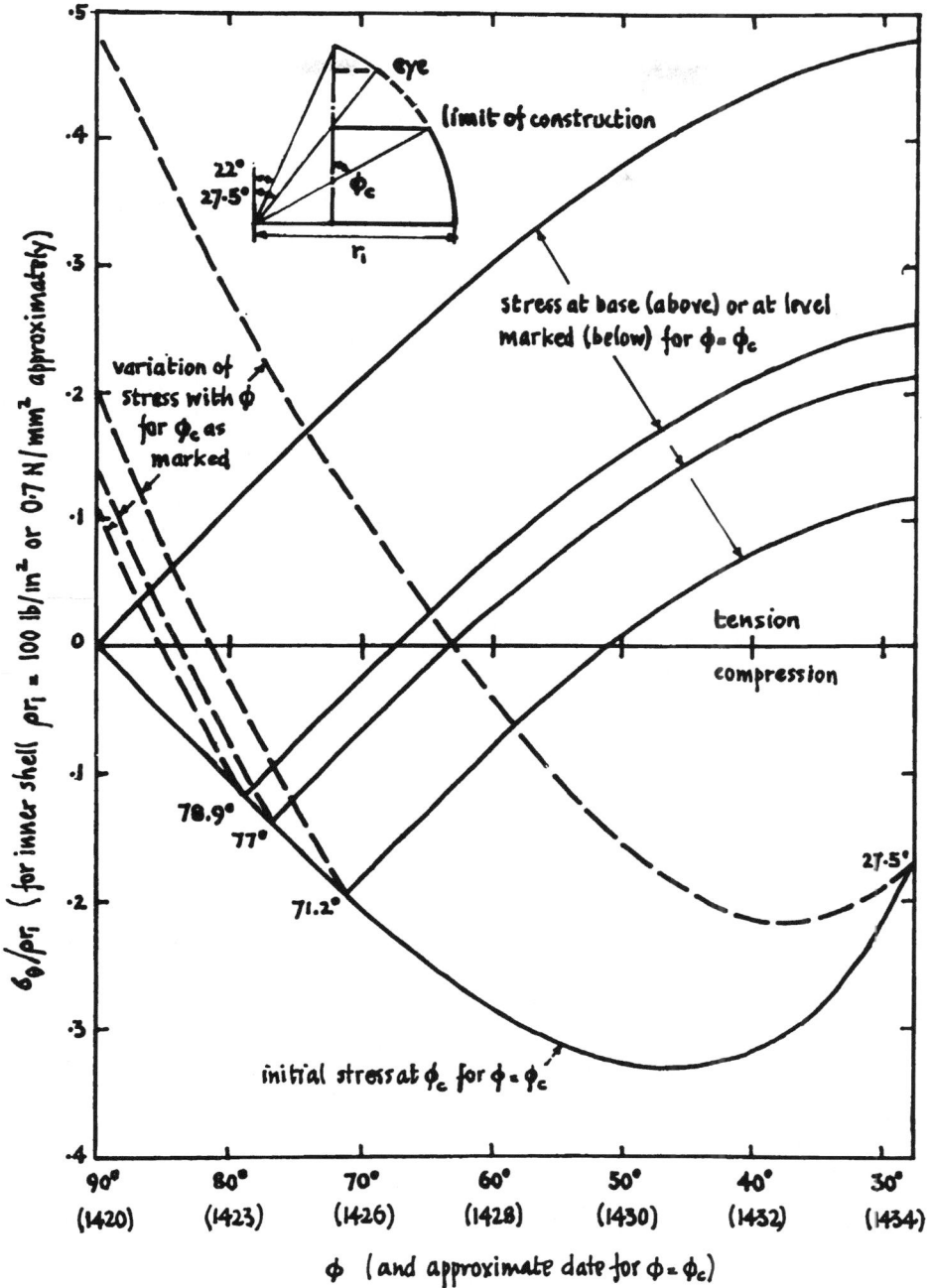

Fig. 5.

The development of hoop stresses $\sigma\theta$ (expressed as $\sigma\theta/\rho r_1$) in a simple statically determinate analogue of the inner shell of the dome.

(*Author*)

BRUNELLESCHI'S DOME OF S. MARIA DEL FIORE

changes of about 0·025 N/mm² each year. (With a supporting centering the stress changes would be indeterminate and would include an unpredictable change at the time of decentering.)

The highly indeterminate actual structure could be considered, conservatively, as if it contained this hypothetical shell wholly inscribed within its own inner shell, as its only active element on which the rest was merely a dead weight. On this basis, and making some allowances for the reductions in thickness and density in the upper parts, and for the fact that much of the lower parts of the outer shell and the ribs could be properly considered as a direct vertical load on the drum, the maximum hoop tensions might be as high as five times those calculated. It is nearer the truth, though, to consider the full widths of both shells as polygonal domes interconnected by and subject to some external load through the ribs, the horizontal arches in the upper part being effectively reinforcements of the outer shell. From this point of view the structure is much more efficiently designed and, ignoring possible local concentrations, the maximum tensions will be perhaps only twice the calculated ones though greater, at any level, in the outer shell than in the inner one because of its greater radius of curvature.

These stresses may be compared with possible ultimate unit tensile strengths of the order of 5 and 0·5 N/mm² for the stone and brickwork respectively, the former taking no account of the weakening effect of the joints and the latter referring primarily to the horizontally coursed brickwork below the start of the herring bone. In the completed structure the frictional forces on the bed joints of the stone should have been sufficient, if all the joints are well lapped, to develop an overall strength about half that of the individual blocks. The final stresses seem, therefore, to be within the ultimate capacities of the materials, even ignoring the cramps and the continuous iron and timber ties. These will have assisted, however, in resisting the lesser hoop tensions at earlier stages of construction. Taking account also of the very slow rate of construction, which would allow the mortar of the brickwork to develop its own bond and tensile strengths before being subjected to tensile stress, it can be further concluded that the strengths should have been adequate at all intermediate stages of construction. These conclusions ignore the effects of shrinkage and thermal strains, but it is only when movement is restrained that such strains give rise to stress. Restraints should have been slight during construction and will always have been minimised by the interpolation of the drum between the dome and its more rigid substructure.

Turning to present observable fact, however, there are today conspicuous cracks on three of the diagonally aligned faces of the octagon.[1] Each extends from below the drum to near the eye, initially almost on the vertical centreline but subsequently tending to follow the inclined bands of vertical bricks as might be expected. (Plate XIV (*b*)). If these cracks developed with or before the completion of construction they would show the conclusions to be mistaken. Did they do so? The documentary evidence suggests not. Considerable concern was expressed about the cracks in the seventeenth century,[2] but prior to that there is no mention of them despite numerous references to damage to the lantern by lightning.[3] This silence may be contrasted also with the immediate concern and almost panic remedial measures provoked in 1366 by the discovery of cracks in some recently completed aisle vaults.[4] If the cracks do indeed date only from about 200 years after the completion of construction, the probable immediate cause was thermal expansion. Contributory factors could have been deterioration of the masonry of the outer shell, of the timber tie, and perhaps also the swelling of embedded iron cramps and chains, all as a result of ingress of water and general dampness.[5] Once a crack had formed through the thickness of either shell it would, of course, tend

---

[1] Illustrated in *Rilievi e studi*, Plates V–VII.
[2] Guasti, *Cupola*, docs. 382–391, pp. 169–83. See also Giovanni Poleni, *Memorie istoriche della gran cupola del Tempio Vaticano*, Padua, 1748, 101–10; *Rilievi e studi*, 18–24.
[3] Guasti, *Cupola*, docs. 337–339, pp. 117–21. See also *Rilievi e studi*, 16–20.
[4] Guasti, *S.M.F.*, docs. 143–149, pp. 168–74.
[5] Guasti, *Cupola*, docs. 382–383, pp. 169–71.

BRUNELLESCHI'S DOME OF S. MARIA DEL FIORE

to spread fairly rapidly as a new pattern of equilibrium was established with some of the outward meridional forces generated in the upper part being resisted by inward horizontal forces at the base rather than by hoop tensions as previously.

ORIGINS AND DEVELOPMENT OF THE DESIGN

It may be assumed that the octagonal dome of the nearby baptistery of S. Giovanni, constructed in the late twelfth century in a remodelling of an earlier structure (Plate XV) was the immediate prototype for the cathedral dome.[1] Its basic structure is, likewise, a pointed octagonal shell with a central eye closed by a lantern. This shell springs from what is, in effect, the inner wall of a double-walled octagonal drum. To span the greater width of the outer wall and provide a unified external aspect, there is also an outer shell carried, very much in the manner proposed in the 1420 specification, by small barrel vaults spanning between more-or-less radial walls above the extrados of the inner shell (Plate XVI (a)). There is also, at a level corresponding roughly to that in the cathedral dome, a circumferential timber chain between the two shells.

Even apart from a 70 per cent increase in scale, the cathedral dome as built is nevertheless a fundamentally different structure. The baptistery structure is not a true double dome but rather a pyramidal roof of stone superimposed on, partly carried by, and near the top merging with, the inner shell, and the walls which carry this roof, though analogous to the ribs in the cathedral dome, do not appear to have been conceived with any intention of integrating the two shells into a single structure. It is possible to identify any other prototypes?

Manetti and Vasari both place considerable emphasis on the years which Brunelleschi is said to have spent in Rome between his failure in the competition for new bronze doors for the baptistery and his preparation of models for the dome.[2] Today it is difficult to assess precisely what he could have learnt there from studies of ancient remains, since much has since been lost or made less revealing by subsequent adaption or restoration. Something is known, however, about the extent of the losses and it seems unlikely that he could have seen any constructional techniques radically different from those exhibited by the remains we know or know of now. If the basic structural characteristics of the cathedral dome are taken to be the double-shell form with intermediate ribs, the horizontal arches that reinforce the upper part of the outer shell, the reinforcements of both shells by iron cramps and stone, timber and iron chains, and the construction without supporting centering using brickwork in a herringbone bond, the only close precedents to be found in Roman construction are for the use of stone chains and iron cramps.[3] Brunelleschi's special bricks are not unlike Roman ones but the use made of them in Rome was very different. All wide-spanning Roman vaults were constructed of concrete cast in horizontal layers and are of the very massive form best exemplified by the Pantheon dome—essentially hemispherical internally but externally much flatter, and composed only of a single shell, although its immensely thick base is lightened by leaving voids within it.[4] All must have been cast on supporting centering, and the nearest approach to dispensing with this is the adoption of various devices to lessen the load upon it while the concrete is still green. In later structures these devices included an increasing use of embedded ribs of brick and in one instance—the so-called

---

[1] The parallels drawn in this concluding section of the paper are based largely on personal observation, but references to the most relevant literature are given below. For the baptistery see E. Isabelle, *Les édifices circulaires et les dômes*, Paris, 1855, Plate 53; Josef Durm, *Die Domkuppel in Florenz und die Kuppel der Peterskirche in Rom*, Berlin, 1887, Plate II (illustrating also the rather similar dome in Cremona) and A. Nardini Despotti Mospignotti, *Il Duomo di San Giovanni*, Florence, 1902.

[2] Frey, pp. 73–5; Toesca, pp. 18–21; Vasari, *op cit.*, Part II, 300–1 (Gaunt, pp. 274–5).

[3] See, for instance, M. E. Blake, *Ancient Roman construction in Italy* and *Roman construction in Italy*, Carnegie Institution of Washington, 1947 and 1959, *passim*; G. Lugli, *La tecnica edilizia romana*, Rome, 1957, I, 235–42.

[4] Lugli, *op. cit.*, pp. 663–93; William Macdonald, *The architecture of the Roman empire*, Yale University Press, 1965, *passim*; Licht, *op. cit.*

BRUNELLESCHI'S DOME OF S. MARIA DEL FIORE

temple of Minerva Medica (Plate XVI (b))—these ribs have a remarkable superficial resemblance to the ribs and horizontal arches of Brunelleschi's dome.[1] It is even possible that, in the fifteenth century, some had been left free standing by the fall of the concrete infill.[2] In general, though, these ribs became an integral part of the concrete mass and always appear so today. If, in terms of structural function, they have a counterpart in Florence, it is to be found in the inclined bands of vertical bricks of the herringbone bond.[3]

Looking further afield, however, it is possible to find fairly close precedents for most of the characteristics, some even during the Roman period. The empire was widespread and, in the Balkans and Asia Minor, brick rather than concrete, was commonly used for vaults.[4] Here, continuing probably a very ancient tradition seen for instance in the store rooms of the Ramesseum at Thebes (Plate XVII (a)),[5] there are examples of domes or vaults constructed without centering at Spalato, (Plate XVII(b)) Salonika, Aspendos and in the Theodosian walls of Constantinople.[6] None of these exhibits a herringbone bond, but each exhibits a variant of the principle of building up the vault by means of a succession of short arches pitched as flatly as possible. The tradition continued and is very widespread in both Byzantine and Islamic architecture, still being characteristic of the latter where brick has remained the normal material of construction and timber is scarce.[7] With the use of brick is associated also the adoption of a more uniform thickness for all vaults, and usually the need for a separate weathering skin. This skin took various forms but, as early as the eleventh century, the true double dome (with pointed profile) emerged in Persia.[8] By the early fourteenth century it had

[1] G. Giovannoni, *La sala termale della Villa Liciniana e le cupole romane*, Rome, 1904; Guido Caraffa, *La cupola della sala decagona degli Horti Liciniani*, Rome, 1944.

[2] This possibility is suggested by a number of drawings etchings, etc., dating from the early sixteenth to the mid-nineteenth centuries. I have found so far a total of eight such representations by (in order of date) G. B. del Porto, M. van Heemskerk, J. d. Ä. Frankaert, G. B. Piranesi (three), F. J. B. Kobell and E. Isabelle. The best is that by Kobell. This is reproduced in Richard Krautheimer, *Early Christian and Byzantine architecture*, Penguin Books, 1965, Plate 80A, and in my forthcoming book, *Developments in structural form*. For the last see Isabelle, *op. cit.*, Plate 23.

[3] I do not, however, consider that a drawing attributed to G. Guidetti (Uffizi, 1330) and showing a concrete semidome with two intersecting sets of inclined bands of vertical bricks embedded in it represents, as Sanpaolesi suggested, a Roman precedent for the Florentine bond. It seems much more likely, for various reasons, to be a rather ill-considered reinterpretation by the artist of the lozenge-shaped coffering of the semidomes of the Temple of Venus and Rome, in terms of Brunelleschi's system. Nor can I find anything relevant in the sketchbook of Guiliano da Sangallo (Vatican codice Barberiniano latino 4424) referred to in this connection by G. C. Mars, *Brickwork in Italy*, Chicago, 1925, 23.

[4] J. B. Ward-Perkins, "Notes on the structure and building methods of early Byzantine architecture" in *The Great Palace of the Byzantine Emperors. Second Report*, Walker Trust, Edinburgh, 1958, 52–104; William Macdonald, "Some implications of later Roman construction," *Journal of the Society of Architectural Historians*, XVII (1958), 2–8.

[5] The Egyptian process of construction is described in Somers Clarke and R. Englebach, *Ancient Egyptian masonry*, Oxford University Presss, 1930, 181–3.

[6] G. Niemann, *Der Palast Diokletians in Spalato*, Vienna, 1910, 75–6, and Plate XIII; E. Herbard and J. Zeiller, *Spalato, le palais de Dioclétien*, Paris, 1912, 89–93, and Plates X and XII; J. B. Ward-Perkins, *op. cit.*, 89–95; idem, "The aqueduct of Aspendos," *Papers of the British School at Rome*, 23 (1955) 115–23.

[7] See, for instance, André Godard, "Voutes Iraniennes: les cupoles," *Athàr-é Iran*, IV (1949) 259–325. To avoid possible confusion it seems desirable, however, to distinguish between different principles of centerless construction—breaking up the continuous horizontal rings into shorter elements that can be constructed independently in succession, pitching the bricks at a flatter angle than the radial one, obtaining adhesion to what has already been completed by the use of a gypsum of fast-setting cement mortar, and mechanically keying each brick or unit to the existing work. These principles are not mutually exclusive, but we are concerned here largely with the first. The method used at the cathedral of St John the Divine was based on the third. See also G. R. Collins, "The transfer of thin masonry vaulting from Spain to America," *Journal of the Society of Architectural Historians*, XXXVII (1968) 176–210. The well-known Italian system using terracotta tubes was based on the last. See, for instance, Paolo Verzone, "Le cupole di tubi fittili nel V e VI secolo in Italia," *Atti del I Congresso nazionale di storia dell'architettura*, Florence (1936), 7-11; and G. Angelis d'Ossat, Nuovi dati sulle volte costruite con vasi fittili', *Palladio*, 5 (1941), 241–51.

[8] David Stronach and T. C. Young, "Three Seljuk tomb towers," *Iran*, IV (1966), 1–20.

BRUNELLESCHI'S DOME OF S. MARIA DEL FIORE

reached a truly monumental form in structures like the mausoleum of Oljeitu at Sultaniya.[1] Here, as in Florence, the inner shell is stepped back at a number of points to reduce its thickness and weight and it is covered by a lighter outer shell of uniform thickness carried by the inner one by means of radial and circumferential ribs. Circumferential timber ties are usually built into the bases of these domes. Finally, there is at least one fairly close parallel to the circumferential iron chains set above "chains" of stone blocks cramped together in the western exedrae of the church of St. Sophia in Constantinople.[2] The chains, composed there of long iron bars, probably date from the mid-sixth century.

I cannot say how many of these latter precedents were known to Brunelleschi. At best, however, they could only have helped to point the way towards a possible solution of the problems posed by the decisions of 1367. Even a synthesis of known forms and techniques on an accustomed scale (as seen elsewhere in Brunelleschi's architecture) demands considerable imagination. If, for example, the dome of the mausoleum of Oljeitu was a second prototype alongside the baptistery dome, it would still have been a major achievment merely to combine the two forms in the manner seen in the cathedral without any change in scale (the mausoleum dome being, like all Persian domes, circular in plan) and to devise a method of construction that did not rely on centering for temporary support. To simultaneously almost double the scale would have introduced problems of an altogether new order.

That the final design and the full elaboration of the technique of erection were not arrived at without careful preparation and continued thought and experiment is shown both by what is known about preparatory studies and by the progressive evolution of the design as construction proceeded. Apart from the masonry models of the dome itself, Brunelleschi tried out the principle of vaulting without centering in a small cupola in a nearby church.[3] Later he made further models of the circumferential chains and other details and Gelli records that, for this purpose, he had a level site prepared large enough to work on the design at full scale.[4]

The way in which the design evolved shows that he must have acquired, perhaps partly through the study of Roman remains,[5] a keen awareness of the general pattern of stresses within a circular dome—of the tensile hoop stresses in the lower part, the corresponding compressive ones above, the way in which the latter made it possible to leave an open eye of any diameter at the top, and the outward thrusts which would be exerted at the base if the tensile hoop stresses led to radial cracking.[6]

---

[1] The best account of this development is in D. N. Wilbur, *The architecture of Islamic Iran*, Princeton, 1955, 61-7. For the mausoleum of Oljeitu see also André Godard, "The mausoleum of Oljeitu at Sultaniya" in A. U. Pope, *A survey of Persian art*, London, 1938-39, II, 1103-18. By the early fifteenth century development had proceeded to the point where the outer shell became bulbous, as in the Gur-i-Mir at Samarkand, inspired by perhaps earlier wooden double domes.

[2] R. L. Van Nice, *Saint Sophia in Istanbul, an architectural survey*, Washington (1966) Plate 21.

[3] Frey, pp. 87-8; Toesca, pp. 37; Vasari, *op. cit.*, Part II, p. 310.

[4] G. B. Gelli, *Vita d'artisti di Giovanni Battista Gelli* in *Archivio storico Italiano*, 5th ser., XVII (1896) 54. "Pippo . . . fecie fare uno ispianata in su renaio d'Arno circha d'un mezo miglio per ogni verso et quivi disegniata in terra questa cupola quanto ella haveva a esser grande appunto, et fatto un punto nel mezo disegnio tutte le pietre . . . che tiravano . . . et colta la misura della grandeza et qualita loro che ve n'era di varie sorte che incastravano l'una nell' altra, ne facie alchuni modegli di rape et mettendovi la misura comincio a farla lavorare di quella maniera a scharppelini et con quelle comincio a voltare detta cupola . . . ."

[5] Particularly from the characteristic patterns of radial cracks extending up and down from the springings of nearly all large domes or semidomes, but stopping short of any open eye at the top. Today these cracks are most apparent in some of the remains of the Baths of Trajan, having mostly been patched or covered with stucco elsewhere. They are equally pronounced in the Pantheon though, as can be seen in the hidden chambers just above the springing line of the dome. Alberto Terenzio, "La restauration du Pantheon de Rome," *La conservation des monuments d'art et d'histoire*, Paris, 1933, 280-5, gives a detailed drawing of these cracks as observed during a restoration (Plate XXVI).

[6] This insight may be contrasted with some of the ideas expressed much later when a proposal to add iron chains was being debated. See Alessandro Cecchini, "Due discorsi sopra la cupola di S. Maria del Fiore" in G. B. Nelli, *op. cit.*, pp. 77-103; and the summary of the debate in Poleni, *loc. cit.*

BRUNELLESCHI'S DOME OF S. MARIA DEL FIORE

The use of stronger materials at the base and of various chains were obviously intended to prevent such cracking and outward thrusts. The thickness of the inner shell was, almost certainly, determined so that it would contain within it the circular one analysed above.[1] Initially the outer shell seems to have been regarded structurally as little more than an unavoidable extra load on the inner one, but the replacement of the small barrel vaults by the horizontal arches of the 1426 amendment seems to have been directed at making it more self supporting. At the same time an attempt was made to bind the two shells together for mutual support in much the same way as parallel arch ribs and even ribs of different radii were interconnected in early iron bridges and with equally little understanding of the strains that might result.[2]

Brunelleschi's greatest problem would have been to match the tensile strengths of the structure to the stresses generated. In tackling it, the models and small-scale tests would have been of little assistance without some intuitive recognition that the stresses would increase in proportion to the radius of curvature even if geometrical similarity was maintained.[3] A dual approach seems to have been adopted. On the one hand the stresses were minimised by reductions in thickness and density in the upper parts and by the adoption of a pointed profile. On the other care was taken to provide as much strength as possible where it was most needed and a possibility was left open of adding additional hoop reinforcement here if, as construction advanced, it appeared to be necessary. This last, in the light of events, I take to be the significance of the reference, in 1420, to unexecuted iron chains above the timber ones. As has been shown, tensile stress would develop only slowly at the level of the one executed timber chain and an iron chain here could serve a useful purpose, if at all, only at a fairly late stage of construction. The individual timbers would have had considerable tensile strength—probably in the range 2–4 MN depending on the extent of defects and on moisture content—but the joints, if the present ones can be taken as a guide, would have been much weaker. It seems quite likely that, since the chain remained visible and accessible and there would have been no means either of determining its effective strength or calculating the strength that it ought to have had, it was simply kept under obseravtion as construction proceeded above it with the intention of adding the iron chain if and when signs of excessive strain appeared.[4] The tightness of the bolts would have been one rough index of the load transmitted. If this was the case, the chain might indeed be regarded as having served as "a tool of observation and test," as Prager suggested.[5]

Taking into account also those aspects of the total construction process which have not been considered here, the final impression is of a highly systematic and unusually comprehensive empirical

---

[1] Indirect confirmation of this is provided both by the stated reason for introducing the horizontal-arch reinforcements to the outer shell in 1426 and by the opening sentence of the passage in Alberti, *De re aedificatoria* already referred to: "*Angularem quoque testudinem sphericam modo per eius istius crassitudinem rectam sphericam interstruas, poteris attollere nullis armamentis.*"

[2] Here, again, there is an echo in this passage in Alberti, "*Sed istic nexura potissimum opus est; qua huius imbecillae partes partibus illius firmioribus arctissime illigentur,*" though there is no direct reference to inner and outer shells as in Leoni's translation (*loc. cit.*). It is possible that, with a less rigid interconnection, the inner one would not have cracked.

[3] Since Brunelleschi left no evidence of his ideas other than his buildings, it is possible only to speculate about any such recognition. Piero Sanpaolesi, "Ipotesi sulle conoscenze matematiche statiche e meccaniche del Brunelleschi," *Belle Arti* (1951), 25–54, discusses Brunelleschi's possible contacts with current theoretic thinking and R. J. Mainstone, "Structural theory and design before 1742," *Architectural Review*, CXLIII (1968), 303–10, gives a more general discussion.

[4] Prager, pp. 495–6. I do not wholly agree with Prager's reasoning.

[5] It is interesting to note, in this connection, that telltales of bronze and marble were inserted at numerous points in 1695 when considering whether the radial cracks called for the insertion of new chains and that, when such chains were proposed, it was stipulated that the bars coming from the foundry should be rigorously tested. No strength criteria were stated, however. Nor was any analysis of the structural action given to justify the proposals. (Guasti, *Cupola*, doc. 391, pp. 177–83.)

BRUNELLESCHI'S DOME OF S. MARIA DEL FIORE

approach which it is difficult to parallel until much more recent times.[1] Some reference has already been made to the impact made on contemporaries. Most obviously it was simply that of the demonstration of a new or reawakened human power. But Brunelleschi's achievement was so woven into the newly emerging patterns of Renaissance thought that the full impact was both wide and profound.[2]

DISCUSSION AND CORRESPONDENCE

In the discussion and in correspondence subsequently arising from the paper Mr. Mainstone amplified some of the points made about the method of construction, the uses of tensile reinforcement, the structural behaviour of the dome, and Brunelleschi's possible understanding of this.

The use of weighted ropes tied back behind the forward edge of the working face and hung over the bricks on this edge as they were set in place was suggested as an alternative method that might have been adopted to retain the bricks in place while the mortar hardened. This alternative to fixed or temporary formwork had been discussed by John Fitchen in his book *The construction of Gothic Cathedrals*, Oxford (1961), 180–7. Fitchen was referring there to the relatively short courses of the webs of Gothic ribbed vaults and based his discussion on a description of the method published in the *Journal of the Royal Institution of Great Britain* and elsewhere in 1831. It was not, however, the method that Mr. Mainstone had observed with Professor Fitchen when they had together inspected the construction of the nave vaults of the neo-Gothic cathedral of St. Peter and St. Paul in Washington, D.C., by courtesy of the resident architect Mr. Howard Trevillian, in 1966. There was no evidence at all for it having been used by Brunelleschi and it would be difficult to account for the upright bricks of the herringbone bond if it had been used. He thought that one of the procedures he had suggested, or something similar, was more likely. He admitted, nevertheless, in relation to his second suggestion, that it would be stretching the meaning of *ansas* in the passage by Alberti to read it as referring directly to the upright bricks. As had been pointed out, Alberti was not even referring directly to the cathedral dome. What seemed most significant about this passage was that it did fairly clearly describe at least the principle of making a limited use of light temporary formwork tied back to the part of the structure just completed.

Nothing was known of the form of the iron chains embedded in the dome of S. Maria del Fiore. The selection of documents published by Guasti did not even include any reference to their having actually been set in place. Only a careful study of the full records (which do include sufficiently full accounts of the purchase of materials to settle such matters) could show whether they were so set and give some indication of their likely size. Such study had been rendered difficult by the damage done to the records by the flood in 1966 and by their removal from Florence for repair. Professor Saalman would, however, be publishing in due course a further selection of those that now seemed important. It seemed to Mr. Mainstone more likely that these chains, if they were set in place, followed the sides of the octagon than that they were continuously curved like the wooden chain.

f they had been curved in this way, he would have expected the associated stone chains to have been similarly curved, which seemed most unlikely. Certainly they did not appear at the surface in

---

[1] The first parallel that comes to mind is Smeaton's reconstruction of the Eddystone Lighthouse in 1756–59. See John Smeaton, *A narrative of the building and a description of the construction of the Edystone Lighthouse with stone*, London, 1793 (second edition). Smeaton, of course, went further than Brunelleschi in conducting preliminary experiments akin to the sort of preparatory testing that might be undertaken today. Brunelleschi relied more on observation in the course of construction as a guide to future action (*perche nel murare la praticha insegnera quello che ss'ara a seguire*), but he seems to have done so in a much more deliberate manner than had been the case previously.

[2] The direct influence within the limited field of dome construction is, of course, to be seen in a long succession of double and multiple domes from that of St Peter's onwards. As a tailpiece, we may note that among the documents published by Guasti (*Cupola*, doc. 350, p. 130) is a free pass into the dome for Michelangelo and two associates, presumably in connection with his studies for the dome of St. Peter's.

BRUNELLESCHI'S DOME OF S. MARIA DEL FIORE

the places where one would expect to see them if they were curved. The timber chain was probably envisaged more in the same way as the arch-like reinforcements of the outer shell (c in Fig. 1), and its more-or-less circular plan form had the practical advantages of standardising and considerably simplifying the jointing of the 24 lengths of timber.

Still less was known about iron chains or any other kind of tensile reinforcement that might perhaps exist in the drum, since there is not even, among the published documents, any specification calling for such reinforcement. He would expect to find some cramping together of the masonry but nothing further. Certainly Brunelleschi had been in no position to call for any chains there. He clearly attempted, moreover, to contain the potential outward thrusts of the dome within its own masonry rather than to rely on any tensile reinforcement of the drum. A final answer must again await a study of the full records.

With adequate tensile reinforcement of the dome itself, there would, of course, have been no need for such reinforcement of the drum. Its presence would then, he thought, have been advantageous in allowing the dome some freedom to "breathe" without the sort of restraint that could lead to cracking.

His dismissal of earthquakes as a serious threat to the stability of the dome was based on a considered judgement of a good deal of evidence accumulated in recent years about the seismicity of different areas and the nature and structural effects of earthquakes. Florence had indeed experienced earthquakes, in 1695 for instance, but never, as far as was known, of the degree of severity that had been experienced further south. Also, as had been pointed out, it was lightning rather than earthquakes that the records mentioned as a cause of damage. As far as structural effects were concerned, a dome was, in itself, one of the forms best fitted to stand up to the momentary shifts in the effective strength and direction of gravity to which an earthquake might be likened. When collapses occurred as a result of these momentary shifts, they were usually attributable to failures of the supports rather than primary failures of the domes themselves. The risk of such failure would have been enhanced by the massive proportions given to the cathedral dome. But this was the only way in which he thought earthquake loading could have been important—and he had been considering the dome only and not the whole structure in this part of his structural assessment.

In support of his belief that the cracks at present seen in the dome did not develop until well after the completion of construction, he could offer no evidence beyond that given in the paper. It might be worth commenting briefly on the positions of the cracks, however. They all occurred over (and extended into) the very substantial diagonally aligned supports (Plate 2) rather than over the principal arches spanning between these supports. Since it was these arches that were most likely to spread, this might, at first sight, appear paradoxical. The explanation, he thought, was that when an arch spreads, other than by slipping of its voussoirs, it is over the haunches that the spandrels must crack to accommodate themselves to the spread, and not over the crown. What had probably happened here was that the cracks had been somewhat further displaced from the crowns of the principal arches on account of the weakening effects of the circular windows in the drum. Each crack passed, as might be expected, through one of these windows.

Finally, his remarks about Brunelleschi's "keen awareness of the general pattern of stresses within a circular dome" were not intended to imply the sort of understanding on which the analysis given in the Paper was based. Brunelleschi almost certainly had no clear abstract concept of a "stress" at all. Even Leonardo was still groping his way towards such a concept. The "awareness" referred to was more a general awareness of the two characteristic actions of a domical structure—one of which we called "compression" and the other "tension", but which Brunelleschi would have thought of in terms of their characteristic manifestations of crushing and pulling apart or tearing. It was a matter of understanding where, and roughly in what direction, each was likely to arise.

Plate XII

S. Maria del Fiore, Florence. (*Author*).

PLATE XIII

(a) Longitudinal Section.

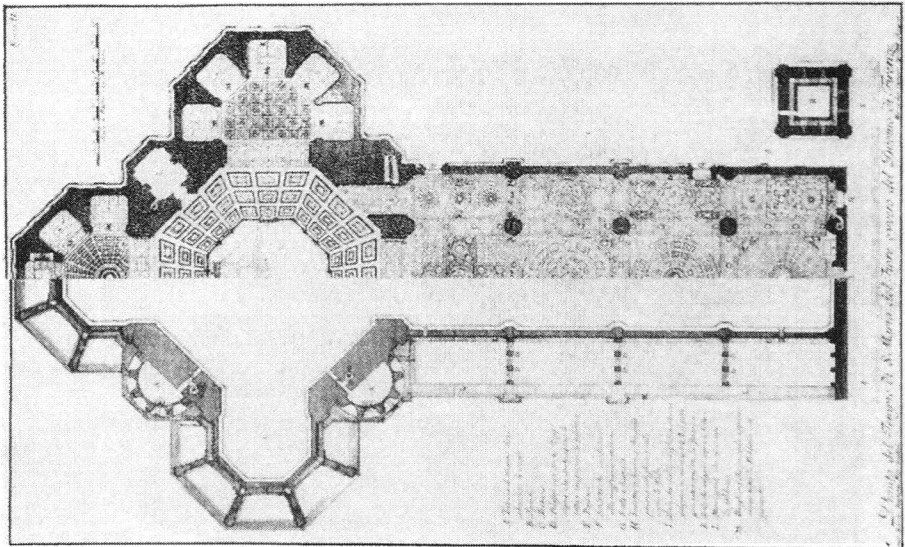

(b) Half plans at floor level (above) and at the level of the springings of the principal arches (below).

S. MARIA DEL FIORE (*Sgrilli*).

PLATE XIV

(a) View between the two shells from the point marked "b" in Fig. 1 looking into the passageway cut through one of the corner ribs.

(b) Detail of the crack running up one of the diagonal faces of the dome.

(c) View between the two shells of the dome looking upwards towards the point marked "b" in Fig. 1.

S. MARIA DEL FIORE (*Author*).

PLATE XV

Baptistery, Florence. Section (*Isabelle*).

PLATE XVI

(a) Baptistery, Florence: looking up between the two shells at the middle of one of the sides (*Author*).

(b) Temple of Minerva Medica, Rome.   Detail of the dome (*Author*).

PLATE XVII

(a) Ramasseum, Thebes: brick vault of a storeroom (*Author*).

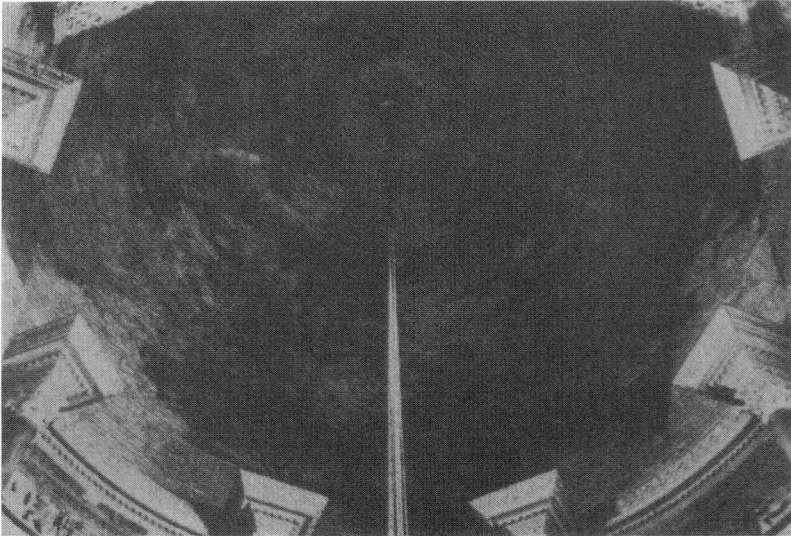

(b) Mausoleum of Diocletian, Spalato: detail of the dome (*Author*).

16

A comment on the function of the upper flying buttress in French Gothic architecture

John F. Fitchen

Agood deal has been written, over the years, about the Gothic structural system [1], culminating in the controversy that arose in the 1930's over the interpretation of the play of forces in the vault [2]. What have been given less consideration are the nature and functions of the flying buttress, that unique and distinctive feature of the mature ribbed vault system. Students of medieval vaulting, to be sure, understand the essential function of the flying buttress: it acts as

---

1. Because of the complexity of the subject, much that has been written on medieval construction, and particularly the Gothic structural system of ribbed vaulting, is confusing or incomplete or downright erroneous. Some of the books are mere classifications of different kinds of vaults, or historical chronologies of vaulting types and practices, or attempts to establish the priority of one country's technical innovation over that of another. Much of the more recent writing deals with the controversy over the function of the ribs in Gothic vaulting.

For scope of coverage and thoroughness of understanding of the totality of the building problem throughout the Middle Ages, however, VIOLLET-LE-DUC's ten volume dictionary is still unequalled. (E. VIOLLET-LE-DUC, *Dictionnaire Raisonné de l'Architecture Française du XIe au XVIe siècle*, Paris, Librairies-Imprimeries Réunies, 1854-1868, 10 vol.) In English, undoubtedly the most accurate observer and reporter was REV. ROBERT WILLIS, of whose writings the highly technical article on medieval vault construction is certainly the most informed and trustworthy account of the nature and the design of various English vaults, gained from first-hand investigation. (ROBERT WILLIS, *On the Construction of the Vaults of the Middle Ages*, in: *Transactions of the Royal Institute of British Architects*, vol. I, part 2, 1842, 1-46. There is a 1910 reprint of this article.) WATSON's monograph on the Glasgow Cathedral's double choir is valuable as an account of the actual erectional procedures and changes in a complex ribbed vault structure, documented by the unequivocal evidence of the rib moldings. (THOMAS LENNOX WATSON, *The Double Choir of Glasgow Cathedral: A Study of Rib Vaulting*, Glasgow, 1901.) The first five chapters of MOORE's book on Gothic architecture, second edition, although confined to the French Gothic of the Ile de France area, is an original and convincing exposition of Gothic construction at its purest, where the integration between structure and esthetics is well understood and effectively set forth, both pictorially and in the text. (CHARLES HERBERT MOORE, *Development and Character of Gothic Architecture*, 2nd ed., New York, Macmillan, 1906.) KUBLER's illuminating article is unique in that it discloses the actual intentions of the Gothic designer, through chapters written by a successful Spanish architect of the XVI Century on the structural computations and methods of construction for Gothic ribbed vaults. (GEORGE KUBLER, *A Late Gothic Computation of Rib Vault Thrusts*, "Gazette des Beaux-Arts," series 6, vol. XXVI, Jul.-Dec. 1944, pp. 135-148.)

2. *Vide*, e.g., the collection of articles published in 1939 by the Centre International des Instituts de Recherche, as *Recherche #1: Le Problème de l'Ogive*. KUBLER, *Op. cit.*, gives an excellent review of the arguments and counter-arguments in this controversy.

70   GAZETTE DES BEAUX-ARTS

a diagonal strut, receiving the thrusts of the vault at the point where they are collected following the lines of the ribs, and transferring them across the space above the side-aisle roof, to the deeply projecting buttress which grounds the thrusts. It is therefore recognized that the flying buttress bridges the side-aisle gap between the point at which a group of vault thrusts are concentrated, on the one hand, and the point outside the building, on the other hand, where these pressures can be taken care of [3].

However, in the fully developed Gothic structural system there are two tiers (sometimes even three) of flying buttresses. Anyone who is acquainted with even the most rudimentary knowledge of statics needs no more than a quick glance at the transverse section of such a building to see that only the lower of the two tiers of flying buttresses is engaged in transferring the thrusts of the high nave vault. In the developed Gothic system this lower strut is set at a very definite level: the intrados of the flying buttress arch where it abuts the clerestory wall is at the level of the top bed of the tas-de-charge, the highest of the corbel stones above the spring of the transverse and diagonal ribs (A in fig. 1). The upper of the two flying buttresses is set too high to receive any of the vault thrust, a fact which many keen observers have noted.

The presence of the upper tier of struts has been accounted for in three ways: 1) on esthetic grounds, 2) as a rain-water conduit, and 3) as a structural compromise. The claims and the degree of justification for each of these explanations need to be reviewed.

The explanation on esthetic grounds argues that the functional (lower) flying buttresses have to be set too low, for structural necessity, to appear effective in the design as seen from the exterior, and hence the medieval builders added a higher tier, for looks. It was even noted that sometimes (as disclosed in Reims Cathedral when the roof burned off during the World War I bombardment) thin membrane walls had been constructed transversely above the vault [4], so that the upper flying buttresses on one side merely pushed against their mates on the opposite side of the building.

This interpretation of the presence of the upper tier of flying buttresses is totally at variance with the articulated interrelationships of Gothic structure. Gothic structure is characterized by the essential nature of each member in the structural system, by the integration of the structure in all of its parts, and by the leanness if not downright attenuation of each member, where each had its job to do and was located and sized accordingly, and no duplication was admitted just " for looks. "

---

3. In many cases flying buttresses were added to earlier Romanesque naves and choirs when the vault thrusts came to push apart the clerestory walls: they were resorted to as more or less clumsy shores to forestall further movement. In mature Gothic construction, however, the flying buttresses were an integral part of the design from the start, and it is about these that this article is concerned.

4. *Vide*, KIMBALL and EDGELL: *A History of Architecture*, 2nd ed., New York, Harper, 1918, p. 287, fig 144. for a photographic view of these membrane walls from above.

UPPER FLYING BUTTRESS 71

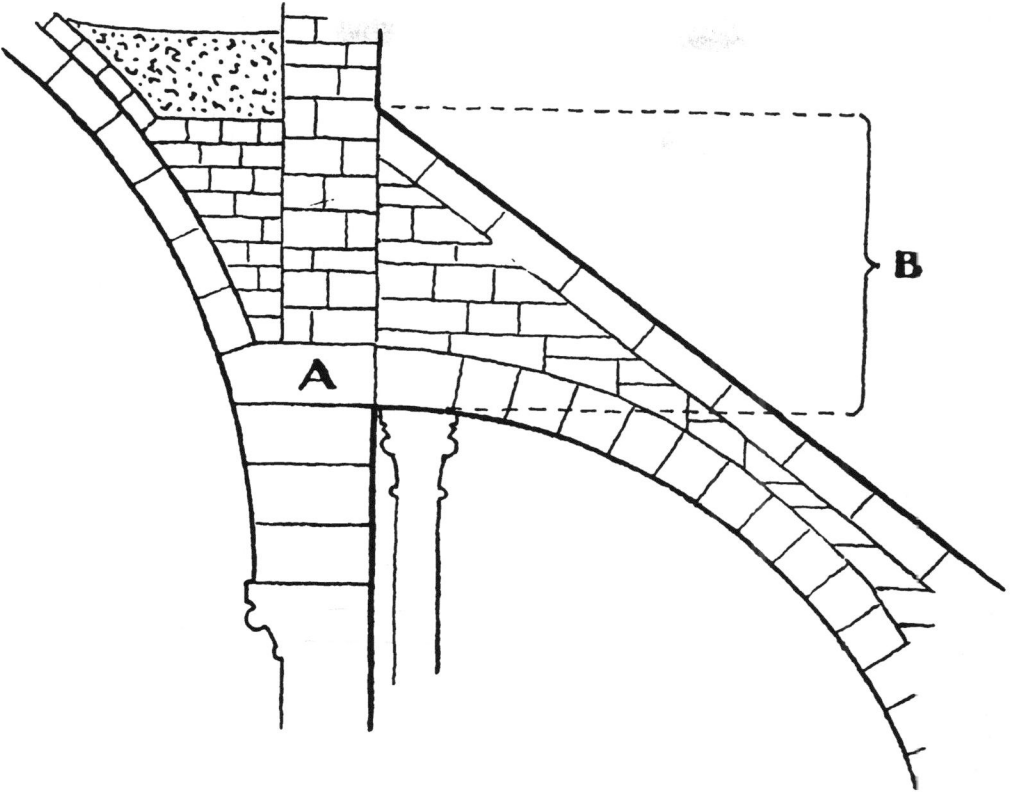

FIG. 1. — See page 70 of the text.

Actually, one of the greatest achievements of the Gothic builders was the masterful way in which they coordinated the solution of both their esthetic and their structural problems. The decoration is never an afterthought, never applied as an independent veneer or mere superfluity; it is always an integral part of the building's fabric, essential to the structure. Thus the stone tracery of a 40-foot diameter window, whose size was dictated by the desire for maximum light, had to support the considerable weight of its mosaic of stained glass, on the one hand, and on the other, had to resist the considerable lateral pressure of air, either as direct wind or as suction on the lee side of the building, during storms. Again, the highly ornamental stone-canopied statues atop the buttresses were useful in weighting the masonry from above, and adding their increment of stability to the buttresses' mass.

Not only is the decoration structural, but conversely, the Gothic structure is decorative. More than that, it reveals its every feature, its every function, in the separation and articulation of the membering. Within, the structural pier is no longer a simple square or cylindrical member but a decorative bundle of differentiated supports, individually expressive of their function and collectively contributing to the sense of upwardness. On the exterior, the flying buttresses themselves relieve the flat continuity of the building's envelope by their open-work silhouette of tense upward leaps. And yet, though all is revealed to the attentive and informed eye, the ensemble effect of this astoundingly complex structure is one of straightforward clarity which the layman can grasp and respond to. Each part is subordinated to the whole, and is related so felicitously and appropriately in size and position to every other part, that one is conscious of an all-pervading unity, both of purpose and of effect.

So evident is the medieval builder's integrity of structure throughout, so undeniable his insistence on integral function, that serious students of this architecture have sought to account for the role of the upper flying buttress in other terms than just for looks. This need for finding a more compatible explanation was apparently satisfied by a practical consideration; namely, provision for channelizing the discharge of rain-water which fell upon the nave roof.

In his articles on eaves, troughs and water conduits, Viollet-le-Duc presents the conditions and the means, both pictorially and in the text, by which the medieval builders sought to conduct to the ground whatever rain-water fell on the roofs[5]. Formerly, in the Romanesque period, there was no gutter at the eaves of the nave roof, so that the rain which struck against the entire area of the high roof tumbled directly upon the side-aisle roofs below. This created damage of various sorts and the need for constant repair. The medieval builders came to provide a masonry gutter at the top of the nave walls, but at first the water was discharged from this only through spouts—the gargoyles—at intervals, onto the lower, side-aisle roofs. After the new Gothic system of construction brought about the essential use of the flying buttress, however, the idea was conceived of utilizing the sloping top of this feature as an open channel to carry the rain-water swiftly but unspectacularly down and away from the clerestory walls[6]. In order that there might be no damaging fall of water, the top of the upper flying buttress—so the argument runs—came to be set at the gutter level, to take its discharge directly.

At first glance, this was an attractive theory for explaining the high setting of the upper flying buttress. The systematic provision for thus collecting, channeling, distributing, and discharging the rain-water seemed to be strikingly in keeping with the systematic organization of the Gothic structural system itself. Choisy states the

---

5. VIOLLET-LE-DUC, *Dictionnaire Raisonné de l'Architecture Française*, article *Cheneau*, III, 219 ff., and, especially, article *Conduite*, III, 505 ff.

6. *Vide*, VIOLLET-LE-DUC, *Op. cit.*, III, 507, fig. 4 B.

UPPER FLYING BUTTRESS   73

conditions as clearly as any[7], and shows how the form and setting of the upper-most slope of the flying buttress units may have been modified with this in mind.

But a considerable number of developed Gothic buildings do not conform to this theory[8]. Surely this practical purpose is not sufficient to account for the existence of the uppermost buttress, even though it be admitted that its setting may have been influenced by this consideration. Even Choisy culminates his series of examples with that of the XIV Century system of Cologne whose two parallel struts, united by tracery to the flying buttress arch in each tier, constitute a scheme over which he exclaims: " It is the most complete realization of the rigid shore[9]. " In other words, that writer winds up his account of the practical consideration of convenient drainage by emphasizing the structural function, without accounting for this emphasis in the case of the upper tier of flying buttresses.

Choisy, along with others, also claims a structural reason for the two tiers of flying buttresses. He mentions the fact that it was difficult to determine exactly at what level the line of pressure from the vaults should be met, but that the medieval builders ordinarily got around this uncertainty by spreading the amount of masonry impinging against the clerestory wall[10]. This spreading could come about naturally due to the strut's divergent boundaries: an arc below, and a straight slope above (fig. 1 at B). But Choisy thinks that the medieval builders looked upon this spread as insufficient in the case of the great cathedrals[11]. He says they considered the vaulting conoid, within the clerestory wall, as a solid which the pressure of the ribs tried to drive outward: in which case the upper and lower flying buttresses gave stability to this block by backing it up, above and below, in the fashion of a yoke.

But again, this explanation is unsatisfactory when the actual situation is analyzed. To be sure, the lower part of the vaulting conoid is ordinarily solid, filled with masonry so as to prevent the excessively thin shell of the vault proper from deforming. This solid filling, however, was carried up only high enough to take care of any bursting action at the haunch, a level considerably below the point at which the upper flying buttress pressed against the clerestory wall in the great majority of examples. Choisy's theory of the yoke action of the two tiers is untenable, then, because the lower end of the yoke normally receives the entire vault thrust where it is concentrated, while the upper end is well above the solid part of the vaulting conoid and, of course, far above the line of pressure of any vault thrusts.

Choisy's statement that it is difficult to determine the precise point at which the vault thrusts are concentrated, is as true today as it was in medieval times. But

7. Auguste Choisy, *Histoire de l'architecture*, Paris, Librairie Georges Baranger (1899), II, pp. 336-337, 308-310.
8. In the nave of Amiens Cathedral, the top of the upper flying buttress meets the clerestory wall some 2¼ meters (about 7½ feet) below the level of the gutter; in Paris, 2¼ m; in Reims, 3 m. The churches of Mantes, Nogent-le-Roi, and the north side of Bourges Cathedral are other examples.
9. Choisy, *Op. cit*, p. 310, line 9.
10. Choisy, *Op. cit.*, p. 307, lines 4-9.
11. Choisy, *Op. cit.*, p. 307, lines 10-18.

that these builders had an uncannily accurate sense of what forces were at work, where their effect was critical, and how to handle them, seems abundantly evident. For example, the vault over the choir at Notre-Dame, Paris, has a span of over forty feet, yet the thickness of this stone shell is only four inches [12]. We could scarcely hope to do better than this today, in spite of elaborate theories and formulae, various means for accurately testing the strength of materials, and complete tables of weights, coefficients of expansion, and other physical properties, none of which data were at the disposal of the medieval builders. Realizing, as they must have, the difficulty of ascertaining the exact point at which the vault thrusts were concentrated, they gradually evolved a combination of techniques which took care of this ticklish problem conclusively and yet with a considerable factor of safety to cover occasional unforeseen eventualities.

Two of the most significant elements in this combination of techniques were the solid filling of the vaulting conoid in the lower part of its funnel, and the tas-de-charge device. The latter, like so many of the features of medieval building, came about through a plurality of causes. First, the transverse and the diagonal ribs of the mature Gothic construction needed to come so close together at their common springing that some of their moldings were mutually absorbed, and there was no room for each rib to be a separate piece of stone. Second, by combining these ribs into a single stone at each of the courses immediately above the springing, a far more stable foundation was provided from which the individual separate ribs could rise. Furthermore, these single-stone courses had level beds and hence no thrust, even though each projected out over the one below in following the curve of the rib arches. The corbelling which resulted from this inward shelving brought about a significant reduction in the span of the vault arches. Thus it made the process of erection somewhat less difficult, probably reduced slightly the thrust of the vault, and, most of all, managed to confine more closely the area in which the arch action of the vault operated, by eliminating the lower part of the vault from that arch action. Lateral confinement was achieved by the warping of the lower portions of the cross vaults, a feature known as the plowshare twist, which Moore has explained and accounted for so clearly [13]. It is evident, then, that the solid filling of the lower part of the conoid, above, the corbelled courses of the tas-de-charge, below, and the plowshare twists on either side, circumscribed the vault thrust and channelized it, as it were, to that sharply limited area of the clerestory wall at which the slender flying buttress could be sure to receive it [14].

---

12. Viollet-le-Duc. *Op. cit.*, article *Construction*, IV. 108. footnote 1.

13. Charles Herbert Moore, *Development and Character of Gothic Architecture*, 2nd ed., New York, 1906, pp. 130-133.

14. Perhaps it would be more accurate to say that the medieval builders became so aware of the nature, the direction, and the application of the high vault thrusts that they were able to pare down the mass of the solid portion of the vaulting conoid to its most irreducible leanness, and hence to its maximum lightness. In this respect a modern counterpart in steel construction is the truss, whose relatively slender web members concentrate and channelize those stresses that the solid-webbed plate girder takes care of more heavily and much less efficiently.

These devices and adjustments make it clear that the medieval builders evolved a very precise scheme for placing the inner, clerestory end of the lower flying buttress so that it could function with sure and maximum effectiveness in receiving the vault thrust and transferring it to the buttress proper. The very measure of this structural effectiveness in the lower flying buttress makes the occurrence of the upper flying buttress, in mature Gothic construction, a matter for the most careful investigation and explanation.

Most serious writers on the structural system of Gothic churches have been preoccupied with the problem of the vaulting and its methods of support: with the means by which the medieval builders spanned space in stone above their lofty interiors. This emphasis upon the thrust and equilibrium of arch-generated forces is certainly the most all-engrossing problem of Gothic structure due to its compound, interrelated complexities and its technical difficulties. But in their understandable preoccupation with the basic problem of the vaulting, well-informed observers and writers alike seem to have lost sight of the fact that even here the complexity is compounded by the fact that there are other active forces at work within the structure; that the arch-generated pressures are not the only ones that have to be provided for if the building is to be kept from collapse. Chief among these other forces affecting the design of the structure is that of wind.

Apparently, with the exception of Viollet le-Duc, no writer on medieval architecture has made any but the most perfunctory remarks about the effect of wind on the aspiring structures of the Middle Ages [15]. Viollet-le-Duc's comments on the occurrence of wind pressure and what the medieval designers did to counteract it are confined almost exclusively to the stone gable-ends of timber roofs [16], on the one hand, and more particularly to the slender wooden spires above the crossing of transept and nave [17], on the other. Neither of these has to do with the matter under discussion here.

Certainly no thorough or systematic analysis has been undertaken for the purpose of showing what the Gothic builders did to secure their increasingly light, increasingly lofty cathedrals against lateral swaying and collapse due to the force of wind. Yet both their loftiness and their lightness of construction made these buildings particularly vulnerable to the action of wind. And it must be remembered that

15. E.g., Pol Abraham, *Viollet-le-Duc et le Rationalisme Médiéval*, Paris, Vincent Fréal, 1934, mentions the wind action only once, in a two-line note, p. 19 To be sure, Prof. Conant, who has done so much painstaking and original work on medieval architecture in many countries, and whose writings and drawings are both illuminating and explicit, published a revealing statement a number of years ago: " After the first construction at Chartres it must have been felt, and rightly, that the slenderness of the piers between the windows compromised the stability of the wall above the windows, particularly since so huge and steep a roof is subject to enormous pressures from the wind. Consequently an upper range of flying buttress arches was added to the buildings. " (Kenneth John Conant, *Observations on the Vaulting Problems of the Period 1088-1211*, " Gazette des Beaux-Arts, " series 6, vol. 26, Jul.-Dec. 1944, 133-134). Apparently, however, Prof. Conant has not developed this two-sentence observation, significant though it is, elsewhere in his writing.
16. Viollet-le-Duc, *Op. cit., Pignon*, VII, pp. 130-133.
17. *Ibid.*, article *Flèche*, V, pp. 446-453 ff.

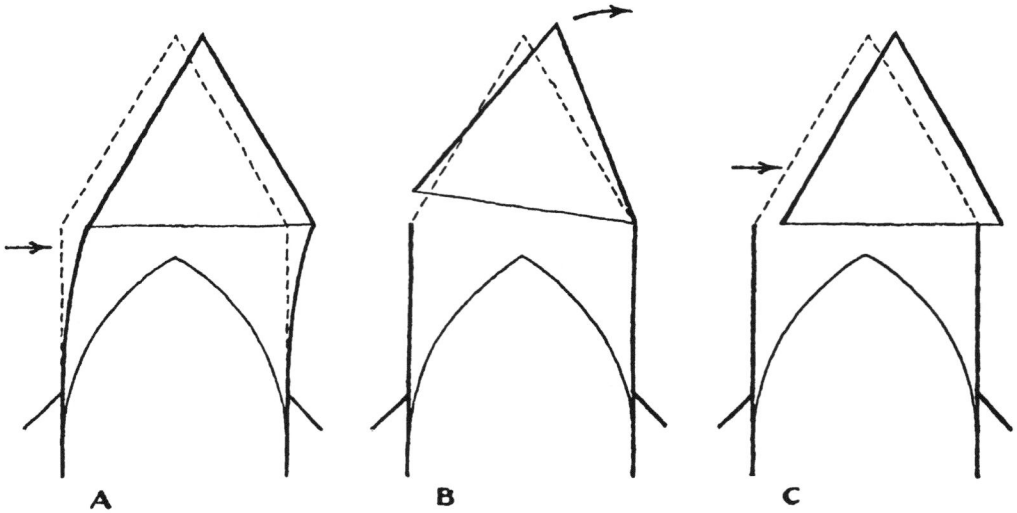

FIG. 2. — See page 77 of the text.

wind, unlike the static forces of a building's structure, is variable not only in force but, even more seriously, in the direction from which that force is applied.

That the consideration of wind action is no idle one is evident in the fact that the great Gothic churches towered above all the other buildings of the town. Indeed, it will be remembered that the Archbishop of Reims took delight in climbing to the eaves of his cathedral to sweep the surrounding town with a glass, with dire penalties for the owner of any man-made structure which projected above the level of the cathedral parapet. Of course, most of the houses and other buildings reached to only a fraction of this height, so that physically the cathedral dominated the town and its skyline from every approach, both at a distance and within the town itself. In such cities as Chartres and Reims this upward-jutting mass of the cathedral, as it soars high above the tree-tops and all the other buildings, is strikingly evident even today [18]. In medieval times, as now, the towering walls and the high-pitched roof had to withstand the unbroken sweep of winds buffeting directly against it, or sucking away from it on the lee side, or eddying in a twisting and wracking motion around rooftop, apse, and transept. That the intricate equilibrium of these glass-walled buildings has not been overthrown or permanently disarranged by the wind's violence over seven centuries is eloquent testimony to the fact that the medieval builders were well aware of the problem and took positive steps to meet it.

---

18. In this connection, see the revealing photograph of Amiens Cathedral, in : STURGIS AND FROTHINGHAM, *History of Architecture*, New York, 1916, III, 56, fig. 63.

UPPER FLYING BUTTRESS   77

There are numerous accounts of various kinds of destruction suffered by medieval buildings. One reason for collapse was lack of repair and general neglect of the fabric of the building. The ravages of rain-water accounted for another serious source of trouble. Another was settlement, due either to faulty construction or to the superimposition of loftier structures on former foundations adequate only for the earlier, lower buildings. Again, with the Gothic builders striving for ever greater height with more and more glass area, the consequent attenuation of the supporting piers put too much weight on too slender supports, and the building collapsed, as at Beauvais. Lightning has accounted for repeated damage, and even total destruction, in medieval churches throughout the intervening centuries, although the builders sought to protect the timbers of roof and spire by a thick sheath of lead.

Any references to destruction by windstorm alone are extremely rare. Robert Willis, who combined exhaustive scholarship with acute and perceptive observation, has listed some of the destruction caused by windstorm in the case of English churches [19]. The brevity of this list, together with the almost complete lack of instances in the generally taller churches of France, testify to the fact that the medieval builders were well aware of the problem and that their provisions for meeting it more than kept pace with the increasing demands of greater height and slenderer construction. We can only surmise that these builders were early aware of danger from this source, and that their techniques for dealing with the problem developed along with their solutions to other structural problems and were never outstripped by the accelerating demands of a dynamic building program.

The problem of providing resistance to wind became more and more serious and acute as the Gothic buildings became loftier, with slenderer supports and higher-pitched roofs. When a wind blows against the side of a Gothic church, one or more of three situations tends to develop, due to the combination of direct pressure on the windward side and suction on the lee side (fig. 2): Either 1) the clerestory walls tend to bend with the wind (diag. A), especially above the level at which the vault-required flying buttresses furnish their abutment; or 2) the triangular prism of the roof tends to rotate on the axis of the top of the lee wall (diag. B); or 3) the whole roof tends to slide off the wall tops to leeward (diag. C). Of course, all three tendencies are to some extent in operation at once and have to be provided against. However, it is convenient, in understanding the possible consequences of these actions, to separate them.

In each of these situations the top of the clerestory wall is subject to lateral pressure. The Gothic tendency to make this clerestory wall as thin as possible,

19. ROBERT WILLIS, *The Architectural History of Chichester Cathedral*, Chichester, 1861, I-XXIV, " Introductory Essay on the Fall of the Tower and Spire, February 21st, 1861. " The list of examples is in connection with his very accurately observed and exciting eye-witness account of the central tower's collapse during a gale.

coupled with its height above the vault-resisting (lower) flying buttress, greatly diminished its ability to resist the overturning force of the wind. Hence the condition, stresswise, at the top of the lee wall was similar, under wind action, to the constant pressure of the vault, lower down: in both cases a lateral thrust, acting high up at right angles to a very thin strip of masonry, had to be met and grounded. This was provided, in both cases, by flying buttresses: the lower tier met the vault thrust, the upper tier took care of the wind load, the pressure exerted by the wind against the upper portion of clerestory wall and the high-pitched roof.

But there was a further difficulty. The masonry of Gothic structures is based throughout on the arch principle, and arches, by their fundamental nature, exert a thrust. This is true of all parts of an arch; and hence the half arches which provided for the bridging of space in the case of the flying buttresses created a crown thrust at the point at which they impinged against the clerestory wall. This counter thrust at the crown was of no special concern where it was met by a steady pressure in the opposite direction, as in the case of the vault-resisting (lower) flying buttress. But where a flying buttress functioned as a wind brace, this crown thrust was of major concern.

On still days, when there was no lateral pressure of wind against wall and roof, there was no need for the upper flying buttress; but it had to be there to function whenever the wind did blow. The critical situation was in the event of storms. For when the wind was blowing, the crown thrust on the windward side of the building added its structural pressure to the live load of the wind against roof and wall.

This condition accounts for the use of the transverse membrane wall above the vault in the mature Gothic construction of Reims Cathedral [20]. This was a device for neutralizing the opposing crown thrusts of the upper tier of flying buttresses, and at the same time tying the top of one thin and lofty clerestory wall to the other, at intervals, for general stability.

It should be remembered that the steep-pitched Gothic roof was perched high on those narrow, continuous strips of masonry that constituted the tops of clerestory walls, themselves underpinned only by the slender stilt-like piers of the nave continuing upward between the very tall clerestory windows. In a structure of this sort, where the great majority of the lofty wall was window, and only a minimum skeleton of stonework existed, the only way, in masonry, to provide a sufficiently stable support for the even loftier roof was to oppose the crown thrust of the upper tier flying buttress on one side to that of the other, opposite. At Reims the transverse membrane walls furnished this cross-bracing in stone, thereby allowing the two thin wall-tops to act together in a sort of skeletonized box construction.

Sometimes the clerestory walls were thick enough to abut the crown thrusts of the upper tier of flying buttresses sufficiently by themselves. In other cases the

---

20. *Vide*, note 4, above.

upper tier was set a number of feet below the gutter level and hence the inward push of these upper flying buttresses was much less dangerous. In any case, the heavy tie-beams of the timber roof at the clerestory wall-tops could be counted on to provide the strutting effect when the transverse membrane walls were absent (always assuming that the wind was never strong enough to prevent the roof from bearing solidly on both wall-tops), and to augment their function when the membrane walls were present [21].

The loftiness of the structure, alone, made one or both of these cross-strutting arrangements advisable even on windless days, in view of the thinness of both the vault web and the clerestory walls. In stormy weather, moreover, it was absolutely essential to stiffen the superstructure by linking the resistance of both clerestory walls against lateral wind pressure. Separately, these two narrow strips of masonry were much too feeble in themselves to withstand the powerful lateral pressures they were called upon to sustain under wind action. It was imperative to augment their individual resistances through united action, giving both increased strength and stiffness. The cross-bracing provided this kind of consolidation in the superstructure. Bracing was accomplished, with typical Gothic rationality and economy of means, by a skeleton construction consisting of 1) the transverse membrane walls above the vault and/or the heavy timber cross-ties, which served as compression members between the clerestory wall-tops at intervals, 2) the opposing diagonal struts of the upper flying buttresses outside the clerestory walls, and 3) the thin yet deeply salient buttresses proper, the only really stable elements in the whole structural complex,

---

21. No one, apparently, has so far taken the trouble to check on the effect of temperature changes upon the width of those cracks which are so often observed in medieval church vaulting and have afforded the basis for so much comment and speculation by architectural historians in recent years. It would be interesting to know to what extent these cracks open and close as a result of winter cold and summer heat, respectively. The matter, if investigated, might conceivably make a modest contribution toward the clarification of just what does take place in the action of Gothic vaults. In any case, it is certain that both wood and stone undergo expansion and contraction due to temperature changes, and that the coefficient of expansion is not the same for wood as for stone masonry.

What makes the point a pertinent one here is the fact that the heaviest of all the wood members of any Gothic roof is the principal tie-beam which, in French Gothic cathedrals at least, spans directly across from the top of one clerestory wall to the other. As this member expands and contracts it must unquestionably push apart and draw together, at different times, the clerestory wall-tops on which it bears. Since the coefficient of oak, longitudinally, is .00027 per unit length per hundred degrees of temperature Fahrenheit, it can readily be understood that this movement of the clerestory wall-tops, now apart now toward each other, can regularly amount to an appreciable fraction of an inch. Actually, for a seventy degree variation in temperature (say from 20° to 90°), the change in length of the 47-foot tie-beams at Reims amounts to just over a tenth of an inch.

This movement might well account for the oft-recurring fracture between the transverse vault and the clerestory wall; indeed, the medieval builders may well have made the formeret of the vault as independent of the clerestory wall as they did because they either sensed, or had actually observed, this movement. The effect of such movement could also manifest itself elsewhere in the vault, particularly where the vault proper adjoins the solid mass of the vaulting conoid's lower portion.

In this connection, see ALBAN D. R. CAROE, *Old Churches and Modern Craftsmanship*, London, Oxford University Press, 1949. CAROE is undoubtedly the modern writer most conversant with the conditions and the practices of medieval materials and workmanship, at least in England, and his book is a mine of practical information of the most experienced and knowledgeable sort. With regard to temperature changes, he says (p. 9), " Such changes have effects which, though chiefly seasonal, are also to a certain extent cumulative."

which acted as anchors on either side of the building. Together, these three groups of elements furnished a slender and lofty but adequately rigid superstructure which provided the even loftier mass of the roof with a substantial framework by way of foundation.

It has long been recognized that the two most persistent and all-pervading motivations affecting the structure of medieval church architecture were the desire for maximum height and the desire for maximum light. The ceaseless search of the medieval builders for more and more light led to larger and larger window areas and consequently less and less supporting masonry: the structure had to become skeletonized. The equally compelling and avid search for means by which lofty height could be achieved necessitated the most accurate and finished masonry throughout, on the one hand, as well as a skeleton construction precisely and rigorously designed to take care of both the intensified and the additional stresses which that increased height imposed.

Like all other structural parts of the building, the Gothic roof and its supports were affected by these twin desires for maximum height and light. It has been remarked above that, as the naves became more and more lofty, the window areas increased at the expense of the masonry supports [22]. Along with the nave's increased height and the reduction in the amount of supporting stonework, there was a corresponding increase in the height of the roof itself. This additional height, which was the consequence of making the slope of the roof of ever steeper pitch, intensified two serious problems. First, by making the triangular cross-section of the roof larger in area, it increased the amount of timber-work and hence the weight of the roof itself, as well as adding greatly to the extent of the weather surfaces which were covered with lead. Second, the more nearly vertical planes of the roof slopes received far greater wind loads, as the pitch increased from less than 45 degrees in Romanesque times to 60 or even 65 degrees in the later Gothic era.

Viollet-le-Duc accounts for the steep slope of French Gothic roofs by the following sequence of related considerations. In the desire for maximum light, the diameter of the wall piers separating the clerestory windows was reduced to a minimum, yet their height was increased. Consequently, the strip of masonry which constituted the top of the clerestory wall was made very narrow, both to accommodate its thickness to the diameter of the supports below, and to reduce the load on these supports. But the thickness of this wall-top had to furnish most of the width of the rather generous eaves gutter, together with its protecting parapet, so that very little was left as a masonry rim on which the roof itself could be poised. This was too narrow a pavement to receive the doubled rafter ends until these rafters were

---

22. In Beauvais Cathedral, for instance, even the " blind story " was glazed to provide additional light in the interior. The extreme attenuation of the supporting masonry throughout, caused the roof over the XIII Century choir to fall shortly after it was built. The medieval builders erected it again, however, with double the original number of piers, and it has continued to stand ever since.

tilted up at a much steeper angle than before—from 40 to 50, to 60 and even 65 degrees [23].

This explanation is doubtless true enough, though such steep slopes would unquestionably not have been arrived at had there not also been an impelling desire for the esthetic effect of aspiring verticality, that religion-generated characteristic that permeated the spirit of Gothic church architecture everywhere. It is necessary to make the point of esthetic compulsion, as well as that of structural convenience, because the latter, by itself, engendered serious complications. Some of these complications have already been mentioned. Three of them—the increased weight of the roof, its tendency to rotate under wind action, and the lateral pressure on clerestory walls and roof alike—need further analysis and explanation.

No useful data on the weight of medieval roofs appear to have been published. In fact, there are scarcely any references whatsoever to this matter. Price [24], a most thorough and perceptive observer, gives an estimate for the weight of the oak timbering in all the roofs making up the ensemble of Salisbury Cathedral, in England, as 2641 tons. Massé states that the lead alone on the roof of the nave of Chartres Cathedral weighed 458,164 livres, or about 247 tons [25]. Moles records that the lead covering the roofs of Paris Cathedral weighs about 231 ½ tons [26]. Such figures as these are not as helpful as one could wish in ascertaining the total weight of the nave roof, and hence the load which the tops of the clerestory walls are called upon to support.

In lieu of trustworthy published data, however, a fair approximation of the weight of a specific nave roof can be arrived at by scaling the individual timbers in the accurate drawings of Viollet-le-Duc, assuming a unit weight for the kind of wood employed, and computing the combined weight of timber work and lead covering. Such a procedure [27], for the nave roof of Reims Cathedral, gives a figure of about 2 ½ tons per lineal foot, or over 57 tons per 23-foot bay, to be supported half on one clerestory wall, half on the other.

It has been noted above that the continuous band of masonry at the top of the clerestory wall was made as narrow as possible. The consequent reduction in weight of masonry on the piers below was partially cancelled out by the increased weight of timber-work, due to the higher pitch and hence additional amount of material employed in both the structure and the covering of the roof. Actually, from the standpoint of wind alone, it was advantageous for these high, steeply-pitched roofs

---

23. Viollet-le-Duc, Op. cit., article Charpente, III, 9.

24. Francis Price, A Series of .. Observations .. upon .. the Cathedral-Church of Salisbury, London, 1753. A note on plate 13, opposite p. 67, reads: " There is on the whole of this pan, in the several roofs, 2641 tons of oak... "

25. H. J. L. J. Massé, The City of Chartres, Its Cathedral and Churches (" Bell's Handbooks to Continental Churches "), London, 1900, p. 21. Taking the Bourbon Standard of 1.079 # for the " old French Livre " gives 247.18 tons.

26. Antoine Moles, Histoire des Charpentiers, Paris, Librairie Gründ, 1949, p. 277: " The whole timber-work of Notre Dame, Paris, supports 1236 sheets of lead, 5 mm thick, weighing 210,000 kilograms. "

27. Vide, Appendix A for computations.

to be relatively heavy. For there is a definite relationship between the height of an object of a given base and its weight, when it comes to its stability in resisting the overturning action of wind. It is evident, for example, that the heavier the triangle that represents the roof in cross section (fig. 3, B), the less likely it is to rotate about one of its lower angles. Where the roof pitch is not more than 45°, the shape itself of the roof can largely be counted on to take care of this overturning tendency of the wind. But where the roof pitch is much above this angle, the weight of the roof acting as a rigid unit becomes increasingly significant. For, in Gothic cathedral roofs, the double plates which constitute the wooden bearing for the entire roof structure are not anchored into the masonry of the clerestory walls; instead, they merely rest upon the top of the thin stone walls. Thus, in a great many Gothic buildings, it is the friction between these plates (weighted as they are by the great roof) and the masonry of the clerestory wall-tops which alone prevents the whole roof from sliding off to one side as a result of wind pressure.

In some but by no means all Gothic cathedral roofs there are diagonal brackets or knee-braces keyed into the great tie-beam near its ends and angling down to the inner face of the clerestory wall [28]. The actual function of these brackets is not so much to help diminish the effective span of the tie-beam, but to resist its wind-imposed movement [29]. The knee-braces accomplish this wind-resisting function by abutting their lower ends against the side of the masonry wall at a lower level than the top, so that more of the wall can be engaged in resisting the lateral pressure. Sometimes these brackets abut the inside of the clerestory wall at just the level at which the upper flying buttresses abut it on the outside. In this situation the tie-beam plus knee-brace combination exactly fulfills, in wood, the function of the membrane walls of masonry, as found at Reims.

Taking the figures given above for the approximate weight of the nave roof of Reims, and making reasonable assumptions for wind pressure, it is possible to compute the probable maximum pressures acting upon the top of the clerestory walls on account of wind action [30]. First, it is evident that each upper flying buttress on the lee side has to help the thin clerestory wall to withstand a lateral pressure of over sixteen tons created by wind action against the roof alone. With the thin strip of this wall-top poised high on relatively slender piers at the nave, triforium, and clerestory window levels, this is no inconsequential force. It is further added to, as has been remarked above, by two other increments. One is the crown thrust of the upper flying buttress on the windward side; the other is the wind pressure against those portions of both wall and clerestory window, on the windward side,

28. MOLES, *Op. cit.*, p. 274 (quoting H. DENEUX, *Ancienne et Nouvelle Charpente de la Cathédrale de Reims*, Reims, Matot-Braine, 1927) mentions the stone corbels which, in the original fire-destroyed roof, received the lower ends of these brackets, in Reims Cathedral.
29. A longitudinal movement with respect to the beam itself, a transverse movement with respect to the clerestory walls.
30. *Vide* Appendix B for wind pressure discussion and computations.

UPPER FLYING BUTTRESS   83

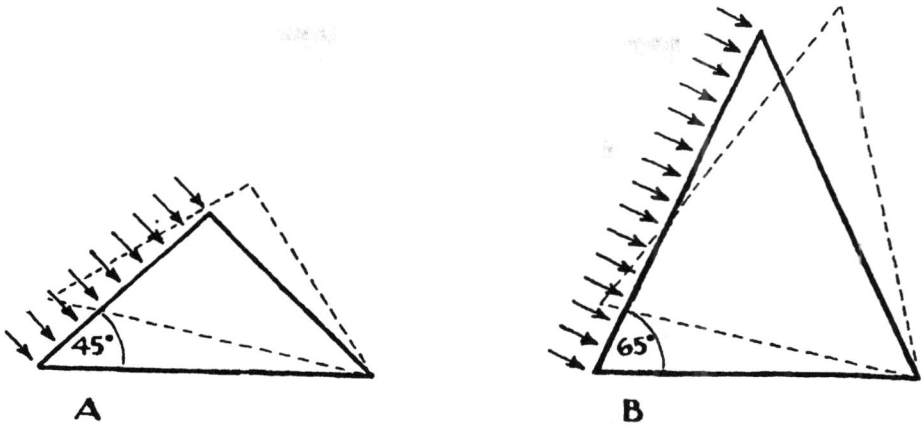

FIG. 3. — See page 82 of the text.

which rise above the level of the lower or vault-abutting flying buttress. It has already been demonstrated that these two additional pressures are transmitted against the upper strip of clerestory wall on the lee side by means of the membrane cross-walls of masonry and/or the great timber tie-beams.

For the moment it may be assumed that, up to the level of the top of the lower flying buttresses, that tier of sloping struts not only handles the vault thrusts but also takes care of much of the pressure exerted by the wind against the lower portions of the clerestory windows and walls. Therefore it is primarily the portion of window and walling *above* this lower tier level whose resistance to wind pressure receives the benefit of lateral bracing provided by the upper tier of flying buttresses on the lee side. There may be as much as 26 feet of vertical rise here between the top of the lower flying buttress and the gutter level, or as much as 32 feet to the wall-top roof plates [31]. This obviously makes a considerable surface area subject to the force of the wind [32].

The upper tier of flying buttresses not only helps to resist this lateral pressure but also, thanks to the cross-bracing provided by the membrane walls and/or timber ties described above, even relieves much of the wind pressure from having to be braced by the lower tier. This lower tier, of course, has no direct or positive cross bracing, and hence is ill-equipped to handle the variable wind pressure now from

31. In Paris Cathedral.

32. At Reims, with the roof plate approximately 30 feet above the top of the lower flying buttress where it meets the clerestory wall, the surface area per bay would be about 23 × 30, or 690 square feet. This area, subject to a 28 #/sq.ft. wind pressure, would need to sustain a horizontal force of some 19300 #, or nearly ten tons.

one side now from the other. The upper tier, by furnishing a box-like frame of relatively rigid members outside and above the vault, allows the lower tier to concentrate more fully on its proper and indispensable function of meeting the primary vault thrusts. And even more important because of the roof's towering height, this upper tier on both sides, along with its cross-bracing above the vault, provides a stable superstructure for the support of the great roof, as has been demonstrated above.

It is not here maintained that each upper flying buttress on the lee side receives the entire maximum designated potential wind load of som twenty-six tons of horizontal pressure [33]. For one thing, the timber cross-beams have been shown to be effective in tying both wall-tops together in a box-like framework, thereby engaging both of them in a cooperative resistance to wind action. Hence, whether the wind comes from one side or the other, the not inconsiderable stability encountered in the clerestory walls, loaded as they are by the weight of the roof, is active in resisting the pressure of the wind. It is evident, also, that in buildings of cathedral size the wind does not press against the entire area of roof and clerestory wall with a uniformly maximum force. The areas momentarily less stressed help to stiffen somewhat the stability of those areas of greater stress adjacent to them. And there is also the anchoring effect of western towers and projecting transepts [34]. Furthermore, it is undoubtedly true that a great deal of stability at the vault level is given by the transverse vaults which intersect the longitudinal vault at each bay. Even when their shells are excessively thin, their narrow pointed tunnel shapes give these shells a surprising degree of rigidity, making them powerfully resistant to compression in the direction of their axes [35]. Thus at each bay there is a stiffening cross vault which, to be sure, contributes its own share to the outward thrust against the lower flying buttresses, but whose direction and shape are most effective in resisting an inward wind pressure at the heads of the clerestory windows. Hence these cross vaults brace the two clerestory walls, from within, between the points at which a pair of flying buttresses impinge against them from both sides, without. And they add to the rigidity of the superstructure by filling out a complex of elements no one of which is very thick or powerful by itself, but whose combination is eminently effective.

---

33. This figure represents the 16-ton horizontal component of the wind load against the roof, plus the 10-ton wind pressure against the upper portion of the clerestory walls. It edos not include the crown thrust of the windward flying buttress, nor the horizontal wind load against the lower portion of the clerestory wall.
34. The long and lofty cathedral of Bourges, however, has no transept.
35. The fact that these cross vaults undoubtedly are active in resisting the lateral displacement of the masonry superstructure under wind pressure may account for some of the fissures that are to be seen in many medieval vaults. This would be most apt to be true near the center of the bay where the compression-resisting tunnel shape of the transverse vault in completely cut away on account of its intersection with the longitudinal vault. Over the centuries, the compression which these transverse vaulting compartments would undergo, now from one side now from the other, would involve some degree of lateral movement, not only in the vault itself but throughout the whole masonry superstructure. It is doubtful whether the precise shape and dimensions of the vault as originally built would be entirely recovered after each slight dislocation, and thus fractures might develop to a point where each portion of the structure adjusted itself to the others in a position and fashion dictated by the stresses it was subject to.

UPPER FLYING BUTTRESS   85

All this does not deny the evident role of the upper flying buttress in taking a major and indispensable part in resisting the pressure of the wind against roof and clerestory wall. The contribution of the other elements of the box-like framework described above is largely that of cross-bracing the superstructure so that all of its parts act together in achieving stability high up within the structure. But it must be remembered that this superstructure, no matter how effectively interconnected at the top, is poised high on slender stilt-like piers, with no positive anchorage at the base that can resist the tensile stresses of an overturning action. Hence the compression struts of the upper tier of flying buttresses, braced against the buttresses proper, are essential to the stability of the whole structure under wind action Without this upper tier, the soaring naves and steep-pitched roofs of the great XIII Century Gothic cathedrals would certainly never have survived the infrequent but inevitable tempests of seven intervening centuries.

*
**

To summarize, we can say that the great churches of the mature Gothic system of construction in France normally have two tiers of flying buttresses. Those in the lower tier are set very precisely at a level where they can most effectively receive the collected vault thrusts and transfer these active forces to the stable buttresses beyond the side aisles. The purpose of the upper tier is not to take care of any vault thrusts but to meet the lateral pressures exerted against these lofty buildings by the force of the wind.

Since the wind may come from any direction, the high thin tops of the clerestory walls are made to act together by means of cross bracing. This cross bracing is provided by transverse membrane walls of masonry above the vault and/or the heavy transverse timbers that act as tie-beams at the base of the roof construction. The upper tiers of flying buttresses on both sides act as lateral struts in this framework by bracing it against the outer buttresses. Such a framework maintains the stability of the clerestory walls, with their large window areas, against pressure on the windward side and suction on the lee side.

In thus providing for the stability of the masonry superstructure, this framework of buttresses, flying buttresses, and cross bracing furnishes a surprisingly rigid structure, in spite of its height above ground and the slenderness of its members, upon which the lofty roof is set. Were it not for the stability assured by this framework, the great planes of the high-pitched roof—so much higher above ground and therefore so much more vulnerable to the sweep and force of the wind—would be in danger of swaying or shifting laterally, thereby bringing about the destruction of all below.

APPENDIX A

During the Middle Ages oak was almost universally the kind of wood employed for structural purposes, whether for the ½" thick split lath strips that supported the lead roof covering, or for the huge tie-beams set across from one clerestory wall-top to the other. VIOLLET-LE-DUC states specifically that the wood timbering in Reims Cathedral is oak [36].

It should perhaps be noted that in medieval practice, because of hand adzing instead of power sawing, long structural timbers such as rafters followed the natural taper of the tree. Thus Moles gives the size of the rafters of Lemoyne's roof at Reims as tapering from 0.22 m square (8 ⅝") to 0.18 m (7 ⅛") in a length of 17.30 m (about 57 feet) [37]. In his exhaustive and invaluable documentary history, SALZMAN gives many instances of this tapering of wood members [38]. Hence the scantling sizes used in the computations herewith appear to be reasonable averages. Since the lath strips were always spaced about an inch apart, this item is figured at 4/5 of the area of the two roof slopes.

There is no unanimity of expert opinion on the weight of a cubic foot of oak, due to many variable factors. For American species, one authority gives 50 pounds per cubic foot for white oak, 45 for red oak [39]. Since Salzman considers the weight of English oak at 55 # [40] it would seem to be a conservative figure, for the purpose of these computations, to use 50 pounds per cubic foot as the weight of the oak in the nave roof of Reims.

As for the lead, VIOLLET-LE-DUC says that the XIII Century sheets of lead on the roof of Chartres Cathedral were about 4 mm (⁵/₃₂" ±) thick [41]. A linear foot strip is figured at 15" on account of the rolled seam, with a 4" lap in every 8 feet down the slope.

---

36. *Op. cit.*, article *Bois*, II, 214.
37. ANTOINE MOLES, *Op. cit.*, p. 274.
38. L. F. SALZMAN, *Building in England Down to 1540*, Oxford, Clarendon Press, 1952, pp. 211-222.
39. HERBERT F. MOORE, *Textbook of the Materials of Engineering*, 6th ed., New York, McGraw-Hill, 1941, p. 225.
40. SALZMAN, *Op. cit.*, p. 242.
41. VIOLLET-LE-DUC, *Op. cit.*, article *Plomberie*, VII, 211.

The schematic diagrams of the roof membering are based on the section drawings given by VIOLLET-LE-DUC [42].

Computations, Nave Roof of Reims Cathedral:

Truss.

1)	604×8×8	38650
1ᵃ)	362×8×(3+3)	17376
2)	2×564×8×9	81216
3)	2×564×8×9	81216
4)	2×652×6×9	70416
5)	92×6×6	3312
6)	188×6×8	9024
7)	280×9×9	22680
8)	140×6×8	6720
9)	2×290×6×7	24360
10)	2×24×6×6	1728
11)	2×48×6×6	3456
12)	2×48×6×6	3456
13)	2×26×6×6	1872
14)	2×26×6×6	1872
15)	2×36×6×6	2592
16)	2×30×6×6	2160
17)	560×9×12	60480
18)	2×276×8×8	35328
18ᵃ)	2×92×8×(4+4)	11776
19)	2×32×6×6	2304
20)	2×34×6×9	3672
	Total.............	485670

Longitudinal.

1)	104×8×8	6656
2)	104×8×8	6656
3)	2×112×6×6	8064
4)	2×112×8×8	14336
5)	2×104×8×8	13312
6)	2×112×6×8	10752
7)	2×112×6×8	10752
8)	2×112×8×8	14336
9)	2×112×8×8	14336
	Total.............	99200

Sway Bracing.

10)	2×84×5×5	4200
11)	2×56×5×5	2800
12)	2×2×44×5×5	4400
13)	2×2×56×5×5	5600
	Total.............	17000
	To transfer :	601870

---

42. *Op. cit.*, article *Charpente*, III, 19, fig. 14.

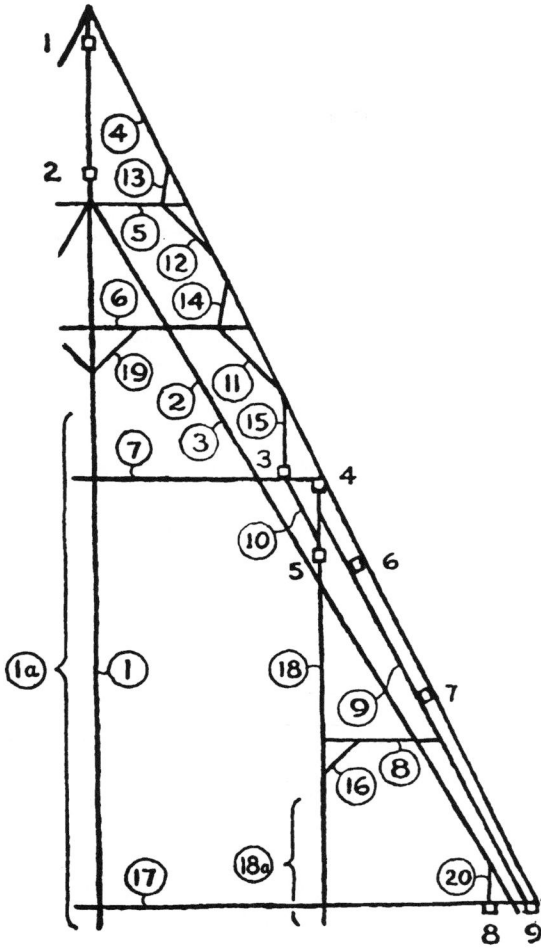
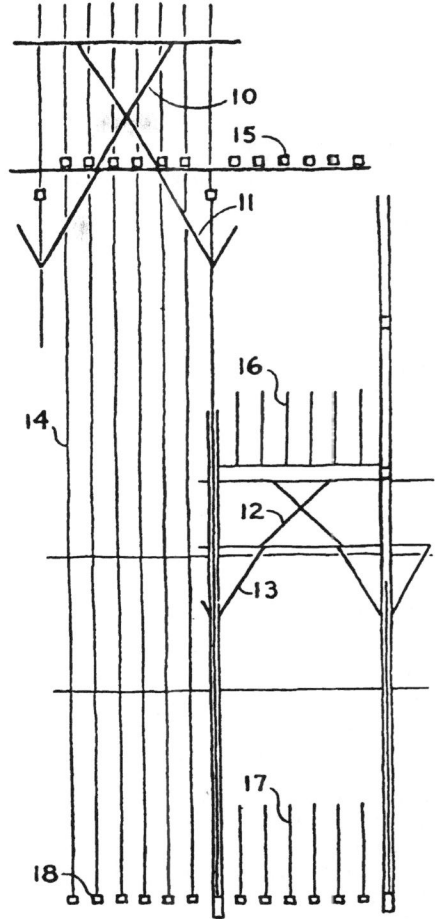

601870

Rafters, Ties, Posts, Bases.

14)	$6 \times 2 \times 652 \times 6 \times 9$	422496
15)	$6 \ \times 76 \times 6 \times 6$	16416
16)	$6 \times 2 \times 36 \times 6 \times 6$	10800
17)	$6 \times 2 \times 56 \times 6 \times 9$	36288
18)	$6 \times 2 \times 44 \times 6 \times 9$	28512

Total.............. 514512

Lath.

$2 \times 670 \times 112 \times \frac{1}{2} \times 4/5$ 60032

Total for Wood: cubic inches... 1176414

or 680.8 cu.ft.

$680.8 \times 50 = 34040 \#$ of oak per carpentry bay.

Lead.

2×672 plus $2 \times 7 \times 4''$ lap, $\times 15 \times 5/32 = 3281\frac{1}{4}$ cu. in.

$3821\frac{1}{4} \times 9.33 = 30614$ cu. in., or 17.716 cu. ft.

$17.716 \times 706 \# = 12507\frac{1}{2} \#$ of lead per carpentry bay.

Total Weight of Roof per Masonry Bay
34040 plus 12507 ½ equals 46547 ½ divided by 9.33 equals 4989 # per lin. ft.
4989×23 equals 114747 # total, or over 57 tons.

APPENDIX B

For a review of literature dealing with wind loads on buildings, *vide*: WHITTEMORE, COTTER, STANG, and PHELAN, *Strength of Houses,* a U. S. Department of Commerce *Building Materials & Structures Report BMS 109,* issued April 1, 1948 [43]. Over a period of ten years all available information on wind loads, including wind tunnel data, was carefully studied by Subcommittee # 31 of the American Society of Civil Engineers. This Subcommittee's six published reports, the Final Report dated 1940, constitute the substantial basis for most of the design recommendations, with respect to wind, in American Building Codes. The Subcommittee recommends a minimum standard wind load in pounds per square foot for the design of buildings, including tall buildings, and makes definite recommendations for wind load assumptions depending on position with respect to direction of wind, slope of roof, height above ground, and size and position of openings relative to wind direction.

It is not here proposed to go into the precise computation (by formulas which involve weight of unit volume of air, height of barometer in inches of mercury, absolute temperature, and the like) of the wind load on the nave roof of Reims Cathedral. To illustrate the thesis of this paper it will perhaps be sufficient to make approximate computations based on available but incomplete information, since precise data covering many centuries are lacking.

Essential data for figuring the velocity pressure (the pressure exerted by the wind) include information on maximum wind velocities as well as height of building, slope of roof, etc.

Lists have been prepared by the Météorologie Nationale, in Paris, which tabulate wind velocities from sixteen compass directions, as recorded at airpots throughout France over a period of years. Examples of the highest recorded values, in meters per second and in corresponding miles per hour, with their directions and the years of observation, are as follows:

---

43. Pp. 22 ff.

UPPER FLYING BUTTRESS   89

Amiens	1924	WSW	24 m/s	54 mph.
Beauvais-Tille	1926-35	NE	24	54
Caen	1945-47	SSW	24	54
Calais-St.-Inglevert	1926-35	WSW	32	72
Chartres-Champhol	1926-35	WSW	25	56
Lille	1945-49	S, SSW, WSW, W	20	45
Nancy	1926-35	SW	32	72
Orléans	1926-35	WSW	24	54
Paris-Le Bourget	1926-35	SW, W	24	54
Paris-Mont Valérien	1926-35	SSW	27	60
Reims	1936-39	W	20	45
Rouen-Rouvray	1945-47	W	25	56
St.-Quentin-Roupy	1936-39	W	22	49
Strasbourg	1926-35	SW, WNW	23	51
Valenciennes	1936-39	SSE	21	47

Only the tables for Caen, Calais, Nancy, and Paris—Mont Valérien record velocities of over 50 mph for winds from approximately north and south, these being the directions which would be most critical for nave roofs. However, the figures are neither conclusive nor comparable in this respect, since readings were taken at quite different heights above ground; e.g., as low as 12 meters (39 feet) at Rouen, and a maximum of 31 meters (nearly 101 feet) at Calais.

Quite different is the situation in many American cities, where winds of these and higher velocities blow far more frequently than in France, and where winds of near hurricane violence or more have occurred in various regions of the country. Thus *The World Almanac* [44] records such maximum velocities as the following, in miles per hour:

Boston, Mass.	73
Buffalo, N. Y.	73
Hatteras, N. C.	90
Galveston, Texas	71
Key West, Fla.	84
Miami, Fla.	123
Mobile, Ala.	87
New York, N. Y.	81
North Head, Wash.	95
Omaha, Nebraska	73
Pensacola, Fla.	91
Savannah, Georgia	71
Tatoosh Is., Wash.	84
Mt. Washington, N. H.	188

44. New York, 1949, p. 794.

These official figures are for an average speed for a five-minute period; gusts of a few seconds' duration have been recorded at 231 mph on Mt. Washington, and at 99 mph on the roof of the 454-foot-high Whitehall Building in New York City at a time when the Weather Bureau's low altitude Central Park station in the same city recorded only 46 mph [45].

Very occasionally, but nonetheless certainly, high winds have occurred in France. VIOLLET-LE-DUC speaks of one tempest which created considerable damage in 1860, before wind velocities were measured and tabulated [46]. Up to 1921 the strongest wind that had been officially recorded in Paris had a velocity of only 18 m/s (40 mph) at 21 meters (69 feet) above ground, but 42 m/s (94 mph) at 305 meters (1001 feet) atop the Eiffel Tower. These velocities were recorded on September 6, 1899, with a southerly wind, but there is no report on the duration of this velocity. On November 12, 1894, another 42 m/s wind was recorded at the top of the Eiffel Tower, and there is indication that winds of more than 50 m/s (112 mph) have been recorded [47].

From the data available, certain general facts appear. In summary they are as follows: 1) Winds blow harder, and thus create greater

45. *Vide* "The New Yorker," Vol. XXV, No. 8, April 16, 1949, pp. 21, 22.
46. VIOLLET-LE-DUC, *Op. cit.*, article *Flèche*, V, 453, Note 1.
47. Vide "Meterologische Zeitschrift," Vol. 39, No. 10, Oct. 1922, pp. 331, 332.

velocity pressure, high up than near the ground. 2) As the wind's force is greater against the lofty tops of buildings than against the portions nearer the ground, so too the thinner the building or part thereof the greater its instability, due to low-pressure suction on the lee side. 3) Data on the duration of maximum velocities in high winds do not appear to be standardized in France, on account of the variety of instruments in use there to record wind velocities. 4) Winds in France blow generally less strongly than those in America, but exceptionally high winds do occur in France at rare intervals. 5) Whatever exceptional winds may have occurred in France during the past seven hundred years, there is no record of their having blown down any Gothic church or cathedral in that country. Any destruction has been from other causes.

Because exceptional winds do blow, however rarely, in France, and because the possibility of their occurrence during so many centuries would appear to be undeniable, it would seem reasonable to use current American design figures in investigating the wind action on the nave roof of Reims Cathedral. These unit figures, as given by the National Board of Fire Underwriters' *National Building Code* [48], are for wind pressures acting horizontally and are graduated as follows:

Height, less than 50 feet... lb/sq ft	20
50-99 foot height	24
100-199 foot height	28
200 feet and above	30

Roofs with slopes greater than 30 degrees are to withstand pressures, acting inward normal to the surface, equal to those specified above, and applied to the windward slope only.

The entire area of the nave roof of Reims Cathedral falls within the 100-199 foot height category. The height along the slope is about 56 feet; multiplying this by a carpentry bay of 9'-4" gives 523 square feet; this multiplied by the wind pressure at 28 #/sq. ft. gives 14630 # normal to the roof slope, of which the vertical component is 6190 #, the horizontal 13240 #. The timber tie-beams help to distribute the action of this horizontal force against both wall-tops at either side of the building. But it is the masonry bay of some twenty-three feet, not the carpentry bay, that determines the spacing of the flying buttresses along the building's flanks. Hence the upper flying buttress on the lee side may be called upon to help resist a wind load against the roof alone of some 36060 #, of which 15250 # is the vertical component, 32640 # the horizontal. Although each of the two clerestory walls-tops has to support at least 57374 # on windless days, this dead load of the roof structure alone is augmented, under the wind loads considered here, to where the windward side has to support 68710 # (over 34 ¼ tons) and the leeward 61185 # (over 30 ½ tons). These vertical loads per masonry bay are quite apart from the 16-ton horizontal pressure of the wind-loaded roof on the wall-tops.

---

48. 1949, p. 67.

17

The high roofs of the east end of Lincoln Cathedral

N.D.J. Foot, C.D. Litton and W.G. Simpson

INTRODUCTION

The work of Hewett[1] has shown that the oak roofs over the entire length of Lincoln Cathedral are largely complete and contemporary with the building they cover, as are those of the minor transepts. Only the principal transepts, the Chapter House and some of the roofs over aisles and chapels have been replaced in recent times (Fig. 1). Lincoln probably has more of its original roofing surviving than any cathedral in the country.

This paper is largely concerned with those roofs to the east of the central tower which comprise the earliest and the latest of the medieval roofs as well as a fair sample of post-medieval repairs and replacements. The roof of the nave[2] is very similar to that of St Hugh's Choir (Fig. 3) both in its original design and in the character of the post-medieval work, while the western chapels have roofs of a design very similar to that of the Chapter House

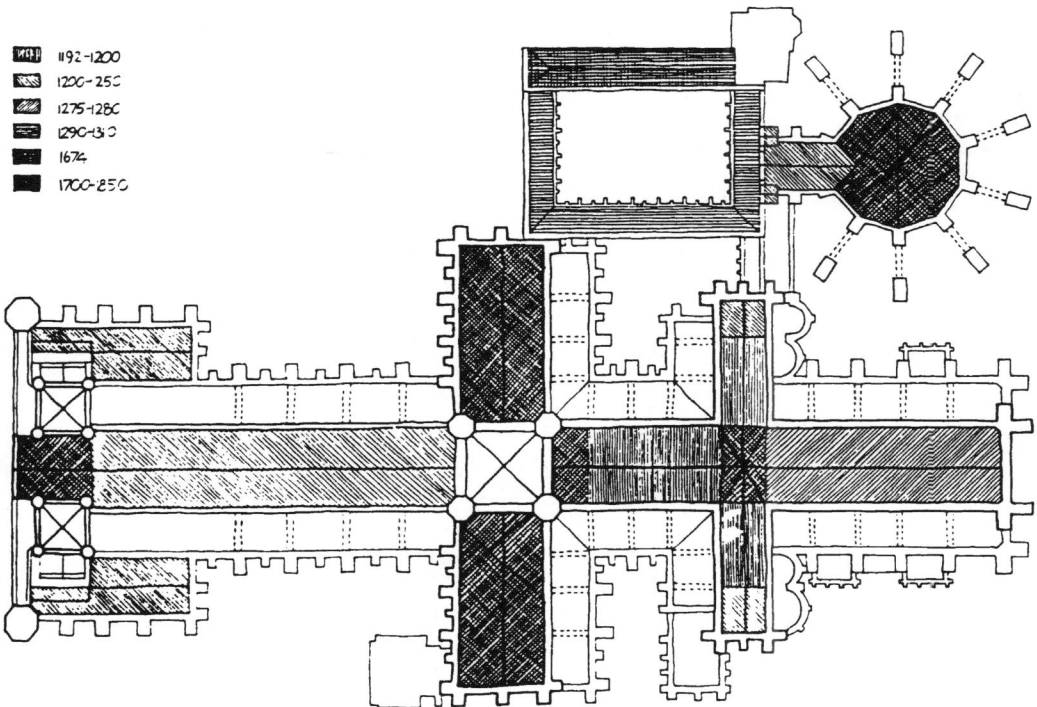

Key:
- 1192–1200
- 1200–25C
- 1275–128C
- 129C–13C
- 1674
- 1700–85C

FIG. 1. Roof plan of the cathedral

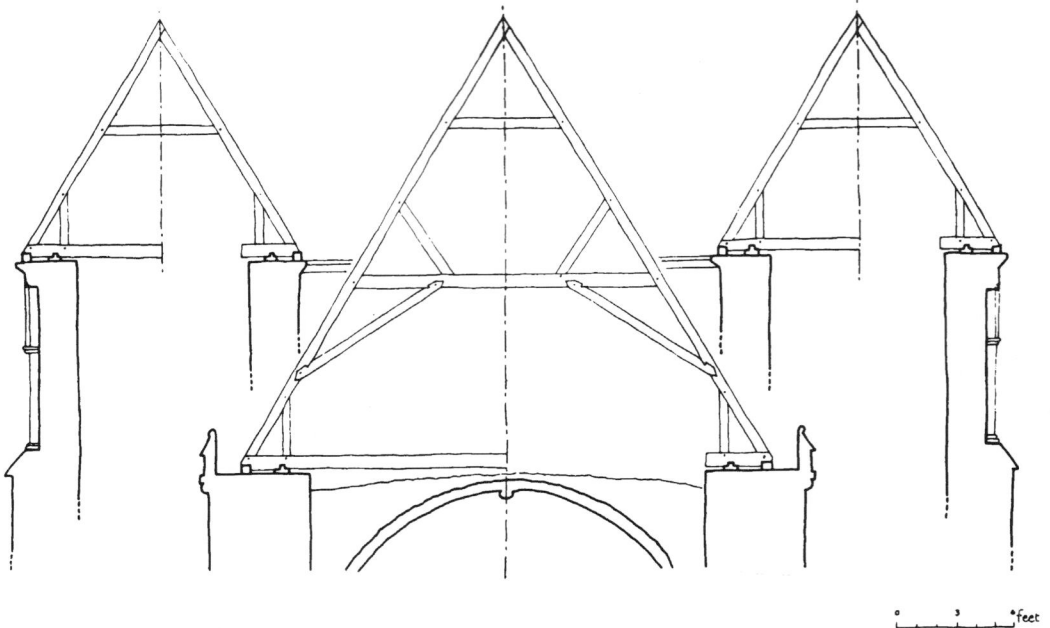

FIG. 2.   Chapter House vestibule and stair towers

vestibule (Fig. 2). It is intended therefore that this paper should serve as an interim report on the results of the detailed architectural study of the roofs throughout the Cathedral and of the analysis of the timber of which they are constructed. This survey was begun in 1979 at the start of an extensive programme of repairs by the Cathedral authorities. These works involved the stripping of the lead covering and of the underlying battens and the replacement of split or decayed timber. It meant that the roof space was opened up to natural light and easy access was afforded to all parts by the carpenters' ladders and scaffolding, while the discarded timber provided ample material for tree-ring analysis.

THE ROOFS OF ST HUGH'S CHURCH

Bishop Hugh of Avalon initiated the rebuilding of the Cathedral in 1192 and by the time of his death eight years later much of the stone building east of the central tower was probably complete — i.e. the presbytery, later demolished and replaced by the Angel Choir — the choir transept, the choir and perhaps even a substantial part of the central tower — later rebuilt after its collapse c.1238. To the first building period belong the roofs of the choir transepts and the choir. Surprisingly, they are all of slightly different designs.

The Choir Transept

The plan of the choir transept is of three bays, the inner of which is the aisle of the church. Its roofing consists of twenty-eight individual frames over the south transept and twenty-nine

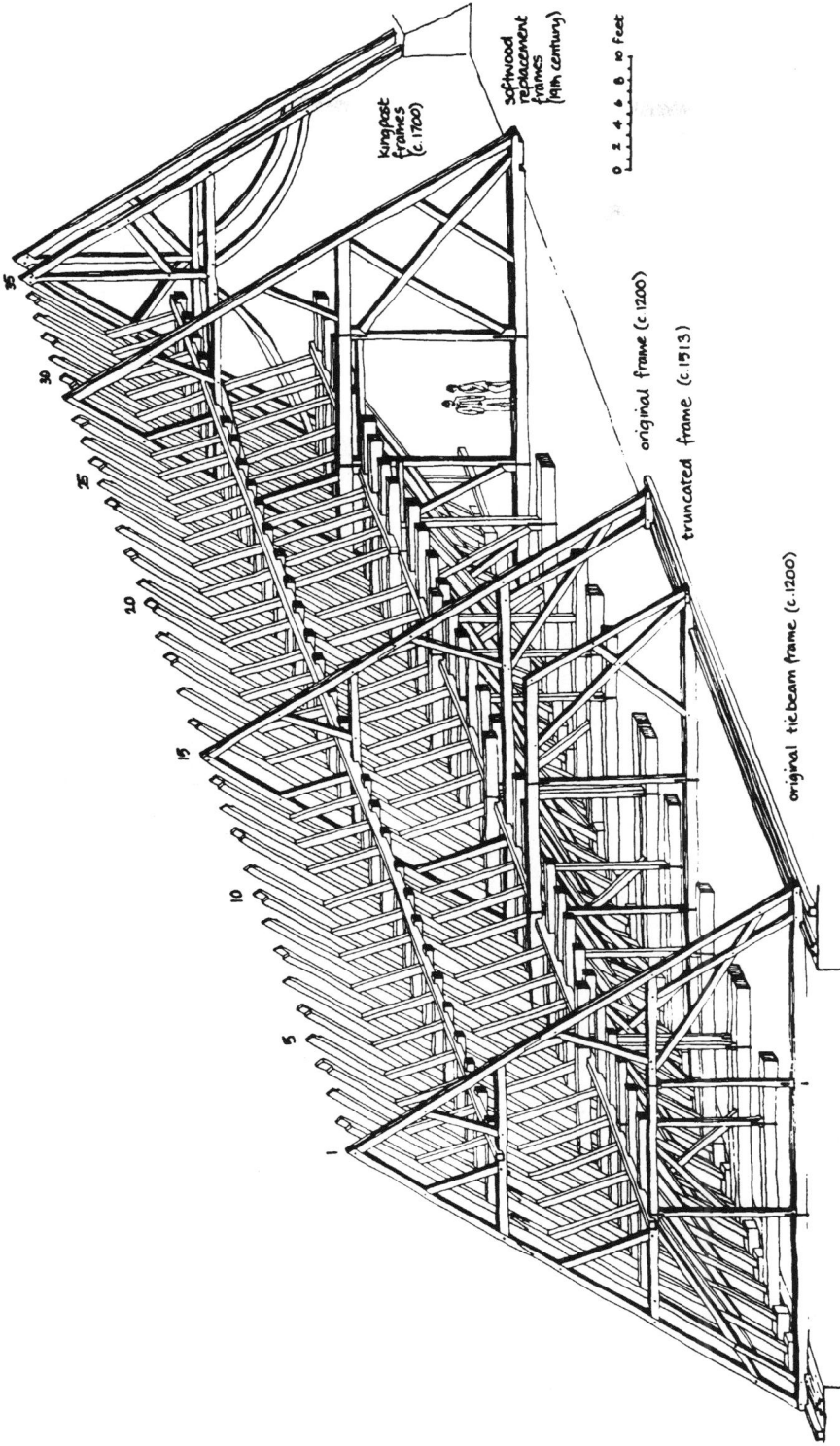

FIG. 3. St Hugh's Choir: projection

FIG. 4a.
South-east transept:
inner bays

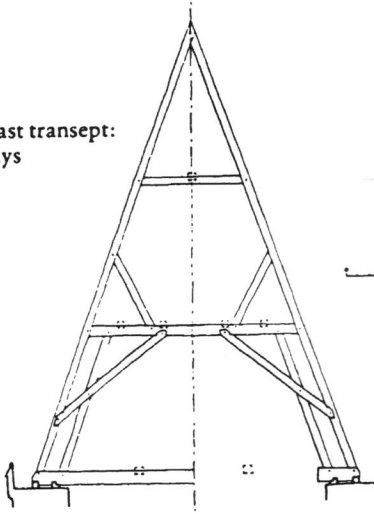

FIG. 4b.
South-east transept:
outer bay

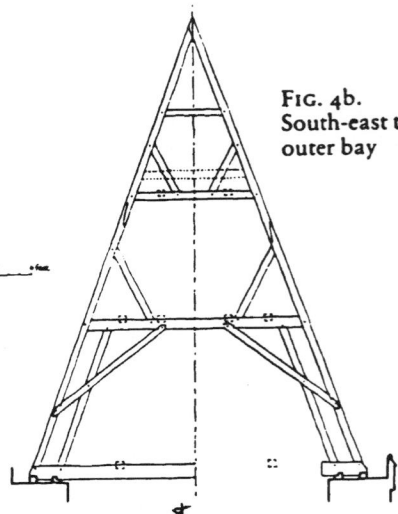

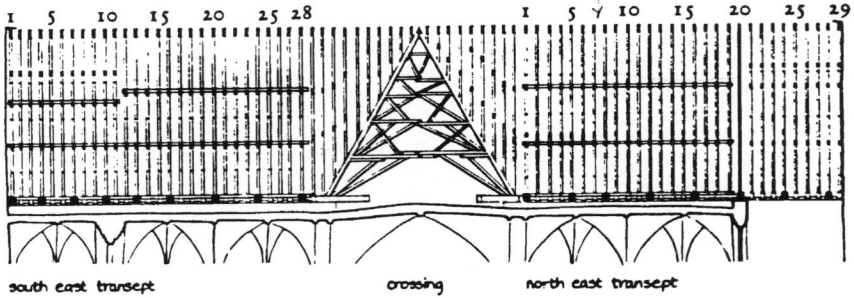

FIG. 4c. Section through east transept

FIG. 4d.
North-east transept:
inner bays

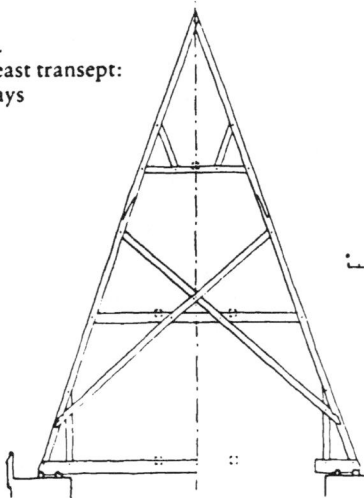

FIG. 4e.
North-east transept:
outer bay

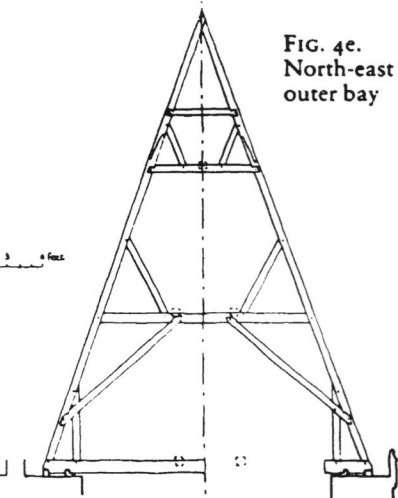

over the north corresponding therefore to about ten frames for each bay of the substructure (Fig. 4c). The pitch of the roof is 70°, the width between wall centres 28 ft, and the height 40 ft. In the north-east transept the frames over the inner two bays are scissor-braced (Fig. 4d). There are two collar-beams, the upper having struts to the rafters and the lower having the scissor-arms obliquely half-jointed across its north face. An ashlar-piece supports the foot of the rafter and the end of the scissor-arm passes across just below its top to fasten in the north face of the rafter with an open notched-lap joint. All other joints are mortice and tenon. The ashlar-piece and rafter foot are based on a sole-piece or, at every fourth frame, on a tie-beam. These, as in all other of the medieval roofs, rest on wall-plates which here are double, the inner having a broad fillet cut along the top which slots into a trench in the soffit of the sole-piece or tie-beam.

In the roof over the inner two bays of the south-east transept the horizontal components of the roof are exactly as described in the north-east transept (Fig. 4a). However, here it is the lower collar that has struts, and soulaces with a notched-lap joint at either end on the north face passing across truncated secondary rafters. These together, carry out the functions of the scissor-braces and ashlar-pieces in the north transept. Carpenters' marks are found in these roofs only in the middle bay of the north transept.

The Choir

The roof of St Hugh's Choir consists of thirty-six individual frames of which twelve at the west end are later replacements (Figs 3, 5a and 5b). It seems likely however that the original number was the same. The pitch of the roof is 60°, its width between wall centres 46 ft, and its height 40 ft. The horizontal components are exactly as described in the choir transept roofs. The tie-beams are at every third frame. Both collars here have struts to the rafters, the extra support perhaps being required because this roof, unlike the transept, has few rafters of a single timber. Most are of two pieces scarfed approximately at mid-length using a scissor-splay-stopped joint (Fig. 6a) — a type that has only been previously recorded at the Blackfriars, Gloucester.[3] The lower collar is supported by truncated secondary rafters and soulaces with open notched-lap joints at either end as in the south transept with the addition, on account of the greater roof span, of queen-posts (Figs 7a and 8). These were originally jointed at either end with mortice and tenon which is used, apart from the soulaces, throughout the roof. The notched-lap joints are on the west face of the frame and there is a consistent series of carpenters' marks throughout the roof.

The Presbytery

The roof of the present presbytery (Angel Choir) contains over eighty pieces of re-used timber which were very probably part of the roof of St Hugh's presbytery. Certainly their dendrochronology is consistent with this conclusion (see Table II, 1). They suggest that the main part of the roof had passing-brace construction since collar-beams with two obliquely cut half-joints occur commonly[4] (Pl. XA). The re-used timber is restricted to the thirty frames which cover the three western bays of the Angel Choir and the quantity is such as to suggest that as far as possible all the timber from the roof of St Hugh's presbytery was re-used in its replacement. It is hoped that by making measured drawings of its components it will be possible to reconstruct its original form. As with the transept roof, carpenter's marks appear only sporadically on its timbers.

FIG. 5a. St Hugh's Choir

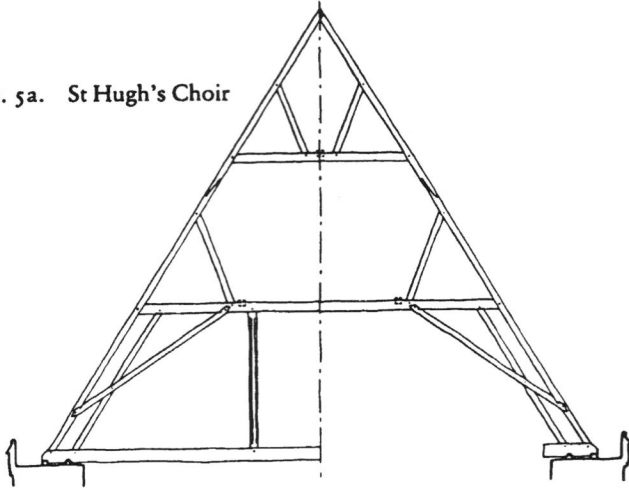

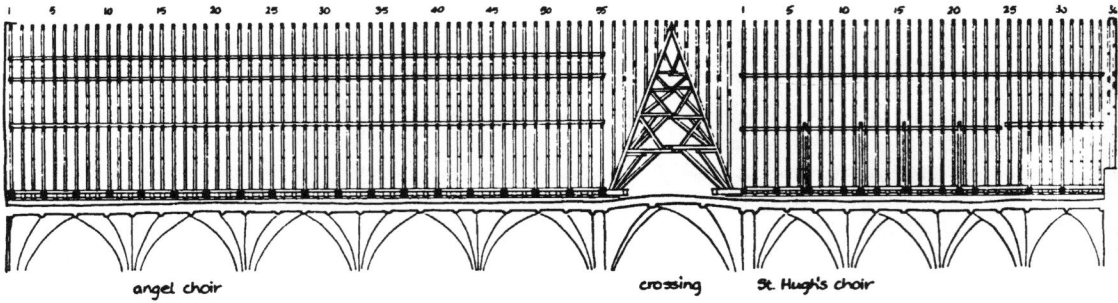

FIG. 5b. Section through choirs

angel choir   crossing   St. Hugh's choir

FIG. 5c. Angel Choir

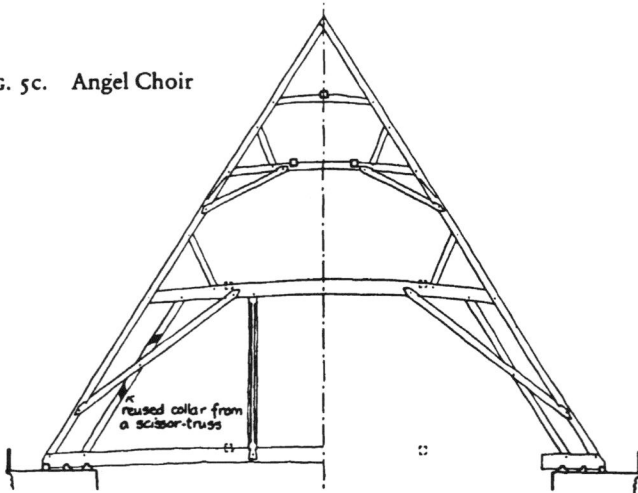

reused collar from
a scissor-truss

N. D. J. FOOT, C. D. LITTON, AND W. G. SIMPSON 53

THE LATER MEDIEVAL ROOFS

The Principal Transept

Hewett has described the roofs of both arms of the principal transept and suggested that the south transept roof (Fig. 9) if not original in its entirety, is at least rebuilt to its original design.[5] Both roofs appear to be built almost entirely of pine although there is some evidence for the re-use of both oak and pine in the southern roof. There is no evidence for the use of pine in the construction of the Cathedral roofs before Essex' use of it for the Chapter House in 1762 and in fact the documentary evidence is quite clear that both roofs were constructed in the 1830s and 1840s, perhaps reusing timber wherever possible (see below p. 66). Hewett's theory that the roof of the south transept is basically of the original design, though attractive, needs therefore to be regarded with some caution.

A certain amount can be deduced about the design of the original roofs from the Cathedral architects' surveys which were carried out before *c.*1830.[6] They strongly suggest that the medieval roofs survived up to this date, though clearly much repaired. In both roofs reference is made to rafters, tie-beams, sole-pieces, collar beams and queen-posts. Reference to 'cross-braces' in the south roof might be taken to mean scissor-braces and as queen-posts are also mentioned, a roof like that of Salisbury choir transept is a possibility and chronologically acceptable.[7] Sir Robert Smirke says in his report of 1826 that the timber of the north roof is 'of slight scantling and bad workmanship'.[8] As will be shown this aptly summarises the character of the Angel Choir roof and the general impression that emerges is that the main transept roofs were probably similar or related in their design to others in the Cathedral.

Another feature of the roofs of the principal transept referred to in Cathedral architects' reports of the late 18th and early 19th century was that their frames were leaning away from the central tower (racking). This applied also to the roofs of the nave and the choir. In the roofs of St Hugh's church the only features giving any vertical stability above wall-plates were the battens and their lead covering. All the roofs now have collar-purlins but none can indubitably be shown to be original. Many are of re-used timber and those in the choir transepts are of pine and so are most probably no earlier than 18th century.

The Angel Choir

The roof of the Angel Choir is of the same pitch and dimensions as that of St Hugh's Choir and is made up of fifty-five frames so that each bay of the masonry structure is covered by eleven frames of roofing. They are basically of the same design as those of St Hugh's Choir but with a number of distinctive features. The rafters are of two pieces scarfed using splayed joints with taced and under-squinted abutments (Fig. 6b). The scarf-joints are invariably between the lower and middle collars. The collars and the tie-beams which are now much altered by repairs, were originally slightly cambered and tapered at the ends. There is an extra collar and soulaces below both the middle and lowest collars. Hewett gives this roof collar-purlins, one lapped over the upper collar and two over the middle one (Pl. XID).[9] The notched-lap joints of the soulaces are on the west face of the frame and there is a consistent series of carpenters' marks throughout the roof (Fig. 5b and c). The queen-posts, though jointed into the collars everywhere with mortice and tenon, have lap-dovetail joints with square, housed shoulders at their bases in the western half of the roof (Fig. 7d, Pl. Xc). This is the only occurrence of that joint in the roofs but it is found elsewhere in contemporary contexts, as for example, at Kersey Priory, Suffolk.[10] In the eastern half of the Angel Choir roof the notched-lap joint is used in this position as also in the western part of the nave roof

(Fig. 7b). At the east end of the nave, mortice and tenon joints are used at both ends of the queen-post, as in St Hugh's Choir roof (Fig. 7a).

Additions to the Choir Transept

The only medieval high roofs at the east end not yet described are those over the end bays of the choir transept (Fig. 4b, c and e). A number of the Cathedral's architectural historians, of whom James Essex was the first[11] have put forward the idea that these were originally of only one storey and so roofed at about triforium level.[12] Later, it was suggested, the bays were raised to the full height of the rest of the transept. An alternative, perhaps preferable theory, is that it was originally intended to have towers terminating the transepts as seen at Exeter and Ottery St Mary.[13] This would explain the thickness of the wall dividing the north end bay from the rest of the transept. At some stage after the roofs over the two inner bays were completed it was decided not to proceed with towers, and the roofs were extended a further ten frames. At the south end, the original south wall and the first storey vault were demolished and replaced with an arch decorated in 13th-century style. Whatever explanation is preferred, a change of design is apparent, both in the masonry of the end bays of the transept and in their roofs. The timbers are of lesser scantling than those over the two inner bays, and this is most apparent in the rafters which are scarfed using splayed joints with taced and under-squinted abutments, as in the nave and Angel Choir roofs (Fig. 6b). At the south end the frame design (Fig. 4b) is much as in the main part of that roof except that there are three collar-beams, only the lowest of which corresponds in level with that of the original roofing. There are also re-used timbers.

At the north end (Fig. 4e) the added roof also has three collars. The upper two have lap-joints into the rafters, as in the roof of the Consistory Court at the west end of the Cathedral. One frame has scissor-bracing in pine which is certainly a later addition. All the other frames have struts on the bottom collar and soulaces with ashlar-pieces, a combination found only in the smaller roofs of Lincoln at the west end and in the Chapter House vestibule (Fig. 2). A single timber from the latter has given a tree-ring date which suggests that it was built shortly after c.1234 and a similar or slightly later date would be acceptable for these additions.[14]

THE TIMBER OF THE MEDIEVAL ROOFS

Samples of timber for study have been obtained by taking cross-sections with a chain saw from beams being replaced by the carpenters. This has given ideal samples and plentiful material for study. However, the quantity of samples and the limited financial resources available has left little time for other work and it is only with the recent purchase of a core-extraction kit that it has been possible to supplement the routine 'rescue' sampling with 'research' sampling and so to begin to answer problems such as those posed in the previous section.

About a hundred and fifty samples have been taken from the roof of St Hugh's Choir and over a hundred, up to the time of writing, from the Angel Choir. These were analysed in the Tree-Ring Dating Laboratory at Nottingham University. The results are important not only for the dates obtained, but also for the information they have provided about the sources of the timber and the use the medieval carpenter made of the available resources. They show over a period of less than a century a decline in the quality of timber available compelling the carpenter to practise greater economy in its use (see Fig. 10).

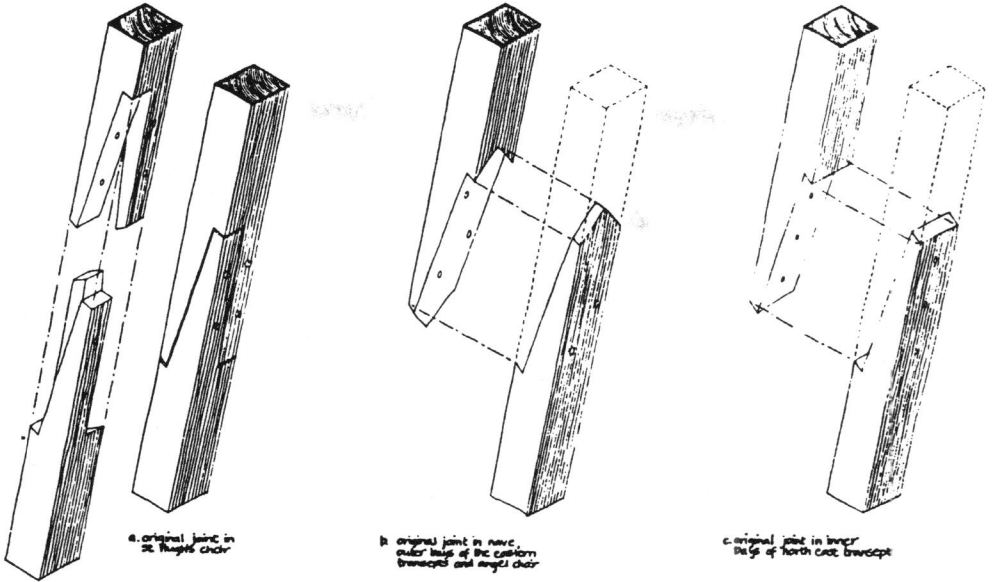

FIG. 6.   Rafter splice joints

a. original joint in
   St Hugh's choir

b. original joint in nave,
   outer bays of the eastern
   transepts and angel choir

c. original joint in inner
   bays of north east transept

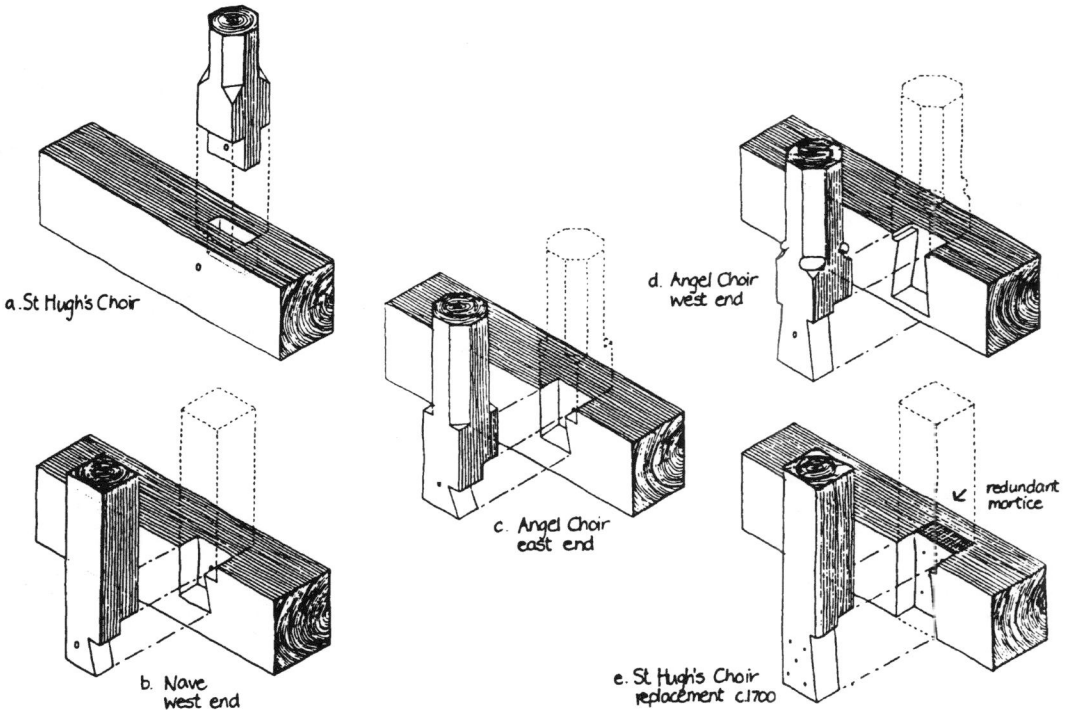

a. St Hugh's Choir

b. Nave
   west end

c. Angel Choir
   east end

d. Angel Choir
   west end

redundant
mortice

e. St Hugh's Choir
   replacement c.1700

FIG. 7.   Queen post joints

FIG. 8.   St Hugh's Choir: details of joints

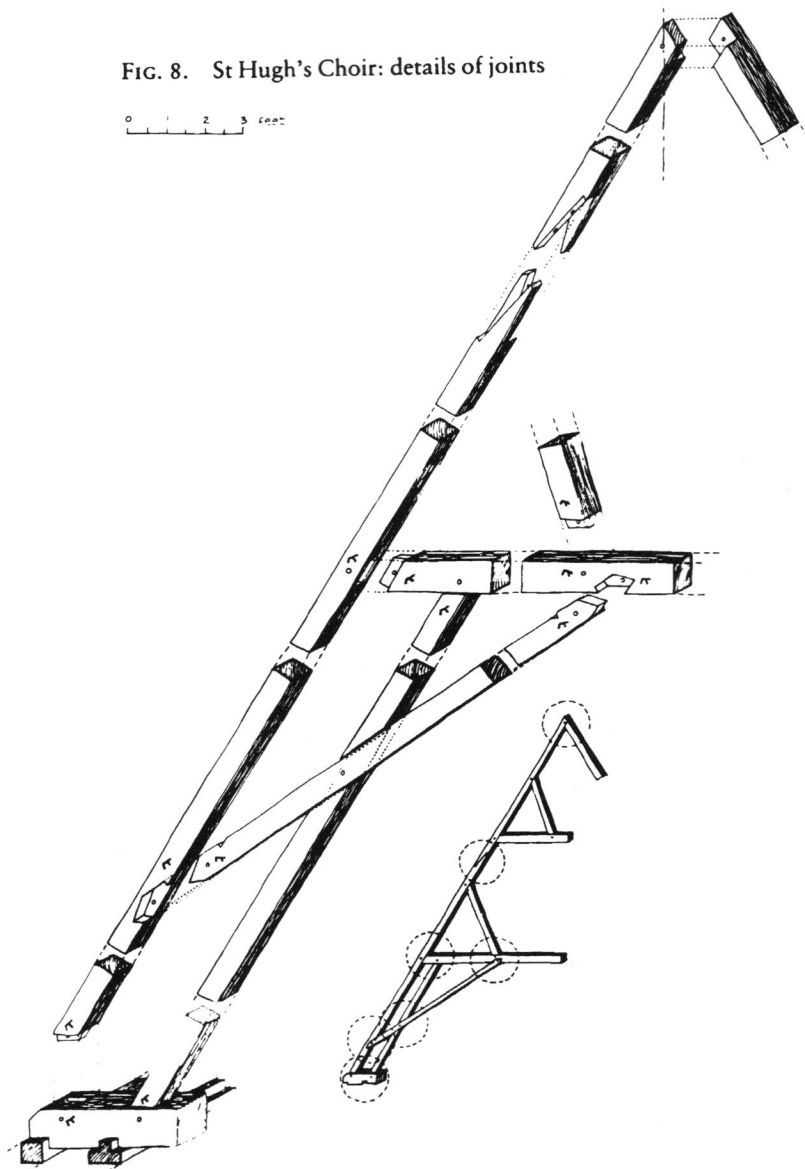

The oakwood of St Hugh's Choir was at least 130–260 years old when it was felled. Its dendrochronology extends from 882 to 1195–6.[15] If it is assumed that it all came from a single woodland then it was already in existence in the early 10th century and many of its trees were already mature at the time of the Domesday Survey.[16] The boundary between heartwood and sapwood on all those samples where it was present, ranged in date from 1158 to 1189 (see Table I). The method of estimating the average felling date of a group of

N. D. J. FOOT, C. D. LITTON, AND W. G. SIMPSON   57

FIG. 9.   South-west transept

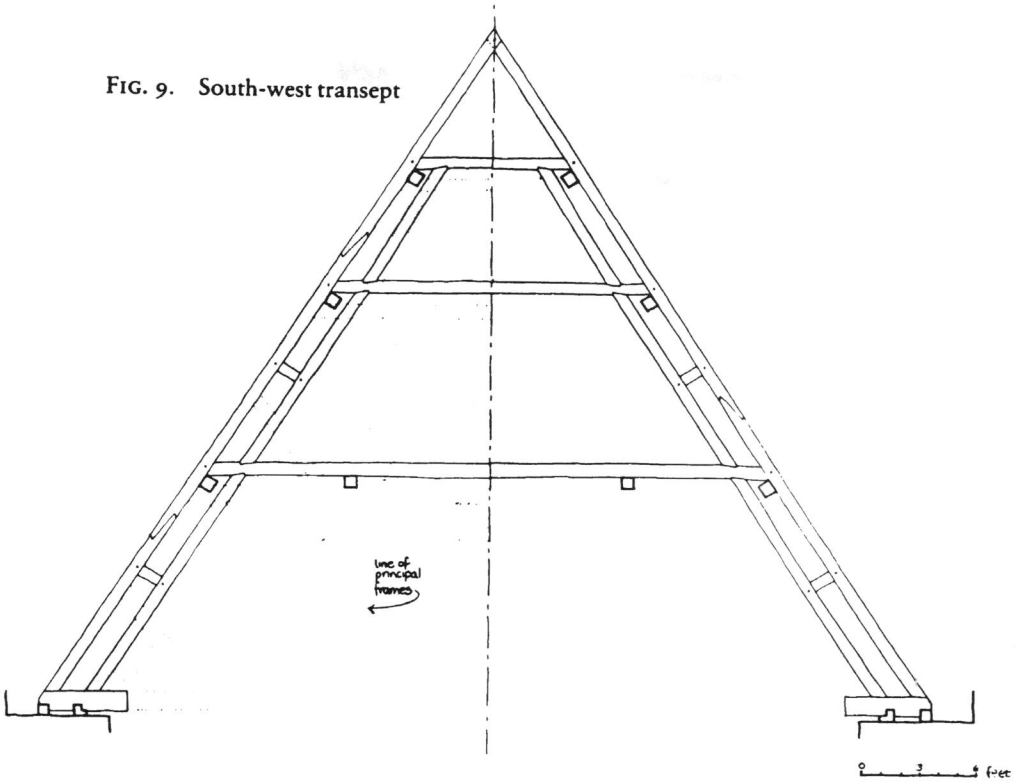

line of
principal
frames

timbers, for which there is no reason to suppose felling did not take place within the space of a few years, is to add thirty years to the average date of the last heartwood rings.[17] It has been shown as a result of statistical evaluation that thirty rings is the average sapwood growth to be expected on a mature oak with a 95% confidence interval of between twenty and fifty rings.[18] The result for St Hugh's Choir timbers is a felling date of 1202–3 (see Table II, Summary, I1) which, assuming the timber was used 'green' as was usual, should be very close to the construction date. However, where a very large project was involved as in this instance it is perhaps unrealistic to expect all the felling to have taken place in a single year. The historical evidence together with the dendrochronological evidence makes it virtually certain that none of the timber was felled before 1192. There is no evidence for the re-use of timber. The only sample (no. 148, see Table I) with complete sapwood on it gave a felling date of 1195–6. This is, however, probably too early a felling date to be generally applicable. For example, the sample (142) from a comparatively young tree with heartwood/sapwood boundary at 1189 showed a greatly increased ring width in the last five years of the heartwood (average ring width, 2.9 mm). If this growth pattern was continued into the sapwood then thirteen or fourteen rings (i.e. a thickness of about 4 cm) would be a reasonable estimate. Because of the inclusion of this particular timber in the roof its construction date is unlikely to be any earlier than the averaged felling date of 1202–3, and the possibility remains that the felling date of some of the timbers may have been later.

The re-used timber in the roof of the Angel Choir is so homogeneous, with similar growth rates and patterns with that from St Hugh's Choir, that it very probably all came from the same source, for the statistical quality of the match between the two chronologies has the very high t-value of 14.4.[19] It is possible, however, that older trees were selected first, for of the ten samples of re-used timber that have yet been dated all seem to have started growing in the 10th century and most of them in the early part of that century (see Table II, 1).

No timber in the choir transept has yet been sampled but the indications are that it is of similar age and character to the earliest of that in the choirs. The rafters over the two inner bays of the south transept (seventeen couples) have all been sawn (Pl. XIc) from huge oaks over 40 ft in usable length. The same is true also in the inner bay of the north transept (nine couples) but those of the middle bay are scarfed (ten couples) just below the upper collar (Fig. 6c), as are most of those in St Hugh's Choir. The tie-beams too are expensive in timber since straight pieces 28 ft long were required — those over the major roofs are 46 ft long. Together with the rafters they are the longest members in the roof.

The first mention of timber for the Cathedral is in a letter of King John dated January 1209 requesting that the canons of Lincoln be allowed to take from the forest the timber that they have acquired and also the lead they had purchased for the building of their church.[20] This must have been roofing material. The use of the word *maeremium* in the original document indicates that timber suitable for heavy construction was involved and its association with lead leaves little doubt of it. This implies that there was building ready to be roofed. Unfortunately the dendrochonology of St Hugh's Choir roof is not sufficiently precise to make it clear if that was the one in question. It is, however, possible that some of the timber (e.g. Table I, Sample nos 09, 68, 78, 90, 114 and 142) was felled c.1209 and used in that roof together with the last of the timber from earlier felling(s) (e.g. Table I, sample nos 53/61, 98, 106, 120, and 148). Alternatively it may have been used for lesser roofs over aisles and chapels. As nothing is known of the quantities involved further speculation is pointless as sampling in the east end of the nave roof and elsewhere should eventually give further information.

The concern of the king in the matter implies that the forest (*foresta*) referred to was royal forest. The nearest available source of lead to Lincoln was in the royal forest of High Peak, north-west Derbyshire — one of the royal forests, like Dartmoor, where trees were thin on the landscape. Lead could have been bought at Lenton Fair (Notts.) as the monks of Lenton Abbey had a tythe of lead produced from this source.[21] Sherwood Forest was the nearest and most likely source of the timber.

The timber might have been felled before 1209 for in his letter the king is pleading for the canons to be allowed to take what already belonged to them. This implies some delay in the progress of the work. A likely cause was a dispute between the clergy and royal officials in connection with the Interdict which had been declared about nine months earlier.[22] Evidence of a break in the construction of the nave at Wells Cathedral has been observed in both the masonry and the roofing and has been convincingly attributed to effects of the Interdict.[23] Another cause of delay, or at least slow progress, could have been the lack of a bishop at Lincoln to give direction and encouragement to the work for much of the first decade of the 13th century. William of Blois held office no more than three years (1204–6) and although his successor Hugh of Wells was appointed (May) and consecrated (December) in 1209, work presumably ceased when he sought refuge on the Continent and may not have resumed until the end of the Interdict in 1213. Even then Bishop Hugh would still have had over twenty years of cathedral building ahead of him although the only documentary record of it is the opening of a new quarry in the ditch of Lincoln Castle by royal grant in 1231.[24] At his death in 1235 another substantial part of the Cathedral must

N. D. J. FOOT, C. D. LITTON, AND W. G. SIMPSON   59

TABLE I

St Hugh's Choir roof. Analysis of timber samples. (All have base of sapwood)

Sample nos	Frame nos	No. of heartwood rings	Average ring in mm	Centre ?	Sap rings	Dates Start	End†
	East End						
82	1	114	2.0	No	13†	1063	1176
131	2	115	1.3	No	—	1058	1172
09	4	164	1.5	No	—	1015	1178
122	5	120	1.2	Yes	—	1052	1171
34	9	98	1.9	No	—	1079	1176
142	9	122	1.6	Yes	—	1068	1189
120	10	224	1.0	Yes	—	938	1161
117	12	96	1.4	Yes	—	No match‡	
106	13	123	1.1	No	17†	1042	1164
113	13	110	1.3	Yes	—	1068	1177
114	14	85	1.5	Yes	—	1097	1181
137	15	99	1.3	Yes	—	1078	1176
100	Wall-plate	177	1.0	No	—	998	1174
98	16	179	1.2	No	—	980	1158
148	16	198	0.7	No	35/6*	963	1160
94	16	231	1.1	Yes	—	935	1165
06	17	130	1.3	Yes	—	1039	1168
90	17	116	1.7	No	—	1066	1181
65	18	150	1.6	Yes	—	No match‡	
86	19	219	0.8	No	—	956	1174
78	19	159	1.2	No	—	1023	1181
73	19	108	1.4	Yes	7†	1070	1177
68	19	146	1.4	No	—	1033	1178
60	21	101	2.2	No	—	1077	1177
56	21	163	1.2	No	—	1009	1171
53/61	22	162	1.5	No	—	997	1158
76/79	23	110	1.8	No	—	1067	1176

* Complete sapwood
† Sapwood rings are not included in the dendrochonology which extends up to
   heartwood/sapwod boundaries only
‡ Identified as original timbers due to carpenters' marks

60   LINCOLN

TABLE II
Angel Choir roof. Analysis of timber samples

1. The reused timbers

Sample nos	Frame no.	No. of heartwood rings	Av. ring in mm	Centre ?	Sapwood ?	Dates From	To
161	51	190	1.0	No	Base?	980	1169
170	50	134	1.5	No	No	1024	1157
190	44	147	1.5	No	No	939	1085
191	42	178	1.1	No	No	949	1126
193	42	107	1.1	No	?	950	1056
194	42	133	1.4	No	No	954	1086
214	42	114	1.0	No	No	955	1068
215	38	167	0.9	No	No?	955	1121
219	36	243	0.8	No	No	912	1154
220	35	201	0.8	No	Base?	963	1163

2. The tie beams ('transversar')

157	52	167	1.3	Yes	No	1022	1188
200	46	176	1.9	Yes	?	1073	1248
202	43	130	1.7	No	No	1109	1238

Summary — Conclusions

No.	No. of samples	Average no. rings heartw'd	Average ring width	Average sapwood start	Estimated sapwood Total	Estimated age of trees (yrs)	Estimated felling years	Royal Grants	Comments
I 1	25	142+	1.3	1172/3	30	130–260	1195–1202+	—	St Hugh's Choir
II 2	10	170+	1.1	c.1166	30–35	over 200	c.1195	—	Angel Choir reused
II 3	3	158+	1.6	c.1248?	30?	about 200	c.1277–79	1276	Angel Choir tie beams

have been nearing completion for in his will dated the same year the Fabric Fund of the Cathedral received the sum of one hundred marks and all the timber that was felled on his estates at that time, reserving to his successor only the right of redeeming it for fifty marks.[25] Whether Bishop Grosseteste did so is not known but in April of the same year, in another royal grant to the Cathedral, one hundred trees from Mansfield Wood in Sherwood Forest were given.[26] Repair of damage done by the collapse of the central tower and the roofing of the Chapter House, the principal transept and nave would probably have taken all this timber (see p. 54). Nearly twenty years later there were two further grants from Sherwood Forest of fifty oaks in 1252,[27] and sixty oaks in 1254 on the accession of the new bishop, Henry Lexington.[28] Some of this was probably used to complete the roofing of the nave and western chapels, and dendrochronology suggests that some was probably left over for the roof of the next great building. This must have started sometime after 1256 when the Dean and Chapter received licence from the king to demolish part of the city wall in order to extend the Cathedral eastwards.[29] Twenty years later, in 1276, just four years before the consecration of the new Angel Choir there is a grant of '50 oaks in Sherwood Forest whereof 6 shall be *transversar*' for the new works of the king's gift'.[30] Oliver Rackham has suggested (pers. comm.) that *transversar*' here indicates the need for trees of a particular quality for a special purpose.

Tables I and II give particulars of the age, growth rates and dendrochronology of the timber used in the roofs of St Hugh's and Angel Choirs. These data, together with the distributions plotted in figure 10 show that at least three different qualities of timber were used in the Angel Choir. The re-used timber is contemporary and very probably from the same source as that of St Hugh's Choir (see p. 58). The different character of these timbers felled in late 12th and early 13th centuries from those felled in the later 13th century and used new in the Angel Choir roof is clearly shown in figure 10, which compares their radial measurements and their age (number of rings). The contrast between the two groups is in fact even greater than is here apparent, for the data for the early timbers are minimum figures since, because of the greater size and age of the trees used, many have lost rings in their reduction to the required dimensions, whereas those for the later timbers are maximum figures, the radii extending in all cases from the centre of the tree to the base of the sapwood. It is estimated that the earlier timbers came from trees 130–260 years old and the later timbers from trees 85–120 years old. As most of the latter have lost their outer rings of sapwood through insect attack and decay only 65–90 rings of their growth remains. Such short ring-sequences are difficult to match into a dendrochronology and further samples will be required before this can be done with confidence. The original (i.e., new) timber in the Angel Choir roof was identified as being of late 13th-century date by its lack of any indication of re-use and from the original carpenters' marks on the members sampled. Three of the later 13th-century timbers it will be noticed are grouped with the earlier timbers (Fig. 10). These samples are the only ones yet taken from tie-beams in the roof (Table II, 2). They match together to form a dendrochronology of 226 years from 1022 to 1248.[31] If it is assumed that the final ring is at the base of the sapwood this would allow a felling date of c.1278. From this date and the distinctive character of the timber there can be little doubt that these tie-beams are the *transversar*' referred to in the grant of timber from Sherwood Forest in 1276.

The re-used timber in the Angel Choir roof, like that of St Hugh's Choir is comparatively straight-grained and knot-free with few waney edges, while the later timber is more often knotted, crooked and shows the full circumference of the trunk indicating that the carpenters were getting the required length from their timber only with difficulty (Pl. XIB). They also had the problem of constructing a rigid structure from timbers some of which would warp and shrink and others of which would not. A means devised to counteract these potentially destabilizing effects was to pair the re-used timbers when used in a vertical position. Thus, for example, a soulace of re-used timber on the south side of a frame is matched by one of re-used timber on the north side also. Such preliminary dating results as have been obtained for the new timber used in the roof suggest that seasoned timber — presumably from the royal grants from Sherwood Forest in the 1250s — may have been used as well as 'green' timber of the 1276 grant. This mixing of timber of different felling dates and the use of timber from comparatively immature trees which has split and shrunk and shed its sapwood since it was built into the roof, and been patched and repaired in recent centuries, contributes to a general impression of 'slight scantling and bad work-manship' which Sir Robert Smirke also noted in the original roof of the principal transept. The fact that this roof has survived is a tribute not only to generations of Keepers of the

62   LINCOLN

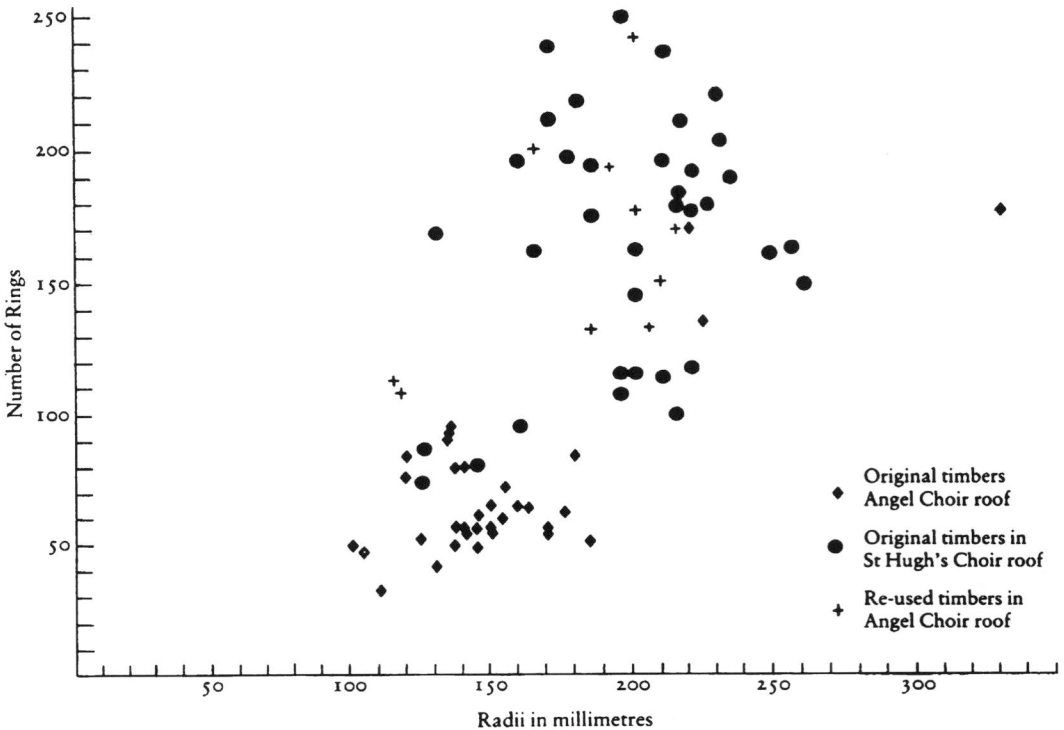

FIG. 10

Fabric but also to the master who, *c.*1200 designed a roof which has stood the test of time, and with minor modifications was suitable, not only for the earliest choir and the latest, but for most of the Cathedral's high roofing.

REPAIRS AND ADDITIONS TO THE MEDIEVAL ROOFS

For over two hundred years after the consecration of the Angel Choir in 1280 there is no evidence for any major work in the medieval roofs of the Cathedral. Nevertheless it is probable that routine maintenance and minor repairs took place. The lead, particularly on the steeply pitched east transept roofs, would not have remained in good order for so long. It was presumably following a stripping of the roof covering of St Hugh's Choir in preparation for re-leading that the four curious frames truncated above the bottom collar were inserted (Fig. 11). It is difficult to see how it could otherwise have been done. They are also in the nave roof and are spaced at six frame intervals as opposed to four over the choir. Four samples, from a brace and tie-beams of those in St Hugh's Choir roof have given a tree-ring date for the felling of the timber of *c.*1513.[32] Apart from their truncated form, the frames differ from the original tie-beam frames in the bracing of the basal angles. The passing brace construction used with secondary rafter and soulace in the medieval roof is here used with struts and simple braces which in the nave roof are curved. Mortice and tenon joints are used throughout and the occurrence of other joints (as in the queen-posts) or ironwork are, as will be shown, the result of later repairs.

N. D. J. FOOT, C. D. LITTON, AND W. G. SIMPSON   63

Indication of the function of these frames may perhaps be found in the collar-beams which invariably have their soffits about level with the tops of the lowest collars of the original frames. Thus short purlins lapped over the original collars abut against the faces of the intrusive collars and so help to prevent racking of the roof frames. Their squat form and solid tie-beams give them a much lower centre of gravity than the medieval frames and so make them anchor-points against which the latter are held by the ends of the purlins. If this theory is accepted it provides the earliest secure evidence for the introduction of purlins into any of the roofs. Further progress on this problem will involve more extensive sampling for dating purposes of the purlins themselves. Even then the interpretation of the results will be fraught with difficulty as purlins seem often either to be made from young trees, re-used timber or pine.

There is no evidence for further work in the roofs until the late 17th century. As in the medieval period the results obtained by dendrochronology are complemented by documentary evidence of varying quality and detail. Work resumed following the Restoration after

the Dean, Chanter, Chancellor and sub-Dean out of zeal to God's Glory and their good affection to the Church did declare — on May 11th, 1661 — that they would give £1000 towards the repairs of the cathedral church.[33]

Major new works were undertaken also with the building of the fine new library presented by Dean Honeywood and designed by Sir Christopher Wren in 1674.[34] In the high roofs of the Cathedral, new frames (nos 35 and 36) were erected over the west end of St Hugh's Choir (Figs 3 and 12). These are two king-post frames — there may once have been more — and five samples taken from their collars, rafters and curved braces gave a date of c.1682 for the felling of their timber. They provide the first evidence for the use of iron (other than nails which are found occasionally in even the earliest roofs) for the bridled scarfs of the rafters are secured with forelock bolts. Other repairs done with oak felled c.1700 were, the replacement of some of the lead-covered parapet boards along the eaves of the Angel Choir or eastern transept, and of some of the queen-posts of the St Hugh's Choir roof (Fig. 7e).[35] The latter were tenoned into the original mortices in the collars, but the joint into the tie beam was a lap-joint cut just in front of the original mortice and supported by an iron stirrup secured with a forelock bolt.

Repair works recorded in documentary sources from the 18th century onwards consisted of small scale operations like those just described interspersed with major works. The first of the latter was undertaken in the roof of the Angel Choir and for it, oak felled in 1703–5 was used.[35] The work involved the replacement of rotted wall-plates, sole-pieces, rafter-feet and ends of collar-beams and was probably completed within the first decade of the century. Then, similar repairs were carried out in the roof of St Hugh's Choir. Here the felling dates of the timber span a much longer period, from at least 1713–25, and this makes it difficult to know if the work progressed slowly, the trees being felled as required, or quickly, using timber which had been in store for varying lengths of time.[36] Perhaps the latter is more likely for there is documentary evidence for fund-raising in 1723 when a circular letter was sent to the prebendaries appealing for subscriptions towards repair of the Cathedral.[37] Three years later a general subscription was raised[38] and Bishop Reynolds gave the Dean and Chapter authority to use the medieval bishop's palace as a source of building material for the same purpose.[39]

From about the middle of the 18th century there is a fairly comprehensive series of architects' surveys, fabric accounts and contractors' bills. This is fortunate because about this time imported pine, which it would be difficult to date by dendrochonology, was generally used for roof repairs. In mid-19th-century accounts it is called Crown Memel or 'best Baltic timber'. James Essex, son of a joiner, became architect to the Cathedral c.1760 and seems to have first used 'fir' which was shipped up the Trent to Gainsborough in 1762 for the rebuilding of the roof of the Chapter House the same year.[40] The ironwork, particularly the stirrups and forelock bolt fastenings used here (Pl. XB) are readily distinguishable from the lighter type used in late 17th and early 18th centuries.[41] They can be found also in the roofs of the Angel and St Hugh's Choirs.

Work in St Hugh's Choir in Essex's time is recorded on a small lead plaque nailed to the north side of frame 23 naming J. Hayward, C[lerk] o[f] W[orks] (?) and dated 1765. Documentary confirmation of this is given, without being very clear about the extent of the work, in the copy of Essex's 1761 survey of the fabric in the Lincoln Archives Office[42] where he has added dates of completion for the repairs he proposed. It is unfortunately not clear if the ten massive frames (Fig. 3, nos 25–34; fig. 13) over the westernmost bay of St Hugh's Choir belong to this time. Much of their ironwork looks comparable to that of the Chapter House, but there are also screw-threaded nuts and bolts. The earliest evidence for their use in the Cathedral roofs comes in a general report of the Clerk of Works, William Hayward, dated 1799 in which he recommends their use to tighten loose joints in the roof of the north-west transept.[43] They are found in their most primitive, and so presumably earliest, form in the heightened roof of the Chapter House which belongs to the early years of 19th century (Pl. XB).[44] So the frames in pine in St Hugh's Choir roof probably belong to the 19th century and were the last major repairs there until those of 1979–80.

The roof over the eastern crossing has not yet been studied in detail but it certainly has oak members, which must belong to the earliest and/or latest phase of medieval building, as well as softwood timbers and elaborate iron strapping, some of which must have been inserted by Essex. In his survey of 1764 he has a sketch of the crossing roof supported on a central post which rests on the top of the stone vault beneath.[45] He recommended that this be removed and the weight taken by raking struts midway up the roof.

The roof of the Angel Choir has been extensively repaired over the last 250 years. Since much of the new timber has been softwood it is only possible to learn the details by study of the documents and the typology of the ironwork. Correlation of these sources suggests the general course of the repairs and additions to the roof. As indicated above, one of Essex's major concerns in his two surveys of the Cathedral was with the stresses laid upon the walls and vaults by roof frames 'running from the perpendicular' (racking) and with the injudicious attempts of his predecessors to correct the fault. Attempts had been made to counter the racking by bracing the sole-piece frames against the centres of the tie-beams. The extra weight upon them and their weakness from rotting at the ends caused them to sag which made their joints with the queen-posts spring. So they became ineffective in their function of holding the roof-line together laterally and its spreading outwards caused stress upon the wall-tops. This in turn put stress upon the vaults, as did the insertion of props, under the tie-beams to prevent them sagging, which rested on the vaults. Essex's solution to the problem was that:

In all parts of the roof the [tie] beams may be restored to their original use by hanging them to the collar-beams as they were at first and to hinder the rafters from running braces [i.e. raking shores?] should be placed under them so as to press upon the ends of the [tie] beams and not upon the middles of them, or near it, as many do now.[46]

These observations were occasioned by his examination of the roof of the north-west transept, but he stated that they were equally applicable to all the high roofs. In his survey of 1764 he recorded what had been done in the intervening three years:[47]

N. D. J. FOOT, C. D. LITTON, AND W. G. SIMPSON   65

The roofs over the East end and over the nave which have been repaired are done very well and made secure. The smith however has overloaded the roofs with iron. Many of the king (*sic*) posts have heavy straps at the bottoms and for the same quantity of iron he could have put them also at the tops where they are much needed now on account of the additional weight of iron below.[48]

The above remarks and quotations from Essex's surveys give an insight into his scholarly interest and thorough understanding of the Cathedral and its maintenance requirements. In the Angel Choir, Essex's contribution to the repair and maintenance of the roof must have been the introduction of raking shores in softwood, the use of softwood for repair or replacement of existing timber and the securing of queen-posts to collar and tie-beams with iron stirrups and forelock bolts.

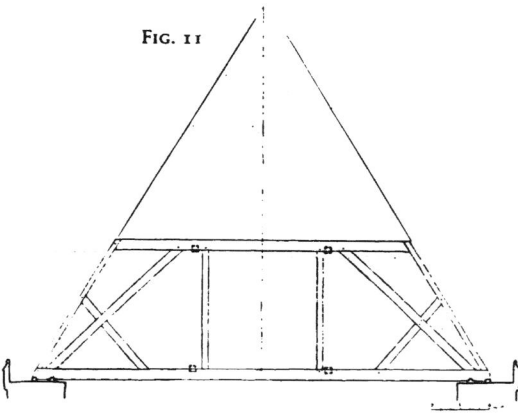

FIG. 11.
St Hugh's Choir: truncated frame

FIG. 12.
St Hugh's Choir: frames 35 and 36 against the Tower (*c.*1700)

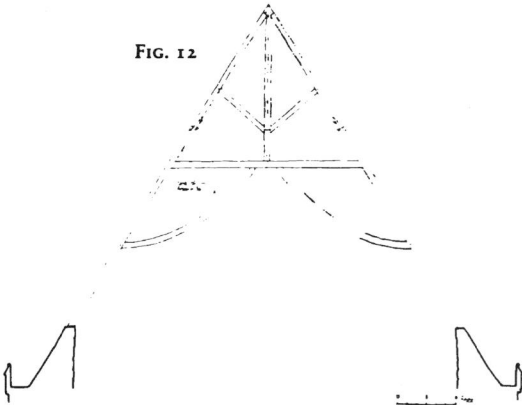

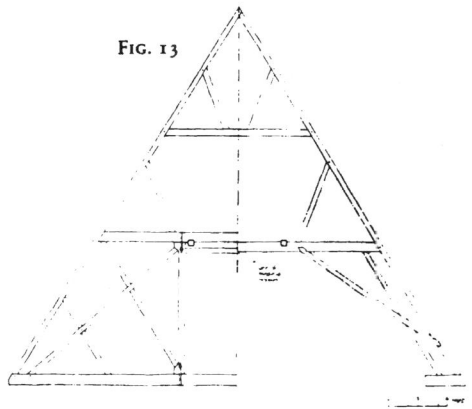

FIG. 13.
St Hugh's Choir: softwood frames

FIG. 11

FIG. 12

FIG. 13

66   LINCOLN

The next occasions when the Angel Choir roof features prominently in the Cathedral archives were in Charles Hollis's report of 15 September 1825 and Sir Robert Smirke's report of the following year.[49] They noted that rotting of the timbers due to damp affected particularly the sole-pieces, wall-plates, tie-beams and rafters in the area of the eaves angle. Some of the collar-beams and queen-posts were also decayed and loose from the framing. The replacement of rotted timber with new pieces scarfed onto the old was advocated. Hollis wrote:

As most of the tie-beams require to be replaced it would be injudicious to trust so extensive a building to so many pieced and repaired tie-beams. It would be proper to introduce some new beams and truss framing and connect the rafters by purlins.[50]

This was very much what was done, presumably following his advice and under his supervision. The rotted ends of the tie-beams were cut off and either a new piece of softwood was scarfed onto what remained of an original oak tie-beam or two pieces of original tie-beams were scarfed, bound with iron stirrups and bolted together with screw-threaded nuts and bolts to make up the length of one (Pl. XD). Only the tie-beam of frame 52 survived this drastic operation intact — until it too had its south end replaced in recent repairs. This work left eight major frames without tie-beams and these were rebuilt in softwood below the bottom collar. The same design was used as for frames nos 25–34 in St Hugh's Choir (Figs 3 and 13).

Eastward racking in the roof was still a serious problem, the rafters being (in 1826) between 6 in. and 3 ft out of perpendicular.[51] Purlins, which Hollis advocated, were helpful but 'struts and braces placed to correct the evil (have) introduced other faults' — as Essex found.[52] The medieval builders expected that the frames of the roof would be kept in place by the horizontally laid battens connecting the rafters and by their own weight and that of the covering lead. The great advantage of having stabilising properties built into the covering skin was that the weight was equally distributed over the whole structure and dangerous stresses would not be generated. Unfortunately the skin was not up to this task — at least not over long periods of time. What was needed was a covering having the strength and flexibility of a clinker-built boat. A means to achieve this was found probably sometime in the 19th century by laying the battens diagonally across the rafters and in opposite directions on either slope of the roof.[53] At Lincoln battens laid diagonally in the same direction on the lower roof slopes and horizontally above are found in the eastern transept, the east end of the Angel Choir and in the south-west transept. In the latter roof it seems quite likely that the battens are of the same date as the roof itself (see p. 53) and a board inscribed in chalk 'James Smith, 1847' nailed to the rafters just below the middle collars of the east end of the Angel Choir roof suggests work in progress there at about the same time. It is not known when or where this technique was first introduced, but it is found also in the roofs of the nave and south-west transept of Westminster Abbey where Sir Gilbert Scott was consulting architect from 1849–78.

DISCUSSION AND CONCLUSIONS

Dendrochronological study of the re-used timbers in the roof of the Angel Choir and the original timbers of the roof of S. Hugh's Choir has shown that they were probably felled at about the same time and came from the same source. Superficial examination of the timber of the two inner bays of the eastern transept suggests that it too was of the same period. If it is accepted that the re-used timbers were part of the roof of St Hugh's presbytery then it had

N. D. J. FOOT, C. D. LITTON, AND W. G. SIMPSON   67

scissor-braced frames. It would have been the first roof of the church to be erected and the order of construction of the existing roofs can be deduced from the joints and associated carpenters' marks which are on one face only of the frames: that facing the direction in which the work was progressing. So it can be deduced that the south-east transept roof was built from the south towards the crossing, the north-east transept from the crossing northwards and St Hugh's Choir from the crossing towards the central tower. The roofs were probably built in this order. The crossing would have required some of the longest and most substantial timbers for the valley rafters and horizontal cross-members. Such timbers were also used in the roof over the two inner bays of the south-east transept and the inner bay of the north-east transept where every rafter was a single piece of timber. The notched-lap joints have the same profiles in these bays. It is only in the middle bay of the latter that the rafters are of two pieces scarfed together and the use of carpenters' marks and other slight differences of construction from the inner bay must indicate a different team of carpenters at work here. Whether or not it was roofed before work started on St Hugh's Choir is not certain, but here again another team of carpenters was involved for not only are the frames of a different design but the carpenters' marks are quite different.

The tree-ring dates for all those timbers from St Hugh's Choir roof which have trace of sapwood are given in Table I. They are not arranged in chronological order of their final ring-dates as is usual, but grouped frame by frame in sequence. It can be seen that the dates are entirely random in that there is no trend of early to later dates from east to west. This suggests that the timbers may have come from a yard where they had been accumulating over a number of years. The only timber with sapwood surviving complete was sample no. 148, from a tree well over two hundred years old felled in 1195–6, which was used in frame 16 right in the middle of the roof. Probably the last tree to be felled — a comparative youngster of 140 years growth — is represented by no. 142 which came from frame 9 at the east end of the roof. The chances of this having had only six or seven rings of sapwood and thus the same felling year as no. 148 are extremely small. Thirteen sapwood rings is about the minimum that can be expected. Clearly the Choir, although well-advanced in construction by the time of St Hugh's death in November 1200, cannot have been roofed and vaulted until some years later. At the consecration of the church in that year only the presbytery, the south-east and perhaps the north-east transepts can have been finished, for it is assumed that construction of the vaults would have closely followed completion of the roof which would have kept the weather off them while the mortar was setting. Their tie-beams might have formed a working platform and means of access from above.

The survival of two-thirds of the original roof of the choir with comparatively few repairs is in marked contrast to its western third which was entirely rebuilt in late 18th–early 19th century. On this evidence alone it can be confidently concluded that the western twelve frames (nos 25–36) of the original roof were destroyed in the collapse of the tower c.1238, and that their replacements, being less well constructed than the originals, in turn themselves needed replacing. That this is the correct explanation can be seen in the crown of the vaults beneath. To the east of frame 26 the upper surfaces have a mortar rendering as elsewhere over the original vaults of the east end, but to the west there is no rendering and their rough, uncoursed rubble construction can be seen. The roof and then the vaults would have suffered the severest damage from the falling masonry and the clerestory walls above the vaults rather less. From the eastern crossing the latter are of squared ashlar in random courses with a continuous chamfered plinth along the inner edge. The extent of damage done to the walls is indicated by its absence to the west of frame 29. The actual point of change is not easily seen as it is hidden by a modern walkway to a door giving out onto the roof parapet. These conclusions are consistent with ones reached from study of the underside of the vaults, clerestory windows and other masonry details of the western bay of the choir.[54]

There seems to be some indication in the early 13th-century roofs of the east end of a tendency to use more substantial timbers and older trees in the earliest work (see Table II.1). It is not yet known if there is a continuous decline in the age and quality of the timber used in the successively later roofs but figure 10 makes this contrast between the earliest and the latest roof very clearly. That suitable timber for the roof of the Angel Choir was difficult to obtain is shown by the special mention of the requirement of trees of sufficient size for tie-beams. It can also be seen in the roof itself in the re-use of timber, the quantity of waney edges, roundwood and timbers running out (Pl. XIB) and the need for additional soulaces and a third collar (Fig. 5c) to the frame designed originally for St Hugh's Choir roof. Here the carpenters were having to make do with timber of smaller dimensions than that to which they had been accustomed, and perhaps high demand on timber resources over the previous century had a lot to do with this state of affairs.

As the high roofs of Lincoln survive almost complete it can be seen how minor and comparatively few were the changes of design made over the period of about seventy-five years. Apart from those just mentioned, the only others of any consequence were to the rafter scarfs (Fig. 6) and the jointing of the queen-posts to the tie-beams (Fig. 7). Changes in the queen-post joints show attempts at the early solution of a problem which was noted by James Essex. The queen-posts were subject to tension and in the original (St Hugh's Choir) design hung literally upon the two pegs holding the tenons in their mortices. As the timbers seasoned *in situ* and the frames settled, the collars and tie-beams sagged at differential rates, the tension increased and the pegs gave way. It was probably the realisation of this eventuality which led to the adoption of the lap-joint in this position in the western half of the nave roof and the eastern half of the Angel Choir roof, where the cambering of the tie and collar-beams would have to some extent alleviated the stressful tendencies. Finally the lap-dovetail with square, housed shoulders was adopted in preference to the lap-joint in the western end of the roof. It is of interest to note that these members where they appear in French roofs are termed *clés pendantes*.[55]

In the analysis of the roof of the presbytery at Westminster Abbey it was observed that, in common with continental church roofs of 13th century, it had a pitch which was steeper than Romanesque predecessors.[56] This is nowhere more apparent than in the roofs of the eastern transept at Lincoln. The particular reason for their steepness must be that these minor transepts were not as wide as other parts of the Cathedral, yet it was desired that their roof lines should be level with those of choir and presbytery. The reason for the greater pitch of Gothic roofs is not known. As McDowall points out, buildings were generally not bigger than before but the thrust upon the walls would be reduced, following the reduction in the number of tie-beams to lighten the roof load — a development which is seen in continental churches.[57]

Although there is supporting evidence for this development in the roofs of continental churches there is very little evidence in this country. Such as it is, Waltham Abbey roof of mid-12th century on Hewett's reconstruction[58] has tie-beams to every frame, but is otherwise more comparable in design to other roofs which according to Hewett are either contemporary[59] (East Ham) or earlier[60] (Chipping Ongar) and apparently have no tie-beams at all. The same basic design on a larger scale is found in the early 13th-century roofs over the nave of Wells Cathedral. Here by reason of the greater span, additional supporting timbers were required. The addition of soulaces below the lowest collar together with the ashlar-pieces and the rafters, make a seven-canted void. There is also an upper collar and a post between it and the lower one, and presumably there were originally tie-beams at regular intervals. Hewett had Wells Cathedral particularly in mind, when he suggested that the Waltham Abbey roof was basic to a long series of developments traceable in English high roofs.[61] A later form of the design is found at Lincoln in the roof of the Chapter House vestibule (Fig. 2) which has a tree-ring date of $c.1234$. The frames have tie-beams every fourth couple and span 26 ft. There are two collars, with V-struts to the rafters on the lower, and below, soulaces and ashlar-pieces framing a seven-canted void. At the west end of this roof smaller, contemporary roofs of almost the same design are found over the stair turrets, except that here, since the spans are less, the soulaces are omitted, so that ashlars, rafters and collar frame a five-canted void. These roofs are of the same design as those over the West Range at Ely and the nave of the parish church at Little Hormead, Herts. which, if contemporary with the buildings they cover, should date to the end of the 12th century.[62] At the west end of the Cathedral the Consistory Court ($c.1250$) has a roof spanning 30 ft of a similar but slightly modified design to that over the Chapter House vestibule.

The roofing tradition outlined above has its beginnings in Romanesque building, perhaps even before the Conquest. All the roofs so far discussed are church roofs designed to sit atop stone walls. It is perhaps unfortunate, therefore, that much of the discussion about the introduction of continental forms of diagonal bracing into Britain has involved early domestic buildings also, and so obscured the fundamental importance of church roofs which was first recognised by Fletcher and Spokes.[63] They have summarised the continental evidence and described three types of diagonal (or 'passing') braces which appear in France in the second half of the 12th century. They are scissor braces, secondary rafters and lateral longitudinal braces. The first two forms were used contemporaneously in the east end of the Cathedral (Fig. 4).

Roofs with secondary rafters comparable to those at Lincoln are found in primitive form over the north transept of Noirlac Abbey, the nave of Lisieux Cathedral and later in the roof over the chapel of the hospital of St John at Angers,[64] founded by the English king Henry II. Most features of the roof of the south-east transept at Lincoln are found in the 12th-century roof over the nave of the church at Hermonville (Marne).[65] The principal differences are that the secondary rafters pass across V-struts in the base angles rather than soulaces, as at Lincoln, and it has the low pitch (42°) of a Romanesque roof. Frames with sole-pieces alternate with tie-beam frames (Table III).

Scissor-bracing on the Continent may not occur quite so early or so often. It is found in the roofs at Lisieux and Angers just mentioned. The latter is of great size with tie-beams 59 ft (18 m) long of two timbers scarfed together. They are spaced at every five frames and are supported at third intervals along their length by two queen-posts (*clés pendantes*) which are halved across them and across the scissor-arms and are lap-jointed into the rafters.[66]

These two types of bracing, which occur in north-east France either separately or even at Angers, with queen-posts, together in a single roof, do not seem to be found in English roofs over stone buildings before the late 12th century when they are found at Lincoln. It is doubtful if there is any lateral longitudinal bracing at Lincoln in any of the original roofs. Its function, like purlins, was to link the frames together and keep them upright. Deneux observed that although wall-plates were normal there was not much evidence for either lateral longitudinal bracing or purlins in French roofs before the 13th century, although many of those roofs without had one or both added later.[67] It is not unexpected, therefore, that there should be no evidence of either in the earliest Lincoln roofs, but perhaps rather more surprising that there is little firm evidence of either in the later medieval roofs also. A useful summary of the principal characteristics of French roofs of the 11th and 12th centuries is given by Fletcher and Spokes.[68]

TABLE III

Earliest Continental and English roofs on stone buildings

	FRANCE, W. GERMANY, BELGIUM			ENGLAND		
	Romanesque tradition	Early Gothic (passing braces) Secondary rafters	Scissor braces	Romanesque tradition	Early Gothic (passing braces) Secondary rafters	Scissor braces
1100	Soignes Cathedral (nave)[57]			Chipping Ongar (chancel)		
				East Ham (chancel)		
1150	Soignes Cathedral (choir)[57] Paris, Montmartre (nave)[65]	Noirlac Abbey		Waltham Abbey (nave)		
		Lisieux Cathedral Angers (St Johns Hospital Chapel)	Troyes, La Madeleine (nave)[63] Tournai Cathedral[63]	Adel Church (chancel)[81]		
		Hermonville (nave)[63] Blaton Church (nave)[63]		Little Hormead Church		
1200		*Mantes, Notre-Dame[65]		Ely, West Range Wells Cathedral (nave)	Lincoln Cathedral (choir & transept)	Lincoln Cathedral (presbytery & transept)
		*Etampes, Notre-Dame choir[63]	Bayeux Cathedral[7] *Tours Cathedral[63]	Lincoln, Chapter House	Lincoln (nave)	
		*Grossenbuseck[80] (chancel)		Roydon Church (nave)[5]	Beverley (choir)	
1250		Marburg (nave)[4] (St Elizabeth Church)		Lincoln, Consistory Court	York (north transept)	*Salisbury (transept) Ely (presbytery) *Westminster Abbey
1300	*Paris, St Leu-St Gilles[65]		†Bourges Cathedral[7]			*Merton Coll, Oxford York, Chapter House

* Indicates roofs which have contemporary lateral longitudinal bracing and/or purlins

† With ridge-pieces only

N. D. J. FOOT, C. D. LITTON, AND W. G. SIMPSON   71

The Lincoln roofs are perhaps a few decades later than the French examples cited. Their remarkable features are the very steep pitch of the transept roofs, the contemporary but discrete use of scissor bracing and truncated secondary rafters in those roofs, evidence for at least two teams of carpenters at work, and the appearance (over St Hugh's Choir) of a roof design with secondary rafters and queen-posts which was to be used in the Cathedral for the next seventy-five years. The width of the eastern transept roofs is such that there would have been no difficulty in roofing them with a traditional type of Romanesque roof without passing-braces. The nave and the choir were presumably considered too wide for a roof of this type. It was used over the nave at Wells but there the span was at least 5 ft less than Lincoln. The significance of the variety of the roofing of the east end of Lincoln Cathedral is probably that innovation and experiment were in progress. The roofs are perhaps the earliest with passing-braces in the country and built at a time when carpenters had yet to reach a satisfactory solution of the problems of building higher roofs which combined the advantages of the new continental technology with the old form of Romanesque truss.

The Lincoln design of roof truss seems to have been seldom imitated elsewhere. The only obvious example is at Beverley where there are many points of comparison with Lincoln in the masonry detail also.[69] It may be significant too that some of the roofing there is built of timber from Sherwood Forest. A grant of twenty oaks to the provost of Beverley is recorded in 1240;[70] there was another of forty oaks to him in 1244;[71] and a grant of sixty oaks 'to the treasurer of St John's Church for its building' was made in 1252.[72] The earlier grants may have been to the provost personally and the later one has been taken to refer to the choir, but it is possible that all three were used in the Minster, the latest and largest being intended for the greatest area of roofing over the principal transept.[73] The roof over the north transept at York Minster, of probably mid-13th-century date, is known from a drawing by A. Pugin.[74] It had a design which, like Angers, combined secondary rafters with scissor braces and is in some respects comparable also to the present roof of the south-west transept at Lincoln, which gives credence to Hewett's views on it (see p. 53). The roof over the north-east transept of Salisbury Cathedral c.1240 has scissor-braces with queen-posts but there similarity ends, for here there are original and more advanced features, not found anywhere at Lincoln, such as secret notched-lap joints and side purlins.[75]

By contrast the scissor-braced design rejected at Lincoln seems to be at the head of a long succession of almost identical designs:[76] (Ely presbytery, c.1245 and Blackfriars, Gloucester, c.1255) or derivative designs[77] (Salisbury, c.1240 and Westminster Abbey, N. transept and presbytery, c.1269), which continue to the end of the century[78] (Merton College Choir, Oxford, c.1294) and into 14th century[79] (York, Chapter House vestibule), (Table III). Blackfriars, Gloucester is the only other building where the rather weak scissor-splay-stopped scarf joint used in the roof of St Hugh's Choir has been found.

The roof construction of St Hugh's Choir, was original and innovative when first built, but the Lincoln carpenters repeated the design with few significant modifications of their own or adoption of new developments taking place elsewhere. Other master carpenters, among them builders of scissor-braced roofs, were more innovative or readier to accept new technology as the example already given of the north-east transept roof of Salisbury serves to show.

ACKNOWLEDGEMENTS

We are grateful to the Dean of Lincoln, the Very Reverend, the Hon. Oliver Fiennes who has permitted us access to the Cathedral roofs; to Reg Godley and Peter Hill, successive Clerks of Works, and their staff for their assistance with problems of sampling and surveying; to Keith Murray, consultant

72   LINCOLN

architect to the Cathedral; to Professor M. W. Barley, Mr A. Cameron, Dr P. W. Dixon and Dr Oliver Rackham for helpful discussion of the results; to Dr C. Salisbury for his photographic record of the roofs and to Dr R. Laxton for his help and encouragement in many aspects of the work, both academic and administrative; to Arthur Robinson and John Bloor of the Works Department, University of Nottingham for their assistance in the preparation of samples and maintenance of equipment and to Mrs Ruth Robinson for the typing of this manuscript.

SHORTENED TITLES USED

HEWETT (1974) C. A. Hewett, *English Cathedral Carpentry* (London, 1974)

HEWETT (1980) C. A. Hewett, *English Historic Carpentry* (London, 1980)

SMITH (1974) J. T. Smith, 'The Early Development of Timber Buildings', *Archaeol. J.*, 131 (1974)

McDOWALL *et al.* (1966) R. W. McDowall, J. T. Smith, and C. F. Stell, 'Westminster Abbey. The Timber Roofs of the Collegiate Church of St Peter at Westminster', *Archaeologia*, 100 (1966)

LAXTON *et al.* (1985) R. R. Laxton, C. D. Litton, and W. G. Simpson, 'Tree-Ring Dates for East Midlands Buildings', *Vernacular Architecture*, 16 (1985)

REFERENCES

1. Hewett (1974).
2. Ibid., fig. 9.
3. W. J. Blair, J. T. Mumby and O. Rackham, 'The thirteenth century roofs and floor of the Blackfriars priory at Gloucester', *Med. Archaeol.*, 22 (1978), 105–22.
4. Smith (1974), 238–63.
5. Hewett (1980), figs 139 and 234. The terms used for the carpentry joints and structural ironwork and reinforcement in this article are those used by Hewett.
6. LAO, D & C. The Ark, no. 22.
7. Hewett (1980), fig. 79. McDowall *et al.* (1966), 167.
8. LAO, op. cit., note 6.
9. Hewett (1974), fig. 19.
10. Hewett (1980), fig. 278.
11. British Library, *Add. Ms.* 6769, 122.
12. F. Bond and W. Watkins, 'Notes on the Architectural History of Lincoln Minster from 1192 to 1255', *RIBA Journal*, 18 (nos. 2–3; 1910), 34–50, 84–97.
13. Rev. E. Venables, 'Notes of an examination of the Architecture of the Choir of Lincoln Cathedral with a view to determining the chronology of St Hugh's work', *Archaeol. J.*, 32 (1875), 229–38.
14. R. R. Laxton, C. D. Litton, W. G. Simpson and P. J. Whitley, 'Tree-Ring Dates for some East Midland Buildings', *Trans. Thoroton Soc.*, 86 (1982), 73–8.
15. Ibid., No. 13a.
16. H. C. Darby, *Domesday England* (Cambridge 1977), fig. 62 gives some idea of the local sources from which this timber may have come.
17. R. R. Laxton, C. D. Litton and W. G. Simpson, 'Tree-Ring Dates for East Midland Buildings: 2', *Trans. Thoroton Soc.*, 87 (1983), 40.
18. M. G. L. Baillie, 'Some thoughts on Art-Historical Dendrochonology', *J. of Archaeol. Science*, 11, no. 5 (1984), 381–3. M. K. Hughes, S. J. Milson and P. A. Leggett, 'Sapwood Estimates in the Interpretation of Tree-Ring Dates', *J. of Archaeol. Science*, 8, no. 4 (1981), 381–90.
19. Laxton *et al.* (1985), 40. For the significance of t-values see M. G. L. Baillie, *Tree-Ring Dating and Archaeology* (Croom Helm 1982).
20. *Cal. Patent Rolls I* (1201–16), 88. The translation of the text which has been followed here is that given by G. A. Poole, 'The Architectural History of Lincoln Minster', *AASR*, 4 (1857–8), 18 and 39.
21. VCH *Nottinghamshire*, 2, 91. We are grateful to Mr Alan Cameron of the Manuscripts Department, University of Nottingham for this reference and also for helpful discussion of the text.
22. A useful discussion of the Interdict and its effect, particularly on the sees of Wells and Lincoln may be found in J. A. Robinson, *Somerset Historical Essays* (London 1921), 141–59. For a good, more recent account of its wider implications see W. L. Warren, *King John* (London 1978), 164 ff.

23. Hewett (1974), 15 and Hewett (1980), 69.
24. *Cal. Close Rolls* (1227–31), 472.
25. C. W. Foster, *The Registrum Antiquissimum of the Cathedral Church of Lincoln*, 2 (LRS, v. 28, 1933), 72.
26. *Cal. Close Rolls* (1234–37), 77. We are grateful to Dr Rackman for bringing this source to our notice.
27. Ibid. (1251–3), 36.
28. Ibid. (1253–4), 296.
29. *Cal. Patent Rolls* (1247–58).
30. *Cal. Close Rolls* (1272–9), 277.
31. Laxton *et al.* (1985).
32. Op. cit., note 17.
33. British Library, *Add. Ms.* 32639, f. 256.
34. The contract between Dean Honeywood and William Evison, the builder, dated June 1674 is published in full in *Wren Society Publications*, 17, 76–77.
35. Laxton *et al.* (1985).
36. Op. cit., in note 14.
37. LAO, A/4/10, no. 45.
38. T. Cocke, 'James Essex, Cathedral Restorer', *Architectural History*, 18 (1975), 12–22.
39. T. Ambrose, *The Bishop's Palace, Lincoln*, Lincolnshire Museums Information Sheet, Archaeology, 18 (Lincoln 1980).
40. LAO, Bj/1/13–16. British Library *Add. Ms.* 6772, 264ᵛ–265.
41. Hewett (1980), figs 241–2.
42. LAO, D & C. The Ark, no. 22.
43. Ibid. For a history of the development of these fastenings see 'Nuts and Bolts', *Scientific American*, 250, no. 6 (June, 1984), 108–15.
44. Rev. E. Venables, 'Architectural History of Lincoln Cathedral', *Archaeol. J.*, 40 (1883), 397n.
45. British Library *Add. Ms.* 5842, 341.
46. LAO, D & C. The Ark, no. 22. 7.
47. LAO, D & C. The Ark, no. 22, 9. and British Library *Add. Mss.*, 6769, 121–7.
48. Op. cit., note 45.
49. LAO, D & C. The Ark, no. 22.
50. Ibid.
51. Ibid.
52. Ibid.
53. McDowall *et al.* (1966), 158, 166.
54. J. Baily, 'Some preliminary observations on the history of Lincoln Minster in the years 1195–1255' (unpublished Ms. Leeds School of Architecture, no date), 53. B/E, *Lincolnshire* (Harmondsworth 1964), 97.
55. Smith (1974), 251.
56. McDowall *et al.* (1966), 173.
57. J. M. Fletcher, 'Medieval Timberwork at Ely', in *Medieval Art and Architecture at Ely Cathedral. BAA CT* (1979), fig. 2, 62.
58. Hewett (1980), fig. 46.
59. Ibid., figs 31 and 32.
60. Ibid., fig. 30.
61. Ibid., 57, 72 and fig. 71, 83.
62. J. M. Fletcher and F. W. O. Haslop, 'The West Range at Ely and its Romanesque Roof', *Archaeol. J.*, 126 (1969), 171–6. C. A. Hewett and A. V. B. Gibson, 'The Nave Roof of the Parish Church of St Mary, Little Hormead', *Herts. Archaeol.*, 3 (1973), 126–7.
63. J. M. Fletcher and P. S. Spokes, 'The origin and development of crown post roofs', *Med. Archaeol.*, 8 (1964), 157.
64. Ibid., 158–66. Smith (1974), fig. 2.
65. H. Deneux, 'L'évolution des charpentes du XIe au XVIIIe siècle', *L'Architecte*, n.s. 4 (1927), 50, fig. 72.
66. Smith (1974), 243, fig. 2; 251, fig. 6.
67. Ibid., and op. cit. in note 65, 53 (6).
68. Op. cit. in note 63, 158–9.
69. L. Hoey, 'Beverley Minster in its 13th century context', *J. Soc. Architect. Historians*, 43, no. 3 (October 1984), 209–24.
70. *Cal. Close Rolls* (1237–42), 181.
71. *Cal. Close Rolls* (1242–7), 182.
72. *Cal. Close Rolls* (1251–3), 63.
73. Smith (1974), 245n.

74. Ibid. It was published as Pl. xxviii in John Britton's *History and Antiquities of York Cathedral* (London 1819) and is reproduced in P. Kidson, P. Murray and P. Thompson, *A History of English Architecture* (Penguin 1969), 91, fig. 33.
75. Hewett (1980), 91–2.
76. Op. cit. in note 57, fig. 1 and op. cit., note 3.
77. Hewett (1980), 91–2; McDowall *et al.* (1966).
78. Op. cit. in note 63, fig. 49c.
79. Hewett (1974), and E. A. Gee, *York Minster: The Chapter House* (Royal Commission on Historic Monuments, 1974), fig. 41.
80. C. A. Hewett, 'The dating of French timber roofs by Henri Deneux: an English summary', *Trans. Ancient Monuments Soc.*, n.s. 16 (1968–9), 97, fig. A.
81. W. H. Lewthwaite and R. D. Chantrell, 'The Norman Roof of Adel Church, near Leeds', *AASR*, 19 (1887), 100–20.

Xᴀ. Angel Choir roof. The middle collar of frame 38 is a re-used collar from a scissor-braced roof. The joint-bed is too wide for the head of the soulace and has been reduced by insertion of a pad

Xʙ. Chapter House roof: part of the double decagonal ring-beam at the base of the roof. The lower timber of James Essex' building (1762) is pierced by a forelock bolt and the upper timber of *c.* 1800 by a screw-threaded bolt secured with washers and nut

Xc. Angel Choir roof: base of queen-post, frame 52, with lap-dovetail joint to the tie-beam. The iron stirrup is 19th-century

Xᴅ. Angel Choir roof: original queen-post with lap-joint to an original (later inverted) tie-beam, frame 7. The inverted (filled) joint-bed is original and the upper (actual) joint-bed and iron stirrup are a result of 19th-century repairs

XIA. St Hugh's Choir: view over the top of the vaults looking south showing the line between the original work (left) and the rebuilt vaulting following the collapse of the tower c. 1238 (right). The tie-beam of frame 27 runs across the top right corner of the picture

XIB. Angel Choir roof: secondary rafter on the south side of frame 46 showing its roundwood profile

XIC. North-east transept roof: pit-saw marks on the face of the lower collar, frame 11

XID. Angel Choir roof: the middle collars showing their cambering. Note the purlin (left) which is a re-used timber

18

Viollet-le-Duc and the flèche of Notre-Dame de Paris: Gothic carpentry of the 13th and 19th centuries

Lynn T. Courtenay

I. THE RESTORATION OF THE FLÈCHE OF NOTRE-DAME[†]

The flèche [spire] of Notre-Dame de Paris by Viollet-le-Duc, models of which were recently on display at the Hôtel de Sully (Fig. 1), is one of the early, influential examples of gothic timber spires reconstructed in the 19th century.[1] In the year of its completion, 1860, Viollet-le-Duc described with enthusiasm his construction 'en charpente et plomb' and celebrated simultaneously the achievement of Parisian carpenters of the 13th century as well as his own 'gothic' re-creation (Fig. 2). He also firmly stated that his monumental spire was 'entièrement construite en bois de chêne de Champagne' and not of iron, as was erroneously

[†] Author's original English text: Lynn T. Courtenay, 'Viollet-le-Duc et la flèche de Notre-Dame de Paris: La charpente gothique au xiiie et xixe siècle', *Journal d'Histoire de l'Architecture*, 2, [1989], 53-68.

Research support was provided by funds from the Alfred P. Sloan Foundation in conjunction with the Sloan Seminar at Princeton University, 1988. I am grateful to the staff of the Centre Recherches Monuments Historiques, especially: Francoise Bercé, Daniel Bontemps and Madame Fresnault, as well as to Bernard Fonquernie, architecte en chef, at Notre-Dame de Paris.

[1] *Le Mont-Saint-Michel, l'archange, la flèche*, Exposition Catalogue, C.N.M.H.S., Paris, 1987. The exhibition included two models (c. 1860) of Viollet-le-Duc's timber framing, but neither is illustrated in the catalogue. In the same work see Jean-Michel Leniaud, 'Les flèches au XIXe siècle', 17-29.

Figure 1: Model of the framing of Viollet-le-Duc's flèche of 1857, C. R. M. H. (author's photograph)

believed.[2] Thus in 1857, when the new iron industry was developing rapidly in France — and a date before

[2] E.E. Viollet-le-Duc, 'La Flèche de Notre-Dame de Paris,' *Gazette des Beaux-Arts [GBA]*, VI (1860), 37.

2

which iron had already been used in several major gothic restorations, such as Alavoine's controversial spire of Rouen Cathedral or, the new, main-span roof of Chartres Cathedral[3] — Viollet-le-Duc chose without hesitation to continue tradition by rebuilding the flèche in timber'. He did so, despite the fact that he had used iron trusses and reinforcement ubiquitously in the restoration of Notre-Dame de Paris.[4] But for Viollet-le-Duc, the central flèche of the Cathedral of Paris (a timber structure of considerable historical and aesthetic importance) was quite another issue. The original carpentry could only be replaced by a new work of timber! For, in addition to the influential legacy of his senior colleague J.-B. Lassus, a pioneer in the study of

---

[3] The medieval roof of the nave of Chartres Cathedral, seen by Viollet-le-Duc before its destruction by fire on June 4, 1836, was replaced by cast and wrought iron trusses, completed in 1841 by Mignon, 'constructeur de la charpente en fer et fonte'. Abbé M. J. Bulteau, *Monographie de la Cathedrale de Chartres*, 2nd ed. vols., Chartres, 1887; I, 271. The timber and lead spire of Rouen by Becquet (1542-62) burned in 1822. For the new cast iron flèche by Alavoine see: Jean-Philippe Desportes, 'Alavoine et la flèche de la Cathedrale de Rouen', *Revue de L'Art*, 13 (1971), 48-62; *Le Gothique Retrouvé*, Exposition Catalogue, C.N.M.H.S., Paris, 1980, 331. On Viollet-le-Duc's reaction to the Rouen flèche see: 'Viollet-le-Duc et l'architecture metallique,' *Viollet-le-Duc*, Exposition catalogue, C.N.M.H.S., Paris, 1980, 255, fn 48; and M. Bideault, 'La flèche de la cathedrale de Rouen, par Alavoine,' *Ibid.* 331-332.

[4] Viollet and Lassus used iron chaining in the choir masonry; iron framing in the western towers, the roofs of the nave and choir triforia, and the main roof of the Nouvelle Sacristie. Viollet-le-Duc, *Journal des Travaux*, archives C.R.M.H. entries: April 18, 1849: 'charpente en fer du comble de la tour sud', f. 221; Sept. 27, 1854: iron framing of nave triforium, f. 252; March 23, 1857: choir triforium 'combles en fer', f. 284. Viollet-le-Duc also replaced the iron chaining in the 12th-century masonry courses of the choir but added a third course to the original two: *Ibid.* July 24, 1844, f. 42. Cf. also E. E. Viollet-le-Duc, *Dictionnaire raisonné de l'architecture française du XI au XVI siècle*, 10 vols., Paris, 1854-1868, II (s.d.) 'Chainage', 400-401. For Viollet-le-Duc's general use of iron in medieval restoration, at St.Denis and Pierrefonds, see the general discussion in *Viollet-le-Duc*, 1980, 248-256.

Figure 2: Exterior view of the base of Viollet-le-Duc's flèche (author's photograph)

medieval monuments, who had erected several 'gothic' timber spires, including the Ste. Chapelle in Paris, Viollet himself had a profound interest in historic carpentry.[5] Yet this aspect of Viollet-le-Duc's interests is seldom discussed, even though his work on medieval carpentry clearly deserves articulation within the wider context of 19th-century attitudes towards the Gothic and particularly the major repairs undertaken at the Cathedral of Paris between 1844 and 1864.

In addition to rebuilding the central spire, which Viollet-le-Duc and Lassus had planned in their initial *Projet* of 1843 (Fig. 3), Viollet-le-Duc's historical research on timber construction, published mainly

---

[5] For the work of Lassus see the excellent monograph: J.M. Leniaud, *Lassus* (Société française d'archéologie), No 212 (1980), 77-86; especially 84-85. Leniaud's figs. 25-27 illustrate Lassus' two proposals and the framing system for the spire for the Ste. Chapelle in Paris. Cf. Leniaud, 'Les flèches au XIXe siècle', 21-29.

3

Figure 3: Lassus and Viollet-le-Duc, *Projet* for the new flèche for Notre-Dame, transverse section, Jan. 1843. C. R. M. H. Inv. no. 1840 (author's photograph)

in the *Dictionnaire* III (1858), manifests his considerable understanding of the technology of medieval carpentry, especially the innovative framing of the 13th century, such as the nave and choir roofs of Notre-Dame (much of which remain today) and also the base of the 13th-century flèche that survived until Viollet dismantled it in 1858. Thus, he had the benefit of seeing the remains of the original support for the flèche both in situ and in context. Viollet was impressed by the ingenious seating of the framing (discussed below). And, although he was acutely aware of the problems of violent wind storms that had destroyed so many similar structures (and which had indeed weakened the structure of the flèche of Notre-Dame), Viollet was nonetheless eager to accept the challenge of erecting a new 'gothic' spire using timber and lead, the traditional building media of the 13th century. Since the survival of a critical portion of the original timberwork of the flèche is a key factor in Viollet's restoration, this

archaeological information seen in the context of the nave and choir roof carpentry is vital to an understanding of Viollet's 'gothic' flèche of 1860. We must, therefore, return to the point where Viollet-le-Duc began in 1857, viz. to the design of an unknown master carpenter of the 13th century

II. THE FLÈCHE OF THE 13TH CENTURY

Aesthetically and symbolically, the tall, slender spire of Notre-Dame de Paris expresses the spiritual aspirations of an age of piety as well as the technical ascendency of early gothic building. As was generally the case, the crossing tower at Notre-Dame above the juncture of the nave and choir served as a bell tower and contained six small bells. It existed, albeit in damaged condition, from its construction in the early 13th century until its mutilation in the French Revolution in 1792-93 and subsequent removal in the early 19th century by Etienne-Hippolyte Godde, architect at Notre-Dame de Paris from 1813-1830.[6] But even before the damage sustained in the Revolution, earlier accounts of 1744 and 1821 indicate that several of the main shores were rotten and that the flèche, whose height was approximately 32 meters to the weather cock, inclined two and a half feet to the southeast.[7] Godde, who

---

[6] The accounts that describe the destruction of the original flèche of Notre-Dame vary in certain details. According to Gilbert (1821), the flèche was destroyed in 1793 and the lead given to the Revolutionary government. A.P.M. Gilbert, *Description historique de la basilique metropolitaine de Paris*, Paris, 1821, 141. Viollet-le-Duc, however, indicates that the flèche was taken down by Godde, architect at Notre-Dame 1813-1830 and that he did not have the funds to restore it. Viollet-le-Duc, *GBA*, 35; and, a similar version is recounted in: Geneviève Viollet-le-Duc, 'La flèche de Notre-Dame de Paris,' *Monuments Historiques de la France*, nouv. ser. XI (1956), 43-44. The authors of the Exhibition catalogue 1980, however, give 1792 as the date for the destruction of the original flèche. *Viollet-Le-Duc*, 1980, 75.

[7] Citing J. Du Bruel's *Théâtre des Antiquités de Paris*, 1612, Viollet reports that the height of the original flèche from the roof ridge to the weather cock was 105 feet (32 meters), Viollet-le-Duc, *GBA*, 36; a second source gives approximately the same dimension (104 ft): M. C. P. Gueffier (Abbé de Monjoy), *Description historique des curiosités de l'église de Paris*, Paris; C. P. Gueffier, père, Librairie, 1763, 42. The 1744 reference is cited in M. Aubert, *Notre Dame de*

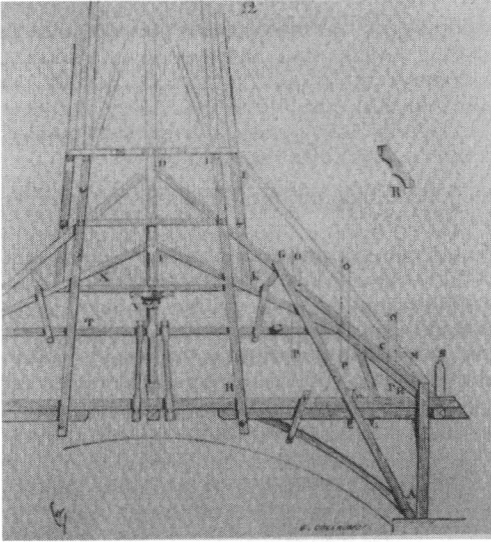

Figure 4: Viollet-le-Duc, Elevation of the 13th century flèche, *Dictionnaire*, V, Fig. 12 (author's photograph)

apparently covered the remaining base of the structure with lead sheathing, fortunately did not remove the critical seating of the timber structure below the level of the main roofs (c.12 meters) thus leaving this valuable archaeological evidence for future inquiry.

Until Viollet's restoration in 1858-60, however, Notre-Dame remained without its crossing spire, as can be seen in several views of Paris painted in this era.[8] Nonetheless, considering the usual fate of medieval spires, it is indeed remarkable that such a perilous structure should survive unaltered for nearly 500 years. Most medieval spires, for example, Reims and Amiens, had long since perished from violent storms and by fires ignited by bolts of lightning, as at Rouen in 1822.[9] And, it is even more extraordinary

that such an important example of early Gothic timberwork was also recorded, not only extensively by Viollet-le-Duc (Fig. 4) but also by Emile Leconte (Fig. 5).[10]

While the precise dating of the campaigns of construction of the Cathedral of Paris awaits definitive analysis, it appears likely that the erection of the flèche occurred in the opening decades of the 13th century and is thus contemporary with the nave roof of c. 1200; Bishop Maurice de Sully provided £100 to purchase lead for this roof in his will of 1186.[11] A thorough investigation of 1) the *inner* upper wall of the nave parapet (which at this point reveals no traces of Viollet's alleged fire or major rebuilding) and more importantly, 2) the relationship between the lower roof framing and the sexpartite vaults (which were certainly not rebuilt in 1230-1240) suggests that the nave roof and the lower portion of the original flèche *predate* the remodeling of the second quarter of the 13th century — a reconstruction that may have resulted from combined structural and aesthetic concerns.

Viollet, however, claimed that a fire in the upper regions of Notre-Dame was an important stimulus for the remodeling of the clerestory and nave flyers which took place in the second quarter of the 13th century. Moreover, he argued: 1) that this fire also destroyed 'les anciens combles' (but which ones?), 2) that the bishops of Paris were therefore forced to undertake new work on the cathedral c. 1230-1240; and 3) that this reconstruction involved an effort to gain more light by enlarging the clerestory windows. It is impossible now to know the extent of this fire, especially since Viollet's evidence has long since dis-

---

Paris, Nouvelle edition, 1950, 38-39; and M. Aubert, 'Les architects de N.-D. de Paris,' *Bull. Monumental* 67 (1908), 438.   Gilbert (1821), *Description historique*, 141.

[8]   Among the examples in the Musée Carnavalet, Paris, is the painting by J. B. Corot, *Le quai des Orfèvres et le pont Michel*, 1833.

[9]   The destruction of Robert Becquet's late Gothic spire of Rouen by lightning is described in: Desportes, 'Alavoine', 1971, 48. For the destruction of other

major gothic spires see: Viollet-le-Duc, 'Cathédrale', *Dict.*, II, 323-326.

[10]   The drawings of Leconte appear in the pirated edition: E .E. Viollet-le-Duc and J. B. Lassus, *Monographie de Notre-Dame de Paris et de la nouvelle sacristie*, Paris; A. Morel, 1856, p 1. 62. The identical plate is also found in E. Leconte, *Notre-Dame de Paris, recueil contentant 80 planches … et une notice archéologique … par Celtibere*, Paris, 1841-43; nouv. ed. 1853. pl. 50.

[11]   See Victor Mortet, *Étude historique et archéologique sur la cathédrale et le palais épiscopal de Paris*, Paris, 1888, 45-46.

appeared, though he only reported charred stone in the area of the original oculi.[12]

Despite this extensive work dated to the second quarter of the 13th century, Viollet-le-Duc asserted that the flèche unquestionably belonged to its early decades. He and other observers noted the wooden capital (Figs.4 & 5) that crowned the first stage of the elevation of the central king-post (poinçon). According to Viollet, in his definitive analysis of the original flèche in the fifth volume of the *Dictionnaire*, this 'chapiteau ... sculpté dans le poinçon central donnait la date exacte de cette flèche (commencement du XIIIe siècle).', and he also reports that in dismantling the original base of the framing, he removed the capital and conserved it in the 'magasins du chantier'; its whereabouts are presently unknown.[13]

---

[12] *Dict.* II, 292-293; *Revue de L'Architecture* (1851), 12. Viollet-le-Duc also mistakenly thought that the nave vaults were raised at this time. *Dict.* III, 267. On the issues of the general construction sequence of Notre Dame see: W. W. Clark and R. Mark, 'The First Flying Buttresses: A New Reconstruction of the Nave of Notre-Dame de Paris,' *Art Bulletin*, LXVI (March, 1984), 47-64; and also C. Bruzelius, 'The Construction of Notre-Dame in Paris', *Art Bulletin*, LXIX (Dec. 1987), 540-569. While both these studies emphasize the tallness and thinness of the 12th-century construction of Notre-Dame, both fail to point out the fact that the nave vaults are 1 to 1.5 meters taller than those of the choir, as measured from the crowns of the extrados of the vaults to the base of the tie-beams. In fact, no accurate measurements exist for the respective heights of the keystones of the vaults for either the nave or the choir. What is certain is that the choir roof was re-erected on a newly constructed parapet wall 2.7 meters higher than the remaining 12th-century masonry, whose exterior decoration conforms with that of the nave. Hence, the framing of the base of the original flèche would have spanned two different levels of masonry connected by a ramped-up central crossing vault (rebuilt several times). Cf. R. Mark, 'Reconstructions of Notre-Dame: Archaeology, Technology and Culture, in *Paris, Center of Artistic Enlightenment* (Papers in Art History from The Pennsylvania State University), IV (1988), 77-93.

[13] Viollet-le-Duc and Lassus first refer to the wooden capital on the king post of the flèche in 1843: 'Un chapiteau fort curieux, taillé dans le poinçon qui existe encore au centre de la souche de cette flèche, suffit pour fixer d'une manière précise l'epoque de sa

The original thin timber spire of Notre-Dame may be the earliest of its type with the possible exception of the Cathedral of Arras.[14] It is distinct from the majority of 12th and 13th-century crossing towers, which were either mainly of stone, as for example, the church of Semur-en-Auxois, or the Cathedrals of Laon and Strassbourg, or those whose construction combined a timber spire upon a visible, stone base, as formerly at Rouen, Mont-Saint-Michel, St. Bénigne, Beauvais, and Amiens, whose present, 16th-century spire is the oldest remaining in France.[15] Rather, the flèche of Paris was a complete structure entirely of timber, with its own base, storeys, and pyramid roof. It emerged directly from the intersection of the timber framing of the nave and choir roofs. As Viollet-le-Duc observed, this was a new conception in monumental architecture of the late twelfth and early 13th centuries — an innovation perhaps of master carpenters at Paris who daringly mounted a spire atop the roof of the tallest gothic cathedral of the 12th century.

The distinctive feature of the carpentry of the original flèche of Notre-Dame was its base. As Viollet-le-Duc aptly described, it was a 'quatre-pied' sprung from the crossing piers *well below* the top of the parapet wall. In plan, the lower framing (Fig. 5) consisted of: (1) two large diagonal frames on massive tie-beams that spanned the crossing and (2) four base-

---

construction, ainsique celle de la charpente, évidemment du XIIIe siècle. *Projet*, 1843, 14. Other references to this capital occur in: Viollet-le-Duc, *GBA*, 36; *Dict.* V, 449; and *Revue générale de L'Architecture* (1851), 13, in which Viollet-le-Duc surprisingly dates the captial to 1245. However, in all other references to the flèche, he consistently dates the construction prior to this date. Cf. also: *Dict.* II, 293 and *Dict.* III, 15-16. Gilbert (1821) refers to this region of the assembly as 'un poinçon taillé en forme de pilastre gothique,...', *Description historique*, 136.

[14] An early drawing of the Cathedral of Arras indicates the possibility of a crossing spire similar in design and scale to that of Notre-Dame. The flèche at Arras is dated by Bony to c. 1190. J. Bony, *French Gothic Architecture*, Berkeley, 1983, 136.

[15] On the flèche of Amiens see: G. Durand, *Monographie de l'Eglise Notre-Dame, Cathédrale d'Amiens*, 3 vols, Paris, 1901-03; I (1901), 512-524. For a general discussion of medieval spires and particularly a comparison between the structure of Amiens and Notre-Dame see: Viollet-le-Duc, *Dict.* V, 466-472.

6

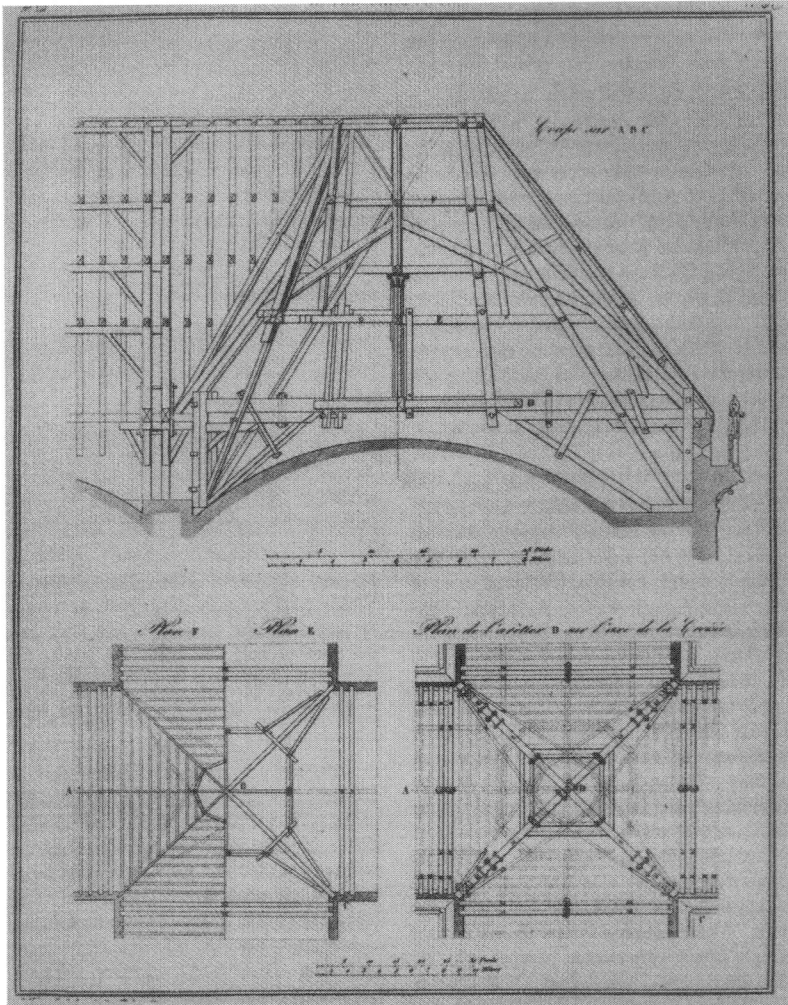

Figure 5: Emile Leconte, Elevation and plans of the 13th century flèche. *Monographie*, Pl. 62 (photograph, Princeton University)

tied frames inclined at about 45°, which circumscribed the crossing and composed the last triangular frames of the adjacent roofs.

At the intersection of the diagonal ties was a central post, which formed the spine of the flèche towards which all the major upright members of the framing inclined. Especially ingenious was the integration of the adjacent roof frames and the long buttresses of the flèche which rose from the base ties to support the octagon. Originally, the inclined timbers (CD & ML in Fig. 4) located in the valleys between the respective roofs, formed a projecting angle buttress with the lower diagonal (CD) being also one of the two prinicipal rafters of the adjacent roof frame; each principal was then strongly reinforced by the brace (AG) which connected it to the footing on the masonry pier below (at A). Additional reinforcement to the octagon was achieved by horizontal ties at points of intersection along the main diagonals (see plans,

7

Figure 6: Notre-Dame nave roof, detail of tiebeam brackets (author's photograph)

Fig. 5). Finally, the success of the entire framing system, which Viollet-le-Duc praised, was also a function of the shape of the octagonal pyramid, which decreased markedly in mass as it ascended and whose frames and angle posts all leaned towards the central post.

As indicated above, the framing design of the flèche was of necessity closely integrated with the main-span roofs around the crossing. It is thus perhaps not surprising, although not commented upon, that certain technological aspects of this ingenious carpentry should also appear in the nave and choir, as is indeed the case. For example, in both nave and choir roofs (and it must be remembered that the choir roof was re-erected and raised c. 1235) one finds: (1) prominent brackets beneath the tie-beams; (2) hangers, or suspended posts to counter deflection of the tie-beam subjected to superimposed loads along the length of the beam; (3) large clasp-like, tension connections to join members both vertically and horizontally; and, (4) timber members of light section in proportion to the scale of the overall framing. While Viollet does

not discuss these features directly in his text, they are clearly depicted in the flèche elevations of both Leconte (Fig. 5) and Viollet-le-Duc (Fig. 4) and can thus be understood from these drawings and in the general context of the existing medieval roofs.

In both the flèche assembly and in the main-span roofs, large brackets function as structural and constructional members of the assembly that patently concern the stabilization of the great tie-beams supporting the main diagonal frames as well as the central king-post of the flèche (Fig. 4) and have an analagous function beneath the tie-beams in the nave (Fig. 6). The bracket beneath each tie beam of the original flèche, whose configuration differs somewhat in Leconte's and Viollet's elevation drawings, was composed of: (1) a horizontal beam (an extended sole-piece) beneath the tie, whose outer end rested on the masonry parapet where it was trenched over a wall plate; (2) a vertical wall post extending c. 4.5 meters down to the masonry pier; and, (3) a diagonal strut which completed the inner side of the triangular bracket. In both drawings the tie-beam and bracket

8

beneath are joined by a short timber clasp, a type of connection seen at other levels of the flèche structure as well.

The constructional advantage of erecting large brackets for positioning the great ties is that these stabilizing consoles, firmly established on the masonry, might also have served as supports for an erection platform in addition to giving continued support to the tie-beam.[16] Moreover, the extra bracing would also provide insurance against rotation of the tie-beam due to the inevitable shrinkage of the timber and opening of the joints or, more importantly, the asymmetrical loading from wind pressure and suction on the flèche — the persistence of which Viollet believed could result in total structural failure.[17] In Viollet's opinion, however, rotation of the frame at the base of the flèche and hence localized torsion in the upper storeys was prevented mainly by the interlocking system of braces, especially the long braces (CD in Fig. 4), which cross at D and which are also the principal rafters of the adjacent roof frame. Viollet-le-Duc noted the considerable amount of coupling of timbers by horizontal clasps in the structure as it rises to the level of the roof ridges, and he interprets this clearly as the carpenters' efforts to resist wind pressure.[18] Moreover, the framing was so constructed that each of the upright angle posts of the octagon discharge their load onto two, perhaps even three of the masonry piers, well below the wall head, i.e. c. 4.5 meters beneath the base of the roof framing. Given the fact that Notre Dame was a daringly tall, thin building at the end of the 12th century and that the original flèche reached an overall height of approximately 77 meters from the ground, it can be assumed, as Viollet-le-Duc suggests, that the designer of the framing, combined the adjacent roofs, a bracing system of crossed shores, and reinforcement of the base tie-beams to offset the danger of torsion due to wind action. And, it might be argued that a similar concern for stabilization of the tie-beams and longitudinal bracing in the nave, indicates the level of awareness of the medieval builders with regard to wind loading

which might cause the rotation of the ties and potential lengthwise deformation (racking).

III. THE RECONSTRUCTION PROPOSALS FOR THE FLÈCHE: 1843 AND 1857

When Viollet-le-Duc designed his new flèche in October of 1857 (a few months after the death of Lassus on July 15) a considerable amount of the vital restoration of the Cathedral of Paris (begun in April, 1844) had already been accomplished. While the cumulative work of Viollet-le-Duc and Lassus involved an extensive rebuilding of the fabric, including completely new guttering for all the roofs, they left intact the medieval carpentry of the main-span roofs (apart from the transepts) and the base of the original flèche whose demolition only took place in the late summer and autumn of 1858.[19] In designing the new flèche for the crossing, Viollet-le-Duc faced both a challenging aesthetic and structural enterprise in the mounting of a tall, wind resistant structure sprung elegantly from the intersection of the main-span roofs. As was Lassus earlier, Viollet-le-Duc was committed professionally and intellectually to a design based on a similar, visual scale and appearance of the 13th-century spire as it had been recorded in various descriptions and pre-Revolution drawings, especially the detailed sketch of c. 1788 by Garneray.[20]

Most important in the initial *Projet* (Fig. 3) of 1843 (probably by Lassus and which closely resembles the external appearance of the Garneray drawing) are: (1) the salient, but *unadorned* angle buttresses in the valleys of the main-span roofs; (2) the short, solid octagonal base above the roof ridges that supports a single, arcaded open storey; and (3) the acutely pointed pyramid that rises above the gables that crown the open storey. This design for the new flèche is consistently shown in several exterior views included in the 22 drawings submitted in 1843, but no plan for the internal framing is known to exist — although one might conjecture a system of framing similar to that used by Lassus at the Ste. Chapel le (which incidently was much admired by Viollet-le-Duc).[21] Yet shortly after

---

[16] On the use of bracket structures in roof framing see: L.T. Courtenay, 'Where Roof Meets Wall: Structural Innovations and Hammer-beam Antecendents', *Annals of the New York Academy of Sciences*, 441 (1985), 89-124.

[17] Viollet-le-Duc, *Dict.*, V, 446.

[18] *Ibid.*, 450.

[19] Viollet-le-Duc, *Journal de Travaux*, passim and especially April-June, 1857, ff. 284-269.

[20] A good illustration of the Garneray drawing can be found in: Pierre du Columbier, *Notre-Dame De Paris, Mémorial de la France*, Paris, 1966, 223.

[21] Leniaud, *Lassus*, 84-85 and Viollet-le-Duc, *Dict.* III, 58.

Lassus' death, on October 29, 1857, Viollet-le-Duc submitted a new proposal which was considerably more elaborate than the original and which indicates the direction of Viollet-le-Duc's own, mature aesthetic preferences.

In contrast to the 1843 design, the 1857 version of Viollet-le-Duc (Fig. 7) exhibits two major changes: (1) the addition of a second open storey, which thus heightens the middle stage of the octagon and (2) an elaborate, stepped treatment of the angle buttresses (inspired by archaeological details of the original spire) along which figures of the Apostles descend. The latter is clearly the most visually dramatic addition; and indeed, this treatment of the angle buttresses is a unique design of Viollet-le-Duc which brings a further iconographic dimension to the structure. Moreover, the descending Apostles empha-size visually the close structural relationship between the spire and the adjacent roofs. In fact, Viollet stressed in his discusssion of the original framing that the base was technologically the most important part of the flèche. And thus, it is not surprising that in devising the structure of the flèche of 1857, Viollet should have taken his lead from the ingenious support system of the early 13th century that had survived so long.

In describing his construction, Viollet-le-Duc indicated his profound respect for the principles upon which the medieval design was based, but at the same time he believes that as a restorer he ought to improve the system and to benefit by 'the perfections of modern industry'.[22] Nonetheless, he clearly adopted the med-ieval system of the two diagonal frames, four circum-scribing inclined frames and a central post, all of which are supported likewise on a 'quatre-pied' seated on the crossing piers about 4.5 meters below the p-arapet (Figs. 8 & 9). He also used the original and ingenious idea of the 13th-century master carpenter of integrating the main rafters of the adjacent roof frames with the buttressing of the octagon. Having employed the same support system and geometry for the base of the new flèche, the octagonal pyramid emerges at the level of the roof ridges as in the 13th-century framing. In Viollet's working drawings, the detailed measure-ments and sketches indicate the proportional adjust-ments necessary to accomodate the irregularity of the crossing, a parallelogram 14,75 x 12,75 meters. As a result, the octagon is irregular, although this is dif-

Figure 7: Viollet-le-Duc, 'Flèche in timber and lead', Oct. 29 1857, C. R. M. H. Inv. no. 2137 (author's photograph)

ficult to perceive visually, because of Viollet-le-Duc's skillful manipulation of scale and decorative elements.

While certainly concerned with the aesthetic quality of his flèche and with its ornamentation (which is a separate topic in itself), Viollet was most absorbed by the essential problem of stability under wind-loading. He described the flèche as the arm of a lever whose base must be immovable in high winds if the

10

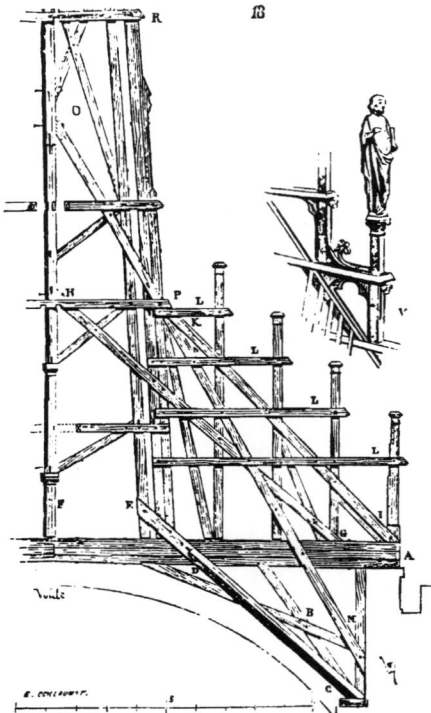

Figure 8: Viollet-le-Duc, Elevation of the flèche of 1857, *Dictionnaire*, V, Fig. 18 (author's photograph)

structure is to survive. He remarked that his timber structure was considerably taller than any building in Paris in 1860. Furthermore, his reason for concern was based on the fact that until the structure was actually tested, i.e. built and observed, no routine methods were available to predict how this complex, framed pyramid would behave in the worst conceivable, climatic conditions, (although by 1860 there were certainly means of measuring wind pressure).[23] Viollet

thus set out to strengthen or modify members of the original structure which, from his vantage point of having looked at a great many medieval buildings, he considered to be weaknesses. Specifically he increased the cross-sectional dimensions of major members; eliminated vertical tension connections; and secured most of the mortise and tenon joints by large iron pins, a screw nut and iron pin with a head that could be tightened after assembly. Nonetheless, Viollet-le-Duc still retained horizontal ties, traditional timber joints, halving for passing braces and mortise-and-tenon connections for major vertical members in compression, e.g. the base of the king post (although evidence of shrinking of the timber and retraction of the tenon is clear).

In addition to changes in jointing, Viollet-le-Duc greatly increased the cross-section of the diagonal tie beams and hence eliminated the brackets, which indeed were no longer needed to reinforce the powerful ties composed of three massive timbers scarfed obliquely and bolted (Fig. 10). Also he added an extra post and braces to the angles of the timber octagon above the roof ridges. In fact, the entire framing, which is also taller (44.5 meters to the weather-cock versus the original 31.7 meters), has greater bulk than its medieval predecessor. This was intentional, since by increasing the section of major members, Viollet insured his framing against local weaknesses caused by general decay and joints cut into the timbers which might reduce their strength. Furthermore the additional weight of the framing, which is also a function of its greater height, provides a comfortable (and conservative) margin of safety that was found to be eight and a half times the weight needed for overall stability in the worst wind conditions.[24] But, this caution — more

---

E. Knowles Middleton, *Invention of Meteorological Instruments*, Baltimore, 1969. I am grateful to Robert Mark for directing me to this source.

[24] Information on maximum wind speed of 115 km/hr. at a height of 45 meters recorded in the Paris region was obtained from: R. Mark, *Experiments in Gothic structure*, Cambridge, Mass., 1982, 21-22. Using information concerning the weight and dimensions of the flèche given by Viollet-le-Duc and a worst wind of 115 km/hr, it was calculated from the equation given in Mark, *Experiments*, 22 that the total force on the flèche was 23.6 tons. And, although accounts vary considerably, a figure of 12 meters was used for the distance between the roof ridge and the base of the framing at the top of the wall. The margin

---

[23] Viollet-le-Duc, *Ibid*. Although there existed a variety of wind-measuring devices prior to the mid-19th century, such as the elegant anemograph of d'Ons-en-Bray of 1734, the major development of precision instruments with sufficient durability came towards the end of the 19th century, especially with the work of W.H. Dines of Cambridge on the rotation anemometer and also the Parisian instrument makers, Richard Frères who employed the electromagnetic anemograph in the 1820's. For more detail, see: W.

Figure 9: Carpentry of the crossing at the base of Viollet-le-Duc's flèche (author's photograph)

it seems than provided by the medieval builders — also acts as an insurance against the natural mutability and deterioration of wood, as was evident to Viollet-le-Duc in the medieval framing he so carefully studied.

The entire timberwork, constructed with oak from Champagne and executed by M. Bellu, weighs 500 tons, including the ironwork; the lead covering is an additional 250 tons thus totaling 750 tons. The timber framing rises 36.5 meters (the exposed 'sail' area) from the roof ridge to the base of the iron cross (8 meters tall and the only significant iron construction in the entire structure). The overall height of the flèche from the ground is 96 meters to the cock, perched

above the cross.[25] Given the increased scale of the 19th-century carpentry in comparison to that of the 13th century, how then did this tall structure of wood actually behave under wind-loading? Fortunately, for the history of building technology, we have a vivid account of a violent wind storm and Viollet-le-Duc's direct observations of structural behavior shortly before the completion of his flèche.

IV. CONCLUSION: THE TRIAL OF THE FLÈCHE:

On February 27, 1860, the stability of the newly-restored timber spire was put to test by the gales of a violent storm that wrought destruction to a large number of chimneys in Paris. At this time, the carpentry of the flèche was finished; the great scaffold had just been removed on January 13; and only the completion of the

of safety was determined by comparing the overturning moment of 571 m-tons (24.2 meters x 23.6 tons) with the righting moment of 4875 m-tons (6.5 m x 750 tons), thus giving an overall stability factor for Viollet's flèche of 8.5. I am indebted to Robert Mark for his assistance in this calculation.

[25] Weights and measurements for the new flèche are found both in: Viollet-le-Duc, *Dict.* V, 461 and in *GBA*, 38.

12

Figure 10: Notre-Dame, detail of diagonal composite tiebeam and reinforcement of Viollet-le-Duc's flèche (author's photograph)

juncture of the lower framing with the roofs and the leading of the lowest portions remained.[26] Shortly thereafter, Viollet-le-Duc wrote with satisfaction and no doubt relief that even while the storm raged at its worst, the oscillation of the cross at the top of the flèche was no more than twenty centimeters.[27] Presumably, Viollet himself with the help of a worker took this alleged measurement by ascending the external ladder whereupon with measuring rod and plumb line in hand, he recorded the deflection. This test evidently assured Viollet that his timber assembly functioned as designed: to distribute the loading onto at least two of the main piers and for the base to remain rigid.

Since Viollet does not give the wind velocity nor mention an anemometer, he likely had no means of obtaining this information for an elevation of 96 meters. Nevertheless, it is possible from modern meteorological information and the weights and dimensions of the flèche provided by Viollet to calculate the maximum, horizontal wind pressure on the flèche at 23.6 tons — a pressure which, given the dead-load of the framing, is well below that which would cause overturning.[28] Thus, Viollet's flèche of Notre-Dame de Paris is a completely stable structure even in the worst wind conditions recorded, and it remains today a significant and influential example of 19th-century, 'Gothic' structural carpentry. Moreover, the essential design principles of Viollet's construction are identical to those of the 13th century, i.e. the creation of a vertical cantilever supported on a rigid, four-footed base sprung from the main piers of the crossing, well below the level of the top of the wall. And, given the less accurate technology of predicting the behavior

[26] Viollet-le-Duc, *Journal de Travaux*, ff. 330-31.

[27] Viollet-le-Duc, *Dict.*, V, 453, fn. 1.

[28] See above, fn. 24.

of tall structures in 1860 (nearly a generation before the brilliant engineering acomplishments of Gustave Eiffel), it should be kept in mind that Viollet-le-Duc, while benefiting from a great deal of hindsight, had little more scientific expertise in predetermining the reactions of the flèche under extreme wind-loading than the medieval builders of 1200. Consequently, both relied on acute observation of the actual framing as well as the experience with previous structures of similar design — as Viollet-le-Duc instructs: 'Les assemblages des charpentes du moyen âge méritent d'être scrupuleusement étudés; ils sont simples, bien proportionnes à la force des bois ou à l'objet particulier auquel ils doivent satisfaire.'[29]

---

[29]   Viollet-le-Duc, *Dict.* III, 57.

19

Building the tower and spire of Salisbury Cathedral

Tim Tatton-Brown

Although — or because — the 19th century saw great archaeological interest in the standing buildings of medieval England, the early legislation to protect British ancient monuments expressly excluded church buildings that were still in use. A new measure, just come into force, gives cathedral archaeology a formal place, and makes this a timely moment to see what kind of work which the archaeology of a standing cathedral can now amount to.

In the last decade or so cathedral archaeology in England has awakened after nearly a century of inactivity, caused partially by the completion of the huge 19th-century restoration campaigns, and partially by the passing of the Ancient Monuments Act in 1913, which exempted ecclesiastical buildings in use from its provisions.

During the last 15 years, many Deans and Chapters have appointed archaeological consultants voluntarily and involved them with major new restorations. In 1990 the *Care of Cathedrals Measure* finally completed its passage through General Synod and Parliament and received the Royal Assent. Among its provisions is the obligation on every cathedral to set up a Fabric Advisory Committee and to appoint an archaeological consultant. In the last year or so those cathedrals without an archaeological consultant or Fabric Committee have been filling those gaps, as the Measure comes into force on 1 March 1991. From this date, all Deans and Chapters lose their unique exemption and have to seek approval for changes to the fabric or character of their cathedrals. This is a major change, and in return cathedrals hope that they will be eligible for government grants.

Cathedral archaeology really got under way in the 1840s with the exceptional work of the Revd Professor Robert Willis (Tatton-Brown

1989), but a century earlier the Clerk of Works at Salisbury Cathedral, Francis Price, was already undertaking what we would today call 'archaeological' investigations (Price 1753). His description and drawings of the spire are the starting-point for the present paper.

'Modern' cathedral archaeology — the detailed investigation of the standing fabric of the building — only really started again with Dr Warwick Rodwell's study of the west front of Wells Cathedral in the later 1970s. In the last decade, major new work has taken place at a number of cathedrals (among them Chichester, Exeter, Lichfield, Oxford, Winchester and York), and a summary of Dr Rodwell's own work at Lichfield was published in this journal two years ago (Rodwell 1989a). Many old techniques for studying cathedrals have been revived and developed (for example, the comparing of mouldings with the help of computer programs), while at the same time modern dendrochronological work has allowed totally independent dates to be given to roofs, doors, etc. At the same time the study of masons' marks, building stone and carpentry (Hewett 1980; 1985) has advanced greatly. Most important of all has been the very detailed recording of areas of fabric using a methodology that has been developed below ground, in archaeological excavations, over the last half-century. Much remains to be done, but later this year a series of essays will show what has been achieved (Tatton-Brown & Munby in press).

As an example of the riches in store for the cathedral archaeologist, this paper looks at one of the greatest and most exhilarating of all high medieval Gothic structures, the spire of Salisbury Cathedral. It attempts to demonstrate what has already been found out as well as what might be discovered from a new archaeological study.

The Salisbury tower and spire

The central tower and spire of Salisbury Cathedral is one of the architectural and engineering masterpieces of the Middle Ages.

Completed perhaps nearly seven centuries ago, it is the tallest surviving medieval building in Europe. Yet neither archaeologists nor architectural historians have studied the structure in detail. Architects, clerks-of-the-works, engineers and others have tried over the years to understand why it is still standing and how they themselves can attempt to make it safer.

The process of adding 'belts and braces' to the structure has continued unabated since at least the 15th century, and we are now in the middle of the most costly and large-scale strengthening and restoration of all time. The Dean and Chapter propose to spend over £5 million on the tower and spire in the next few years, and most of the tower and a section of the spire has since August 1989 been covered in scaffolding.

This is an ideal time to carry out a full measured survey of the structure, and to make a detailed analysis. This paper surveys earlier work, and brings together what is known at present about how and when the tower and spire were built (FIGURE 1).

Earlier published work

As is well known, the first written survey of Salisbury Cathedral was undertaken by Sir Christopher Wren in 1668–9, commissioned by his friend Bishop Seth Ward after the spire had been struck by lightning. This report is a masterly survey for its time. It was Wren who first recognized that the tower and spire were a later addition to the cathedral, not planned by the original masons. After a brief description of the fabric, Wren goes on to look at faults in the structure and to suggest some remedies (Wren c. 1669), though details of what was actually done

in the following years are not well documented. In 1671 'all the fines from the fabric lands, amounting to £4200, were spent on repairs of the cathedral fabric, to which the prebendaries also contributed £500 by a tax on their prebendal incomes' (Edwards 1956: 194). The 'clerk of the fabric' of a few years later, Thomas Naish, may also have carried out various repairs to the spire and other parts of the building during his long term of office (1679–1714). He was 'clerk of the fabric' until his death (Colvin 1978: 404), and made a survey in 1691, which was published in the Civil Engineers and Architects Journal (1843). In 1717, his son William became his deputy, succeeding his father in 1727 (Slatter 1965). During William's time and that of his assistant and successor Francis Price (1737–53), various documented works were undertaken. We know this from a quite remarkable book by Francis Price himself which was published in the year of his death, 1753. Price was well-known in his day for his standard work The British Carpenter. The Salisbury book, A Series of Particular and Useful Observations made with Great Diligence and Care upon that Admirable Structure, the Cathedral Church of Salisbury, was, Howard Colvin (1978) has pointed out, 'the first serious attempt to describe and analyse the structure of a major Gothic building'. It is indeed remarkable for its time, and in many ways has not been superseded.

The description of the spire fabric, for example, is a model of careful writing by a man who knew his building very well. In documenting the repair work of Price's time the book is illustrated by Price's own quite excellent drawings, not pretty views but scaled plans and sections. Though it was reprinted (anonymously) in 1774 with some additions and 'an account of Old Sarum', both editions are today rare books.

No such detailed analysis of the fabric of the building has been undertaken since Francis Price's time, though excellent shorter works have been published. Next in chronological order were Professor Robert Willis' two lectures of 1849 (Willis 1849 [1973]), given to the General Meeting at Salisbury of the Archaeological Institute. Willis, the greatest Medieval architectural historian of the 19th century, had little to add to Price's analysis of the tower and spire (which he probably never ascended). He did, however, add considerably to the docu-

76

TIM TATTON-BROWN

FIGURE 1. *General view from the air of the tower and spire of Salisbury cathedral from the northeast. (© British Crown Copyright/MoD. Reproduced with the permission of the Controller of Her Britannic Majesty's Stationery Office.)*

mented history of the building, and it is Willis who first associates the contract of Richard de Farley (sic) of 1334 with the building of the tower and spire (quoting Dodsworth 1814). Price had suggested that the tower and spire might have been started after the 'rehallowing of the cathedral in 1280' (quoting Bishop Godwin's 'catalogue of bishops' (1616: 279)).

The first analytical drawings of the tower and spire since Price's time were carried out just before the last war by the architect W.A. Forsyth, at a scale of 8 feet to the inch i.e. 1:96. In January 1946 these were published, much reduced, in the *Journal of the RIBA* (Forsyth 1946) with a useful short essay on the structure of the tower and spire. Since this time, no new measured survey has been carried out.

Sir Nikolaus Pevsner wrote (as usual) a superb description for the Wiltshire volume of his *Buildings of England* series (Pevsner 1975; revised as Pevsner & Metcalf 1985), but unfortunately he says unequivocally that the contract between the Chapter and Richard Farleigh for work on the tower and spire dates from 1334. The contract in 1334 does not mention the tower and spire, and it is almost certain that the work was complete well before that date.

In the late 1970s the inside of the tower and spire was visited by Cecil Hewett, and as a result of this detailed (but unmeasured) sketch drawings of the carpentry and ironwork within the stone shell were published (Hewett 1980; 1985), along with brief descriptions, which showed the great importance of these two materials to the fabric. Hewett's drawings do not distinguish the different phases of carpentry, or show how they relate to the surrounding stone shell. Without Cecil Hewett's immense achievements in the study of the evolution of carpentry (carried out in the last 25 years or so), we would not begin to understand the 'amazingly complex' framing inside the tower and spire (to quote Hewett himself). Below, I hope to identify two main phases of framing: original medieval work within the spire, and a later, post-medieval 'timber-tower' within the upper stage of the tower.

For the last two decades the clerk-of-the-works at Salisbury has been Roy Spring. In the tradition of his great predecessor, Francis Price, he has recently published a book on Salisbury Cathedral (in the 'New Bells Cathedral Guides' series). His chapter on the tower and spire

throws many new insights on the history of the fabric, and, as with Francis Price, it shows his intimate knowledge of the structure and of its many restorations gained over many years (Spring 1987: 48–60). Spring also rejects the 'standard' dating of 1334 for the start of new work on the tower.

The most recent article on the tower and spire fabric is by the current consulting engineer to the Dean and Chapter, Peter Taylor (1988), and gives some of the background thinking to the present restoration. Detailed discussion of the foundations (or lack of them!) is followed by notes on the lantern, the tower and the spire with some details of earlier restoration works. Mr Taylor also outlines the main areas of work during the present huge restoration campaign.

As can be seen from this summary no detailed architectural analysis has yet been attempted, let alone any archaeological work.

The building of the tower

By the middle years of the 13th century much of the main fabric of Salisbury Cathedral was nearing completion. It is certainly to this time that we can date the construction of the small and low tower at the main crossing of the cathedral. The top of this lantern, at c. 120 ft above ground level, is only just above the top of the ridges of all the main high roofs, and it is not known what originally capped it, possibly a low pyramidal roof. The lantern was meant to be open to the ground floor, and on all four inside faces an arcade was built with four bays of alternating piers and vertical shafts of Purbeck marble.

The style of the architecture is very similar to that found in the rest of the cathedral. High up in the corners were pairs of small windows that lit the upper area of the lantern. No other windows were possible because the large triangular area in the middle of each side was covered by the ends of the high roofs.

The walling of most of the lantern at this level is remarkably thin. As Wren pointed out, the original builders of the cathedral never conceived of a large tower, let alone a stone spire.

A short way above the top of the pointed arches of the internal arcade, a clear break in the ashlar masonry can be seen. This must mark the top of the original work and the base of the new work for the tower. This is also the level of the top of the elaborate system of iron ties which

TIM TATTON-BROWN

FIGURE 2. *The medieval iron ties, with decorative elements at the top of the lantern tower. In the foreground is one of Shield's large diagonal ties put in for Scott in 1869, while top left are three diagonal struts supporting one corner of the 18th-century gallery. (Salisbury cathedral works department.)*

bind together this area (FIGURE 2). The ironwork, which almost certainly dates from when work on the new tower started, is an extremely sophisticated system of strengthening which was crucial to the stability of the new work. The iron ties are at two levels. An upper level forming a prefabricated and 'pre-stressed' grid is connected with the lower level of inner wall ties by vertical bars and curved braces (with decorative cusps in the middle). The ironwork, which is mostly made up of two-inch square bars, also shows signs of modification during framing in two different areas. The bars continue through the wall at both levels. Examining there on the outside in 1974, Mr Spring found them to be in good condition and protected from rusting by lead (Spring 1987: 50). A detailed examination of the lead should show whether it was poured in molten (as with iron

clamps), and hence whether it was poured in from above before the upper stage of the tower had been started.

This elaborate ironwork obviously impressed Wren and Price very much; Price (Anon.1774: 14) calls it 'the best piece of Smith's work, as also the most excellent mechanism, of anything in Europe of its age'.

When one considers how thin are the main walls of the lantern behind the arcade (only about 2 ft thick), and that the tower and spire above weigh about 5000 tons (Taylor 1988: 7), it is no wonder that all commentators on the structure single out this level as the most dangerous. Here most has been done to strengthen the tower. Most obvious today are Scott and Shields' four tiers of crossed diagonal ties (1869) and Forsyth's bricking-up of the four corner-stair turrets and infilling of the lantern

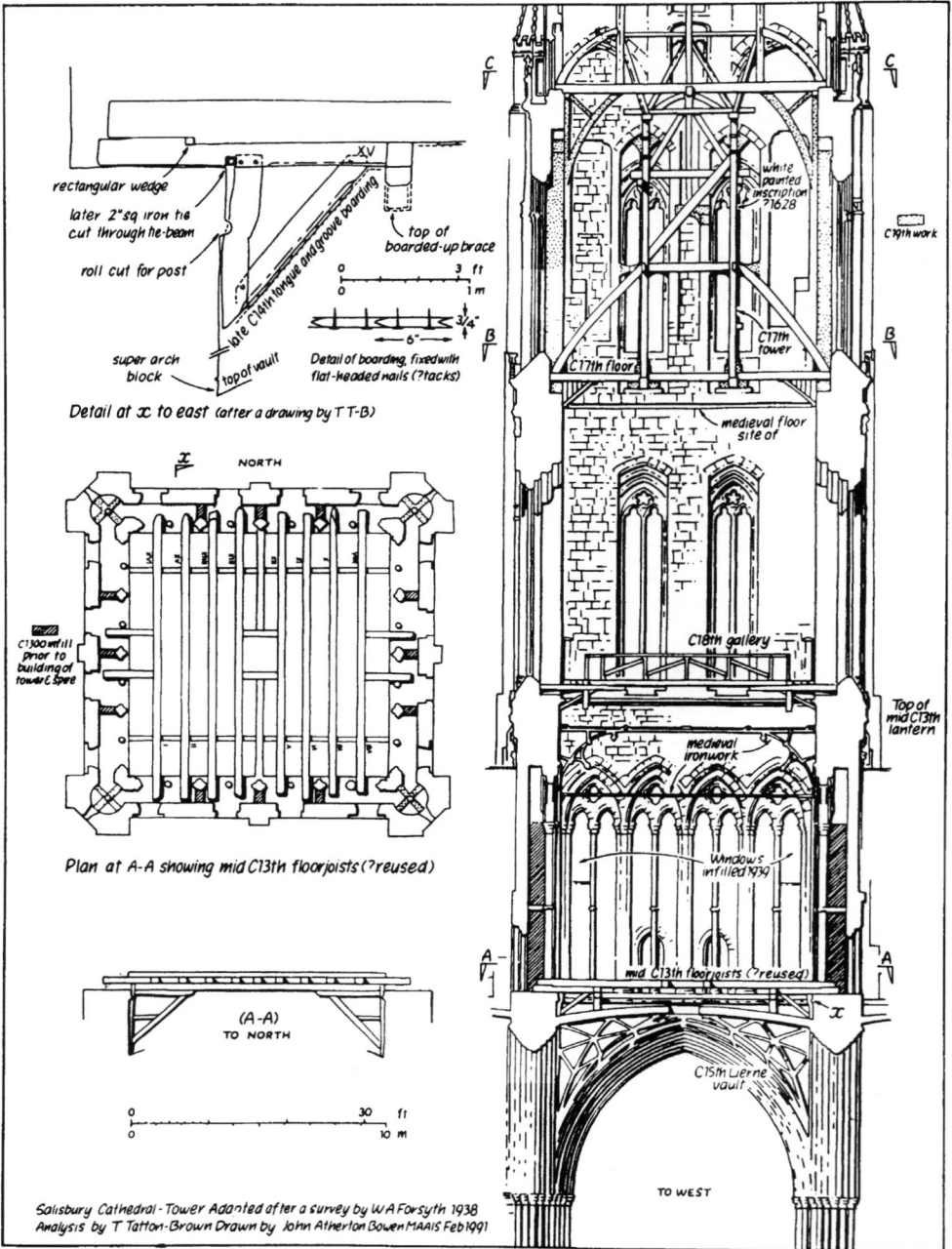

FIGURE 3.   Plan with sectional details of lantern floor joists.

windows (1937–9). The original builders of the tower themselves undertook many works to buttress and strengthen the lantern. A few feet above the bases of the main piers of the internal arcade can be seen heavy iron bands which tie the piers back to the main wall. The tower builders also filled solid the space between these piers and the outer wall, blocking earlier passageways at lantern level. (The iron ties at capital level in the internal arcade may date from the original phase of construction of the lantern in the mid 13th century.)

In blocking wall-passages in the lantern stage, access between the roofs and stair-turrets was made very difficult, if not impossible. It seems very likely, therefore, that the insertion of a floor at the base of the lantern stage was a very early requirement. This floor certainly dates from well before 1479–80, when the vault was put in, because its underside was originally covered with a ceiling of tongued and grooved boarding (FIGURE 3). This was first noticed by Price, but not fully re-exposed until 1971 when repairs were carried out by Mr Spring (1987: 49). The floor and ceiling also effectively cut off the tower works from the cathedral below. Some of the timbers of the floor-framing appear to be re-used. They may have come from the original lantern roof (Spring 1987: 49), and dendrochronology may confirm this. (Provisional results (December 1990) from dendrochronology now suggest a felling date for most of the timbers used in the floor of c. 1251. Two timbers, however, appear to be earlier (seasoned) timber. The tongue-and-groove boarding has been shown by dendrochronology to be of Baltic origin with a felling date of c. 1371)

As well as many iron 'bandages' and new masonry to join the arcade piers to the outer lantern wall, a very large number of new buttresses and flying buttresses were built around the base of the lantern to spread the load of the new tower. Inside the cathedral they can be seen built into the triforium and clerestory arcade at all places where they join the main tower piers. The clerestory windows in this area were also blocked.

Outside the cathedral an elaborate series of three-way flying buttresses can be observed above the triforium roofs at all four corners of the tower. There are also extra buttresses to the nave and transept clerestory walls. The whole system is described by Price, who enumerates

'one hundred and twelve additional supports, exclusive of the strength resulting from the bandages of iron' (Anon. 1774: 60). As the exact order in which these buttresses were inserted, and their dating, is unknown, detailed study is required. It seems likely they were all built at the time that the tower and spire were being erected. The only known exceptions are the two newer buttress foundations to the southwest corner of the lantern (the side of the tower that has settled most), where buttressing was suggested by Wren. The work was apparently started not long afterwards but never completed (Spring 1987: 60). The remains of the buttress foundations and vertical stiffeners can still be seen on the nave south wall and south transept west wall bound up by the lower three bands of the Scott/Shields ironwork. Its bulky form is in sharp contrast to the slender flying buttresses of the Middle Ages.

The first stage of the new tower above the original lantern stage is the start of the new work proper. It is characterized on the exterior by its elegant 'Decorated' architecture with many hundreds of 'ballflowers' around the windows and blind openings (FIGURE 4). This style of decoration is well-known in the west of England in the years around 1300, but no close dating can be given by architectural style alone.

Internally (FIGURE 5) the base of this new first stage never had a floor, though the doors in the stair-towers, the window-seats in the pairs of open central windows and the large roll string-course all suggest that there should have been a floor at this level. In the walls, however, there are no signs of the joist holes for an earlier floor. Instead there is now an elegant gallery which Cecil Hewett (1985: 139–42) says is

ingeniously framed; the great angles of the square void were first spanned by beams at 45 degrees, onto which were laid joists that bisected the angles, into which were chase-tenoned three [in fact, two] common-joists to form a walk-way along each side. A forelock-bolt was used to secure the first two timbers at their crossing; and the internal common-joists were braced from beneath – the braces springing from pockets in the masonry and crossing each other during their ascent. The hand-rails around the gallery thus produced seem to be the earliest 'trussed' timber girder construction yet known, and it is of interest that all the compressive diagonal struts meet square, hewn, outsets on the posts; these later became square abutments. Though much repaired this gallery is mainly of original timbers, and of its initial design.

FIGURE 4. *Aerial view of the outside of the tower from the south. (© British Crown Copyright/MoD. Reproduced with the permission of the Controller of Her Britannic Majesty's Stationery Office.)*

TIM TATTON-BROWN

Capstone

Weather door

9th stage

8th stage

7th stage

6th stage

5th stage

4th stage

3th stage

2nd stage

1st stage

Parapet (8 doors) level

Site of medieval floor

Medieval Ironwork

Mid C13th floorjoists (?reused)
Late C14th boarded ceiling
C15th Lierne vault

180'

404'

TO WEST

100 ft
E
30

Salisbury Cathedral - The Tower and spire showing the medieval
timber scaffold Adapted after a survey by W A Forsyth 1938
Analysis by Tim Tatton-Brown
Drawn by John Atherton Bowen MAAIS Feb 1991

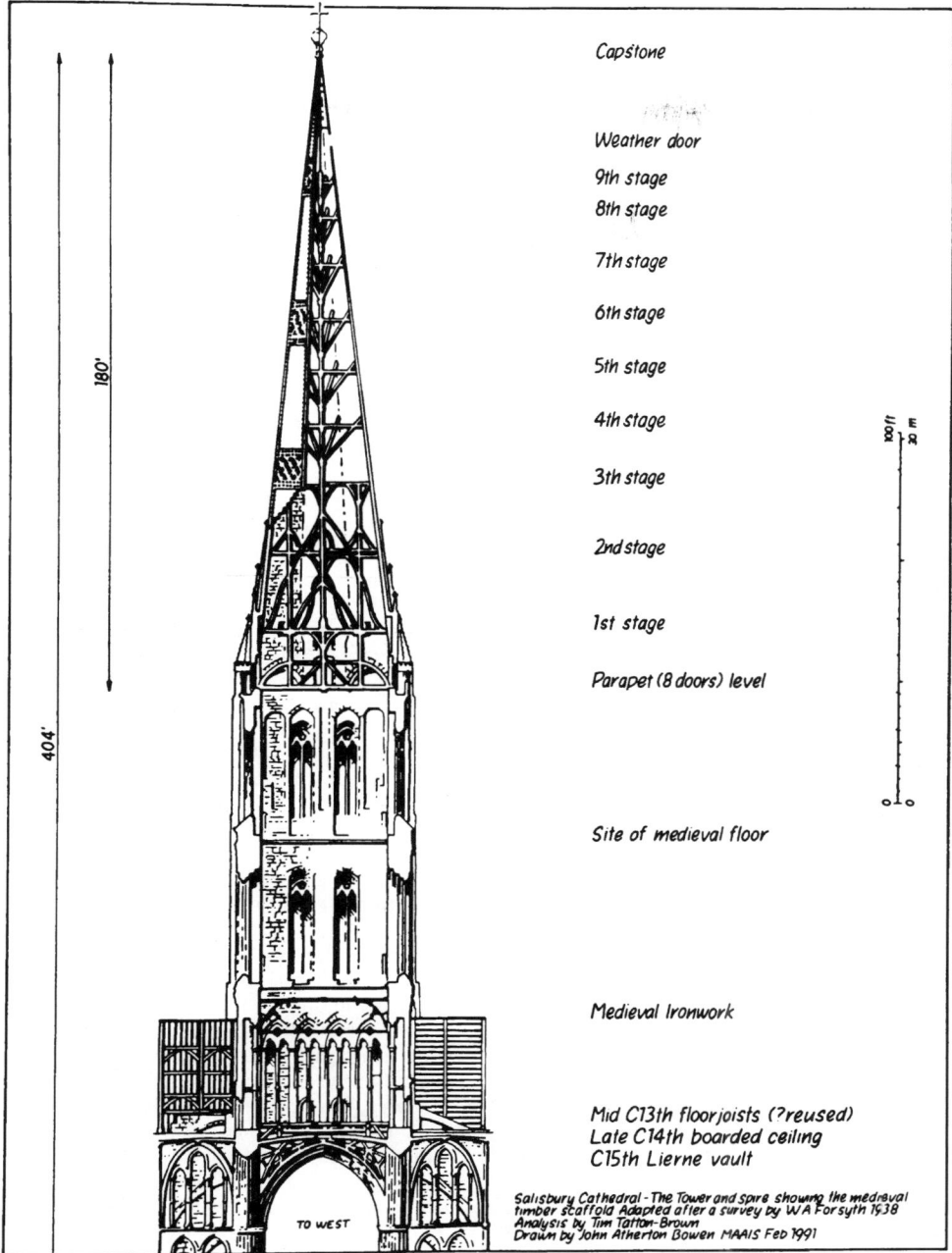

FIGURE 5. *Vertical section through tower and spire, showing medieval timberwork, and key to the different levels.*

In this, one of his masterly descriptions of the carpentry within the tower and spire, Hewett implies that the gallery is of early date. This is most unlikely and an 18th-century date seems a better suggestion; the gallery is not shown in Price's section of the tower and spire dated 1738 (Anon. 1774: plate 6 facing p. 12; FIGURE 6). The gallery also lies at a higher level than would be expected for original medieval work; the main beams that span the angles sit awkwardly on blocks of wood in the space between the window-seats. Some of the timber shows clear signs of re-use, while one has to go up later steps from the medieval doorways in the stair-turrets to reach the gallery.

As the masonry of the two stages of the tower is uniform both externally and internally it was almost certainly built continuously. Half-way up between the upper and lower stages are clear signs of joist-holes for an original medieval floor to the upper chamber. The present floor to this chamber (which, since 1884, has held the four chiming bells) is several feet higher and reached by short flights of steps in the corners. Again this floor has been described and illustrated by Hewett, and it is clear that it was built at the same time as the timber tower which it supports. Hewett (1985: 145) again:

Upon this [floor] a four-posted structure 33 ft. high was built, designed to distribute its weight and that of the superstructure evenly betwixt the carcase and the bridging-joists. The posts were secured by stirrup-irons and forelock-bolts.

This floor and tower are clearly of post-medieval date (perhaps 17th century), and were certainly built, with many renewed timbers, to support the sagging base of the medieval spire-scaffold. (Provisional results from dendrochronology on the tower structure have been difficult to interpret. An early 17th-century felling date seems, however, to be most likely.) The late date of this structure is confirmed by the large red bricks which fill the medieval floor-joist holes and support the later joists at the higher level. This tower scaffold is clearly depicted on Price's section of 1738, and Price himself says (Anon. 1774: 24):

within the last century, a floor and frame of timber was fixed in the tower, about forty feet below the eight doors, as plainly appears, on purpose to uphold the former and prevent that utter destruction, which it threatened 'till that time.

The structure is apparently not mentioned by Wren. It is certainly possible that it was put in during the earlier seventeenth century; a date after 1669 is more likely (FIGURE 7).

At the top of the timber tower four short posts were placed on the middle of each side. These posts had heavily 'jowled' tops so that they could clasp the two cross-beams at the base of the original spire scaffold. These short posts must have been inserted sideways, and once in place were themselves supported by pairs of low braces. To tighten up the whole structure, wedges were put in at the outer ends of the chase tenons of the low braces. The top four tie-beams of the tower, which supported the above-named posts, and braces, were themselves strengthened by cross-bracing below. Towards the ends of these four top tie-beams, four more pairs of braces were seated which went up to brace, from the inside, the lowest stage of the four corner-posts of the original spire scaffold. This very complicated area is made more difficult to understand by the later floor at the base of the spire which masks the connections between the two structures. The area beneath the floor is at present difficult to get to.

All of these timbers were drawn (in two perspective sketches) by Cecil Hewett (1980: 142–3, figures 125–6). Unfortunately his perspectives are not quite right, and he has not differentiated the two periods of carpentry, so his drawings are difficult to 'read'. He has also failed to show the position of the surrounding masonry at the top of the tower. To understand this extremely complex area fully (FIGURE 8), measured plans and sections of all the timbers at the top of the tower need to be made, and all phases of the work need to be differentiated, ideally using dendrochronology as well.

There are further complications: extra long wall-posts and braces (with cross-struts) were inserted at the centre of each side at the top of the tower, also to support from below the two original cross-members at the base of the spire scaffold. These inserted timbers, which look relatively recent, are themselves supported on thick stone 'transoms' put in (probably during G.G. Scott's restoration) between the tops of the inner jambs of the upper tower windows. These masonry additions were put in at the same time as new vertical 'mullions' and 'transoms' were put into the inside of the upper tower window

84   TIM TATTON-BROWN

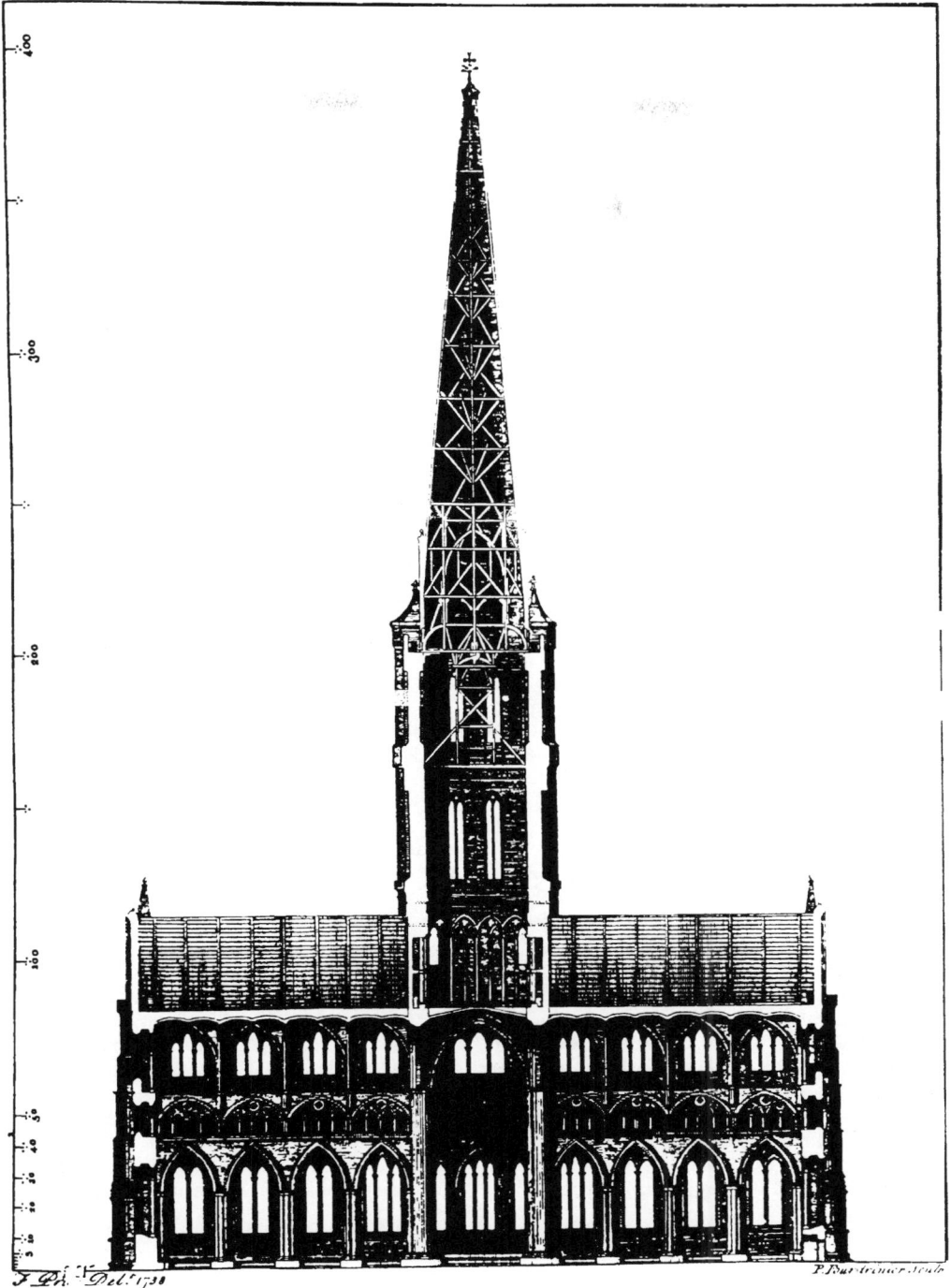

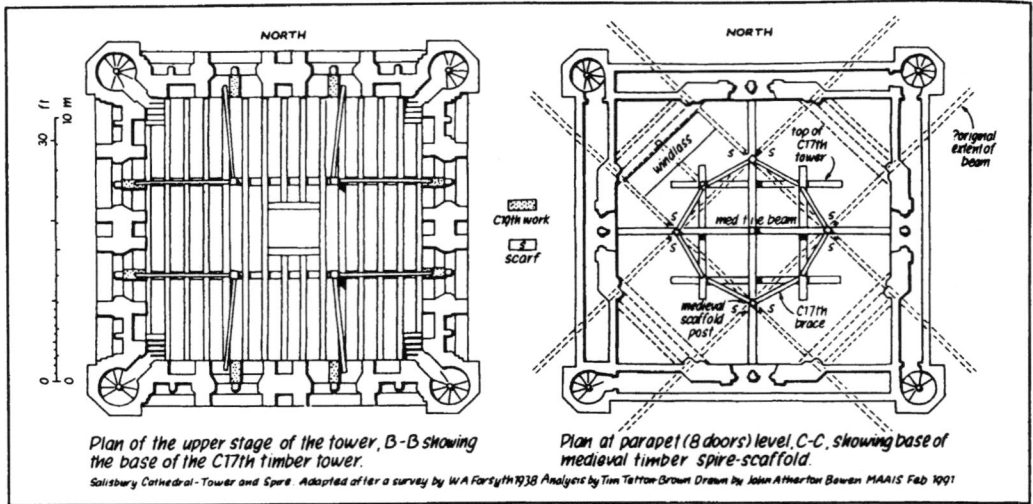

Plan of the upper stage of the tower, B-B showing the base of the C17th timber tower.

Plan at parapet (8 doors) level, C-C, showing base of medieval timber spire-scaffold.

Salisbury Cathedral - Tower and Spire Adapted after a survey by W A Forsyth 1938 Analysis by Tim Tatton-Brown Drawn by John Atherton Bowen MAAIS Feb 1991

FIGURE 7.   Plans of the top of the tower and base of the spire.

openings. At the same time also, the top of the tower walls were re-built and strengthened, apparently with metal. The details of this are not clear.

The building of the spire
The new two-stage tower, added to the lantern, is about 85 ft high. Before the spire was even started the builders were about 220 ft above the ground. To add a 180-ft high stone spire on top, albeit only a thin stone shell, was extraordinarily daring. At St Paul's Cathedral in London a much higher tower, capped by a timber and lead spire, had already been built. Also, at the same time or a little after the Salisbury spire was being erected, the central tower of Lincoln Minster was being heightened to 270 ft so that it too could receive a timber and lead spire. Both the St Paul's and Lincoln spires came down in the 16th century, but it seems that

their tops were perhaps more than 500 ft above the ground (Harvey 1974: 77–80). Timber and lead spires, though they must have contained extraordinarily complicated and sophisticated carpentry, were probably easier to build than stone, but as we know with hindsight they always had a shorter life. (The Chapter House roof of York Minster is a rare surviving example. See also the upper stages and spire of the free-standing bell-tower at Salisbury Cathedral, demolished in 1758 but recorded in section by Francis Price (Anon.1774: plate 10 facing p. 35). The building of such a huge central stone spire at Salisbury was a venture almost without precedent, though it was followed by lesser stone spires at Chichester, Lichfield and Norwich, and by central timber spires (all now gone) at York, Wells and Durham.

Once the top of the tower had been reached, the base of the octagonal spire was constructed

FIGURE 6.   Francis Price's section through the spire and tower, made in 1738, looking east.

TIM TATTON-BROWN

FIGURE 8. *Perspective cutaway of the top of the tower and lower part of the spire.*

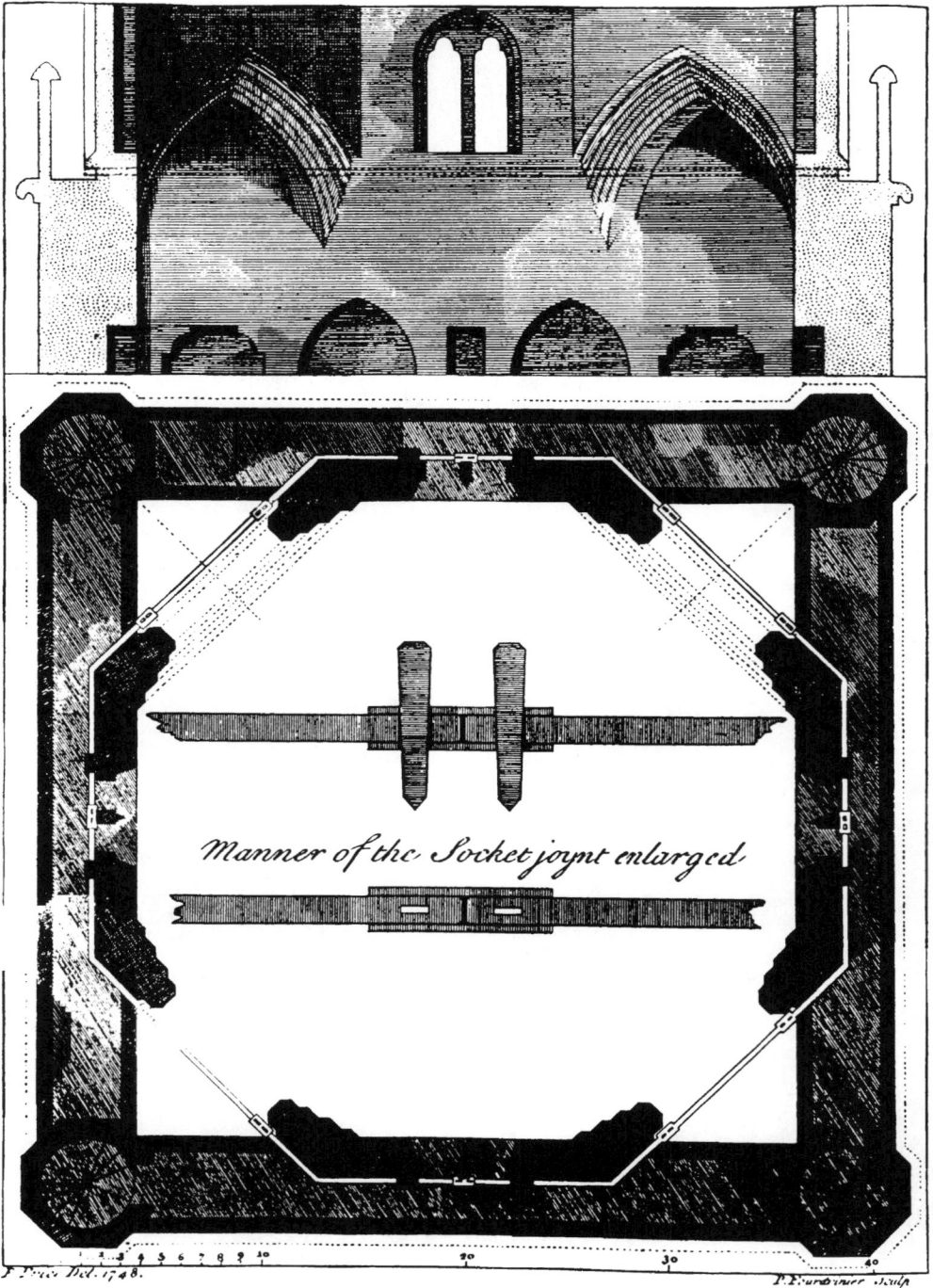

Manner of the Socket joynt enlarged

FIGURE 9. Francis Price's plan and section of the base of the spire of 1748 with newly added iron bandage.

TIM TATTON-BROWN

FIGURE 10. *Aerial view of the outside of the lower part of the spire from the south. (© British Crown Copyright/MoD. Reproduced with the permission of the Controller of Her Britannic Majesty's Stationery Office.)*

on the inner halves of the central sections of the tower walls. To form the octagon, squinch arches were built across the angles of the tower, clearly shown in one of Price's drawings (Anon. 1774: plate 8 facing p. 15; reproduced as FIGURE 9). The internal diameter of the tower and of the base of the octagon is about 33 feet, and the spire wall is only about 2 ft thick at its base (the top of the tower wall is about 5 ft thick). The internal faces of the spire walls rise vertically for about the first 20 ft, while externally the sloping outer face reduces the thickness of the masonry to just 8 inches. The first 25 ft at the base of the spire is a complicated area of masonry as it contains inner pinnacles at the angles (sitting on the broaches above the squinch arches), and at the centre of each side pairs of low doorways with tall lucarnes (dormers) above, capped by crocketed gables with pinnacles. Above the outer corner buttresses (at the top of the stair turrets) are short crocketed pinnacles, and all of these are covered in ballflower (FIGURE 10).

There is a highly-decorated parapet at the top of the tower with a narrow passage behind it which originally connected the pairs of low doorways at the base of the spire with a door into the top of one of the spiral staircases. The wall-top at the base of this passageway has received considerable thrusting over the centuries. This was dealt with in the 1740s by placing an iron bandage around the base of the octagon, as is shown in Price's clear illustration (see FIGURE 9 above). In 1968–9, this iron bandage was removed and replaced by (to quote Peter Taylor 1988: 8):

a stainless steel bar which incorporates turnbuckle screws in the squinch openings. To ensure a positive bearing against the 'square' sides of the tower the bar was encased in an epoxy resin/sand mortar cast against the stone shell. A concrete ring-beam formed in the floor of the gutter, and made structurally continuous around the square tower, enclosed the epoxy resin casing. The turnbuckles may be adjusted at any time should the need arise.

This new bandage will have destroyed any evidence for the earlier ties, including any medieval iron ties. An internal iron 'bandage' does survive a few feet higher immediately above the heads of the eight doors at the base of the spire, as a series of iron bars, set in lead, which were hooked together at the internal corners of the octagon. (Where one of these has 'blown' the internal arrangement can be seen clearly.) Other iron bars, hooked over this, then ran out through the walls of the spire, where their outer plates, covered in lead, can still be seen, damaged and rusted by the weather. This system of iron ties must have been part of the original work, as the molten lead surrounding the iron can only have been poured in liquid from above. The same system was used to join all the Chilmark stone blocks that make up the shell of the spire, and many of the original iron clamps (surrounded in lead) have been revealed over the years during restoration work.

Apart from the iron bandage of the 1740s at the base of the spire, seven other bandages were added externally following Sir Christopher Wren's advice. These were replaced between 1903 and 1914 by copper bands which can still be seen on the outside of the spire (Spring 1987: 28). According to Price (Anon. 1774: 14):

Sir Christopher . . . did most certainly direct the making of others, as time should require, particularly those which, as it were, hoop the spire together; seven of these bandages are applied to that purpose, viz. one below the first network [i.e. band of lozenge decoration], two betwixt the first and second network, and four betwixt the middle and upper network: there is likewise a bandage round the tower itself just below the eight doors, which was probably done by the same advice.

The most extraordinary survival within the spire is the original timber scaffold used to build the spire. This was recognized as such by Francis Price, and briefly described and illustrated by him. Since then only Cecil Hewett (1980 and 1985) has examined the frame closely and made some drawings of it. The engineer Freeman (of Messrs Douglas Fox and Partners) in 1937 proposed that the entire timber frame be removed and replaced by a light steel structure! If this had happened, we would only have had Price's drawings as a rough record of the original. W.A. Forsyth's section only records the upper stages of the spire-scaffold. Although this has been much repaired (and added to), the majority of the original structure survives and merits a detailed study.

The first parts of the spire-scaffold to be built were the two large crossing tie-beams which run north–south and east–west from the centres of the tops of the tower walls (now mostly concealed below the floor). These tie-beams, which

90   TIM TATTON-BROWN

are rather like the cross-trees of a post mill, and were the only original and supports for the whole of the spire-scaffold, are remarkably slim for such a large structure. As we have seen above, they were later supported by a 17th-century timber-tower, and by long 19th-century braces. At this base level (and now below the more recent floor), four diagonal 'plank-like' timbers were also placed. They are not structurally part of the spire-scaffold, but were prob-ably used by the original masons to build the squinch arches at the base of the spire. (They run diagonally from each of the 'eight doors'.) These are supported from below, at their ends by short plates which are in turn on wall-posts and corbels. All of this is below the floor and difficult of access; it needs more careful study.

Before the new floor, there was an earlier floor which Price (Anon 1774: 24) tells us was in very poor condition. He says:

At the eight doors or beginning of the spire, there was a floor of timber originally laid in, connected to the timber-frame within the spire; the beams of this floor being neglected and in time becoming rotten on the south side, it had nothing to bear it, and therefore hung up to the frame of timber above it; and by long continuance in this condition, drew after it the said timber frame to the south side of the spire, which most certainly affect the same. The parts of that frame intended to be strictly level, are declined out of level, nearly four inches to the south; and by this means, the stones about twenty feet above the floor, are scaled or frushed; how long it might continue in this state, is doubtful, but that it was so, is very certain.

The framing within the spire that is erected on the two main cross-beams is a masterpiece of medieval joinery. Although it is purely utilitarian, all the main posts, braces and cross-members have stopped-chamfers along their arrises. This is typical of the medieval period, and allows original medieval timbers to be distinguished from later replacements and additions. The later timbers are also often lighter in colour and more sharply sawn. At this stage, all the original timbers seem to have mortice-and-tenon joints, with only the later timbers being nailed in. Some of the forelock-bolts and iron spikes appear to be original. (Cores have now been taken from quite a large number of timbers in the original frame, and these have all been found to contain very short tree-ring sequences. Clearly young wide-ringed

trees were selected. Even the cross-beams at the base were only about 50 years old. This confirms that the frame is only a scaffold, but makes dating by dendrochronology almost impossible.)

The first stage of the scaffold, about 15 feet high, is an elaborate affair that turns the whole structure through 45 degrees. There is a central post and four others that stand diagonally on the cross-beams half-way between the centre and outer walls. The central post or 'mast' is braced from the cross-beam in four directions, while the outer posts have single external braces and horizontal members set diagonally half-way up the posts (opposite the tops of the braces), which make the whole structure rigid. At the tops of the four outer posts, long horizontal (and fairly slim) timbers have been trenched in. They are halved over each other on top of the outer posts, and diagonal timbers are in turn trenched over them. These diagonal timbers are halved over each other above the central post (see FIGURE 8).

The pairs of horizontal timbers set on the four outer posts all have scarf-joints just outside the crossing area (8 scarfs in all). These scarfs intrigued Cecil Hewett (1980: 145) who writes:

This seems to have been the first scarf ever to possess a bridled abutment; the reasons for its use at this point, or in this construction, are not yet determined, but its effect upon the subsequent development of scarfing was remarkable.

Beyond the scarf-joints short timbers continue until they meet the offset (shoulder) on the inside of the masonry wall where the vertical face changes to a slope, the only point within the spire where direct contact is made between the spire-scaffold and the masonry. It is clear that the short outer pieces of timber were dropped down from above after the masonry offset was completed. After this had been done, another diagonal timber was trenched over it and run back to the central diagonals, where it was attached on the side by a forelock-bolt. When the second stage was built, curved braces were tenoned in on the upper surface of the very ends of the outer timbers to run up to the next stage of four vertical posts. There has clearly been much later compression on the masonry shoulders from the frame (indicated by the scaling and cracks in the masonry beneath the ends of the timbers first noticed by Price). Here arose the need for additional later timbers in the stage above to transfer some of the weight back to the inner posts (FIGURE 8).

The scarfs would not really be necessary here if the carpentry was all integral. However, as a spire-scaffold, it was clearly built stage by stage (to nine stages in all), as the masonry went up around it. The bottom stage would have been particularly elaborate, extending beyond the spire shell-walls so that it could have been used to construct the tall pinnacles around the base of the spire. If the crossing pairs of timbers at the top of the first stage originally extended out beyond the spire walls to flank the pinnacles, then scarfs would have been necessary as the total length of each horizontal member may have been up to 50 feet. If these timbers were extended they would flank perfectly the sides of the inner pinnacles; and if the lower stage of scaffolding was erected to build these pinnacles as well as the spire itself, it would also explain why the main structure was set diagonally on the tower top. All this ties in with the position of the small rectangular openings which open from the first stage of the spire-scaffold into the hollow core of the pinnacles.

Also to be noted is the windlass, at the first stage of the spire-scaffolding, which has been thought medieval (Hewett 1974: 70; Hewett 1985: 191; Backinsell 1980). Yet it, together with its surrounding frame, is clearly a later insertion, most likely put in after the timber tower structure in the upper stage of the tower and the floor at the base of the spire had been rebuilt (perhaps in the 16th or 17th century). The complex hub box of the windlass (using nails), and the way the iron is used, may also suggest a post-medieval date. It may, however, contain elements of an earlier machine. A closer study, perhaps using dendrochronology, is needed before any proper conclusions can be reached.

The second and third stages of the medieval spire-scaffold have been illustrated and very briefly discussed by Hewett (1980: 143–5, figure 127; 1985: 144–5, figure 139). He calls it 'one of the most remarkable works of carpentry known', but again fails to distinguish original and later timbers. Each stage has clearly been built as the masonry octagon went up around it, and quite remarkably the octagons of scaffold boards (the masons' working platforms) also still survive on top of each stage. The second

stage, about 17 ft high, continues up the central mast (with a large base that sits over the crossing timbers at the top of the first stage). There are also four more posts around the central mast (all with chamfered arrises and stops at the top and bottom), each with two tall curved braces from the outside. The tops of the posts support a pair of horizontal timbers (northwest to southeast) which are sandwiched by pairs of double horizontal timbers (northeast to southwest). These timbers then support the second stage of octagonal planking and are themselves braced from below. At a later date, perhaps in 1738 when Price (Anon. 1774: 24) tells us that all the timber work in the spire was 'thoroughly repaired', vertical posts with large jowelled tops have been inserted under the ends of the horizontal timbers for extra support. These posts sit on the shoulders of the braces below, and are themselves supported by diagonal struts from the base of the four original posts (FIGURE 8).

The third stage has four more posts around the central mast, braced in all four directions from below, and from the horizontal pairs of timbers above. Some of these braces (particularly the upper external straight ones) must be secondary, but there are also some repairs and replacements, and full measured drawings are essential before a proper analysis can be given. Another probable later element at the third stage is the bolted-on shorter inner posts, with jowelled tops that support a diagonal cross-timber and straight braces up to the central mast at the next stage. The horizontal timbers at the top of the third stage are this time just two pairs of timbers at right angles to, and sitting on top of, each other. They also support an octagon of scaffold boards.

From the fourth stage to the final ninth stage, each scaffold platform is supported by a series of eight radiating timbers held in place by eight struts from the central mast below (like the ribs of an umbrella), which get steeper and steeper at each stage. The lower ends of the struts are supported by carved capital-like blocks on the central mast, while the upper ends of these struts broaden out so that the slender horizontal timbers can pass through them. They were then adjusted and held in place by wedges, and as Francis Price tells us (Anon 1774: 16):

The said arms and braces may be taken out, and put in at pleasure, consequently capable of easy repair.

TIM TATTON-BROWN

The central mast is crucial to the whole structure, and Price once again is our informant about a major repair undertaken in his time (Anon. 1774: 24–5):

In the year 1738, when all the timber work within was thoroughly repaired, the central piece was found to have been broken in the solid, a little below the weather door, and exactly corresponding with this, a sudden bending in the spire appears, and was taken notice of by Sir Christopher Wren, in his before mentioned survey. This may well be supposed to contribute to the declination [of the spire], and therefore the utmost care was [taken] to restore the original connection of the central piece at this part; and to make all the others secure, wherever they required such a core.

Above the ninth stage of the internal timber scaffolding, where the octagon of masonry is only about five feet across internally, a temporary external scaffold had to be erected. This was fixed to horizontal timbers taken through the thickness of the masonry. When work was complete, this scaffold was taken down and the putlog holes filled with masonry plugs. These plugs, which now have iron handles, are set in weak lime mortar and can be removed when scaffolding of the top of the spire is needed for repair work (Spring 1987: 57). When major repairs to the top of the spire, or the cross or weather vane, were carried out in 1762, 1849, 1903 and 1949–51, the putlog holes were once again used. Access to the outside of the spire is through a small door near the top of the north side called the 'weather door', and from here it is possible to climb to the capstone using iron rings (set in lead) fixed in the masonry. The weather door is approached on the inside by a series of ten small ladders set against the internal spire scaffold (perhaps put in in 1738 when it was repaired).

Between the top of the central timber mast and the cap-stone(s) of the spire, an iron 'machine' (to use Hewett's term) was contrived which locked together the whole structure, and allowed much of the weight of the timber scaffold to be carried down the masonry shell from the top of the spire. The whole arrangement is excellently described and illustrated by Price (Anon. 1774: 16, plate 9; reproduced as FIGURE 11):

This timber frame, though used as a scaffold while the spire was building, was always meant to hang up to

the capstone of the spire, and by that means prevent its top from being injured in storms, and so add a mutual strength to the shell of stone. The central piece of timber is not mortised, to receive the arms which served as floors, but has an iron hoop round it with hooks riveted through; and upon these hooks a flat iron bar is fitted, with a hole in it, which is fastened on the brace: the upper part of the brace is mortised, and the arms tenanted at the end, to slide into and through the mortise in the brace; so that by a key, or wedge on the outside of the brace, the connection is made compleat, the central piece, and the other end of the arm, being provided with iron, as before, renders it the most compleat piece of work imaginable; nor is its connection at the top inferior to it.

Sometime in the century or so before 1738 (see FIGURE 6 above), the new floor and timber tower were built in the upper stage of the tower to support the spire scaffold which was sagging badly at its base. Only at this time, then, did the timber scaffold cease to act like a huge chandelier. When Wren saw the structure in 1668, it had only recently been supported, and he must have seen immediately the deadweight strength and wind damping effect it gave to the structure as a whole. Some time later Wren took down and rebuilt the top of the spire at Chichester Cathedral

and fixed therein a pendulum stage to counteract the effects of the south and south-westerly gales of wind, which act with some considerable power against it, and had forced it from its perpendicularity

to quote James Elmes, an early biographer of Wren, who had himself been called in by Chichester Cathedral in 1813–14 to take down, repair, and reinstate Wren's 'useful piece of machinery'. In his Memoirs of the Life and Works of Sir Christopher Wren (London, 1823), Elmes illustrates and describes this device which he had carefully examined and measured (quoted in Corlette 1901: 40, 42). Sadly it was destroyed by the fall of the spire at Chichester in 1861. It seems very likely that Wren got his idea for this machine from what he saw at Salisbury.

Between 1949 and 1951 the top 23 ft of the spire at Salisbury were completely rebuilt using Doulting stone (Taylor 1988: 7). The ironwork at the top of the spire was modified at this time, but the 'special arrangement of iron brackets and folding wedges' still allows the spire to be tensioned, and 'when conditions demand it, the

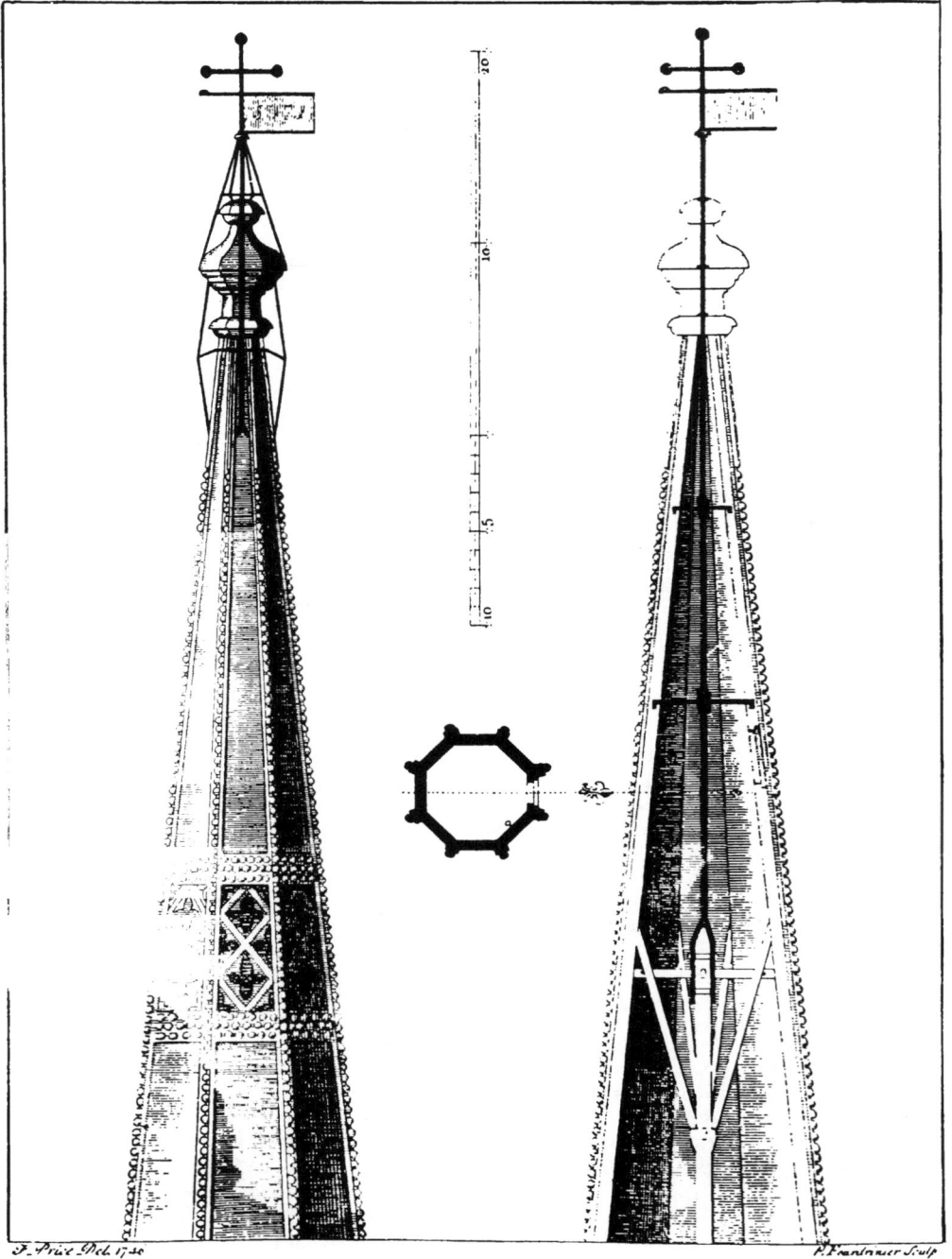

FIGURE 11.   Francis Price's elevation and section of the top of the spire in 1746.

94   TIM TATTON-BROWN

spire can still be "tightened up"' (Spring 1987: 57).

The masonry, timber, iron (and lead) work of the 180-ft spire of Salisbury Cathedral all, therefore, form an extraordinarily integrated whole, which for any age would be considered a masterpiece of architectural design and civil engineering. The fact that it was conceived and built on top of a 220-ft tower some time around 1300 makes it even more remarkable.

The date of the tower and spire
Sir Nikolaus Pevsner in his masterly description of Salisbury Cathedral (Pevsner & Metcalf 1985: 264) says:

In the 1330s the crossing tower was wonderfully heightened and the spire added: the contract for this work dates from 1334, between the chapter and Richard of Farleigh, who must have been a man of some reputation, as he insisted on also carrying on with commitments at Bath Abbey and Reading Abbey. Later he was to be in charge at Exeter.

Unfortunately no such contract for the tower and spire exists.

The contract with the mason, Richard 'de Farlegh', which was made in July 1334, is an agreement

that he should have the custody of the fabric, to superintend, direct and appoint masons, and himself do the necessary work. He should make such stay in Salisbury as the needs of the fabric demanded and, notwithstanding prior obligations at Bath and Reading [Abbeys], he should not neglect or delay the work of the cathedral.

He was to receive 6d. a day when present, and the salary of 10 marks annexed to the office of guardian of the fabric, if he survive Robert the Mason (Dodsworth 1814: 151; Salzman 1952: 45). The work that Richard of Farleigh was superintending at this time was almost certainly the demolition of parts of the old cathedral at Old Sarum and the building of the crenellated close wall (re-using much Old Sarum material) and the three stone gateways (Dodsworth 1814: 146-7). This was the final completion of the work at the new cathedral site and the spire must have been built by this date. From the beginning of the 1330s chapter records are fairly complete, and no mention of the building of the tower or spire is made. Nor, as

Spring (1987: 17) has pointed out, is there any Old Sarum material re-used in the tower and spire.

Pevsner was, in fact, misled into his 1330s date by the entry on Richard of Farleigh in John Harvey's Dictionary of English medieval architects (Harvey 1954: 404-5). In this Harvey mentions the contract, and then goes on to say that Farleigh's work at Salisbury 'must have been the great tower and spire'. He continues:

Preparations for the construction of the upper part of the tower at Salisbury had begun about 1320, and Farleigh's appointment probably marks the start of actual work. The spire was not completed until well into the second half of the 14th century, but the design of tower and spire forms a perfect unity, and must have been the conception of one man.

In the second edition of the Dictionary, Harvey (1987: 106) adds 'probably Robert the Mason VII'. This man Harvey equates with the Robert mentioned in the 1334 contract, who 'is said to have controlled the work of the cathedral for 25 years'. He then says (Harvey 1987: 255):

Robert may have designed the great tower at a date around 1320, taking account of its stylistic dependence upon the Hereford tower then only approaching completion. On grounds of detail, notably the profusion of ballflower ornament, the project is to be dated in the 1320s rather than later, and so is unlikely to be the work of Farleigh.

This is all very tenuous, and just shows that Harvey initially felt that the tower and spire were started in 1334 and completed 'c. 1380'. Later he moved back the 'date of design', and possibly construction, to the 1320s, though he did discuss some of the 'difficulties' in an appendix to his Cathedrals of England and Wales (Harvey 1974: 215-6). Unfortunately, however, no earlier documentary evidence really helps, and Harvey is thrown back on to stylistic comparisons with Gloucester (the ballflower), and

architectonic aspects of the design [which] indicate strong influence from the Lincoln tower completed in 1311 as well as the more obvious assimilation of detail from Lichfield and Hereford.

(Perhaps the closest parallel as a forerunner to the Salisbury spire is the stone spire on the church of St Mary-the-Virgin in Oxford, the University church. In the middle and later years of the 13th century, many Oxford scholars were migrating to Salisbury. Is there a connection here?) His final conclusion (Harvey 1974: 216) is

Design within the decade 1320-30 seems to accommodate all the factors. The spire was probably finished before 1387.

Leaving aside the date of design for a moment, it seems inconceivable that a tower and spire could be designed in the decade 1320–30, and then be built very slowly over the next half-century. The spire itself could have been, and logically it would have been, built in just a few seasons. If the materials for each year's work had been prepared previously, then a decade at most would have been more than adequate. The thin rings of masonry, carefully cut, could have been erected quite quickly with their iron clamps. As the spire only became stable once the internal scaffold was hung from the capstone, to leave it half-built through too many winters would have been very risky. Warwick Rodwell in correspondence makes the following suggestion:

I have wondered whether the principal stages of the scaffolding did not mark yearly breaks, at the lower levels. The main timbers which formed the floor levels could have been allowed to oversail the temporary 'top' of the spire and wedged so that the scaffold (suitably loaded with stone acted as a dead-weight to hold down the vulnerable upper courses over the winter (i.e. on the same principle as the capstone and pendulum, but in a more spread-out version). The void of the spire, and its exposed wall top would be protected from the ingress of water by boarding over. At the beginning of each new season the carpenters would trim back the over-sailing ends of the uppermost scaffold stage, and carry on upwards. I think this might even hazard an estimate at the number of seasons the spire took to build on this basis.

Is it too fanciful to think of the nave floor being used as the place where the lines of the spire were laid out, and where it was prefabricated, with the masonry being laid out in readiness for erection each season?

If this is correct, then the spire and tower would have been designed and built very rapidly, and we should perhaps look for a

completion date for the whole structure by 1330 at the very latest. The dating of and sequences of ballflower decoration has been studied by Richard Morris (1972: 172–83) who suggests a date in the 1320s for the tower and spire. He points out (pers. comm.) that the earliest *datable* example of ballflower is in the Wells Chapter House (1286–1306), but that Salisbury follows the Hereford tower (complete before 1319), and precedes Pershore Abbey tower. He also says that the mouldings of Salisbury compare closely with those of the south aisle of Gloucester Abbey (built 1318–29).

One other piece of evidence might suggest an even earlier date. Up to 1297 almost all the bishops, deans and senior canons were a remarkable group of men who, in effect, presided at Salisbury over a third university (Edwards 1956: 169). Three deans, Masters Robert de Wykehampton (1259–74), Walter Scammell (1274–84) and Henry de Brandeston (1285–87) became in succession bishops of Salisbury (from 1271 to 1287), and, up until the appointment of Peter of Savoy as Dean in 1297, all of them were usually resident. (In a letter of 28 June 1290, Edward I granted all fines levied on the Dean and Chapter and its tenants be given to the fabric of the cathedral as granted by Henry III (Rolls Series: Sarum Charters and Documents (1891): 365)) By the end of the century, the incipient university was disappearing, and the pernicious practice had arrived of giving the deanery and most of the senior canonries to non-resident foreigners (i.e. papal provisions). As Dr Edwards points out (1956: 170), between 1297 and 1379 all six deans were non-resident foreigners (one a kinsman of Edward I and the rest French or Italian cardinals), four foreigners were precentors and three cardinals were treasurers. Only the chancellors were mostly English. The distinguished scholar bishop, Master Simon of Ghent (1297–1315) and later his successor, Master Roger Mortival (1315–1330), tried on many occasions to stop this, but to no avail. (Simon of Ghent's tomb in the cathedral has ballflower on it, while that of his successor, Roger Mortival, has not.) This would perhaps suggest that the 'inception' of the tower and spire should be before 1297 when the whole chapter was still a flourishing resident body. Does bishop Ghent's strong protest ' (Edwards 1956: 170) against these provisions of absentee foreigners

96   TIM TATTON-BROWN

that the dignitaries [i.e. Dean and Residentiary Canons] were intended to be like living cornerstones or pillars of the cathedral, and that their neglect of residence might cause the whole fabric to crash in ruins

suggest that the whole cathedral, including the tower and spire, had been completed by this time, i.e. the beginning of the 14th century?

None of this evidence is conclusive. I would suggest that the tower and spire were built rapidly in the years around 1300. If correct it would mean that the whole of Salisbury Cathedral was built in only about 80 years, a truly remarkable achievement. Even on a longer chronology we can be fairly certain that it was built within about a century (i.e. c. 1220–1320) in a Golden Age at New Salisbury that was never to be repeated.

Acknowledgements. I am extremely grateful to Roy Spring, clerk of the works at Salisbury Cathedral, for allowing me to make numerous visits to both the outside (on a scaffold!) and inside of the tower and spire. He has also given me much useful information on the fabric and lent me the photographs that illustrate this article. John Atherton Bowen has very kindly drawn the illustrations for me, using my rough sketches. I am also very much indebted to Drs Thomas Cocke and Warwick Rodwell for reading and commenting constructively on an earlier draft of this paper; also Dr Richard Morris (of Warwick University) for discussing the ball-flower decoration and mouldings. Just before this paper went to press Gavin Simpson most kindly sent me (December 1990) the provisional results of dendrochronological sampling work within the tower and spire undertaken by him and his colleagues at Nottingham University a few months earlier. This has allowed me to make a few additions in the text, though it is sad that dendrochronological dating of the spire scaffold does not appear to be possible. Finally, my wife Veronica has laboured much in producing a clean word-processed text for me from a very messy manuscript.

References

ANONYMOUS. 1774. A Description of that Admirable Structure the Cathedral Church of Salisbury, with the chapels, monuments, grave stones, and their inscriptions, etc. London. 2nd edition of Price 1753.

BACKINSELL, W. 1980. Medieval windlasses at Salisbury, Peterborough and Tewkesbury. Salisbury: South Wiltshire Industrial Archaeological Society. Historical monograph 7.

COLVIN, H.M. 1978. A biographical dictionary of British architects 1600–1840. London: John Murray.

CORLETTE, H. 1901. The cathedral church of Chichester. . . . London: Bells Cathedral Series.

DODSWORTH, W. 1814. An Historical Account of the Episcopal See and Cathedral Church of Sarum, or Salisbury. Salisbury.

EDWARDS, K. 1956. Salisbury Cathedral, an ecclesiastical history, in Victoria History of Wiltshire III: 156–209. London: Institute of Historical Research.

FORSYTH, W.A. 1946. The structure of Salisbury Cathedral tower and spire, Journal of the Royal Institute of British Architects (January): 85–7.

GODWIN, T. 1616. De Praesulibus Angliae.

HARVEY, J. 1954. Dictionary of English medieval architects. London: Batsford.

1974. Cathedrals of England and Wales. London: Batsford.

1987. Dictionary of English medieval architects. 2nd revised edition. Gloucester: Alan Sutton.

HEWETT, C. 1974. English cathedral carpentry. London: Wayland.

1980. English historic carpentry. Chichester: Phillimore.

1985. English cathedral and monastic carpentry. Chichester: Phillimore.

MORRIS, R.K. 1972. Decorated architecture in Herefordshire. Unpublished Ph.D thesis, London University.

PEVSNER, N. 1975. Salisbury cathedral, in Buildings of England: Wiltshire: 388–418. Harmondsworth: Penguin.

PEVSNER, N. & P. METCALF. 1985. The cathedrals of England: Southern England: 261–86. London: Penguin.

PRICE, F. 1753. A Series of Particular and Useful Observations made with Great Diligence and Care upon that Admirable Structure, the Cathedral-Church of Salisbury. London.

RODWELL, W.J. 1989a. Archaeology and the standing fabric: recent studies at Lichfield Cathedral, Antiquity 63: 281–94.

1989b. Church archaeology. London: Batsford.

SALZMAN, L.F. 1952. Building in England down to 1540. Oxford: Clarendon Press.

SLATTER, D. 1965. The diary of Thomas Naish. Devizes: Wiltshire Record Society. No. 20.

SPRING, R. 1985. Salisbury Cathedral. London: Unwin Hyman.

TATTON-BROWN, T.W.T. 1989. Great cathedrals of Britain: an archaeological history. London: BBC.

TATTON-BROWN, T.W.T. & J.T. MUNBY (ed.). In press. The archaeology of cathedrals.

TAYLOR, P. 1988. The tower and spire of Salisbury cathedral, Associations for Studies in the Conservation of Historic Buildings, Transactions for 1987 12: 3–10.

WILLIS, R. 1849 [1973]. Minutes of lecture on the architecture of Salisbury cathedral, in Architectural history of some English cathedrals II. Chicheley: Paul P.B. Minet.

WREN, C. c. 1669. Survey, printed in part in Price 1753: 1–4 and in Wren (Jr) 1750: 304–6.

WREN, C. JR. 1750. Parentalia: or memoirs of the family of Wren. London.

Index

Please note: Page numbers which appear in italics are references to tables or illustrations.